PSEUDODIFFERENTIAL OPERATORS AND APPLICATIONS

PROCEEDINGS OF SYMPOSIA
IN PURE MATHEMATICS
Volume 43

PSEUDODIFFERENTIAL OPERATORS AND APPLICATIONS

AMERICAN MATHEMATICAL SOCIETY
PROVIDENCE, RHODE ISLAND

PROCEEDINGS OF SYMPOSIA IN PURE MATHEMATICS OF THE AMERICAN MATHEMATICAL SOCIETY
VOLUME 43

PROCEEDINGS OF THE SYMPOSIUM ON PSEUDODIFFERENTIAL OPERATORS AND FOURIER INTEGRAL OPERATORS WITH APPLICATIONS TO PARTIAL DIFFERENTIAL EQUATIONS
HELD AT THE UNIVERSITY OF NOTRE DAME
NOTRE DAME, INDIANA
APRIL 2–5, 1984

EDITED BY

FRANCOIS TRÈVES

Prepared by the American Mathematical Society
with partial support from National Science Foundation grant DMS-8318439

1980 *Mathematics Subject Classification.* 22E30, 32F99, 35-XX, 42B20, 43A80, 47-02, 47D30, 47G05, 58GXX, 81C12, 85B40, 85P05.

Library of Congress Cataloging in Publication Data

Symposium on Pseudodifferential Operators & Fourier Integral Operators with Applications to Partial Differential Equations (1984: University of Notre Dame)

Pseudodifferential operators and applications.

(Proceedings of symposia in pure mathematics; v. 43)

Proceedings of a symposium held at the University of Notre Dame, Apr. 2–5, 1984.

Bibliography: p.

1. Pseudodifferential operators–Congresses. 2. Fourier integral operators–Congresses. 3. Differential equaitons, Partial–Congresses. I. Trèves, François, 1930– . II. American Mathematical Society. III. Title. IV. Series.

QA329.7.S96 1984 515.7′242 85-1419

ISBN 0-8218-1469-9

Table of Contents

Preface vii

Propagation of local analyticity for the Euler equation
By S. ALINHAC and G. METIVIER 1

A functional calculus for a class of pseudodifferential operators with singular symbols
By J. L. ANTONIANO and G. A. UHLMANN 5

Uniqueness in a class of nonlinear Cauchy problems
By M. S. BAOUENDI 17

Propagation of smoothness for nonlinear second-order strictly hyperbolic differential equations
By MICHAEL BEALS 21

Multidimensional inverse scatterings and nonlinear partial differential equations
By R. BEALS and R. R. COIFMAN 45

Nonlinear harmonic analysis and analytic dependence
By R. R. COIFMAN and YVES MEYER 71

On some C^*-algebras and Fréchet*-algebras of pseudodifferential operators
By H. O. CORDES 79

Boundary-value problems for second-order elliptic equations in domains with corners
By G. ESKIN 105

Imbedding $\mathbb{C}^n$ in H_n
By P. C. GREINER 133

On some results of Gelfand in integral geometry
By VICTOR GUILLEMIN 149

The propagation of singularities for solutions of the Dirichlet problem
By LARS HÖRMANDER 157

Application of the microlocal theory of sheaves to the study of $\mathcal{O}_X$
By M. KASHIWARA and P. SCHAPIRA 167

Recent progress on boundary-value problems on Lipschitz domains
By C. E. KENIG 175

Estimates for $\bar{\partial}_b$ on compact pseudoconvex CR manifolds
By J. J. KOHN 207

Real analysis and operator theory
By Yves Meyer 219
Integrability and holomorphic extendibility for rigid CR structures
By L. P. Rothschild 237
Multiple wells and tunneling
By Johannes Sjöstrand 241
The real analytic and Gevrey regularity of the heat kernel for $\Box_b$
By N. K. Stanton and D. S. Tartakoff 247
Fefferman-Phong inequalities in diffraction theory
By M. E. Taylor 261

Preface

This volume gathers a number of papers devoted to microlocal analysis, that is to say, to the precise application of Fourier transforms to analysis on manifolds. Mostly these articles are the written versions of lectures given by their authors at the AMS Symposium held at the University of Notre Dame from April 2 to April 5, 1984. In some instances the version is an abbreviated one; in one or two cases the content of the written paper differs from that of the lecture delivered, but covers a related topic.

Microlocalization is the most powerful tool of linear analysis to have emerged since distribution theory; it is the natural continuation of the latter. Its core is the theory of pseudodifferential and Fourier integral operators. Its central tenet is that the natural context for the study of PDE is the cotangent bundle—a claim that would not have excessively surprised the mathematicians of the nineteenth century or the physicists of the twentieth. In fact, physics has anticipated much of the new developments in the direction of microlocal analysis, often with less rigor and greater daring.

The successes of microlocalization have been truly splendid. They have made the analytic foundation of the Atyah–Singer formula simple, helped us penetrate deeper into the difficult study of uniqueness of solutions, and helped us understand why and when boundary problems like the oblique derivative or the $\bar{\partial}$-Neumann problem could be solved. It has allowed us to assert the existence or nonexistence of solutions to large classes of linear PDE and apply the vivid language of geometrical optics to describe their singularities. That same blend of geometrical optics and hard analysis has led to a greatly improved mathematical theory of diffraction. And hyperfunction theory has revitalized the study of overdetermined systems in the analytic category. Recently, the range of microlocal analysis has widened even more: Its advancing edge now tackles inverse scattering, the analysis of tunneling, the study of Cauchy–Riemann structures. It is even beginning to percolate into that most forbidding realm—nonlinear PDE. Although the truly severe nonlinearities, particularity those connected with shocks, seem at present out of reach, there is reason to believe that the future (and the near future, at that) will witness good progress in this direction.

Each of the topics I have just alluded to were discussed by the lecturers at the Symposium in Notre Dame; most of them, and many more, are studied in the articles in this volume. They bear testimony to the vitality and scope of the Fourier transform method.

FRANCOIS TRÈVES

Proceedings of Symposia in Pure Mathematics
Volume **43** (1985)

Propagation of Local Analyticity for the Euler Equation

S. ALINHAC AND G. METIVIER

The following is the text of a talk given by the first author at the AMS meeting in Notre-Dame, Indiana on the joint paper *Propagation de l'analyticité locale pour les solutions de l'équation d'Euler* (to appear in Arch. Rational Mech. Anal.)

1. Introduction and results. Let $x \in \mathbb{R}^n$, $t \in [0, T]$, and consider the motion of an incompressible nonviscous fluid given by the equations

$$\partial u/\partial t + (u \cdot \nabla) u = \nabla p + f, \quad \operatorname{div} u = 0, \quad u(x,0) = u_0(x),$$

where $u(x, t) = (u_1(x, t), \ldots, u_n(x, t))$ is the velocity, p is the (unknown) pressure, and f and u_0 are given.

Assume that a sufficiently regular solution (u, p) is given globally in the strip $\mathbb{R}^n_x \times [0, T]$. More precisely, assume that, for some $\mu > n/2$ (say $\mu \in \mathbb{N}$), $u \in C^0([0, T], (H^{\mu+1}(\mathbb{R}^n_x))^n)$, $p \in C^0([0, T], H^{\mu+1}(\mathbb{R}^n_x))$, and $u_t \in C^0([0, T], (H^{\mu}(\mathbb{R}^n_x))^n)$. Then the geometry of the fluid lines $x'(t) = u(x(t), t)$ is well defined. If $\phi_{t,s}(x)$ is the solution of these equations, with $\phi_{s,s}(x) = x$, then $\phi_{t,s}\colon \mathbb{R}^n_x \to \mathbb{R}^n_x$ is a diffeomorphism.

Let $\Omega \subset \mathbb{R}^n$ be an open set, and let $\mathscr{C}$ be the tube with basis Ω:

$$\mathscr{C} = \{(x, t), 0 \leqslant t \leqslant T, x \in \phi_{t,0}(\Omega)\}.$$

We prove the following

THEOREM. *If u_0 is analytic in Ω and f in $\mathscr{C}$, then u is analytic in $\mathscr{C}$.*

REMARK. The same theorem holds for u on $U \times [0, T]$, with $u \cdot \nu = 0$ on $\partial U \times [0, T]$.

2. Some related results. (a) The existence of smooth enough solutions for the Euler equation has been established by many authors: see Kato [**12**], Ebin–Marsden [**11**], Temam [**15**], etc.

1980 *Mathematics Subject Classification*. Primary 35L75.

(b) For the case of globally analytic data (on a compact manifold or on a bounded set), existence has been proved by Baouendi–Goulaouic [**4**]–[**6**] and Delort [**10**].

(c) Recently, G. Metivier [**14**] has obtained local existence for analytic pseudodifferential operators and analytic data.

(d) A priori regularity results such as ours have been proved for the Euler equation by Bardos [7], Bardos–Benachour–Zerner [**8**], and Benachour [**9**] in the case of globally analytic data.

(e) For general (nonlinear) hyperbolic equations or systems, propagation of analyticity has been proved by Alinhac–Metivier [**1, 2**].

(f) Generalizations to nonhomogeneous fluids and boundary-value problems have been obtained by Le Bail [**13**].

What is the new feature of the present problem? The problem has a pseudodifferential character. More precisely, let $\pi = 1 - \operatorname{grad} \Delta^{-1} \operatorname{div}$ be the projector on divergence-free fields, orthogonally to gradients; then u satisfies

$$\partial u/\partial t + \pi(u \cdot \nabla u) = \pi f, \qquad u(x,0) = u_0(x),$$

which can be thought of as a (nonlinear) hyperbolic pseudodifferential system (whatever that means). The main difficulty here is that π being nonlocal, zones where u is not smooth contribute to the value of $\pi(u \cdot \nabla u)$ in $\mathscr{C}$. The quantitative control of this contribution will be the crucial step in the proof (Lemma 3.5).

3. Some ideas of the proof. The proof is carried out by taking x-derivatives of the equation and estimating them recursively with the aid of an energy inequality (control of t-derivatives being then given by the equation).

For fixed (forever) μ and any p, α, β, let $|\alpha| = p + 1$, $|\beta| \leqslant \mu$, and $v = \partial_x^{\alpha+\beta} u$. Then

$$\begin{aligned}\partial v/\partial t + u \cdot \nabla v &= \partial_x^{\alpha+\beta}(u \cdot \nabla u) - u \cdot \nabla \partial_x^{\alpha+\beta} u + \partial_x^{\alpha+\beta} \nabla p + \partial_x^{\alpha+\beta} f \\ &= g_1 + g_2 + g_3 = g.\end{aligned}$$

3.1. Classical papers (Friedman, Morrey–Nirenberg, etc.) and experience in the field show that "nested" open sets in Ω (which can be assumed smooth without loss of generality) have to be considered.

If $\delta(x) \geqslant 0$ is locally the distance to $\partial\Omega$, let $\Omega_\delta = \{x \in \Omega,\ \delta(x) \geqslant \delta\}$, $\Omega_{t,\delta} = \phi_{t,0}(\Omega_\delta)$, etc. We will estimate, at time t, the H^μ-norm in x of $\partial_x^{\alpha+\beta} u$ in $\Omega_{t,\delta}$. Let

$$X_p(t,\delta) = \sup_{|\alpha| = p} \|\partial_x^\alpha u(t,\cdot)\|_{H^\mu(\Omega_{t,\delta})}$$

and

$$Y_p = \sup_{1 \leqslant q \leqslant p}\ \sup_{t \in [0,T]}\ \sup_{0 < \delta \leqslant \delta_0} \left\{ \frac{e^{-\lambda(q-1)t}(\varepsilon\delta)^{q-1}}{m_{q-1}} X_q(t,\delta) \right\},$$

where $\varepsilon > 0$ ($\varepsilon \ll 1$) and $\lambda \gg 1$ have to be chosen, and $m_p = ctep!/(p+1)^2$ (see [**1**]).

The final aim of the recursion argument will be to prove $Y_p \leqslant H$, $\forall p$, which implies analyticity in Ω.

3.2. *The energy inequality.* Recall the following classical inequality. If $\partial_t v + u \cdot \nabla v = g$, then

$$\|v(t,\cdot)\|_{L^2(\Omega_{t,\delta})} \leqslant \|v(0,\cdot)\|_{L^2(\Omega_\delta)} + 2\int_0^t \|g(s,\cdot)\|_{L^2(\Omega_{s,\delta})}\,ds.$$

3.3. *Estimates for g_1 and g_3.* Since f is analytic, the estimate of g_3 is straightforward.

The estimate of g_1 can be obtained by using the machinery of [**1**]. The key facts are, of course, that H^μ is an algebra, and the special choice of m_p, which allows us to keep the same ε, λ in the recursion.

We summarize the results in the following lemma.

LEMMA. *There exist C, $\varepsilon_0 > 0$, and $\delta_0 > 0$ such that for all $0 < \varepsilon \leqslant \varepsilon_0$, $\lambda \geqslant 0$, $0 < \delta \leqslant \delta_0$, $t \in [0, T]$, we have ($i = 1$ or 3)*

$$\begin{aligned}&\|g_i(t,\cdot)\|_{L^2(\Omega_{t,\delta})}\\ &\quad\leqslant C(p+1)\left\{X_{p+1}\left(t, \delta\left(1-\frac{1}{p+1}\right)\right) + \left(Y_p^2+1\right)e^{\lambda pt}(\varepsilon\delta)^{-p}m_p\right\}.\end{aligned}\tag{3.3}$$

Of course, all constants are independent of p.

3.4. *Estimates for the derivatives of the pressure.* We have

$$\Delta p = \sum \frac{\partial u_j}{\partial x_k}\frac{\partial u_k}{\partial x_j} - \operatorname{div} f \equiv A,$$

so

$$g_2 = \partial_x^{\alpha+\beta}\nabla p = \Delta^{-1}\partial_{x_j}\nabla\partial_x^{\alpha+\beta-1_j}A = K\partial_x^{\alpha+\beta-1_j}A,$$

where Δ^{-1} means the usual convolution, and $K = \Delta^{-1}\partial_j\nabla$ is an analytic pseudodifferential operator of order 0.

The remarkable fact here is that the type of control of the derivatives of u expressed by the inequalities $Y_p \leqslant H$ ($p \leqslant N$) is preserved by the action of K (even the constants ε, δ, λ are preserved).

This is precisely stated in Lemma 3.5, and it allows us to prove for g_2 the inequality (3.3).

3.5 *The crucial (pseudolocal) lemma.*

LEMMA 3.5. *Let $w \in H^\mu(\mathbb{R}_x^n) \cap C^\infty(\overline{\Omega})$, and let K be an analytic pseudodifferential operator of order* 0.

Set $|\varphi|_{p,\delta} = \sup_{|\alpha|=p}\|\partial_x^\alpha\varphi\|_{H^\mu(\Omega_\delta)}$. Then there exist C, $\varepsilon_0 > 0$, $\delta_0 > 0$, s.t. for $p \geqslant 1$, $0 < \varepsilon \leqslant \varepsilon_0$, $0 < \delta \leqslant \delta_0$, we have

$$|Kw|_{p,\delta} \leqslant C\left\{|w|_{p,\delta(1-1/(p+1))} + (\varepsilon\delta)^{-p}m_p\left([w]_{p-1} + \|w\|_{H^\mu(\mathbb{R}^n)}\right)\right\},$$

where

$$[w]_p = \sup_{\substack{0 \leqslant q \leqslant p \\ \delta \in]0, \delta_0]}} \frac{(\varepsilon\delta)^q |w|_{q,\delta}}{m_q}. \qquad \blacksquare$$

When this lemma is applied to the situation of 3.4, t is to be considered as a parameter.

A very similar result has been obtained by Baouendi–Goulaouic [**6**], who deal with functions holomorphic in certain domains of the complex space.

This lemma is proved by cutting the kernel k of K in p zones concentric about the origin.

3.6. *End of the proof*. Inequality (3.3) for g and the energy estimate 3.2 imply easily (taking into account the analyticity of u_0)

$$Y_{p+1} \leqslant \sup\left\{ Y_p, H_1 + C_1\left(Y_p^2 + 1\right)/(\lambda - \lambda_1)\right\}$$

for some constants C_1, H_1 and all ε, $1/\lambda$ small enough. Appropriate choices of H, ε, λ then give $Y_p \leqslant H$.

References

1. S. Alinhac and G. Metivier, *Propagation de l'analyticité des solutions de systèmes hyperboliques non-linéaires*, Invent. Math. **75** (1984), 189–204.

2. ______, *Propagation de l'analyticité des solutions d'équations non-linéaires de type principal*, Comm. Partial Differential Equations 9 (1984), 523–537.

3. ______, *Propagation de l'analyticité locale pour les solutions de l'équation d'Euler*, Arch. Rational Mech. Anal. (to appear).

4. M. S. Baouendi and C.. Goulaouic, *Problèmes de Cauchy pseudo-différentiels analytiques non-linéaires*, Séminaire Goulaouic-Schwartz 75–76 (No. 13), École Polytechnique, Paris.

5. ______, *Solutions analytiques de l'équation d'Euler d'un fluide incompressible*, Séminaire Goulaouic-Schwartz 76–77(No. 22), École Polytechnique, Paris.

6. ______, *Sharp estimates for analytic pseudo-differential operators and application to Cauchy problems*, J. Differential Equations **48** (1983), 241–268.

7. C. Bardos, *Analyticité de la solution de l'équation d'Euler dans un ouvert de* $\mathbb{R}^n$, C.R. Acad. Sci. Paris **283** (1976), 255–258.

8. C. Bardos, S. Benachour and M. Zerner, *Analyticité des solutions périodiques de l'équation d'Euler en dimension deux*, C.R. Acad. Sci. Paris **282** (1976), 995–998.

9. S. Benachour, *Analyticité des solutions périodiques de l'équation d'Euler en dimension trois*, C. R. Acad. Sci. Paris **283** (1976), 107–110.

10. J. M. Delort, *Estimations fines pour des opérateurs pseudo-différentiels analytiques sur un ouvert à bord de* $\mathbb{R}^n$. *Application aux équations d'Euler*, Comm. Partial Differential Equations (to appear).

11. D. G. Ebin and J. Marsden, *Groups of diffeomorphisms and the motion of an incompressible fluid*, Ann. of Math. (2) **92** (1970), 102–163.

12. T. Kato, *Nonstationary flows of viscous and ideal fluids in* $\mathbb{R}^3$, J. Funct. Anal. **9** (1972), 296–305.

13. Le Bail, Univ. de Rennes, preprint.

14. G. Metivier, *Un théorème de Cauchy-Kowalevski pseudo-différentiel local*, Séminaire Goulaouic-Schwartz 83–84, No. 16, École Polytechnique, Paris.

15. R. Temam, *On the Euler equations of incompressible perfect fluids*, J. Funct. Anal. **20** (1975), 32–43.

Université Paris - Sud, France

Université de Rennes 1, France

Proceedings of Symposia in Pure Mathematics
Volume **43** (1985)

A Functional Calculus for a Class of Pseudodifferential Operators with Singular Symbols

JOSÉ L. ANTONIANO AND GUNTHER A. UHLMANN[1]

0. Introduction and statement of results. Distributions whose wave front set is contained in several intersecting Lagrangian manifolds appear naturally in many situations. For instance, the Schwartz kernel of the parametrices constructed by Duistermaat and Hörmander (see [**DH**]) for pseudodifferential operators P of real principal type have wave front set contained in the diagonal and in the flow-out, from the diagonal intersected with $p = 0$, by the integral curves of H_p. Also in the case of operators with double characteristics, several intersecting Lagrangian manifolds appear due to the interaction of the different flows (see [**GU**, **MU**, **M**, **U**]).

In this paper we extend the symbol calculus developed in Guillemin–Uhlmann [**GU**] for the case of two Lagrangian manifolds intersecting cleanly to a functional calculus under certain restrictions. Of particular importance is the case of a functional calculus for pseudodifferential operators with singular symbols. We now describe our results more precisely.

Let X be a C^∞ manifold of dimension n, $\Delta = \Lambda_0$ the diagonal in $T^*(X) \times T^*(X)$, p a symbol of real principal type (i.e., p is a real-valued canonical C^∞ function homogeneous of degree 1 on $T^*X \setminus 0$), $dp \neq 0$ on $p = 0$, dp and the one-form are linearly independent on $p = 0$. Let Λ_1 be the Lagrangian manifold obtained as the flow-out from $\Delta \cap \{p = 0\}$ by the integral curves of the Hamiltonian vector field associated to p. In this case the compositions of the different Lagrangians do not generate new manifolds since $\Lambda_0 \circ \Lambda_0 = \Lambda_0$, $\Lambda_0 \circ \Lambda_1 = \Lambda_1$, $\Lambda_1 \circ \Lambda_0 = \Lambda_1$, $\Lambda_1 \circ \Lambda_1 = \Lambda_1$. Notice also that the intersection of Λ_1 with itself is clean. Thus, the question naturally arises of whether the

1980 *Mathematics Subject Classification*. Primary 35S99; Secondary 58G15.

[1] The author was partially supported by NSF grant DMS 8402581. The author is an Alfred P. Sloan Research Fellow.

composition of two operators in $I^{\cdot}(X \times X; \Lambda_0, \Lambda_1)$, as defined in [**GU**], is in the same class. We prove

THEOREM 0.1. *Let* $A \in I^{p,l}(X \times X; \Lambda_0, \Lambda_1)$, $B \in I^{r,s}(X \times X; \Lambda_0, \Lambda_1)$ *be properly supported. Then* $A \circ B \in I^{\tilde{p},\tilde{l}}(X \times X; \Lambda_0, \Lambda_1)$, *with* $\tilde{p} = p + r - n/4$, $\tilde{l} = l + s - 1/2$.

Theorem 0.1 is proved in §2. In §3 we compute the symbol of the composition on Λ_0, Λ_1 away from the intersection $\Sigma = \Lambda_0 \cap \Lambda_1$. $A \circ B$ is a pseudodifferential operator on $\Lambda_0 - \Sigma$. We have

$$\sigma(A \circ B)|_{\Lambda_0 - \Sigma} = \sigma(A)|_{\Lambda_0 - \Sigma}\, \sigma(B)|_{\Lambda_0 - \Sigma}. \tag{0.1}$$

Also $A \circ B$ is a Fourier integral operator on $\Lambda_1 - \Sigma$. Let $(x, \xi, z, \zeta) \in \Lambda_1 - \Sigma$. Then, microlocally,

$$\begin{aligned} &\sigma(A \circ B)|_{\Lambda_1 - \Sigma}(x, \xi, z, \zeta) \\ &\qquad = \int \sigma(A)|_{\Lambda_1}(x, \xi, y(t), \eta(t))\, \sigma(B)|_{\Lambda_1}(y(t), \eta(t), z, \zeta)\, dt, \end{aligned} \tag{0.2}$$

where $(y(t), \eta(t))$ denotes the (maximally extended) bicharacteristic curve joining (x, ξ) and (z, ζ). (0.2) does not take into account Maslov contributions and half-densities. Also, (0.2) is, at the moment, a formal expression because of the singularities of $\sigma(A)$ and $\sigma(B)$ on Σ. However, as we shall see in §3, the singularities of $\sigma(A)|_{\Lambda_1}$ and $\sigma(B)|_{\Lambda_1}$ in (0.2) occur at different t's, and one can make sense of (0.2) in the sense of distributions. In the case

$$A \in I^{p-n/4}(X \times X; \Lambda_1) \subseteq I^{p,l}(X \times X; \Delta, \Lambda_1),$$
$$B \in I^{r-n/4}(X \times X; \Lambda_1) \subseteq I^{r,s}(X \times X; \Delta, \Lambda_1),$$

formula (0.2) was obtained by Duistermaat and Guillemin (see [**DG**]). In §1 we briefly review the symbol calculus of [**GU**]. In §4 we give an application. We prove

THEOREM 0.2. *There exist elliptic* $\mathcal{U}, \tilde{\mathcal{U}} \in I^{\cdot}(\mathbf{R}^n \times \mathbf{R}^n; \Delta, \Lambda_1)$ *(in the sense described in* §1) *such that*

$$\left(\frac{\partial}{\partial x_1}\frac{\partial}{\partial x_2} + A\right)\mathcal{U} = \tilde{\mathcal{U}}\frac{\partial}{\partial x_1}\frac{\partial}{\partial x_2} \quad \textit{microlocally near } \Delta \cap \{\xi_1 = 0\}.$$

Here A is a classical pseudodifferential operator of order 0 *in* $\mathbf{R}^n$, $n > 2$, *and* $p = \xi_1$ *in this case.*

We recall that operators of the form $(\partial/\partial x_1)(\partial/\partial x_2) + A$ are a microlocal model for operators with double involutive characteristics of product type satisfying the Levi condition [**U**]. In §4 we also give further examples of operators in $I^{\cdot}(X \times X; \Delta, \Lambda_1)$, namely, pseudodifferential powers of operators with simple, real characteristics.

Theorem 0.1 was obtained independently by Jiang and Melrose (see [**JM**]).

1. Symbol calculus (see [GU] for details). Let X be a smooth manifold of dimension $n > 2$, Λ_0, Λ_1 two conic Lagrangian submanifolds of $T^*X\setminus 0$ intersecting cleanly in a submanifold of codimension 1 on each Λ_i, $i = 0, 1$, and

$$T_\lambda(\Lambda_0) \cap T_\lambda(\Lambda_1) = T_\lambda(\Lambda_0 \cap \Lambda_1) \quad \text{for all } \lambda \in \Lambda_0 \cap \Lambda_1. \tag{1.1}$$

A basic example of such an intersecting pair in $T^*(\mathbf{R}^n)$ is given by

$$\begin{aligned} \tilde{\Lambda}_0 &= \{(0, \xi) \in T^*\mathbf{R}^n \setminus 0\}, \\ \tilde{\Lambda}_1 &= \{(x, \xi) \in T^*\mathbf{R}^n \setminus 0 \,|\, x' = 0, \xi_1 = 0\}. \end{aligned} \tag{1.2}$$

Here (x, ξ) are the standard coordinates in $T^*(\mathbf{R}^n)$, $x = (x_1, x') \in \mathbf{R} \times \mathbf{R}^{n-1}$, and $\xi = (\xi_1, \xi') \in \mathbf{R} \times \mathbf{R}^{n-1}$. Let $\mathbf{R}^m = \mathbf{R}^n \times \mathbf{R}$. Let z, x, s be coordinates on $\mathbf{R}^m$, $\mathbf{R}^n$, and $\mathbf{R}$, respectively, and let ξ, σ be the dual variables of x, s.

$S^{p,l}(m, n, 1)$ denotes the space of all C^∞ functions in (z, ξ, σ) compactly supported in z and satisfying

$$\left|\left(\frac{\partial}{\partial z}\right)^\alpha \left(\frac{\partial}{\partial \xi}\right)^\beta \left(\frac{\partial}{\partial \sigma}\right)^\gamma a\right| \leqslant C_{\alpha,\beta,\gamma}(1 + |\xi|)^{p-|\beta|}(1 + |\sigma|)^{l-|\gamma|} \tag{1.3}$$

uniformly in (z, ξ, σ).

Let $a_r(z, \xi, \sigma)$ be a smooth function on the set $\xi \neq 0$, homogeneous of degree r in ξ, and let $a_{r,s}(z, \xi, \sigma)$ be a smooth function on the set $\xi \neq 0$, $\sigma \neq 0$, bihomogeneous of degree (r, s) in (ξ, σ). We shall say

$$a_r \sim \sum_{s=l}^{-\infty} a_{r,s} \quad \text{if } \rho(\xi, \sigma)\left(a_r - \sum_{s=l}^{-N} a_{r,s}\right) \in S^{r,-(N+1)}, \tag{1.4}$$

where ρ is a smooth function that is zero near $\xi = 0$ and $\sigma = 0$ and one outside a compact set. $S^{p,l}_{cl}(m, n, 1)$ denotes the subspace of $S^{p,l}(m, n, 1)$ consisting of all symbols that admit an asymptotic expansion of the form

$$a \sim \sum_{r=p}^{-\infty} a_r, \tag{1.5}$$

with the a_r's as in (1.4). Here $\sim$ means

$$\rho(\xi)\left(a - \sum_{r=p}^{-N} a_r\right) \in S^{-(N+1),l}(m, n, 1),$$

where ρ is a smooth function that is zero near $\xi = 0$ and one outside a compact set.

$I^{p,l}(X; \Lambda_0, \Lambda_1)$ denotes the space of oscillatory integrals with singular symbols as defined in [GU]. A microlocal model is $I^{p,l}(\mathbf{R}^n; \tilde{\Lambda}_0, \tilde{\Lambda}_1)$. We say $\mu \in I^{p,l}(\mathbf{R}^n; \tilde{\Lambda}_0, \tilde{\Lambda}_1)$ if $\mu = \mu_1 + \mu_2$, with $\mu_1 \in C_0^\infty(\mathbf{R}^n)$, and μ_2 can be represented by means of the oscillatory integral

$$\mu_2 = \int e^{i[(x_1 - s)\xi_1 + x'\xi' + s\sigma]} a(s, x, \xi, \sigma)\, ds\, d\xi\, d\sigma, \tag{1.6}$$

with $a \in S^{p',l'}_{cl}(m, n, 1)$, $p' = p - n/4$, $l' = l - 1/2$.

We have the following facts from [**GU**].

PROPOSITION 1.1. *Let* $\tilde{\Lambda}_0$ *and* $\tilde{\Lambda}_1$ *be as in* (1.2), *and let* $\tilde{\Sigma} = \tilde{\Lambda}_0 \cap \tilde{\Lambda}_1$. *Then* $\mathrm{WF}(\mu_2) \subseteq \tilde{\Lambda}_0 \cup \tilde{\Lambda}_1$. *Moreover, near* $\tilde{\Lambda}_0 - \tilde{\Sigma}$, μ_2 *is microlocally in the space* $I^{p+l+n/4}(\mathbf{R}^n; \tilde{\Lambda}_0)$, *and its leading symbol is* $a_{p,l}(z, \xi, \sigma)|_{z=0, \sigma=\xi_1}$, *omitting half-densities and Maslov contributions. Near* $\tilde{\Lambda}_1 - \tilde{\Sigma}$, μ_2 *is microlocally in the space* $I^{p+n/4}(\mathbf{R}^n; \tilde{\Lambda}_1)$, *and its leading symbol is*

$$\left(\int a_p(z, \xi, \sigma) e^{is\sigma}\, d\sigma\right)\Bigg|_{\substack{x'=0\\ \xi_1=0\\ s=x_1}}, \tag{1.7}$$

omitting half-densities and Maslov contributions.

PROPOSITION 1.2. *With* l *fixed,* $\bigcap_p I^{p,l} = C_0^\infty(\mathbf{R}^n)$. *With* p *fixed,* $\bigcap_l I^{p,l} = I^p(\mathbf{R}^n; \tilde{\Lambda}_1)$.

DEFINITION 1.1. Given $u \in I^{p,l}(X; \Lambda_0, \Lambda_1)$, let $\sigma_0(u) = \sigma(u)|_{\Lambda_0 - \Sigma}$ and $\sigma_1(u) = \sigma(u)|_{\Lambda_1 - \Sigma}$. $\sigma_0(u)$ is called the principal symbol of u.

The main result is

PROPOSITION 1.3. *The following sequence*

$$\begin{aligned} 0 \to I^{p,l-1}(X; \Lambda_0, \Lambda_1) + I^{p-1,l}(X; \Lambda_0, \Lambda_1) &\to I^{p,l}(X; \Lambda_0, \Lambda_1) \\ &\to S^{p,l}(X; \Lambda_0, \Sigma) \to 0 \end{aligned}$$

is exact, where $S^{p,l}(X; \Lambda_0, \Sigma)$ *is the subspace of* $R^{l-1/2}(\Omega_0 \otimes L_0; \Lambda_0, \Sigma)$ (*as defined in* [**GU**]). $R^{l-1/2}$ *is, intuitively, "the space of smooth functions on* $\Lambda_0 - \Sigma$ *that have a singularity of order* l *at* Σ", *and consist of elements that are homogeneous of degree* $p + l + n/4$.

Let $X = \mathbf{R}^n \times \mathbf{R}^n$, let $\Lambda_0 = \Delta$ be the diagonal, and let Λ_1 be the flow-out from $\Lambda \cap \{\xi_1 = 0\}$ by the integral curves of $\partial/\partial x_1$. It is easy to check

PROPOSITION 1.4. *If* $A \in I^{p,l}(X; \Delta, \Lambda_1)$ *then* $A^t \in I^{p,l}(X; \Delta, \Lambda_1)$.

DEFINITION 1.2. $A \in I^{p,l}(X; \Delta, \Lambda_1)$ is elliptic if $\sigma_0(A) \neq 0$ on $\Delta - \Sigma$.

2. Proof of Theorem 0.1. Using the invariance of the distributions in $I^{p,l}(X; \Lambda_0, \Lambda_1)$ under conjugation by Fourier integral operators (see [**GU**]), we can assume $X = \mathbf{R}^n \times \mathbf{R}^n$, Λ_0 is the diagonal in $T^*(\mathbf{R}^n) \times T^*(\mathbf{R}^n)$, and Λ_1 is the flow-out from $\Lambda_0 \cap \{\xi_1 = 0\}$ by the integral curves of $\partial/\partial x_1$; i.e.,

$$\Lambda_0 = \{((x, \xi), (x, \xi)) \in T^*(\mathbf{R}^n \times \mathbf{R}^n) \setminus 0\},$$

$$\Lambda_1 = \{((x_1, x', 0, \xi'), (y_1, x', 0, \xi')) \in T^*(\mathbf{R}^n \times \mathbf{R}^n) \setminus 0\},$$

where $x = (x_1, x') \in \mathbf{R} \times \mathbf{R}^{n-1}$, $\xi = (\xi_1, \xi') \in \mathbf{R} \times \mathbf{R}^{n-1}$. We take the Schwartz kernel of A and B to be

$$k_A = \int e^{i[(x_1 - y_1 - s)\xi_1 + (x' - y')\xi' + s\sigma]} a(s, x, y, \xi, \sigma)\, ds\, d\xi\, d\sigma, \tag{2.1}$$

with $a \in S_{cl}^{p',l'}(m, n, 1)$ a product type symbol in ξ, σ; $m = 2n+1$, $p' = p - n/4$, $l' = l - 1/2$, $s, \sigma \in \mathbf{R}$ and

$$k_B = \int e^{i[(y_1 - z_1 - t)\eta_1 + (y' - z')\eta' + t\tau]} b(t, y, z, \eta, \tau)\, dt\, d\eta\, d\tau, \tag{2.2}$$

with $b \in S_{cl}^{r',s'}(m, n, 1)$ a product type symbol in (η, τ), m as above, $r' = r - n/4$, $s' = s - n/2$, $t, \tau \in \mathbf{R}$. We first cut off a near $\xi = 0$, $\sigma = 0$ and b near $\eta = 0$, $\tau = 0$. Let $\chi \in C^\infty(\mathbf{R}^n)$ be a homogeneous function of degree 0 for $|\xi| \geqslant 1$ such that $\chi = 0$ near $\xi = 0$, $\chi = 1$ for $|\xi| \geqslant 1$, and let $\tilde{\chi} \in C^\infty(\mathbf{R})$ be a homogeneous function of degree 0 for $|\sigma| \geqslant 1$ s.t. $\tilde{\chi} = 0$ near $\sigma = 0$, $\tilde{\chi} = 1$ for $|\sigma| \geqslant 1$. Then

$$k_A = \int e^{i[(x_1 - y_1 - s)\xi_1 + (s' - y')\xi' + s\sigma]} \chi(\xi)\tilde{\chi}(\sigma) a(s, x, y, \xi, \sigma)\, ds\, d\xi\, d\sigma + \tilde{A} \quad \text{with } \tilde{A} \in I^{p-n/4}(\mathbf{R}^n \times \mathbf{R}^n; \Lambda_1) \tag{2.3}$$

and

$$k_B = \int e^{i[(y_1 - z_1 - t)\eta_1 + (y' - z')\eta' + t\tau]} \chi(\eta)\tilde{\chi}(\tau) b(t, y, z, \eta, \tau)\, dt\, d\eta\, d\tau + \tilde{B} \quad \text{with } \tilde{B} \in I^{r-n/4}(\mathbf{R}^n \times \mathbf{R}^n; \Lambda_1). \tag{2.4}$$

Thus we may assume that a (resp. b) in (2.1) (resp. (2.2)) vanishes near $\xi = 0$, $\sigma = 0$ (resp. $\eta = 0$, $\tau = 0$), since the terms $A \circ \tilde{B}$, $\tilde{A} \circ B$ can be shown to be in the appropriate class by a similar argument to the one used below. We have

$$k_{A \circ B} = \int e^{i(\phi_1 + \phi_2)} a \cdot b\, dy\, ds\, d\xi\, d\sigma\, dt\, d\eta\, d\tau, \tag{2.5}$$

where

$$\phi_1 = (x_1 - y_1 - s)\xi_1 + (x' - y')\xi' + s\sigma,$$
$$\phi_2 = (y_1 - z_1 - t)y_1 + (y' - z')\xi' + t\tau.$$

Let $0 < \varepsilon < 1$, and let χ_1 (resp. χ_2) be a C^∞ homogeneous function of degree 0 in (ξ, η) such that $\chi_1 = 1$ for $|\eta| \leqslant \frac{1}{2}\varepsilon|\xi|$, $\chi_1 = 0$ for $|\eta| \geqslant \varepsilon|\xi|$ if $|(\xi, \eta)| = 1$ (resp. $\chi_2 = 1$ for $|\xi| \leqslant \frac{1}{2}\varepsilon|\eta|$, $\chi_2 = 0$ for $|\xi| \geqslant \varepsilon|\eta|$ for $|(\xi, \eta)| = 1$). Integrating by parts in the y-variable, we can easily show that the term

$$\int e^{i(\phi_1 + \phi_2)} \chi_i a \cdot b\, dy\, ds\, d\xi\, d\sigma\, dt\, d\eta\, d\tau$$

is smoothing, $i = 1, 2$. We then assume that the amplitude in (2.5) satisfies

$$\tfrac{1}{2}\varepsilon \cdot |\eta| \leqslant |\xi| \leqslant 2|\eta|/\varepsilon \quad \text{on supp}\,(a \cdot b). \tag{2.6}$$

Now making the change of variables $s = u + r$, $t = -u + r$, and $y_1 = -u + v$, we obtain

$$k_{A \circ B} = 2\int e^{i[(u+r)\sigma + (-u+r)\tau]} D\, du\, dr\, d\xi\, d\sigma\, d\tau, \tag{2.7}$$

where

$$D = \int e^{i[(x_1 - v - r)\xi_1 + (x' - y')\xi' + (v - z_1 - r)\eta_1 + (y' - z')\eta']} a \cdot b\, dv\, dy'\, d\eta_1\, d\eta'. \tag{2.8}$$

In order to be able to apply the stationary phase method, we introduce polar coordinates in ξ. Writing $\omega = \xi/|\xi| \in S^{n-1}$, $\xi = |\xi|\omega$, and making the change of variables $\eta = |\xi|\bar{\eta}$ in (2.8), we get

$$(2.9)\ D = \int e^{i|\xi|[(x_1-v-r)\omega_1+(z'-y')\omega'+(v-z_1-r)\bar{\eta}_1+(y'-z')\bar{\eta}']}$$
$$\cdot a \cdot b(u+r, -u+r, x, (-u+v, y'), z, |\xi|\omega, |\xi|\bar{\eta}, \sigma, \tau)\, dv\, dy'\, d\eta.$$

Observe that $\bar{\eta} \leqslant 2/\varepsilon$ on supp $a \cdot b$ ((2.6)). We are now in position to apply stationary phase in $(v, y', \bar{\eta})$. We obtain

$$(2.10) \qquad D \sim (2\pi/|\xi|)^n e^{i[(x_1-z_1-2r)\xi_1+(x'+z')\zeta']}$$
$$\cdot d(u+r, -u+r, x_1, (-u+v, y'), z, \xi, \sigma, \tau),$$

where $d \in S_{cl}^{p'+r', l', s'}(n,1,1)$ is a product type symbol of order $p'+r'$, l', s' in ξ, σ, τ, respectively; $\sim$ in (2.10) means that

$$\left(D - \sum_{j=1}^{N} d_j\right) \in S^{p'+r'-(N+1), l', s'}(n,1,1),$$

where $d \sim \sum d_j$ as in §1.

Let us now consider

$$(2.11) \qquad F = \int e^{i[(u+r)\sigma+(-u+r)\tau]} d\, du\, d\tau.$$

Let $\chi_1(\sigma, \tau)$, $\chi_2(\sigma, \tau)$ be defined in a similar way as above. Then

$$\int e^{i[(u+r)\sigma+(-u+r)\tau+(x_1-z_1-2r)\xi_1+(x'-z')\xi']} \chi_i\, d\, du\, d\tau$$

is an element of $I^{p+l}(\mathbf{R}^n \times \mathbf{R}^n \Lambda_1)$. Therefore, we may assume

$$(2.12) \qquad \tfrac{1}{2}\varepsilon|\tau| \leqslant |\sigma| \leqslant 2|\tau|/\varepsilon \quad \text{on supp } d.$$

Then we apply the stationary phase Lemma to F in u, $\bar{\tau}$, with $\bar{\tau} = \tau/|\sigma|$. Putting everything together we find that

$$(2.13) \quad k_{A \circ B} = \int e^{i[(x_1-z_1-2r)\xi_1+(x'-y')\xi'+2r\sigma]} c(r, x, z, \xi, \sigma)\, d\xi\, d\sigma\, dr,$$

where $c \in S_{cl}^{p', r'}$ is a classical product type symbol in the (ξ, σ) variables ending the proof of the theorem.

3. Computation of the principal symbol of $A \circ B$. First we compute the principal symbol of $A \circ B$ on $\Delta - \Sigma$, where A, B, $A \circ B$ are pseudodifferential operators. We have

$$\sigma(A)|_{\Delta-\Sigma} = a_{p',l'}(0, x, x, \xi, \xi_1)|_{\xi_1} \neq 0,$$
$$\sigma(B)|_{\Delta-\Sigma} = b_{r',s'}(0, z, z, \zeta, \zeta_1)|_{\zeta_1} \neq 0$$

(see §1). From the proof in §2 it is easy to see that the highest order of homogeneity of c (as in (2.13)) in ξ is $a_{p',l'} \cdot b_{r',s'}$. Therefore, we conclude that

$$(3.1) \qquad \sigma(A \circ B)|_{\Delta-\Sigma} = \sigma(A)|_{\Delta-\Sigma} \cdot \sigma(B)|_{\Delta-\Sigma}.$$

We next compute the principal symbol of $A \circ B$ on $\Lambda_1 - \Sigma$. We first show that we can extend the C^∞ function on $\Lambda_1 - \Sigma$, $\sigma(A)|_{\Lambda_1 - \Sigma}$, to a symbol-valued conormal distribution associated to $\Sigma \subseteq \Lambda_1$. Let

$$f^A_{p'}(s, x, y, \xi) = \int e^{is\sigma} a_{p'}(s, x, y, \xi, \sigma)\, d\sigma,$$

where $a \sim \sum_{j \leqslant p'} a_j$ as in §1. Then $f^A_{p'} \in C^\infty(\mathbf{R}^n_x \times \mathbf{R}^n_y \times \mathbf{R}^n_\xi; \mathscr{S}'(\mathbf{R}_s))$. In fact, $f^A_{p'}$ is a conormal distribution associated with $s = 0$. We also have (see §1)

$$\sigma(A)|_{\Lambda_1 - \Sigma} = f^A_{p'}\,(x_1 - y_1, (x_1, x'), (y_1, x'), (0, \xi'))|_{x_1 \neq y_1}.$$

Then we extend $\sigma(A)|_{\Lambda_1 - \Sigma}$ to Λ_1 as an element of

$$C^\infty\Big(\mathbf{R}^{n-1}_{x'} \times \mathbf{R}^{n-1}_{\xi'}; \mathscr{D}'\Big(\mathbf{R}^2_{(x_1, y_1)}\Big)\Big)$$

—in fact, as a conormal distribution associated with $x_1 = y_1$. Clearly, $(\sigma(A)|_{\Lambda_1}, \psi(x_1, y_1))$ is a homogeneous function of degree p' in ξ' depending smoothly on x', y' $\forall \psi \in C^\infty_0(\mathbf{R}^2)$. Let

$$f_A = \int e^{is\sigma} a(s, x, y, \xi, \eta)\, d\sigma, \quad f_B = \int e^{it\tau} b(t, y, z, \eta, \tau)\, d\tau. \tag{3.2}$$

We now write

$$k_{A \circ B} = \pi_* \tilde{K} \tag{3.3}$$

where π: $\mathbf{R}^n_x \times \mathbf{R}^n_z \times \mathbf{R}_s \times \mathbf{R}_t \to \mathbf{R}^n_x \times \mathbf{R}^n_z$ is the projection. $\tilde{K}$ is formally given by

$$(\tilde{K}, \phi) = \int e^{i(\psi_1 + \psi_2)}(f_A f_B, \phi)\, dy\, dx\, dz\, d\xi\, d\eta \tag{3.4}$$

$\forall \phi \in C^\infty_0(\mathbf{R}^n_x \times \mathbf{R}^n_z \times \mathbf{R}_s \times \mathbf{R}_t)$, where

$$\psi_1 = (x_1 - y_1 - s)\xi_1 + (x' - y')\xi', \quad \psi_2 = (y_1 - z_1 - t)\eta_1 + (y' - z')\eta'.$$

Formally,

$$(f_A f_B, \phi) = \int f_A f_B\, \phi\, ds\, dt \quad \forall \phi \tag{3.5}$$

as above.

PROPOSITION 3.1. *$(f_A f_B, \phi)$, as in* (3.5), *is a symbol-valued symbol in ξ, η; i.e.,*

$$\left|D^\alpha_x D^\beta_z D^\gamma_\xi D^\delta_\eta (f_A f_B, \phi)\right| \leqslant C_{\alpha,\beta,\gamma,\delta}(1 + |\xi|)^{p' - |\gamma|}(1 + |\eta|)^{r' - |\delta|}$$

uniformly on compact subsets in (x, z).

PROOF. We write

$$(f_A f_B, \phi) = \int e^{i(s\sigma + t\tau)} a \cdot b\, \phi\, ds\, dt.$$

We develop a (resp. b) in a Taylor series around $s = 0$ (resp. $t = 0$) and obtain

$$(3.6)\qquad (f_A f_B, \phi) = \int \sum_{j=0}^{N} \sum_{k=0}^{N} \frac{d^k}{ds^k} a \Big|_{s=0} \frac{d^j}{dt^j} b \Big|_{t=0} \widehat{D_s^k D_t^j \phi}\, d\sigma\, d\tau + (R_N, \phi),$$

where the Fourier transform is taken in the s, t variables. The first term in (3.6) satisfies the estimate in the proposition, since now the amplitude is rapidly decreasing in σ, τ, and we use the estimates for a and b as product type symbols. (R_N, ϕ) contains terms of the form

$$(3.7)\qquad \begin{aligned} &\int e^{i(s\sigma + t\tau)} \frac{s^{(N+1)}}{(N+1)!} h_N \frac{d^j}{dt^j} b \Big|_{t=0} \phi\, d\tau\, d\sigma\, dt\, ds \\ &= \int \frac{D_\sigma^{N+1}(e^{is\sigma})}{(N+1)!} h_N \frac{d^j}{dt^j} b \Big|_{t=0} \hat{\phi}\, d\tau\, d\sigma\, ds, \end{aligned}$$

where the Fourier transform is in the t variable. h_N is the remainder term of order $N + 1$ in the Taylor series of a. Integrating by parts in (3.7) in the σ variable, we obtain that, for N large, the integrand in (3.7) is absolutely integrable in σ, τ, and, therefore, the estimate of Proposition 3.1 is solid for terms of the form (3.4). The other terms in (R_N, ϕ) are estimated similarly.

PROPOSITION 3.2. $\tilde{K} \in \mathcal{D}'(\mathbf{R}_x^n \times \mathbf{R}_z^n \times \mathbf{R}_s \times \mathbf{R}_t)$.

PROOF. We may assume $\xi, \eta \neq 0$. Then we can write (3.4) in the form

$$(3.8)\qquad (\tilde{K}, \phi) = \int \Delta_x \Delta_z e^{i(\psi_1 + \psi_2)} (f_A f_B, \phi) \frac{1}{|\xi|^2} \frac{1}{|\eta|^2}\, dy\, dx\, dz\, d\xi\, d\eta,$$

where Δ_x, Δ_z denotes the Laplacian in the x and z variables, respectively. Integrating by parts in the x, z variables, we obtain the result. Thus

$$k_{A \circ B} = \int e^{i(\psi_1 + \psi_2)} f_A f_B\, dy\, ds\, d\xi\, dt\, d\eta$$

is a well-defined distribution in $\mathcal{D}'(\mathbf{R}^n \times \mathbf{R}^n)$. Now in order to prove (0.2) we make the change of variables $s = u + r$, $t = -u + r$, $y_1 = -u + v$. As in §2 we apply the stationary phase lemma in v, y', η. We obtain

$$(3.9)\qquad \begin{aligned} k_{A \circ B} = \int & e^{i[(x_1 - z_1 - 2r)\xi_1 + (x' - z')\xi']} \\ & \cdot f_C(u + r, -u + r, x, (z_1 + r - u, z'), z, \xi)\, du\, dr\, d\xi, \end{aligned}$$

where

$$f_C \sim \left(\frac{2\pi}{|\xi|}\right)^n \sum_{j=0}^{\infty} \frac{(i)^j}{j!} \left(\frac{\partial}{\partial v} \frac{\partial}{\partial y'} \frac{\partial}{\partial \eta}\right)^j (f_A f_B) \Bigg|_{\substack{\eta = \xi_1 \\ v = z_1 + r \\ y' = z'}} |\xi|^{-j}.$$

Thus we obtain, using §1 (particularly (1.7)),

$$\begin{aligned}
&\sigma(A\circ B)((x_1,x'),(z_1,x'),(0,\xi'),(0,\xi')) \\
&\quad = \int \sigma(A)((x_1,x'),(z_1+t,x'),(0,\xi'),(0,\xi')) \\
&\qquad \cdot\sigma(B)((z_1+t,x'),(x_1,x'),(0,\xi'),(0,\xi'))\,dt
\end{aligned} \tag{3.10}$$

locally, disregarding Maslov contributions and half-densities. This proves (0.2).

Notice that we are computing $\sigma(A\circ B)$ on $\Lambda_1 - \Sigma$—i.e., at points such that $x_1 \neq z_1$. The singularity of $\sigma(A)$ in the integrand in (3.10) occurs when $t = x_1 - z_1 \neq 0$; that of $\sigma(B)$ occurs when $t = 0$. Therefore (3.10) makes sense interpreted as a distribution in the t variable. Formula (3.10) makes sense in the whole Λ_1 if we can multiply the distributors in question at $t = 0$. This happens, for instance, if $\sigma(A)$ or $\sigma(B)$ is smooth or locally integrable in the t variable.

4. Applications.

A. We shall first prove Theorem 0.2. Let E_1 be the parametrix for $\partial/\partial x_1$ constructed in [**DH**]. $E_1 \in I^{0,-1}(\mathbf{R}^n\times\mathbf{R}^n;\Delta,\Lambda_1)$ (see [**MU**]). First we are going to find $\mathscr{V}\in I^{0,0}(\mathbf{R}^n\times\mathbf{R}^n;\Delta,\Lambda_1)$ such that

$$\frac{\partial}{\partial x_2}\mathscr{V}-\mathscr{V}\frac{\partial}{\partial x_2}+AE_1\mathscr{V}=0 \quad \text{microlocally near } ((0,\tilde\xi),(0,\tilde\xi)) \quad \text{with } \tilde\xi=(0,0,\dots,1). \tag{4.1}$$

We first try to eliminate the singularities on the left side of (4.1) on the diagonal. We write

$$\mathscr{V}\sim\sum_{j\geqslant 0}\mathscr{V}_j \quad \text{with } \mathscr{V}_j\in I^{0,-j}(\mathbf{R}^n\times\mathbf{R}^n;\Delta,\Lambda_1) \tag{4.2}$$

and choose $\mathscr{V}_0 = \mathrm{Id}$. The transport equation for $\sigma_0(\mathscr{V}_1)$ is

$$D_{x_2}\sigma_0(\mathscr{V}_1) = -\sigma_0(AE_1). \tag{4.3}$$

We can solve (4.3) with $\sigma_0(\mathscr{V}_1)\in S^{0,-1}(\mathbf{R}^n\times\mathbf{R}^n;\Delta,\Sigma)$, since $AE_1\in I^{0,-1}(\mathbf{R}^n\times\mathbf{R}^n;\Delta,\Lambda_1)$, and integrating along the x_2 variable does not affect the homogeneity in ξ or the singularity in ξ_1. Then using the symbol calculus (see §1), we have

$$\frac{\partial}{\partial x_1}\mathscr{V}_1-\mathscr{V}_1\frac{\partial}{\partial x_1}+AE_1\mathscr{V}_0\in I^{0,-2}+I^{-1,-1}.$$

Inductively, we find $\mathscr{V}_j\in I^{0,-j}$, solving

$$D_{x_2}\sigma_0(\mathscr{V}_j) = -\sigma_0(AE_1\mathscr{V}_{j-1}+\mathscr{U}_{0,j+1}) \quad \text{with } \mathscr{U}_{0,j}\in I^{0,-j}. \tag{4.4}$$

Using the functional calculus we have $AE_1\mathscr{V}_{j-1}\in I^{0,-j}$, so that we can solve (4.4) with $\sigma_0(\mathscr{V}_j)\in S^{0,-j}$. Thus we obtain, with $\mathscr{V}\sim\sum_{j\geqslant 0}\mathscr{V}_j$,

$$\frac{\partial}{\partial x_2}\mathscr{V}-\mathscr{V}\frac{\partial}{\partial x_2}+AE_1\mathscr{V}\in I^{0,-\infty}(\mathbf{R}^n\times\mathbf{R}^n;\Delta,\Lambda_1)+I^{-1,-1}. \tag{4.5}$$

Now $I^{0,-\infty} \subset I^0(\Lambda_1)$ (see §1). The next step is to eliminate $I^{-1,-1}$. We keep $\mathscr{V}$ as in (4.5) (the notation becomes cumbersome here!).

Proceeding in exactly the same fashion as before, we can construct $\mathscr{V}^j \in I^{-1,-1-j}$ s.t.

$$\frac{\partial}{\partial x_2}\mathscr{V} - \mathscr{V}\frac{\partial}{\partial x_2} + AE_2\mathscr{V} \in I^{-1,-\infty} + I^{-1,-2}, \tag{4.6}$$

with $\mathscr{V} \sim \sum \mathscr{V}^j$. Now we eliminate $I^{-1,-2}$. Continuing with this tedious process, which we spare the reader, we obtain $\mathscr{V} \in I^{0,0}$ s.t.

$$[\partial/\partial x_2, \mathscr{V}] + AE_1\mathscr{V} = B \quad \text{with } B \in I^0(\mathbf{R}^n \times \mathbf{R}^n; \Lambda_2). \tag{4.7}$$

Now we eliminate the singularities on Λ_2. We construct $T \in I^0(\mathbf{R}^n \times \mathbf{R}^n; \Lambda_1)$ such that

$$[\partial/\partial x_2, T] + AE_1 T = -B \quad \text{microlocally}. \tag{4.8}$$

We have

$$\sigma\left(\frac{\partial}{\partial x_2}T - T\frac{\partial}{\partial x_2}\right)\bigg|_{\Lambda_1} = \frac{d}{dx_2}\left(\sigma(T)|_{\Lambda_1}\right).$$

Using formula (0.2) and the fact $\sigma(E_1)|_{\Lambda_1} = H(x_1 - z_1)$, with H the Heaviside function, we obtain the transport equation for $\sigma(T)$ on Λ_1:

$$\begin{aligned}\frac{d}{dx_2}\sigma(T) + \int \sigma(A)((x_1, x'), (z_1 + t, x'), (0, \xi'), (0, \xi'))\\ \cdot \int_{s \leqslant t} \sigma(T)((z_1 + s, x'), (z_1, x'), (0, \xi'), (0, \xi'))\, ds\, dt\\ = -\sigma(B)\end{aligned} \tag{4.9}$$

(4.8) can be written as

$$\begin{aligned}\frac{d}{dx_2}\sigma(T) + \int \alpha((x_1, x'), (z_1 + t, x'), (0, \xi'), (0, \xi'))\\ \cdot \sigma(T)((z_1 + t, x'), (z_1, x'), (0, \xi'), (0, \xi'))\, dt\\ = -\sigma(B),\end{aligned} \tag{4.10}$$

with α the primitive of $\sigma(A)(\ldots, z_1 + t, \ldots)$ in the t variable.

(4.9) can be solved iteratively for x_2 near 0, using a Picard iteration scheme. Notice that the integration in (4.9) does not occur in the x_2 variable.

We write $T \sim \sum_{j \geqslant 0} T_j$, $T_j \in I^{-j}(\Lambda_2)$, and we solve (4.9) iteratively, using standard procedures, we have then finished the construction of T, as in (4.8), and, therefore, of $\mathscr{V}$, as in (4.1).

Now to prove the theorem we just take $\tilde{\mathscr{V}} = E_1\mathscr{V}(\partial/\partial x_1)$ since

$$\left(\frac{\partial}{\partial x_1}\frac{\partial}{\partial x_2} + A\right)\tilde{\mathscr{V}} = \left(\frac{\partial}{\partial x_2}\mathscr{V} + AE_1\mathscr{V}\right)\frac{\partial}{\partial x_1} = \mathscr{V}\frac{\partial}{\partial x_2}\frac{\partial}{\partial x_1} \quad \text{microlocally},$$

proving the theorem.

We remark that, using Theorem 0.2, one can give another proof of the theorem proved in [**MU**] on the propagation of singularities for operators with double involutive characteristic of product type satisfying the Levi condition.

B. *Applications.* Let $P \in L^1(X)$ be a properly supported, classical, pseudodifferential operator of real principal type, and X a C^∞ manifold. We first study the singularities of complex powers of P.

We define (see [**I**])

$$P^\lambda = \int \frac{t_+^{m-\lambda-1}}{\Gamma(m-\lambda)} U(t) P^m \, dt, \tag{4.11}$$

where $U(t)$ solves

$$(D_t - P)U(t) = 0, \qquad U(0) = \mathrm{Id}, \tag{4.12}$$

and m is chosen sufficiently large in (4.11) so that (4.11) makes sense (see [**I**]). The solution of (4.12) is (mod smoothing) a Fourier integral operator. Thus, we can write

$$k_{P^\lambda} = \int e^{i[\phi(t,x,y,\xi)+t\tau]} a(t,x,y,\xi)(\tau + io)^{\lambda-1} \, d\xi \, d\tau \, dt, \tag{4.13}$$

where a is a classical symbol and ϕ is the solution of the eikonal equation

$$\partial\phi/\partial t = p(t,x,d_{t,x}\phi), \qquad \phi(0,x,\xi) = x \cdot \xi. \tag{4.14}$$

Introducing a cutoff function near $\tau = 0$, we easily see that

$$k_{P^\lambda} \in I^{0,\lambda}(X; \Delta, \Lambda_1), \tag{4.15}$$

with Λ_1 as in the introduction. It is natural then to define

$$k_{P^{B(x,D_x)}} = \int e^{i[\phi(t,x,y,\xi)+t\tau]} a(t,x,y,\xi)\chi(\tau)(\tau + io)^{-b(x,\xi)-1} \, d\xi \, d\tau \, dt, \tag{4.16}$$

with ϕ and a as in (4.13), $B(x, D_x)$ a classical pseudodifferential operator of order 0 with symbol $b(x,\xi)$, which we assume with compact support in x, and $\chi \in C^\infty(\mathbf{R})$: $\chi = 0$ for τ near 0, and $\chi = 1$ for τ large. We have

$$M_- \leqslant b(x,\xi) \leqslant M_+, \tag{4.17}$$

where $M_- = \min(b(x,\xi))$ and $M_+ = \max(b(x,\xi))$.

Now, it is easy to see that $\forall \varepsilon > 0$,

$$\left| D^\alpha_{t,x,y} D^\beta_\xi D^\gamma_\tau \left(a \cdot \chi \cdot (\tau + io)^{-b-1} \right) \right| \leqslant C_{\alpha,\beta,\gamma}(1 + |\xi|)^{-\varepsilon-|\gamma|}(1 + |\tau|)^{-M_- - 1 - |\gamma|}$$

uniformly in t, x, y. Thus we have that the amplitude in (4.16) is a product type symbol, which is, however, not of classical type; thus, $P^B \in I^{-\varepsilon, -M-1}(X; \Delta, \Lambda_1)$ $\forall \varepsilon > 0$. Pseudodifferential powers of operators with simple characteristics and their composition have been used by [**I**] and [**Iw**]. We leave to the future further applications of these powers and the functional calculus developed here.

REFERENCES

[DG] J. J. Duistermaat and V. Guillemin, *The spectrum of positive elliptic operators and periodic bicharacteristics*, Invent. Math. **29** (1975), 39–79.

[DH] J. J. Duistermaat and L. Hörmander, *Fourier integral operators*. II, Acta. Math. **128** (1972), 183–269.

[GU] V. Guillemin and G. Uhlmann, *Oscillatory integrals with singular symbols*, Duke Math. J. **48** (1981), 251–267.

[I] V. Ja. Ivrii, *Sufficient conditions for regular and completely regular hyperbolicity*, Trudy Moskov. Mat. Obsc. **33** (1976), 3–65, English transl. in Trans. Moscow Math. Soc. **1** (1978), 1–65.

[Iw] N. Iwasaki, *Cauchy problems for effective hyperbolic equations* (*a special case*), R.I.M.S., 1983 (preprint).

[JM] L. Jiang and R. Melrose, *Totally bicharacteristic boundary problem for the wave equation* (preprint).

[MU] R. Melrose and G. Uhlmann, *Lagrangian intersection and the Cauchy problem*, Comm. Pure Appl. Math. **32** (1979), 483–519.

[M] G. Mendoza, *Symbol calculus associated with intersecting Lagrangian*, Comm. Partial Differential Equations **7** (9) (1982), 1035–1116.

[U] G. Uhlmann, *Pseudodifferential operators with double involutive characteristics*, Comm. Partial Differential Equations **2** (7) (1977), 713–779.

UNIVERSIDAD NACIONAL AUTÓNOMA, MEXICO

MASSACHUSETTS INSTITUTE OF TECHNOLOGY

Proceedings of Symposia in Pure Mathematics
Volume **43** (1985)

Uniqueness in a Class of Nonlinear Cauchy Problems

M. S. BAOUENDI

In this paper we discuss some recent results in the uniqueness for certain complex first-order nonlinear Cauchy problems. These results were obtained jointly with C. Goulaouic and F. Treves. We refer to [**1**] for more details.

Consider the following nonlinear overdetermined system of first-order differential equations:

$$\partial u/\partial t_j = F_j(t, x, u, u_x), \qquad 1 \leqslant j \leqslant m, \tag{1}$$

$$u(t_0, x) = u_0(x). \tag{2}$$

Here $t \in \mathbb{R}^m$, $x \in \mathbb{R}^n$. The functions F_j are assumed to be holomorphic in a neighborhood of $(t_0, x_0, \zeta_0, \xi_0) \in \mathbb{R}^m \times \mathbb{R}^n \times \mathbb{C} \times \mathbb{C}^n$.

We need to introduce the following complex vector fields. For $j = 1, \ldots, m$ set

$$\mathscr{L}_j = \frac{\partial}{\partial t_j} - \sum_{k=1}^{n} \frac{\partial F_j}{\partial \xi_k} \frac{\partial}{\partial x_k} + \left(F_j - \sum_{k=1}^{n} \xi_k \frac{\partial F_j}{\partial \xi_k} \right) \frac{\partial}{\partial \zeta} + \sum_{k=1}^{n} \left(\frac{\partial F_j}{\partial x_k} + \frac{\partial F_j}{\partial \zeta} \xi_k \right) \frac{\partial}{\partial \xi_k}.$$

We assume that system (1) is in *involution*—i.e., the following commutation conditions are satisfied:

$$[\mathscr{L}_k, \mathscr{L}_j] = 0, \qquad j, k \in \{1, \ldots, m\}. \tag{3}$$

Note that condition (3) implies existence and uniqueness of a local analytic solution u of (1) and (2) for a given analytic Cauchy datum u_0 satisfying

$$u_0(x_0) = \zeta_0, \qquad u_{0,x}(x_0) = \xi_0. \tag{4}$$

We are interested here in the following uniqueness question. Assume that u and $u_\#$ are two (germs of) C^2 complex-valued functions satisfying (1) and (2) near (t_0, x_0) in real space with the same C^2 Cauchy datum u_0 satisfying (4). Can one conclude that $u \equiv u_\#$ in a neighborhood of (t_0, x_0)?

1980 *Mathematics Subject Classification*. Primary 35A07, 35A35, 35N10.

0082-0717/85 $1.00 + $.25 per page

Let us first point out that the answer to this question is not known even when $m = 1$ and u and $u_\#$ are of class C^∞. However, we can prove that such uniqueness holds in each of the following cases:

(a) $n = 1$; i.e., x is a single real variable;

(b) the Cauchy datum u_0 is an analytic function;

(c) the semilinear case—i.e.,

$$F_j(t, x, \zeta, \xi) = \sum_{k=1}^{n} \lambda_j^k(t, x)\xi_k + f_j(t, x, \zeta), \tag{5}$$

where the coefficients λ_j^k are holomorphic near (t_0, x_0), and f_j is holomorphic near (t_0, x_0, ζ_0).

We state the main results of this paper.

THEOREM 1. *Suppose that condition* (3) *holds and $n = 1$. If the Cauchy datum u_0 is of class C^2 near x_0, then there is at most one solution u of class C^2 to the Cauchy problem* (1), (2).

THEOREM 2. *Suppose that condition* (3) *holds and u_0 is analytic in a neighborhood of x_0 (satisfying* (4)*). Then any C^2 solution u of* (1), (2) *is analytic near* (t_0, x_0); *therefore it is unique.*

THEOREM 3. *In the semilinear case* (5), *assuming* (3), *if u_0 is of class C^1 in a neighborhood of x_0, then there is at most one C^1 solution u to the Cauchy problem* (1), (2).

The proofs of Theorems 1–3 rely in a crucial way on the earlier approximation and uniqueness results of Baouendi–Treves [**2**] in the linear case. We shall now give some idea of these proofs and refer to [**1**] for more details.

Without loss of generality we may assume that $t_0 = 0$, $x_0 = 0$. If u is a C^2 function satisfying (4) and defined near (t_0, x_0), and if v is a holomorphic function defined near $(t_0, x_0, \zeta_0, \xi_0) \in \mathbb{C}^{2n+m+1}$, we write

$$\tilde{v}(t, x) = v(t, x, u(t, x), u_x(t, x)).$$

With this notation we define the vector fields

$$L_j = \frac{\partial}{\partial t_j} - \sum_{k=1}^{n} \widetilde{\frac{\partial F_j}{\partial \xi_k}}(t, x)\frac{\partial}{\partial x_k}, \qquad 1 \leqslant j \leqslant m. \tag{6}$$

Note that the coefficients of L_j are of class C^1, whereas the coefficients of $\mathscr{L}_j$ are analytic. An easy computation gives the following lemma.

LEMMA 1. *For every holomorphic function v as above, the following identities hold*:

$$L_j\tilde{v} = \widetilde{\mathscr{L}_j v}, \qquad 1 \leqslant j \leqslant m. \tag{7}$$

Lemma 1 has two immediate consequences:

1. The system of vector fields $(L_1, \ldots, L_m)$ is *integrable*. More precisely, there are C^1 functions $\tilde{Z}_p(t, x)$ satisfying

$$L_k\tilde{Z}_p \equiv 0, \qquad p \in \{1, \ldots, n\},\ k \in \{1, \ldots, m\}, \tag{8}$$

and $\tilde{Z}_p|_{t=0} = x_p$ $(x = (x_1, \ldots, x_n))$.

2. There exist $n+1$ holomorphic functions $H, H_1, \ldots, H_n$, defined near $(t_0, x_0, \zeta_0, \xi_0) \in \mathbb{C}^{2n+m+1}$, satisfying

$$H|_{t=0} = \zeta, \qquad H_k|_{t=0} = \xi_k, \quad 1 \leqslant k \leqslant n,$$

such that the C^2 function u is a solution of (1) if and only if there exist $(n+1)$ C^1 functions $w, w_1, \ldots, w_n$, defined near (t_0, x_0) in $\mathbb{R}^{n+m}$, satisfying

$$w(t_0, x_0) = \zeta_0, \quad w_k(t_0, x_0) = \xi_{0,k}, \qquad 1 \leqslant k \leqslant n,$$

$$L_j w = L_j w_k = 0, \qquad j \in \{1, \ldots, m\}, \; k \in \{1, \ldots, n\}, \tag{9}$$

and

$$u = H(t, x, w, w_1, \ldots, w_n), \qquad u_{x_k} = H_k(t, x, w, w_1, \ldots, w_n). \tag{10}$$

PROOF OF THEOREM 2. Assume that u is a solution of (1) and u_0 is analytic. Replacing u by $u - u_0$, we can assume that $u_0 \equiv 0$. Making use of consequence 2 above, we conclude that

$$w|_{t=0} = u_0 = 0, \qquad w_k|_{t=0} = u_{0,x_k} = 0, \quad 1 \leqslant k \leqslant n. \tag{11}$$

Since the system of vector fields $L_1, \ldots, L_m$ is integrable, using the uniqueness result of **[2]** we conclude from (9) and (11) that $w = w_k = 0$, $1 \leqslant k \leqslant n$. Substituting in (10) yields

$$u(t, x) = H(t, x, 0, \ldots, 0),$$

which proves that u is analytic and therefore unique.

Note that the uniqueness result of **[2]** is proved under the hypothesis that the coefficients of the L_j's and the $\tilde{Z}_k$'s are C^∞. Inspection of the proof shows that the same argument works when the $\tilde{Z}_k$'s are only C^1 and the coefficients of the L_j's are only C^0.

PROOF OF THEOREM 3. In the semilinear case the vector fields L_j defined by (6) are independent of u. Using a slightly easier version of consequence 2, we can see that if $u \in C^1$ satisfies (1) then

$$u(t, x) = H(t, x, w(t, x)),$$

where H is holomorphic near (t_0, x_0, ζ_0) and

$$L_j w = 0, \quad 1 \leqslant j \leqslant n, \qquad w(0, x) = u_0(x).$$

Theorem 3 follows from the uniqueness for the vector fields L_j.

PROOF OF THEOREM 1. It is in the proof of this theorem that the approximation formula of **[1]** is crucial. Here x is one single real variable. If $u \in C^1$ satisfies (1), then, making use of consequence 2, we have

$$u = H(t, x, w, w_1), \qquad u_x = H_1(t, x, w, w_1), \tag{12}$$

with w and w_1 satisfying (9). Note that

$$w(0, x) = u_0(x), \quad w_1(0, x) = u_{0,x}(x).$$

Let $\tilde{Z}$ be the function satisfying (8) (here $n = 1$, $\tilde{Z} = \tilde{Z}_1$). If $g \in C_0^\infty(\mathbb{R})$, $g \equiv 1$ near 0, and supp g is small enough, then inspection of the proof of the approximation formula in [**2**] shows that the sequence of entire functions

$$(13) \qquad P_\nu(z) = \frac{\nu}{\sqrt{\pi}} \int e^{-\nu^2[z-y]^2} g(y) u_0(y)\, dy$$

converges uniformly in $\tilde{Z}(K)$ to W, where K is a compact neighborhood of 0 in $\mathbb{R}^{m+1}$. In addition,

$$(14) \qquad w = W \circ \tilde{Z} \quad \text{in } K.$$

Similarly, replacing u_0 by $u_{0,x}$ in (13), we have, with a different function W_1,

$$(15) \qquad w_1 = W_1 \circ \tilde{Z} \quad \text{in } K.$$

Let $u_\sharp$ be a second C^1 solution of (1) with the same Cauchy datum u_0. Let $Z_\sharp$, $w_\sharp$, $w_{1,\sharp}$, and $K_\sharp$ be the analog of Z, w, w_1, and K for $u_\sharp$ (instead of u). Because the entire functions (13) depend only on u_0 (and not on u or $u_\sharp$), we have

$$(16) \qquad w_\sharp = W \circ \tilde{Z}_\sharp, \quad w_{1,\sharp} = W_1 \circ \tilde{Z}_\sharp \qquad \text{in } K_\sharp.$$

It can be shown that the compact sets K and $K_\sharp$ can be chosen so that $\tilde{Z}(K)$ and $\tilde{Z}_\sharp(K_\sharp)$ have Lipschitz boundaries and that W and W_1 are Lipschitz continuous on $\tilde{Z}(K) \cup \tilde{Z}_\sharp(K_\sharp)$.

Recall that

$$(17) \qquad \begin{aligned} \tilde{Z}(t, x) &= Z(t, x, u(t, x), u_x(t, x)), \\ \tilde{Z}_\sharp(t, x) &= Z(t, x, u_\sharp(t, x), u_{\sharp,x}(t, x)), \end{aligned}$$

where $Z(t, x, \zeta, \xi)$ satisfies

$$\mathscr{L}_j Z = 0, \quad j \in \{1, \ldots, m\}, \qquad Z|_{t=0} = x.$$

Using (12) (and its analog for $u_\sharp$), (14)–(17), and the properties of W and W_1 stated above, we conclude that, for (t, x) small enough,

$$|u(t, x) - u_\sharp(t, x)| \leqslant \tfrac{1}{2}|u(t, x) - u_\sharp(t, x)|,$$

which implies $u \equiv u_\sharp$.

References

1. M. S. Baouendi, C. Goulaouic and F. Treves, *Uniqueness in certain first-order nonlinear complex Cauchy problems*, Comm. Pure Appl. Math. (to appear).

2. M. S. Baouendi and F. Treves, *A property of the functions and distributions annihilated by a locally integrable system of complex vector fields*, Ann. of Math. (2) **113** (1981), 341–421.

3. C. Carathéodory, *Calculus of variations and partial differential equations of the first order*. Part I, Holden–Day, San Francisco, 1965.

PURDUE UNIVERSITY

Proceedings of Symposia in Pure Mathematics
Volume **43** (1985)

Propagation of Smoothness for Nonlinear Second-Order Strictly Hyperbolic Differential Equations

MICHAEL BEALS[1]

Introduction. Let $u \in H^s(\mathcal{O})$, $\mathcal{O} \subset \mathbb{R}^n$, $s > n/2$, be the solution of a nonlinear strictly hyperbolic equation with given singular support on an initial hypersurface or in the past. As is well known, the singular support of u in the future will generally be larger than that of the solution to the corresponding linear problem. These "anomalous singularities" that develop have been shown to arise in two ways. If singularity-bearing characteristics for the linear problem cross, nonlinear singularities can propagate along all forward characteristics issuing from the crossing point (Rauch–Reed [**17**], Lascar [**12**], Beals [**2**]). And, for $n > 2$, if $\Gamma = \{x(s)\}$ is a characteristic that is the projection of the two null bicharacteristics $\Gamma_\pm = \{(x(s), \pm\xi(s))\}$, and if the corresponding linear solution has wave front set containing $\Gamma_+ \cup \Gamma_-$, then nonlinear singularities can "self-spread" from Γ along all possible characteristics (Beals [**3**]).

The nonlinear propagation theory that has been developing over the past several years is an attempt to characterize the location and strength of anomalous singularities. They will always be smoother than the linear singularities that create them: if $u \in H^s(\mathcal{O})$, then the worst such singularities will be of type roughly $H^{2s-n/2}(\mathcal{O})$ (Rauch [**16**], Bony [**7**]). If the order of the equation is greater than two, then nonlinear singularities of this strength do actually appear, even when $n = 2$, as a result of the crossing of singularities (Rauch–Reed [**17, 18**]). On the other hand, if $n = 2$ and the equation is semilinear and of order two, no anomalous singularities appear. In higher dimensions, additional hypotheses on the singular structure of u in the past, or on the initial surface, will again prevent or sharply curtail the appearance of nonlinear singularities: conormal solutions

1980 *Mathematics Subject Classification*. Primary 35L70.
[1] Research partially supported by NSF Grant #DMS-8201281.

(Bony [8, 9], Melrose–Ritter [13]), striated solutions (Rauch–Reed [19]), and angularly smooth solutions (Beals [4]). But without such extra conditions, singularities of strength approximately $2s - n/2$ will appear in a solution to a nonlinear equation of order greater than two.

For second-order equations, however, even if $n > 2$, anomalous singularities of order $2s - n/2$ will not appear. In [3] it is shown that if $\Box u = f(u)$, where f is a polynomial and $\Box$ is the ordinary wave operator, then anomalous singularities will be at worst of order roughly $3s - n$, and solutions with such singularities due to self-spreading are constructed. (In [2] solutions with singularities of this order due to crossing characteristics are shown to exist.) In this paper we extend the $3s - n$ smoothness result to the general second-order, strictly hyperbolic equation, providing complete proofs for the general semilinear equations $p_2(x, D)u = f(x, u)$ and $p_2(x, D)u = f(x, u, Du)$, with f smooth, and an outline of the proof in the quasilinear case. The idea behind the proofs bears certain similarities to the arguments used in the study of conormal or stratified solutions cited above. In those cases, microlocal control over certain derivatives $D_j u$ is assumed in the past; commutators of the D_j with the operator $p(x, D)$ are expressed microlocally in terms of the D_j and of p, and estimates on u in the future are proved in terms of estimates on $D_j(f(u))$ and $p(x, D)u$. In our case, only the operator $p_2(x, D)$ is present, but an algebraic property of $p_2(x, \xi)$ allows estimates on $e_{2k}(D)(f(u))$ for certain microlocally elliptic operators e_{2k} in terms of estimates on $(p_2(x, D))^k u$ in the case of singularities due to self-spreading. Since $(p_2(x, D))^k u \in H^{s-k}(\mathcal{O})$ if $p_2(x, D)u = f(u)$ and $k < s - n/2$ (see Lemma 1.5), it follows that an extra $s - n/2$ derivatives are estimated where e_{2k} is elliptic. A geometric property of the characteristic set of p_2 then allows the conclusion that, for both self-spread singularities and those due to characteristics crossing, extra derivatives in characteristic directions of order $s - n/2$ are controlled, allowing the improvement of the regularity theorem for second-order equations from $2s - n/2$ to $3s - n$.

In §1 the space $\tilde{H}^s(p_2)$ of functions u satisfying $(p_2(x, D))^k u \in H^{s-k}(\mathcal{O})$ for certain k is considered. It is an algebra, and its elements can be microlocalized. Most importantly, if (x_0, ξ_0) is characteristic for p_2, the functions in $\tilde{H}^s(p_2)$ that are microlocally in H^g at (x_0, ξ_0) are shown to form an algebra for $g < 3s - n$ (Theorem 1.12). (The corresponding statement for functions only assumed to be in $H^s(\mathcal{O})$ is true only for $g \leqslant 2s - n/2$.) This property is proved to follow from the algebraic and geometric properties of the second-order differential symbol $p_2(x, \xi)$ mentioned above.

In §2 the algebra result is applied in a simple bootstrap argument using Hörmander's linear theorem, exactly as in Rauch [16] for the $2s - n/2$ case, to prove that microlocal smoothness along null bicharacteristics is propagated up to order $3s - n + 1$ for equations $p_2(x, D)u = f(x, u)$, with $u \in H^s(\mathcal{O})$ (Theorem 2.1). For the general semilinear equation $p_2(x, D)u = f(x, u, Du)$, with $u \in H^{s+1}(\mathcal{O})$, $s > n/2$, an additional argument, along the lines of Beals–Reed [5], is provided. If the equation is differentiated, an extension of Hörmander's linear result to the case in which the lower-order terms have coefficients in $\tilde{H}^s(p_2) \cap H^g_{ml}(x_0, \xi_0)$ allows the completion of the proof of the nonlinear propagation property (Theorem 2.5). Finally, in §3 the analogous algebraic and microlocal results for quasilinear second-order equations are stated, culminating in the description of the propagation of smoothness (Theorem 3.7). Details of the proof for this case will appear elsewhere.

I would like to thank Professor G. Metivier for discussions that led to the present form of Lemma 1.9, which is a considerable simplification of the argument given even for the case of $\square$ in Lemma 1.5(iii) in [**3**].

Notation. $\langle\xi\rangle = (1 + |\xi|^2)^{1/2}$, and $A \approx B$ means that there are constants $c, C > 0$ with $cB \leqslant A \leqslant CB$. $f(x, u, \ldots, D^j u)$ stands for a smooth function of x, u, and derivatives $D^\alpha u$ of order $|\alpha| \leqslant j$. If $\mathcal{O} \subset \mathbb{R}^n$ is open, $H^s(\mathcal{O})$ is the Sobolev space of distributions u with $\phi u \in H^s(\mathbb{R}^n)$ for all smooth functions ϕ with support in $\mathcal{O}$. If $(x_0, \xi_0) \in \mathbb{R}^n \times \mathbb{R}^n \setminus 0$, $H^s_{ml}(x_0, \xi_0)$ is the space of distributions u which are in H^s microlocally at (x_0, ξ_0). $\mathrm{WF}(u) \subset \mathbb{R}^n \times \mathbb{R}^n \setminus 0$ is the wave front set of u, and $\Pi\mathrm{WF}(u)$ stands for the projection on the second factor. For K_1, K_2 cones in $\mathbb{R}^n \setminus 0$, $K_1 \Subset K_2$ means that there is an open cone K with $\overline{K}_1 \cap S^{n-1} \subset K \cap S^{n-1} \subset K_2 \cap S^{n-1}$. $\overline{K}$ denotes the closure of K; K^c, the complement of K; and χ_K, the characteristic function of K.

1. Spaces of functions satisfying a second-order equation. If $u_\pm \in H^s(\mathbb{R}^n)$, $s > n/2$, have wave front sets with $\Pi\mathrm{WF}(u_\pm) = \{\pm\rho\xi_0, \rho > 0\}$, then, in general, $\Pi\mathrm{WF}(u_+ u_-) = \mathbb{R}^n \setminus 0$, and microlocally away from $\pm\xi_0$ the singularity of $u_+ u_-$ is of strength $2s - n/2$. But if $u_\pm$ also satisfy the equations $\square u_\pm = f_\pm(x, u_+, u_-)$, it is not difficult to check that, for $\chi_\pm(\xi)$ with support near the characteristic set $\xi_1 = \pm|\xi'|$,

$$\chi_\pm(\xi)\langle\xi\rangle^s\big(\xi_1 \mp |\xi'|\big)^{s-n/2}\hat{u}_\pm(\xi) \in L^2(\mathbb{R}^n).$$

This extra control over the Fourier transform of $u_\pm$ for ξ near the characteristic set of $\square$ prevents the appearance in the wave front set of $u_+ u_-$ of singularities as strong as in the general case: away from the tangent hyperplane to char$\square$ at $\pm\xi_0$, the singularity of $u_+ u_-$ is of strength $3s - n$ (see [**3**]). In order to reach a similar conclusion for solutions of general second-order equations, it is necessary to define spaces that measure the extra control over behavior of the function microlocally near the characteristic set of the operator. Fix $s > n/2$, $n \geqslant 2$, $\mathcal{O} \subset \mathbb{R}^n$ open, and $p_2(x, D)$ a second-order *differential* operator.

DEFINITION 1.1. $u \in \tilde{H}^s(p_2)$ if $(p_2(x, D))^j u \in H^{s-j}(\mathcal{O})$ for $0 \leqslant j < s - n/2$ and $(p_2(x, D))^k u \in H^{2s-n/2-2k-}(\mathcal{O})$ for $s - n/2 \leqslant k < s - n/2 + 1$. Here $H^{r-}(\mathcal{O}) = \bigcap_{\varepsilon>0} H^{r-\varepsilon}(\mathcal{O})$.

Since the conditions on elements of $\tilde{H}^s(p_2)$ are given in terms of differential, rather than pseudodifferential, operators, it is not difficult to check that this space is an algebra.

LEMMA 1.2. *If* $u \in \tilde{H}^s(p_2)$, $s > n/2$, *and* $g \in C^\infty$, *then* $g(x, u) \in \tilde{H}^s(p_2)$.

PROOF. If u is smooth, it is easily checked using induction, the chain rule, and Leibniz's rule that

$$(p_2(x,D))^j g(x,u) = g\left(x,u,\ldots,D^{\alpha_i}(p_2(x,D))^{a_i}u,\ldots\right), \qquad |\alpha_i| + a_i \leqslant j.$$

For $u \in \tilde{H}^s(p_2)$ and $j < s - n/2$, $D^{\alpha_i}(p_2(x,D))^{a_i}u \in H^{s-j}(\mathcal{O})$, and, by Schauder's lemma, the use of the chain rule and the Leibniz rule is justified, and $g(x,u,\ldots,D^{\alpha_i}p_2^{a_i}u,\ldots) \in H^{s-j}(\mathcal{O})$. If $s - n/2 \leqslant k < s - n/2 + 1$,

$$(p_2(x,D))^k g(x,u) = p_2(x,D) g\left(x,u,\ldots,D^{\alpha_i}(p_2(x,D))^{a_i}u,\ldots\right),$$
$$|\alpha_i| + a_i \leqslant k - 1.$$

For first-order differential operators ∂_1, ∂_2,

$$\partial_1 g(x,u,\ldots,D^{\alpha_i}p_2^{a_i}u,\ldots) = \sum \tilde{g}_i(x,u,\ldots,D^{\alpha_i}p_2^{a_i}u,\ldots) D^{\alpha_i}p_2^{a_i}\partial_1 u$$

and

$$\partial_2\partial_1 g(x,u,\ldots,D^{\alpha_i}p_2^{a_i}u,\ldots) = \sum \tilde{g}_i(x,u,\ldots,D^{\alpha_i}p_2^{a_i}u,\ldots) D^{\alpha_i}p_2^{a_i}\partial_2\partial_1 u$$
$$+ \sum \tilde{g}_{i,j}(x,u,\ldots,D^{\alpha_i}p_2^{a_i}u,\ldots)(D^{\alpha_i}p_2^{a_i}\partial_1 u)(D^{\alpha_j}p_2^{a_j}\partial_2 u),$$

so it follows that

$$(p_2(x,D))^k g(x,u) = \sum \tilde{g}_i(x,u,\ldots,D^{\alpha_i}p_2^{a_i}u,\ldots) D^{\alpha_i}(p_2(x,D))^{a_i+1}u$$
$$+ \sum \tilde{g}_{i,j}(x,u,\ldots,D^{\alpha_i}p_2^{a_i}u,\ldots)(D^{\alpha_i+1}p_2^{a_i}u)(D^{\alpha_j+1}p_2^{a_j}u), \tag{1.1}$$

where $|\alpha_i| + a_i \leqslant k - 1$. By Schauder's lemma (since $H^r(\mathbb{R}^n) \cdot H^\rho(\mathbb{R}^n) \subset H^\rho(\mathbb{R}^n)$ for $|\rho| \leqslant r$, $r > n/2$) and the definition of $\tilde{H}^s(p_2)$, the terms in the first sum on the right side of (1.1) are in $H^{2s-n/2-2k-}(\mathcal{O})$ because $n/2 - 2 < 2s - n/2 - 2k < n/2 < s - (k-1)$. Similarly, the terms in the second sum on the right side of (1.1) are of the form

$$H^{s-k+1}(\mathcal{O}) \cdot H^{s-k}(\mathcal{O}) \cdot H^{s-k}(\mathcal{O}) \subset H^{s-k}(\mathcal{O}) \cdot H^{s-k}(\mathcal{O}),$$

and the next lemma shows that such functions are in $H^{2s-n/2-2k-}(\mathcal{O})$ because $2s - 2k \geqslant n - 2 \geqslant 0$. ■

The following well-known result is frequently used in the rest of §1 (see e.g. **[5]**).

(1.2) If $f, g \in L^2$ and $\sup \int G^2(\xi,\eta)\,d\eta < \infty$ or $\sup_\eta \int G^2(\xi,\eta)\,d\xi < \infty$, then, for $h = \int G(\xi,\eta) f(\xi - \eta) g(\eta)\,d\eta$,

$$\|h\|_{L^2} \leqslant C\|f\|_{L^2}\|g\|_{L^2}.$$

LEMMA 1.3. *For $i = 1, 2$ let $r_i \leqslant n/2$, $r_1 + r_2 \geqslant 0$. If $u_i \in H^{r_i}(\mathbb{R}^n)$, then $u_1u_2 \in H^{r_1+r_2-n/2-}(\mathbb{R}^n)$.*

PROOF.

$$\langle\xi\rangle^{r_1+r_2-n/2-\varepsilon}|u_1u_2(\xi)| = \int G(\xi,\eta) f(\xi-\eta) g(\eta)\,d\eta,$$

where f, g are in $L^2(\mathbb{R}^n)$ and

$$|G(\xi,\eta)| = \frac{\langle\xi\rangle^{r_1+r_2-n/2-\varepsilon}}{\langle\xi-\eta\rangle^{r_1}\langle\eta\rangle^{r_2}}.$$

Consider three cases for the relative sizes of ξ, $\xi - \eta$, and η.

(i) If $|\xi - \eta| \leqslant \frac{1}{2}|\xi|$, then $\frac{1}{2}|\xi| \leqslant |\eta| \leqslant \frac{3}{2}|\xi|$ and

$$|G(\xi, \eta)| \leqslant \frac{C}{\langle\xi\rangle^{n/2+\varepsilon-r_1}\langle\xi-\eta\rangle^{r_1}} \leqslant \frac{C}{\langle\xi-\eta\rangle^{n/2+\varepsilon}}.$$

(ii) If $|\xi - \eta| \geqslant \frac{1}{2}|\xi|$ and $|\eta| \leqslant \frac{1}{2}|\xi|$, then $|\xi - \eta| \leqslant \frac{3}{2}|\xi|$ and, hence,

$$|G(\xi, \eta)| \leqslant \frac{C}{\langle\xi\rangle^{n/2+\varepsilon-r_2}\langle\eta\rangle^{r_2}} \leqslant \frac{C}{\langle\eta\rangle^{n/2+\varepsilon}}.$$

(iii) If $|\xi - \eta| \geqslant \frac{1}{2}|\xi|$ and $|\eta| \geqslant \frac{1}{2}|\xi|$, then

$$|G(\xi, \eta)| \leqslant \frac{1}{\langle\xi\rangle^{n/2+\varepsilon}} \cdot \max\left(\frac{\langle\xi\rangle^{r_1+r_2}}{\langle\xi-\eta\rangle^{r_1+r_2}}, \frac{\langle\xi\rangle^{r_1+r_2}}{\langle\eta\rangle^{r_1+r_2}}\right)$$

$$\leqslant \frac{C}{\langle\xi\rangle^{n/2+\varepsilon}}, \quad \text{since } r_1 + r_2 \geqslant 0.$$

It follows that G satisfies the estimate for (1.2) in each of the three cases, so $u_1 u_2 \in H^{r_1+r_2-n/2-\varepsilon}(\mathbb{R}^n)$ for any $\varepsilon > 0$. ■

If $p_2(x, \xi)$ is strictly hyperbolic, it may be factored microlocally as the product of a first-order elliptic symbol $e_1(x, \xi)$ and a first-order hyperbolic symbol $h_1(x, \xi)$. Functions u in $H^s(p_2)$ have better behavior near the characteristic set of p_2 than the general function does in $H^s(\mathcal{O})$: roughly, $h_1^r(x, D)u \in H^s(\mathcal{O})$ for all $r < s - n/2$. Away from the characteristic set the following property holds.

LEMMA 1.4. *If $u \in \tilde{H}^s(p_2)$, $s > n/2$, and p_2 is microlocally elliptic at (x_0, ξ_0), then $u \in H^{2s-n/2-}_{ml}(x_0, \xi_0)$.*

PROOF. For $s - n/2 \leqslant k < s - n/2 + 1$, $(p_2(x, D))^k$ is a differential operator of order $2k$, microlocally elliptic at (x_0, ξ_0). Since $(p_2(x, D))^k u$ is in $H^{2s-n/2-2k-}(\mathcal{O})$, the conclusion follows from the usual microlocal elliptic regularity theory. ■

The space $\tilde{H}^s(p_2)$ is of interest because it includes solutions of certain semilinear equations with principal symbol p_2. (For the general semilinear equation, see §2.)

LEMMA 1.5. *Let $u \in H^s(\mathcal{O})$, f, $\tilde{f}_\alpha \in C^\infty$, and suppose*

$$p_2(x, D)u = f(x, u) + \sum_{|\alpha|=1} \tilde{f}_\alpha(x, u) D^\alpha u \quad \text{on } \mathcal{O}. \tag{1.3}$$

Then $u \in \tilde{H}^s(p_2)$.

PROOF. By induction for $j < s - n/2$,

$$(p_2(x, D))^j u = \tilde{f}(x, u, \ldots, D^j u), \tag{1.4}$$

since

$$\begin{aligned} p_2(x, D)h(x, u, \ldots, D^{j-1}u) &= h(x, u, \ldots, D^j u) \\ &\quad + \sum_{|\alpha|=j-1} h_\alpha(x, u, \ldots, D^{j-1}u) D^\alpha p_2(x, D)u \\ &= \tilde{f}(x, u, \ldots, D^j u) \quad \text{by (1.3)}. \end{aligned}$$

If $j < s - n/2$, then, by Schauder's lemma, $\tilde{f}(x, u, \ldots, D^j u) \in H^{s-j}(\mathcal{O})$. Now fix k with $s - n/2 \leqslant k < s - n/2 + 1$. As in the proof of Lemma 1.2, from (1.4) it follows that

$$\begin{aligned}(p_2(x, D))^k u &= p_2(x, D) f(x, u, \ldots, D^{k-1}u) \\ &= \tilde{h}(x, u, \ldots, D^{k-1}u) + \sum_{|\alpha|=k} \tilde{h}_\alpha(x, u, \ldots, D^{k-1}u) D^\alpha u \\ &\quad + \sum_{\substack{|\alpha|=k \\ |\beta|=k}} \tilde{h}_{\alpha\beta}(x, u, \ldots, D^{k-1}u) D^\alpha u D^\beta u \\ &\quad + \sum_{|\alpha|=k-1} \tilde{f}_\alpha(x, u, \ldots, D^{k-1}) p_2(x, D) D^\alpha u.\end{aligned}$$

From (1.3), for $|\alpha| = k - 1$,

$$\begin{aligned}p_2(x, D) D^\alpha u &= p_k(x, D) u + D^\alpha(p_2(x, D) u) \\ &= \sum_{|\beta|=k} \tilde{g}_\beta(x, u, \ldots, D^{k-1}u) D^\beta u.\end{aligned}$$

Thus

$$\begin{aligned}(p_2(x, D))^k u = g(x, u, \ldots, D^{k-1}u) &+ \sum_{|\alpha|=k} g_\alpha(x, u, \ldots, D^{k-1}u) D^\alpha u \\ &+ \sum_{\substack{|\alpha|=k \\ |\beta|=k}} g_{\alpha\beta}(x, u, \ldots, D^{k-1}u) D^\alpha u D^\beta u.\end{aligned}$$

As in the proof of Lemma 1.2, by Schauder's lemma and Lemma 1.3, the right side of this equation is in $H^{2s-n/2-2k-}(\mathcal{O})$. ∎

In order to study the propagation of microlocal regularity, it is important that the space $\tilde{H}^s(p_2)$ allow microlocalization. This is indeed the case.

LEMMA 1.6. *Let $u \in \tilde{H}^s(p_2)$, $s > n/2$, and $b_0(x, \xi) \in S^0_{1,0}(\mathbb{R}^n)$. Then $b_0(x, D)u \in \tilde{H}^s(p_2)$.*

PROOF. By induction there are symbols $b_i(x, \xi) \in S^i_{1,0}(\mathcal{O})$ with

$$\begin{aligned}(p_2(x, D))^j b_0(x, D) = b_0(x, D)(p_2(x, D))^j &+ b_1(x, D)(p_2(x, D))^{j-1} \\ &+ \cdots + b_j(x, D),\end{aligned}$$

since

$$\begin{aligned}&p_2(x, D)\Big(b_0(x, D)(p_2(x, D))^{j-1} + b_1(x, D)(p_2(x, D))^{j-2} \\ &\qquad\qquad + \cdots + b_{j-1}(x, D)\Big) \\ &= b_0(x, D)(p_2(x, D))^j + [p_2(x, D), b_0(x, D)](p_2(x, D))^{j-1} \\ &\quad + b_1(x, D)(p_2(x, D))^{j-1} + [p_2(x, D), b_1(x, D)](p_2(x, D))^{j-2} \\ &\quad + \cdots + b_{j-1}(x, D) p_2(x, D) + \big[p_2(x, D), b_{j-1}(x, D)\big].\end{aligned}$$

For $j < s - n/2$ it follows from the definition of $\tilde{H}^s(p_2)$ that

$$(p_2(x, D))^j b_0(x, D)u \in H^{s-j}(\mathcal{O}),$$

and, for $s - n/2 \leqslant k < s - n/2 + 1$,

$$(p_2(x, D))^k b_0(x, D)u = b_0(x, D)(p_2(x, D))^k u + \sum_{j=1}^{k} b_j(x, D)(p_2(x, D))^{k-j} u.$$

The first term on the right is in $H^{2s-n/2-2k-}(\mathcal{O})$, and the remaining terms are in $H^{s-k}(\mathcal{O}) \subset H^{2s-n/2-2k-}(\mathcal{O})$ because $s - k \geqslant 2s - n/2 - 2k$. ∎

We now recall the argument of Rauch [**16**], which is the key to the description of the algebraic properties of functions that are microlocally better in some directions than in others. A proof is included, because, in part, it is different from the usual one, and the analogue of this part will appear when the algebraic properties of the spaces $\tilde{H}^s(p_2) \cap H^g_{ml}(x_0, \xi_0)$ are proved later. Moreover, this proof yields the conclusion that the ordinary spaces $H^s_{\text{loc}} \cap H^r_{ml}(x_0, \xi_0)$ of functions in H^s, which are microlocally in H^r, $n/2 < s \leqslant r$ (see e.g. [**5**]), are algebras for $r \leqslant 2s - n/2$ rather than just $r < 2s - n/2$. This result was first given by Meyer [**14**] using Bony's calculus [**7**].

LEMMA 1.7. *Let K_1, K_2, and K be cones in $\mathbb{R}^n \setminus 0$. Assume that $u_i \in H^{s_i}(\mathbb{R}^n)$ and $\Pi \mathrm{WF}(u_i) \subset K_i$, $i = 1, 2$.*

(1) *If $K \Subset K_2^c$, then $\chi_K(D)(u_1u_2) \in H^{s_1-k}(\mathbb{R}^n)$ if $k \geqslant 0$, $s_1 - k \geqslant 0$, $s_2 + k > n/2$.*

(2) *If $K \Subset K_1^c \cap K_2^c$, then $\chi_K(D)(u_1u_2) \in H^{s_1+s_2-n/2}(\mathbb{R}^n)$ if $s_1 + s_2 > n/2$.*

PROOF. (1)

$$\langle\xi\rangle^{s_1-k}\chi_K(\xi)\widehat{u_1u_2}(\xi) = \int G(\xi, \eta) f(\xi - \eta) g(\eta)\, d\eta,$$

where f, $g \in L^2(\mathbb{R}^n)$ and

$$|G(\xi, \eta)| \leqslant \frac{\langle\xi\rangle^{s_1-k}\chi_K(\xi)\chi_{K_1}(\xi - \eta)\xi_{K_2}(\eta)}{\langle\xi - \eta\rangle^{s_1}\langle\eta\rangle^{s_2}}.$$

If $K \Subset K_2^c$, then $\langle\xi - \eta\rangle \geqslant \varepsilon\langle\xi\rangle$ on the support of G, so

$$|G| \leqslant \frac{C}{\langle\xi - \eta\rangle^k\langle\eta\rangle^{s_2}}$$

and, by (1.2), since $s_2 + k > n/2$,

$$\langle\xi\rangle^{s_1-k}\chi_K(\xi)\widehat{u_1u_2}(\xi) \in L^2(\mathbb{R}^n).$$

(2)

$$\|\chi_K(D_x)u_1u_2(x)\|_{H^{s_1+s_2-n/2}(\mathbb{R}^n)} = \left\| \chi_K(D_x + D_y)u_1(x)u_2(y)\big|_{y=x}\right\|_{H^{s_1+s_2-n/2}(\mathbb{R}^n)}$$

$$\leqslant C\left\|\chi_K(D_x + D_y)u_1(x)u_2(y)\right\|_{H^{s_1+s_2}(\mathbb{R}^{2n})}$$

by the theorem on traces of functions on $\mathbb{R}^{2n}$ restricted to $\mathbb{R}^n$. (See e.g. Stein [**20**].) Let ξ, η be dual to x, y; then

$$\left\|\chi_K(D_x + D_y)u_1(x)u_2(y)\right\|_{H^{s_1+s_2}(\mathbb{R}^{2n})}$$

$$= \left\|\langle\xi, \eta\rangle^{s_1+s_2}\chi_K(\xi + \eta)\chi_{K_1}(\xi)\hat{u}_1(\xi)\chi_{K_2}(\eta)\hat{u}_2(\eta)\right\|_{L^2(\mathbb{R}^{2n})}.$$

If $K \Subset K_1^c \cap K_2^c$ on the support of $\chi_K(\xi + \eta)\chi_{K_1}(\xi)\chi_{K_2}(\eta)$, then $\langle\xi\rangle \geqslant \varepsilon\langle\xi + \eta\rangle$ and $\langle\eta\rangle > \varepsilon\langle\xi + \eta\rangle$. It follows that $\langle\eta\rangle \approx \langle\xi\rangle$, and thus

$$\chi_K(\xi + \eta)\langle\xi, \eta\rangle^{s_1+s_2}\chi_{K_1}(\xi)\chi_{K_2}(\eta) \leqslant C\langle\xi\rangle^{s_1}\langle\eta\rangle^{s_2}.$$

It follows that

$$\|\chi_K(D_x)u_1u_2(x)\|_{H^{s_1+s_2-n/2}(\mathbb{R}^n)} \leqslant C\|\langle\xi\rangle^s\hat{u}_1(\xi)\langle\eta\rangle^s\hat{u}_2(\eta)\|_{L^2(\mathbb{R}^{2n})}$$

$$= C\|u_1\|_{H^{s_1}(\mathbb{R}^n)}\|u_2\|_{H^{s_2}(\mathbb{R}^n)}. \quad \blacksquare$$

It follows immediately from Lemma 1.7 that the space of functions $H^s_{\text{loc}} \cap H^r_{ml}(x_0, \xi_0)$ is an algebra for $n/2 < s \leqslant r \leqslant 2s - n/2$. But in fact more is true: this space is invariant under the action of smooth functions (Bony [7], Meyer [**14**]). This property may be proved in an elementary fashion using just the above result, and the same argument will apply later in the $3s - n$ case.

COROLLARY 1.8. *If $u \in H^s_{\text{loc}} \cap H^r_{ml}(x_0, \xi_0)$, $n/2 < s \leqslant r \leqslant 2s - n/2$, and $f \in C^\infty$, then $f(x, u) \in H^s_{\text{loc}} \cap H^r_{ml}(x_0, \xi_0)$.*

PROOF. The conclusion holds by Schauder's lemma if $r = s$. Now assume that for $\varepsilon = \min(s - n/2, 1)$, $s \leqslant \rho - \varepsilon$, and $\rho \leqslant r$, and that $g(x, u) \in H^s_{\text{loc}} \cap H^{\rho-\varepsilon}_{ml}(x_0, \xi_0)$ for all smooth functions g. It suffices to prove that, for any first-order differential operator D, $D(f(x, u)) \in H^{\rho-1}_{ml}(x_0, \xi_0)$ in order to conclude that $f(x, u) \in H^s_{\text{loc}} \cap H^\rho_{ml}(x_0, \xi_0)$, and the result then follows by bootstrapping. But

$$D(f(x, u)) = g(x, u) + f'(x, u)\, Du.$$

The first term on the right is in $H^s_{\text{loc}} \cap H^{\rho-\varepsilon}_{ml}(x_0, \xi_0)$, and the second term is of the form $v\, Du$ with $v \in H^s_{\text{loc}} \cap H^{\rho-\varepsilon}_{ml}(x_0, \xi_0)$ and $Du \in H^{s-1}_{\text{loc}} \cap H^{\rho-1}_{ml}(x_0, \xi_0)$. It is enough to assume that v and Du are supported near x_0; let K have conic support near ξ_0. Using microlocal partitions of unity, it suffices to prove that

(1.5) $\chi_K(D)(\chi_{K_1}(D)v\chi_{K_2}(D)Du) \in H^{\rho-1}(\mathbb{R}^n)$ for K_1, K_2 with small conic support.

If K_1 and K_2 both have support sufficiently near ξ_0, then (1.5) holds by Schauder's lemma, since $\varepsilon \leqslant 1$. If K_1 has support near ξ_0 and K_2 has support away from K, then (1.5) holds by Lemma 1.7(1), with $k = 1 - \varepsilon$, since $(s - 1) + (1 - \varepsilon) > n/2$. Similarly, if K_2 has support near ξ_0 and K_1 has support away from

K_1, then (1.5) holds by Lemma 1.7(1), with $k = 0$ (and the roles of K_1 and K_2 reversed). Finally, if both K_1 and K_2 are supported away from K, it follows from Lemma 1.7(2) that

$$\chi_K(D)\big(\chi_{K_1}(D)v\chi_{K_2}(D)Du\big) \in H^{s+(s-1)-n/2}(\mathbb{R}^n) \subset H^{\rho-1}(\mathbb{R}^n),$$

since $\rho \leqslant r \leqslant 2s - n/2$. ■

Our goal is to show that for functions in $\tilde{H}^s(p_2)$, extra microlocal smoothness, say in $H^g_{ml}(x_0, \xi_0)$ for (x_0, ξ_0) a characteristic direction, is preserved under the action of smooth (nonlinear) functions for a better range of g, namely $g < 3s - n$. In [3] it is proved that for the usual wave operator $\square$, if $u_\pm \in \tilde{H}^s(\square)$ and $u_\pm$ have sufficiently small conic support near $\pm(1,1,0)$, then (u_+u_-) is microlocally in H^{3s-n-} away from the tangent hyperplane to char$\square$ at $\pm(1,1,0)$. In order to extend this conclusion to the general second-order, strictly hyperbolic operator p_2, it is necessary first to prove an algebraic property about the symbols of such operators. It will then be demonstrated how to use this property in order to obtain the appropriate microlocal estimates.

Let $p_2(x_0, (1,0,0)) = 0$ (where $\xi = (\xi_1, \xi', \xi_n)$). Coordinates can be chosen near x_0 such that, after division by a nonzero function of x,

$$\begin{aligned} &p_2(x,\xi) = \xi_1\xi_n - \textstyle\sum' a_{ij}(x)\xi_i\xi_j, \quad \text{where } a_{ij}(x_0) = 0 \\ &\text{and } \textstyle\sum' \text{ stands for the sum over } (i,j) \neq (1,1), (1,n), (n,1), (n,n). \end{aligned} \tag{1.6}$$

(See e.g. Rauch–Reed [19].)

LEMMA 1.9. *With $p_2(x, \xi)$ as in* (1.6), *let $r(x,\xi) = \Sigma' a_{ij}(x)\xi_i\xi_j$. Then*

$$\begin{aligned} \xi_n + \eta_n &= \frac{p_2(x,\xi)}{\xi_1} + \frac{p_2(y,\eta)}{\eta_1} + \frac{r(x,\xi) - r(y,\xi)}{\xi_1} \\ &\quad + r(y,\xi)\frac{\xi_1 + \eta_1}{\xi_1\eta_1} + \textstyle\sum \tilde{r}_j(y,\xi,\eta)\frac{\xi_j + \eta_j}{\eta_1}. \end{aligned} \tag{1.7}$$

Here $\tilde{r}_j(y, \xi, \eta)$ are first-order polynomials in (ξ, η).

PROOF.

$$\xi_n + \eta_n = \frac{\xi_n\xi_1}{\xi_1} + \frac{\eta_n\eta_1}{\eta_1} = \frac{p_2(x,\xi)}{\xi_1} + \frac{p_2(y,\eta)}{\eta_1} + \frac{r(x,\xi)}{\xi_1} + \frac{r(y,\eta)}{\eta_1},$$

and

$$\begin{aligned} \frac{r(x,\xi)}{\xi_1} + \frac{r(y,\eta)}{\eta_1} &= \frac{r(x,\xi) - r(y,\xi)}{\xi_1} + r(y,\xi)\left(\frac{1}{\xi_1} + \frac{1}{\eta_1}\right) \\ &\quad + (r(y,\eta) - r(y,\xi))\frac{1}{\eta_1}. \end{aligned}$$

Since $r(y,\xi) = r(y,-\xi)$, writing $\eta = -\xi + (\eta + \xi)$, we see that

$$r(y,\eta) = r(y,\xi) + \textstyle\sum(\xi_j + \eta_j)\tilde{r}_j(y,\xi,\eta). \qquad ■$$

We will see that the terms involving $r(x,\xi)$, $r(y,\xi)$, and $\tilde{r}_j(y,\xi,\eta)$ in (1.7) act as small remainders when appropriate microlocal estimates are made. If they were not present, and $\xi_n + \eta_n = p_2(x,\xi)/\xi_1 + p_2(y,\eta)/\eta_1$, then

$$D_{x_n}(uv(x)) = (D_{x_n} + D_{y_n})u(x)v(y)\big|_{y=x}$$
$$= D_{x_1}^{-1}p_2(x,D)u(x)v(y)\big|_{y=x} + u(x)D_{y_1}^{-1}p_2(y,D)v(y)\big|_{y=x}.$$

For $u, v \in \tilde{H}^s(p_2)$ with support microlocally where ξ_1 is elliptic, this would imply estimates on the H^{s+1} norm of uv microlocally where ξ_n is elliptic in terms of the $\tilde{H}^s(p_2)$ norms of u and v.

It is enlightening to examine the handling of the remainder terms in (1.7) in the case where p_2 has constant coefficients in order to prepare the way for the general case. Assume that $\hat{u}$ and $\hat{v}$, respectively, have support on small conic neighborhoods K_+ and K_- of $\pm(1,0,0)$, and K has small conic support away from the hyperplane $\{\xi_n = 0\}$. Write

$$\chi_K(D_x)D_{x_n}uv(x) = \chi_K(D_x + D_y)(D_{x_n} + D_{y_n})\chi_+(D_x)\chi_-(D_y)u(x)v(y)\big|_{y=x},$$

where $\chi_\pm \in S^0_{1,0}(\mathbb{R}^n)$ are supported near $K_\pm$. As in the proof of Lemma 1.7(2), $\langle\xi\rangle \approx \langle\eta\rangle$ on the support of $\chi_K(\xi+\eta)\chi_+(\xi)\chi_-(\eta)$. Thus there is a smooth function χ_0 with compact support in $(0,\infty)$ such that

$$\chi_K(\xi+\eta)\chi_+(\xi)\chi_-(\eta) = \chi_K(\xi+\eta)\chi_0(\langle\xi\rangle/\langle\eta\rangle)\chi_+(\xi)\chi_-(\eta).$$

From (1.7) in the constant coefficient case,

$$\chi_K(\xi+\eta)(\xi_n+\eta_n)\chi_+(\xi)\chi_-(\eta)$$
$$= \chi_K(\xi+\eta)\left(\frac{p_2(\xi)}{\xi_1} + \frac{p_2(\eta)}{\eta_1} + \sum(\xi_j+\eta_j)b_j(\xi,\eta)\right)\chi_+(\xi)\chi_-(\eta).$$

The $b_j(\xi,\eta)$, which are of the form $(r(\xi)/\xi_1\eta_1)\chi_0(\langle\xi\rangle/\langle\eta\rangle)$, with r of order 2, or $(\tilde{r}_j(\xi,\eta)/\eta_1)\chi_0(\langle\xi\rangle/\langle\eta\rangle)$, with $\tilde{r}_j$ of order 1, are symbols in $S^0_{1,0}(\mathbb{R}^{2n})$. Moreover, by (1.6), each b_j contains a term of the form ξ_i/η_1 or η_i/η_1 for $i \neq 1$. Thus the norms of the corresponding operators are of order ε, where $\varepsilon \to 0$ as the supports of χ_+ and χ_- shrink. On the other hand, on the support of $\chi_K(\xi+\eta)$, $\xi_n+\eta_n$ is elliptic, so on that set $|(\xi_j+\eta_j)/(\xi_n+\eta_n)| \leqslant M$ for some fixed constant M. Thus we can write

$$\chi_K(\xi+\eta)(\xi_n+\eta_n)\chi_+(\xi)\chi_-(\eta)$$
$$= \frac{\chi_K(\xi+\eta)(\xi_n+\eta_n)\left(1-\Sigma(\xi_j+\eta_j)(\xi_n+\eta_n)^{-1}b_j(\xi,\eta)\right)\chi_+(\xi)\chi_-(\eta)}{1-\Sigma(\xi_j+\eta_j)(\xi_n+\eta_n)^{-1}b_j(\xi,\eta)}$$
$$= a_0(\xi,\eta)\left(\frac{p_2(\xi)}{\xi_1} + \frac{p_2(\eta)}{\eta_1}\right)\chi_+(\xi)\chi_-(\eta),$$

where

$$a_0(\xi,\eta) = \frac{\chi_K(\xi+\eta)\tilde{\chi}_+(\xi)\tilde{\chi}_-(\eta)}{1-\Sigma(\xi_j+\eta_j)(\xi_n+\eta_n)^{-1}b_j(\xi,\eta)}.$$

If the supports of $\tilde{\chi}_{\pm}$ are small enough, $a_0 \in S^0_{0,0}(\mathbb{R}^{2n})$, and it follows easily that H^s norms of $\chi_K(D_x)D_{x_n}uv(x)$ can be estimated by norms of u, v in $\tilde{H}^s(p_2)$, as in the preceding paragraph.

In the nonconstant coefficient case it is again true that terms in (1.7) like $r_j(y, \xi, \eta)(\xi_j + \eta_j)/\eta_1$ are small compared to $\xi_n + \eta_n$. But operators in $S^0_{0,0}(\mathbb{R}^{2n})$ do not have a satisfactory symbolic calculus, so that care must be taken in handling these remainders. The idea is to express

$$\left(1 - \sum \frac{\xi_j + \eta_j}{\xi_n + \eta_n} b_j(y, \xi, \eta)\right)^{-1}$$

in a geometric series and use the fact that the resulting terms are of the form $a_0(\xi + \eta)b_0(\xi, \eta)$, with $a_0 \in S^0_{1,0}(\mathbb{R}^n)$ and $b_0 \in S^0_{1,0}(\mathbb{R}^{2n})$. Also, the term $(r(x, \xi) - r(y, \xi))/\xi_1$ present in (1.7), which vanishes when $y = x$, will contribute additional lower-order remainder terms when acted on by pseudodifferential operators on the left. But the basic idea—that we may replace $D_{x_n} + D_{y_n}$ with $p_2(x, D_x)D_{x_1}^{-1} + p_2(y, D_y)D_{y_1}^{-1}$ and, hence, bring into play the spaces $\tilde{H}^s(p_2)$ in order to make microlocal estimates—still holds. Precisely, the following result holds.

LEMMA 1.10. *Let χ_K have conic support away from the hyperplane $\{\xi_n = 0\}$. If p_2 is as in* (1.6) *and $\hat{u}$, $\hat{v}$ have sufficiently small conic support near* $(1, 0, 0)$ *and* $(-1, 0, 0)$, *respectively, then for $\sigma > 0$ and any choice of σ_i, σ_j subject to $\sigma_i + \sigma_j = \sigma$,*

$$\|\chi_K(D_x)uv\|_{H^{\sigma+k-n/2}} \leqslant C_{\sigma_i,\sigma_j} \sum_{i+j\leqslant k} \|(p_2(x, D))^i u\|_{H^{\sigma_i - i}} \|(p_2(x, D))^j v\|_{H^{\sigma_j - j}}.$$

PROOF. Consider first the case $k = 1$. For simplicity of notation write (1.7) as

$$\xi_n + \eta_n = p_2(x, \xi)/\xi_1 + p_2(y, \eta)/\eta_1 + (a(x) - a(y))\tilde{b}_1(\xi) + (\xi_i + \eta_i)\tilde{b}_0(y, \xi, \eta).$$

(The last two terms in equation (1.7) are actually summations of terms like $(a(x) - a(y))b_1(\xi)$ and $(\xi_i + \eta_i)b_0(y, \xi, \eta)$.) As in the above discussion for the constant coefficients case, with $a_0(\xi + \eta) = (\xi_i + \eta_i)/(\xi_n + \eta_n)$, it follows that

$$\begin{aligned} &(1 - b_0(y, \xi, \eta)a_0(\xi + \eta))(\xi_n + \eta_n)\chi(\xi, \eta)\chi_+(\xi)\chi_-(\eta) \\ &\qquad = \left(\frac{p_2(x, \xi)}{\xi_1} + \frac{p_2(y, \eta)}{\eta_1}\right)\tilde{\chi}(\xi, \eta) + (a(x) - a(y))b_1(\xi, \eta). \end{aligned} \tag{1.8}$$

Here χ, $\tilde{\chi}$, $b_0 \in S^0_{1,0}(\mathbb{R}^{2n})$, $b_1 \in S^1_{1,0}(\mathbb{R}^{2n})$, and all have support where $\langle\xi\rangle \approx \langle\eta\rangle$.

As above, the operator norm of $b_0(y, D_x, D_y)$ can be made small by taking $\chi_{\pm}$ with sufficiently small support, and the norm of $\chi_K a_0 \in S^0_{1,0}(\mathbb{R}^n)$ is bounded. It follows that the sum

$$q_0(x, y, D_x, D_y) = \sum_{j=0}^{\infty} \chi_K(D_x + D_y)a_0^j(D_x + D_y)(b_0(y, D_x, D_y))^j$$

converges as an operator of order 0, and

$$\begin{aligned}
&\chi_K(D_x + D_y)\left\{1 + \sum_1^\infty a_0^j(D_x + D_y)\big(b_0(y, D_x, D_y)\big)^j\right\} \\
&\quad\cdot\big(1 - b_0(y, D_x, D_y)a_0(D_x + D_y)\big)(D_{x_n} + D_{y_n}) \\
&= \chi_K(D_x + D_y)\left\{1 + \sum_1^\infty a_0^j(D_x + D_y)\Big[a_0(D_x + D_y), \big(b_0(y, D_x, D_y)\big)^j\Big]\right\} \\
&\quad\cdot(D_{x_n} + D_{y_n}).
\end{aligned} \tag{1.9}$$

Write

$$\big[a_0, b_0^{j+1}\big](D_{x_n} + D_{y_n}) = \big[a_0(D_{x_n} + D_{y_n}), b_0^{j+1}\big] - a_0\big[b_0^{j+1}, D_{x_n} + D_{y_n}\big].$$

The second term on the right is an operator of order 0 because $(\xi_n + \eta_n) \in S^1_{1,0}(\mathbb{R}^{2n})$. And since $a_0(\xi + \eta)(\xi_n + \eta_n)$ is a first-order polynomial in (ξ, η) by the definition of a_0, it also is in $S^1_{1,0}(\mathbb{R}^{2n})$. Thus the first term on the right is also an operator of order 0. It is then easily verified that the sum

$$r_0(x, y, D_x, D_y) = \chi_K(D_x + D_y)\sum_1^\infty a_0^j(D_x + D_y)\big[a_0, b_0^j\big](D_{x_n} + D_{y_n})$$

is an operator of order 0 (as long as the norm of b_0 is small enough). Therefore, from (1.8) and (1.9),

$$\begin{aligned}
&\chi_K(D_x + D_y)(D_{x_n} + D_{y_n})\chi_+(D_x)\chi_-(D_y) + r_0(x, y, D_x, D_y) \\
&= q_0(x, y, D_x, D_y)\big(p_2(x, D_x)D_{x_1}^{-1}\tilde{\chi}(D_x, D_y) \\
&\quad + p_2(y, D_y)D_{y_1}^{-1}\tilde{\chi}(D_x, D_y) + (a(x) - a(y))b_1(D_x, D_y)\big).
\end{aligned} \tag{1.10}$$

Now

$$\begin{aligned}
&\chi_K(D_x + D_y)a_0^j(D_x + D_y)b_0^j(y, D_x, D_y)(a(x) - a(y))b_1(D_x, D_y) \\
&\quad = \chi_K(D_x + D_y)a_0^j(D_x + D_y)(a(x) - a(y))b_0^j(y, D_x, D_y)b_1(D_x, D_y) \\
&\quad\quad + \chi_K(D_x + D_y)a_0^j(D_x + D_y)b_{-1,j}(x, y, D_x, D_y)b_1(D_x, D_y),
\end{aligned}$$

where $b_{-1,j} = [b_0^j, a(x) - a(y)]$ is an operator of order -1 with symbol in $S^{-1}_{1,0}(\mathbb{R}^{2n})$. It is then easily checked that

$$\tilde{r}_0(x, y, D_x, D_y) = \chi_K(D_x + D_y)\sum_0^\infty a_0^j(D_x + D_y)b_{-1,j}b_1$$

is an operator of order 0. Moreover,

$$\chi_K(D_x + D_y)a_0^j(D_x + D_y)(a(x) - a(y))f(x, y)\big|_{y=x} = 0.$$

Thus from (1.10),

$$\begin{aligned}
&\chi_K(D_x)D_{x_n}uv(x) = \chi_K(D_x + D_y)(D_{x_n} + D_{y_n})\chi_+(D_x)\chi_-(D_y)u(x)v(y)\big|_{y=x} \\
&= \left\{\begin{array}{l} q_0(x, y, D_x, D_y)\big(p_2(x, D_x)D_{x_1}^{-1}\tilde{\chi}(D_x, D_y) + p_2(y, D_y)D_{y_1}^{-1}\tilde{\chi}(D_x, D_y)\big) \\ + \tilde{r}_0(x, y, D_x, D_y) - r_0(x, y, D_x, D_y) \end{array}\right\} u(x)v(y)\Bigg|_{y=x}.
\end{aligned}$$

Since D_{x_n} is elliptic on supp χ_K, it follows that

$$\begin{aligned}\|\chi_K(D)uv\|_{H^{\sigma-n/2+1}(\mathbb{R}^n)} &\leqslant C\|\chi_K(D)D_{x_n}uv\|_{H^{\sigma-n/2}(\mathbb{R}^n)}\\ &\leqslant C\Big\{\big\|p_2(x,D_x)D_{x_1}^{-1}\tilde{\chi}(D_x,D_y)u(x)v(y)\big\|_{H^\sigma(\mathbb{R}^{2n})}\\ &\qquad+\big\|p_2(y,D_y)D_{y_1}^{-1}\tilde{\chi}(D_x,D_y)u(x)v(y)\big\|_{H^\sigma(\mathbb{R}^{2n})}\\ &\qquad+\big\|\chi(D_x,D_y)u(x)v(y)\big\|_{H^\sigma(\mathbb{R}^{2n})}\Big\}.\end{aligned}$$

And since $\langle\xi\rangle \approx \langle\eta\rangle$ on supp χ, $\tilde{\chi}$, as in the proof of Lemma 1.7(2), the terms on the right are bounded by

$$\begin{aligned}C\Big\{&\|p_2(x,D_x)u\|_{H^{\sigma_1-1}(\mathbb{R}^n)}\|v\|_{H^{\sigma_2}(\mathbb{R}^n)}+\|u\|_{H^{\rho_1}(\mathbb{R}^n)}\|p_2(x,D_x)v\|_{H^{\rho_2-1}(\mathbb{R}^n)}\\ &+\|u\|_{H^{\tau_1}(\mathbb{R}^n)}\|v\|_{H^{\tau_2}(\mathbb{R}^n)}\Big\}\end{aligned}$$

for any choice of $\sigma_1+\sigma_2=\rho_1+\rho_2=\tau_1+\tau_2=\sigma$.

For $\chi(D_x)D_{x_n}^k uv$ with $k>1$, the argument is similar. From (1.8),

$$\begin{aligned}&(1-b_0(y,\xi,\eta)a_0(\xi+\eta))^k(\xi_n+\eta_n)^k\chi(\xi,\eta)\chi_+(\xi)\chi_-(\eta)\\ &\qquad=\left\{\left(\frac{p_2(x,\xi)}{\xi_1}+\frac{p_2(y,\eta)}{\eta_1}\right)\tilde{\chi}(\xi,\eta)+(a(x)-a(y))b_1(\xi,\eta)\right\}^k.\end{aligned}$$

Write $(1-b_0a_0)^k=1-B_0\cdot A_0$, with

$$B_0\cdot A_0=\sum_{1\leqslant i\leqslant k}c_i b_0^i(y,\xi,\eta)a_0^i(\xi+\eta)$$

and

$$\frac{1}{(1-a_0b_0)^k}=1+\sum_{j=1}^{\infty}(A_0\cdot B_0)^j,$$

$$(A_0\cdot B_0)^j=\sum_{1\leqslant l\leqslant jk}c_{j,k,l}a_0^l(\xi+\eta)b_0^l(y,\xi,\eta).$$

Then as above we consider the operator

$$\begin{aligned}&\left(1+\sum_{j=1}^{\infty}\sum_{1\leqslant l\leqslant jk}c_{jkl}a_0^l(D_x+D_y)\big(b_0(y,D_x,D_y)\big)^l\right)\\ &\qquad\cdot\left(1-\sum_{1\leqslant i\leqslant k}c_i\big(b_0(y,D_x,D_y)\big)^i a_0^i(D_x+D_y)\right)\circ(D_{x_n}+D_{y_n})^k\\ &\quad=1-\sum_{j,i,l}c_{jkl}c_i a_0^l\big[b_0^{l+i},a_0^i\big]\circ(D_{x_n}+D_{y_n})^k.\end{aligned}$$

Now,

$$\begin{aligned}\big[b_0^{l+i},a_0^i\big](D_{x_n}+D_{y_n})^k&=\Big[b_0^{l+i},a_0^i(D_{x_n}+D_{y_n})^i\Big](D_{x_n}+D_{y_n})^{k-i}\\ &\quad-a_0^i\Big[b_0^{l+i},(D_{x_n}+D_{y_n})^i\Big](D_{x_n}+D_{y_n})^{k-i}.\end{aligned}$$

The second term on the right is a sum of products of zero-order operators with powers of $D_{x_n} + D_{y_n}$ of order $\leqslant k-1$. And since $(a_0(D_x + D_y))(D_{x_n} + D_{y_n})$ is a sum of terms of the form $(D_{x_j} + D_{y_j})$, the first term on the right is a sum of products of zero-order operators with $(D_{x_1} + D_{y_1})^{\alpha_1} \cdots (D_{x_n} + D_{y_n})^{\alpha_n}$, with $|\alpha| \leqslant k-1$. It follows, as in the derivation of (1.10), that

$$\begin{aligned} &\chi_K(D_x + D_y)(D_{x_n} + D_{y_n})^k \chi_+(D_x)\chi_+(D_y) + r_{k-1}(x, y, D_x, D_y) \\ (1.11) \qquad &= \chi(D_x + D_y)\Big(\sum c_{jkl} a_0^l(D_x + D_y)\big(b_0(y, D_x, D_y)\big)^l\Big) \\ &\circ\Big\{ p_2(x, D_x) D_{x_1}^{-1} \tilde{\chi}(D_x, D_y) + p_2(y, D_y) D_{y_1}^{-1} \chi(D_x, D_y) \\ &\qquad + (a(x) - a(y)) b_1(D_x, D_y)\Big\}^k, \end{aligned}$$

where estimates on $r_{k-1}(x, y, D_x, D_y)|_{y=x}$ follow from those on

$$\chi_K(D_x + D_y)(D_{x_n} + D_{y_n})^{k-1} \chi_+(D_x)\chi(D_y)\Big|_{y=x}.$$

(Again this uses the fact that D_{x_n} is elliptic on the support of $\chi_K(D_x)$ in order to estimate $D_{x_j}^{\alpha_j}$ for $|\alpha_j| \leqslant k-1$.)

The terms on the right side of (1.11) are handled by expanding the quantity raised to the kth power and commuting the powers $(a(x) - a(y))^l$ to the left wherever they appear. The commutators have the appropriate form, since, for example,

$$\begin{aligned} &\Big[p_2^j(x, D_x), (a(x) - a(y))^{k-j}\Big] D_{x_1}^{-j} b_1^{k-j}(D_x, D_y) \\ &\qquad = \sum_{1 \leqslant l \leqslant j} a(x, y) p_2^{j-l}(x, D_x) p_l(x, D_x) D_{x_1}^{-j} b_1^j(D_x, D_y), \end{aligned}$$

where $p_l \in S^l_{1,0}(\mathbb{R}^n)$. Thus these terms are of the form

$$\sum_{1 \leqslant l \leqslant j} \tilde{a}(x, y) p_2^{j-l}(x, D_x) \tilde{p}_l(x, y, D_x, D_y), \qquad \tilde{p}_l \in S^l_{1,0}(\mathbb{R}^{2n}).$$

Finally, the terms with $a(x) - a(y)$ at the leftmost vanish when restricted to $y = x$. The rest of the argument follows exactly as in the case $k = 1$. ■

All of the ingredients are now in place for showing that extra microlocal smoothness for functions in $\tilde{H}^s(p_2)$ is preserved up to order $2s - n$ under nonlinear operations. For products of such functions the basic idea is to microlocalize, handle pairs of characteristic directions that are opposite to each other by using Lemma 1.10, and handle all other pairs of directions by using the arguments in the usual $s_1 + s_2 - n/2$ case.

DEFINITION 1.11. Let $p_2(x, \xi)$ be the symbol of a second-order, strictly hyperbolic differential operator on $\mathcal{O} \subset \mathbb{R}^n$. We say that $u \in \tilde{H}^s(p_2) \cap H^g_{ml}(x_0, \xi_0)$ if $p_2(x_0, \xi_0) = 0$, $u \in \tilde{H}^s(p_2)$ locally near x_0, and $u \in H^g$ microlocally at (x_0, ξ_0). And $u \in \tilde{H}^s(p_2) \cap H^g_{ml}(\Gamma)$ means that $\Gamma \subset \operatorname{char}(p_2)$ and $u \in \tilde{H}^s(p_2) \cap H^g_{ml}(x_0, \xi_0)$ for each $(x_0, \xi_0) \in \Gamma$.

THEOREM 1.12. *If $n/2 < s \leqslant g < 3s - n$ and $u, v \in \tilde{H}^s(p_2) \cap H^g_{ml}(x_0, \xi_0)$, then $vD^\alpha u \in H^{g-|\alpha|}_{ml}(x_0, \xi_0)$ for $|\alpha| \leqslant 1$. In particular, $\tilde{H}^s(p_2) \cap H^g_{ml}(x_0, \xi_0)$ is an algebra for $n/2 < s \leqslant g < 3s - n$.*

PROOF. We can assume that v and u have been multiplied by smooth cutoff functions so as to be supported near x_0. Using a pseudodifferential partition of unity, by Lemma 1.6 it is enough to prove the result assuming that $\hat{v}$ and $\hat{u}$ have small conic supports K_1 and K_2, respectively. Let K be a conic neighborhood of ξ_0. There are three cases to consider for the locations of K_1 and K_2: near ξ_0, away from char(p_2), and near char(p_2) but away from ξ_0.

(i) If K_1 and K_2 both have support near ξ_0, then $v \in H^g(\mathbb{R}^n)$, $D^\alpha u \in H^{g-|\alpha|}(\mathbb{R}^n)$, and $g \geqslant g - |\alpha| \geqslant s - 1 \geqslant 0$. So by Schauder's lemma, $vD^\alpha u \in H^{g-|\alpha|}(\mathbb{R}^n)$. If K_1 has support near ξ_0, but K_2 has support away from K, then $v \in H^g(\mathbb{R}^n)$, $D^\alpha u \in H^{s-|\alpha|}(\mathbb{R}^n)$, and $(s - |\alpha|) + |\alpha| > n/2$. So by Lemma 1.7(1), $\chi_K(D)(vD^\alpha u) \in H^{g-|\alpha|}(\mathbb{R}^n)$. If K_2 has support near ξ_0, but K_1 has support away from K, then $D^\alpha u \in H^{g-|\alpha|}(\mathbb{R}^n)$ and $v \in H^s(\mathbb{R}^n)$. So by Lemma 1.7(1) (with the roles of K_1 and K_2 reversed), $\chi_K(D)(vD^\alpha u) \in H^{g-|\alpha|}(\mathbb{R}^n)$. Therefore, in the remaining cases we can assume that neither K_1 nor K_2 is a conic neighborhood of ξ_0.

(ii) If K_1 (respectively K_2) has support away from char(p_2), then $v \in H^{2s-n/2-}(\mathbb{R}^n)$ (resp. $D^\alpha u \in H^{2s-n/2-|\alpha|-}(\mathbb{R}^n)$) by Lemma 1.4. Thus, by Lemma 1.7(2), if K has support sufficiently near ξ_0, $\chi_K(D)(vD^\alpha u) \in H^{3s-n-|\alpha|-}(\mathbb{R}^n)$.

(iii) It remains to consider the case in which both K_1 and K_2 are small conic neighborhoods of characteristic directions ξ_1 and ξ_2, with $\xi_1 \neq \xi_0$, $\xi_2 \neq \xi_0$. If $\xi_2 \neq -\xi_1$, then $\Pi\mathrm{WF}(uv)$ is contained in the closure of $\Pi\mathrm{WF}(u) + \Pi\mathrm{WF}(v)$ (Hörmander [**11**]). But by the convexity of the two halves of the characteristic cone $\{\xi\colon p_2(x_0, \xi) = 0\}$, which follows from the strict hyperbolicity of p_2 (see Atiyah–Bott–Gårding [**1**]), this set avoids all characteristic directions away from ξ_1 and ξ_2. (See [**3**]; the argument for p_2 is the same as for $\square$.) Therefore $\chi_K(D)(vD^\alpha u) \in H^\infty(\mathbb{R}^n)$ for these pairs ξ_1 and ξ_2 if K has support sufficiently near ξ_0.

Finally, if $\xi_2 = -\xi_1$, fix k with $s - n/2 \leqslant k < s - n/2 + 1$. Then

$$3s - n - |\alpha| = \left(3s - n/2 - |\alpha| - k\right) - n/2 + k,$$

and

$$\sigma = 3s - n/2 - |\alpha| - k > 3s - n/2 - 1 - (s - n/2 + 1) = 2s - 2 > 0.$$

Hence, by Lemma 1.10,

$$\|\chi_K(D) vD^\alpha u\|_{H^{3s-n-|\alpha|}} \leqslant C \sum_{i+j \leqslant k} \left\|(p_2(x, D))^i v\right\|_{H^{\sigma_i - i}} \left\|(p_2(x, D))^j D^\alpha u\right\|_{H^{\sigma_j - j}},$$

where $\sigma_i + \sigma_j = \sigma = 3s - n/2 - |\alpha| - k$. Since $\sigma \leqslant 2s - |\alpha|$, if neither i nor j equals k, we can pick $\sigma_i = s$, $\sigma_j = s - |\alpha|$; by Lemma 1.4 and Definition 1.1,

$(p_2(x, D))^i v \in H^{s-i}(\mathbb{R}^n)$ and $(p_2(x, D))^j D^\alpha v \in H^{s-j-|\alpha|}(\mathbb{R}^n)$. If $i = k$, $j = 0$, pick $\sigma_i = 2s - n/2 - k$, $\sigma_j = s - |\alpha|$, while if $i = 0, j = k$, pick $\sigma_i = s$, $\sigma_j = 2s - n/2 - k - |\alpha|$. In either case, Lemma 1.4 and Definition 1.1 imply that $\|(p_2(x, D))^i v\|_{H^{\sigma_i - i}} \|(p_2(x, D))^j D^\alpha u\|_{H^{\sigma_j - j}}$ is finite for $v, u \in \tilde{H}^s(p_2)$. Therefore $vD^\alpha u \in H^{g-|\alpha|}_{ml}(x_0, \xi_0)$ as long as $g < 3s - n$. ∎

COROLLARY 1.13. *If $u \in \tilde{H}^s(p_2) \cap H^g_{ml}(x_0, \xi_0)$, $n/2 < s \leqslant g < 3s - n$, and $f \in C^\infty$, then $f(x, u) \in \tilde{H}^s(p_2) \cap H^g_{ml}(x_0, \xi_0)$.*

PROOF. If $g = s$ the conclusion holds by Lemma 1.2. Now let $\varepsilon = \min(2s - n, 1)$ and assume that, for fixed ρ with $s \leqslant \rho - \varepsilon$ and $\rho \leqslant g$, it is true that $\tilde{f}(x, u) \in H^{\rho-\varepsilon}_{ml}(x_0, \xi_0)$ for all smooth functions $\tilde{f}$. If it can be concluded that $f(x, u) \in H^\rho_{ml}(x_0, \xi_0)$, the result follows by bootstrapping. But for any first-order differential operator D,

$$D(f(x, u)) = g(x, u) + f'(x, u)\, Du.$$

The first term on the right is in $H^{\rho-\varepsilon}_{ml}(x_0, \xi_0) \subset H^{\rho-1}_{ml}(x_0, \xi_0)$. The second term on the right is of the form vDu, with $v \in \tilde{H}^s \cap H^{\rho-\varepsilon}_{ml}(x_0, \xi_0)$ and $u \in \tilde{H}^s \cap H^\rho_{ml}(x_0, \xi_0)$. Just as in the proof of Corollary 1.8, it follows from Theorem 1.12 that $vD^\alpha u \in H^{\rho-1}_{ml}(x_0, \xi_0)$ for $\rho < 3s - n$. Therefore $D(f(x, u)) \in H^{\rho-1}_{ml}(x_0, \xi_0)$ and $f(x, u) \in H^\rho_{ml}(x_0, \xi_0)$. ∎

2. Propagation of smoothness for semilinear equations. The linear propagation theorem of Hörmander [**10**] may be used directly, along with the results of §1, to handle the simplest type of semilinear equation. The argument is the same as that of Rauch [**16**].

THEOREM 2.1. *Let $\mathcal{O} \subset \mathbb{R}^n$ be an open set, and let $u \in H^s(\mathcal{O})$, $s > n/2$. Assume that $p(x, D)$ is a second-order differential operator that is strictly hyperbolic with principal symbol p_2, that $f \in C^\infty$, and that*

$$p(x, D)u = f(x, u) \quad \text{on } \mathcal{O}. \tag{2.1}$$

If $\Gamma \subset \mathcal{O} \times \mathbb{R}^n$ denotes a null bicharacteristic for p_2 and

$$u \in H^g_{ml}(x_0, \xi_0) \quad \text{for some point } (x_0, \xi_0) \text{ on } \Gamma, \tag{2.2}$$

then $u \in H^g_{ml}(\Gamma)$ as long as $g < 3s - n + 1$.

PROOF. By Lemma 1.5, $u \in \tilde{H}^s(p_2)$. Suppose it has been proved that $u \in H^\gamma_{ml}(\Gamma)$. Then by Corollary 1.13, $f(x, u) \in H^\gamma_{ml}(\Gamma)$ as long as $\gamma < 3s - n$. And by (2.1), (2.2) and Hörmander's theorem, $u \in H^{\min(g, \gamma+1)}_{ml}(\Gamma)$. ∎

COROLLARY 2.2. *With u, p_2 as above, assume that for every $x \in \mathcal{O}$, all characteristics for p_2 through x intersect $\mathcal{O} \cap \{x_n = 0\}$ and are wholly contained in $\mathcal{O}$ between $\{x_n = 0\}$ and x. If $u|_{x_n=0} = u_0$, $u_{x_n}|_{x_n=0} = u_1$, and $\tilde{x}_1$ is a point in the lacuna above the union of the characteristics over* sing supp(u_0, u_1), *then $u \in H^{3s-n+1-}_{\text{loc}}$ at $\tilde{x}$. (See Figure 1.)*

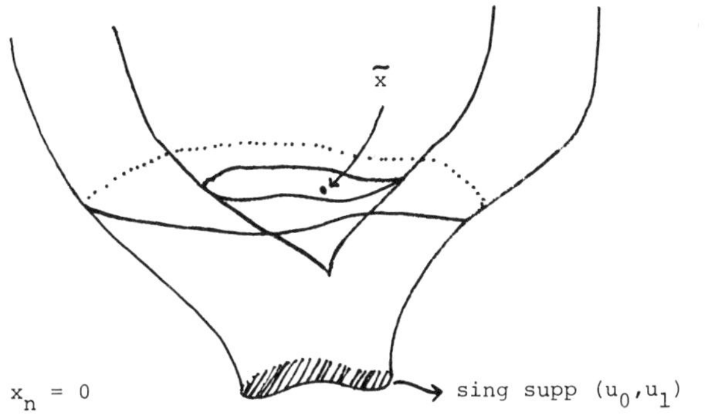

FIGURE 1.

PROOF. If $p_2(\tilde{x}, \tilde{\xi}) = 0$, Γ is the null bicharacteristic through $(\tilde{x}, \tilde{\xi})$, and $(x_0, \xi_0) = \Gamma \cap \{x_n = 0\}$, then u is smooth near x_0, so, in particular, $u \in H^g_{ml}(x_0, \xi_0)$ for all $g < 3s - n + 1$. Therefore

$$(2.3) \qquad u \in H^{3s-n+1-}_{ml}(\tilde{x}, \tilde{\xi}) \quad \text{for all characteristic directions } \tilde{\xi}.$$

From Lemma 1.4, $u \in H^{2s-n/2-}_{ml}(\tilde{x}, \xi')$ for all elliptic directions ξ', so it follows that $u \in H^{2s-n/2-}_{\mathrm{loc}}$ at $\tilde{x}$. And if $u \in H^{\sigma}_{ml}(\tilde{x}, \xi')$, then the same holds for $f(x, u)$ by Corollary 1.8, as long as $\sigma < 2(2s - n/2) - n/2$. By microlocal elliptic regularity, it can then be concluded that $u \in H^{\sigma+2}_{ml}(\tilde{x}, \xi')$ for all elliptic directions ξ'. So a bootstrap argument yields that $u \in H^{2(2s-n/2)-n/2+2-}_{ml}(\tilde{x}, \tilde{\xi}')$ for all elliptic directions ξ'. Since $2(2s - n/2) - n/2 + 2 > 3s - n + 1$, it follows from (2.3) that $u \in H^{3s-n+1-}_{\mathrm{loc}}$ at $\tilde{x}$. ∎

For the general semilinear equation $p_2(x, D)u = f(x, u, Du)$, an argument similar to that given in Beals–Reed [5] is needed: in microlocalizing to apply Hörmander's linear theorem, it becomes necessary to commute a microlocal cutoff operator with a differential operator having nonsmooth coefficients. The "$3s - n$" argument is essentially no more difficult than the "$2s - n/2$" argument given in [5], because the new feature—the interaction of H^s singularities microlocally in the characteristic directions $\pm\xi_1$ producing weaker-than-expected spread singularities—occurs microlocally on a set where the properties of commutators do not come into play. (See the subsequent proof of Theorem 2.5.) As in [5], the argument is begun by differentiation of the original equation to produce a linear equation with nonsmooth coefficients. We will differentiate the equation only once and provide a slightly simpler proof of the propagation theorem than that given in [5], where for certain values of s the equation was differentiated twice. The simplification is achieved through the use of a microlocal partition of unity on the nonsmooth coefficients to separate the pieces whose smoothness is being improved by a bootstrap argument from those whose smoothness is always just H^s.

To begin, we have the analogue of Lemma 1.5 for the solution of a differentiated semilinear equation.

LEMMA 2.3. *Let $u \in H^{s+1}(\mathcal{O})$, $s > n/2$, let $f \in C^\infty$, and suppose*

$$p_2(x, D)u = f(x, u, Du) \quad \text{on } \mathcal{O}. \tag{2.4}$$

If ∂ is a first-order differential operator and $v = \partial u$, then $v \in \tilde{H}^s(p_2)$.

PROOF. Upon differentiation of (2.4), it follows that $p_2(x, D)v = \tilde{f}(x, u, Du, D^2u)$, where $\tilde{f}$ is linear in D^2u. By Schauder's lemma, since $s > n/2$ and $Du \in H^s(\mathcal{O})$, the right side is in $H^{s-1}(\mathcal{O})$. As in the proof of Lemma 1.5, for $j < s - n/2$,

$$(p_2(x, D))^j v = f(x, u, \dots, D^{j+1}u) \in H^{s-j}(\mathcal{O}),$$

while if $s - n/2 \leqslant k < s - n/2 + 1$,

$$\begin{aligned}(p_2(x, D))^k v = g(x, u, \dots, D^k u) &+ \sum_{|\alpha|=k+1} g_\alpha(x, u, \dots, D^k u) D^\alpha u \\ &+ \sum_{\substack{|\alpha|=k+1 \\ |\beta|=k+1}} g_{\alpha\beta}(x, u, \dots, D^k u) D^\alpha u D^\beta u.\end{aligned}$$

Since $g(x, u, \dots, D^k u)$, $g_\alpha(x, u, \dots, D^k u)$, $g_{\alpha\beta}(x, u, \dots, D^k u) \in H^{s+1-k}(\mathcal{O})$ and $D^\alpha u$, $D^\beta u \in H^{s-k}(\mathbb{R}^n)$, Lemma 1.3 yields that

$$(p_2(x, D))^k v \in H^{2s-n/2-2k-}(\mathcal{O}). \quad \blacksquare$$

The following result is an extension of the commutator lemma and the lemma following Theorem 2 in [5].

LEMMA 2.4. *Let K_1, K_2, and K be cones in $\mathbb{R}^n \setminus 0$. Assume that $a \in H^{s_1}(\mathcal{O})$, $v \in H^{s_2}(\mathcal{O})$, $s_i > n/2$, $\Pi\mathrm{WF}(a) \subset K_1$, and $\Pi\mathrm{WF}(v) \subset K_2$. Let $p_1(x, D)$ be a first-order operator, and let $b_0(x, \xi) \in S^0_{1,0}(\mathbb{R}^n)$.*

(1) *If $s_1 = s_2 = g$ and $\delta < \min(1, g - n/2)$, then*

$$[b_0(x, D), a(x)p_1(x, D)]v \in H^{g-1+\delta}(\mathcal{O}).$$

(2) *If $K \Subset K_1^C$ and $\delta < \min(1, s_1 - n/2)$, then*

$$\chi_K(D)[b_0(x, D), a(x)p_1(x, D)]v \in H^{s_2-1+\delta}(\mathcal{O}).$$

(3) *If $K \Subset K_2^C$ and $\delta < \min(1, s_2 - n/2)$, then*

$$\chi_K(D)[b_0(x, D), a(x)p_1(x, D)]v \in H^{s_1-1+\delta}(\mathcal{O}).$$

PROOF. For notational simplicity assume that b_0 and p_1 depend on ξ only; the proof in the general case proceeds in the same fashion. Then

$$[b_0, ap_1]v\hat{\ }(\xi) = \int \hat{a}(\eta)(b_0(\xi) - b_0(\xi - \eta))p_1(\xi - \eta)\hat{v}(\xi - \eta)\, d\eta.$$

$$\langle\xi\rangle^{g-1+\delta}[b_0, ap_1]v\hat{\ }(\xi) = \int G_1(\xi, \eta) f(\xi - \eta) g(\eta)\, d\eta, \tag{1}$$

where $f, g \in L^2(\mathbb{R}^n)$ and

$$|G_1(\xi, \eta)| = \frac{\langle \xi \rangle^{g-1+\delta} |b_0(\xi) - b_0(\xi - \eta)|}{\langle \eta \rangle^{g} \langle \xi - \eta \rangle^{g-1}}.$$

If $|\eta| \geqslant \frac{1}{2}|\xi|$ and $|\xi - \eta| \geqslant \frac{1}{2}|\xi|$, then $|G_1| \leqslant C/\langle \xi \rangle^{g-\delta}$. If $|\eta| \geqslant \frac{1}{2}|\xi|$ and $|\xi - \eta| \leqslant \frac{1}{2}|\xi|$, then

$$|G_1| \leqslant \frac{C\langle \xi \rangle^{-1+\delta}}{\langle \xi - \eta \rangle^{g-1}} = \frac{C}{\langle \xi \rangle^{1-\delta} \langle \xi - \eta \rangle^{g-1}} \leqslant \frac{C}{\langle \xi - \eta \rangle^{g-\delta}}.$$

And if $|\eta| < \frac{1}{2}|\xi|$, then $\xi - \eta$ and ξ are comparable and, hence, from Taylor's theorem, $|b_0(\xi) - b_0(\xi - \eta)| \leqslant C\langle \eta \rangle / \langle \xi \rangle$. Therefore

$$|G_1| \leqslant \frac{C\langle \xi \rangle^{g-1+\delta} \langle \eta \rangle}{\langle \eta \rangle^{g} \langle \xi - \eta \rangle^{g-1} \langle \xi \rangle} \leqslant \frac{C}{\langle \eta \rangle^{g-1} \langle \xi \rangle^{1-\delta}} \leqslant \frac{C}{\langle \eta \rangle^{g-\delta}}.$$

In all cases, (1.2) is satisfied because $g - \delta > n/2$, and thus $[b_0, ap_1]v \in H^{g-1+\delta}(\mathcal{O})$.

(2) $$\langle \xi \rangle^{s_2-1+\delta} \chi_K(\xi) [b_0, ap_1]v^\wedge(\xi) = \int G_2(\xi, \eta) f(\xi - \eta) g(\eta)\, d\eta,$$

with $f, g \in L^2(\mathbb{R}^n)$ and

$$|G_2(\xi, \eta)| = \frac{\langle \xi \rangle^{s_2-1+\delta} \chi_K(\xi) \chi_{K_1}(\eta) \chi_{K_2}(\xi - \eta) |b_0(\xi) - b_0(\xi - \eta)|}{\langle \eta \rangle^{s_1} \langle \xi - \eta \rangle^{s_2-1}}.$$

On supp $\chi_K(\xi)\chi_{K_1}(\eta)\chi_{K_2}(\xi - \eta)$, since $K \Subset K_1^C$, $\langle \xi \rangle \leqslant C\langle \xi - \eta \rangle$, so

$$|G_2| \leqslant \frac{C\langle \xi \rangle^{\delta} |b_0(\xi) - b_0(\xi - \eta)|}{\langle \eta \rangle^{s_1}}.$$

If $|\eta| > \frac{1}{2}|\xi|$, then $|G_2| \leqslant C/\langle \eta \rangle^{s_1-\delta}$, while if $|\eta| < \frac{1}{2}|\xi|$, then

$$|b_0(\xi) - b_0(\xi - \eta)| \leqslant \frac{\langle \eta \rangle}{\langle \xi \rangle} \quad \text{and} \quad |G_2| \leqslant \frac{C}{\langle \xi \rangle^{1-\delta} \langle \eta \rangle^{s_1-1}}.$$

In either case, (1.2) holds because $\delta < 1$ and $s_1 + \delta > n/2$.

(3) $$\chi_K(\xi) \langle \xi \rangle^{s_1-1+\delta} [b_0, ap_1]v^\wedge(\xi) = \int G_3(\xi - \eta) f(\xi - \eta) g(\eta)\, d\eta,$$

with $f, g \in L^2(\mathbb{R}^n)$ and

$$|G_3(\xi, \eta)| = \frac{\langle \xi \rangle^{s_1-1+\delta} \chi_K(\xi) \chi_{K_1}(\eta) \chi_{K_2}(\xi - \eta) |b_0(\xi) - b_0(\xi - \eta)|}{\langle \eta \rangle^{s_1} \langle \xi - \eta \rangle^{s_2-1}}.$$

On supp $\chi_K(\xi)\chi_{K_1}(\eta)\chi_{K_2}(\xi - \eta)$, sinc $K \Subset K_2^C$, $\langle \xi \rangle \leqslant C\langle \eta \rangle$. Therefore

$$|G_3| \leqslant \frac{C\langle \xi \rangle^{-1+\delta}}{\langle \xi - \eta \rangle^{s_2-1}} = \frac{C}{\langle \xi \rangle^{1-\delta} \langle \xi - \eta \rangle^{s_2-1}},$$

and (1.2) holds because $s_2 + \delta > n/2$. ■

It is now a straightforward matter to prove the propagation of smoothness result for the general semilinear equation. For all but one term Lemma 2.4 suffices to allow the argument given in [5] to apply; the one remaining term yields directly to an application of Theorem 1.12.

THEOREM 2.5. *Let $\mathcal{O} \subset \mathbb{R}^n$ be an open set, and let $u \in H^{s+1}(\mathcal{O})$, $s > n/2$. Assume that $p_2(x, D)$ is a second-order strictly hyperbolic differential operator, that $f \in C^\infty$, and that*

$$p_2(x, D)u = f(x, u, Du). \tag{2.5}$$

If $\Gamma \subset \mathcal{O} \times \mathbb{R}^n$ denotes a null bicharacteristic for p_2, and

$$u \in H^g_{ml}(x_0, \xi_0) \quad \text{for some point } (x_0, \xi_0) \text{ on } \Gamma, \tag{2.6}$$

then $u \in H^g_{ml}(\Gamma)$ for $g < 3s - n + 1$.

PROOF. By means of a local partition of unity and an argument on successive pieces of Γ, it can be assumed that there is a first-order differential operator ∂ that is microlocally elliptic near Γ. Upon differentiation of (2.5), it follows (with $v = \partial u$) that

$$p_2(x, D)v = \sum_{|\alpha|=1} a_\alpha(x, u, Du)D^\alpha v + \tilde{p}_2(x, D)u + \tilde{f}(x, u, Du). \tag{2.7}$$

Here $v \in H^s(\mathcal{O})$, $Du \in H^s(\mathcal{O})$, $\tilde{p}_2(x, \xi) = \partial p_2(x, \xi)$, and, from (2.6), $v \in H^{g-1}_{ml}(x_0, \xi_0)$. It suffices to prove that $v \in H^{g-1}_{ml}(\Gamma)$ as long as $g - 1 < 3s - n$. From [5] (or Bony [7]) this property holds if $g - 1 \leqslant 2s - n/2$. So by a bootstrap argument, it is enough to assume that $u \in H^{\nu+1}_{ml}(\Gamma)$ and $v \in H^\nu_{ml}(\Gamma)$ and to prove that $v \in H^{\nu+\delta}_{ml}(\Gamma)$, with $\delta < \min(s - n/2, 1)$ fixed and $\nu + \delta \leqslant g - 1 < 3s - n$.

By Lemma 2.3 and Corollary 1.13, v, $a_\alpha(x, u, Du)$, and $\tilde{f}(x, u, Du)$ are in $\tilde{H}^s \cap H^\nu_{ml}(\Gamma)$. Let $b_0 \in S^0_{1,0}(\mathbb{R}^n)$ have conic support near Γ, be microlocally elliptic near Γ, and satisfy $\{p_2, b_0\} = 0$ (see Nirenberg [15]), so that $[p_2(x, D), b_0(x, D)]$ has order zero. Then, from (2.7),

$$\begin{aligned} p_2(x, D)b_0(x, D)v = &\sum_{|\alpha|=1} a_\alpha D^\alpha b_0(x, D)v + \sum_{|\alpha|=1} [b_0(x, D), a_\alpha D^\alpha]v \\ &+ [p_2(x, D), b_0(x, D)]v + \tilde{p}_2(x, D)b_0(x, D)u \\ &+ \tilde{p}_1(x, D)u + b_0(x, D)\tilde{f}(x, u, Du). \end{aligned}$$

Since $\tilde{p}_1 = [b_0, \tilde{p}_2]$ has conic support near Γ, $\tilde{p}_1(x, D)u \in H^\nu(\mathcal{O})$. And $\tilde{p}_2 b_0 u = q_1(x, D)b_0(x, D)v$ (since $v = \partial u$ and ∂ is elliptic on $\operatorname{supp} b_0$). Absorbing this term in with the other first-order operators acting on $b_0 v$, we have

$$\begin{aligned} p_2(x, D)b_0(x, D)v = &\sum_{|\alpha|=1} a_\alpha D^\alpha b_0(x, D)v \\ &+ \sum_{|\alpha|=1} [b_0(x, D), a_\alpha D^\alpha]v + r(x). \end{aligned} \tag{2.8}$$

By Corollary 1.13, $r(x) \in H^\nu(\mathcal{O})$. Our aim is to show that

$$\tilde{r}(x) = \sum_{|\alpha|=1} [b_0(x, D), a_\alpha D^\alpha]v \in H^{\nu-1+\delta}_{ml}(\Gamma).$$

Decompose a_α and v microlocally as in the proof of Theorem 1.12 and assume, as a result, that $\hat{a}_\alpha$ and $\hat{v}$ have small conic supports K_1 and K_2, respectively. For K with microlocal support near $\xi_0 \in \Pi(\Gamma)$, consider $\chi_K(D)[b_0, a_\alpha D^\alpha]v$.

(i) If K_1 and K_2 both have support near that of K, then $a_\alpha, v \in H^\nu(\mathcal{O})$ and, by Lemma 2.4(1), $[b_0, a_\alpha D^\alpha]v \in H^{\nu-1+\delta}(\mathbb{R}^n)$.

(ii) If K_2 is supported near K while K_1 is supported away from K, then $v \in H^\nu(\mathcal{O})$, $a_\alpha \in H^s(\mathcal{O})$, and, by Lemma 2.4(2), $\chi_K(D)[b_0, a_\alpha D^\alpha]v \in H^{\nu-1+\delta}(\mathcal{O})$. On the other hand, if K_1 is supported near K while K_2 is supported away from K, then $a_\alpha \in H^\nu(\mathcal{O})$, $v \in H^s(\mathcal{O})$, and, by Lemma 2.4(3), $\chi_K(D)[b_0, a_\alpha D^\alpha]v \in H^{\nu-1+\delta}(\mathbb{R}^n)$.

In the remaining cases to be considered, K is supported away from both K_1 and K_2. The special properties of commutators may be ignored, and the terms $b_0(x, D)a_\alpha D^\alpha v$ and $a_\alpha b_0(x, D)D^\alpha v$ in $\tilde{r}(x)$ will be treated individually. Since $\nu - 1 + \delta < 3s - n - 1$, it is enough to show that the remaining terms are in $H^{3s-n-1}(\mathcal{O})$.

(iii) If K_1 is supported away from K, and if K_2 is supported away from char(p_2), then $a_\alpha \in H^s(\mathcal{O})$ and $D^\alpha v \in H^{2s-n/2-1-}(\mathcal{O})$, by Lemmas 1.6 and 1.4. Therefore by Lemma 1.7(2),

$$\chi_K(D)(b_0 a_\alpha D^\alpha v) \in H^{3s-n-1-}(\mathcal{O})$$

and

$$\chi_K(D)(a_\alpha b_0(x, D)D^\alpha v) \in H^{3s-n-1-}(\mathcal{O}).$$

Similarly, if K_2 is supported away from K, and if K_1 is supported away from char(p_2), then $a_\alpha \in H^{2s-n/2-}(\mathcal{O})$, by Lemmas 1.6 and 1.4, and $D^\alpha v \in H^{s-1}(\mathcal{O})$. Thus

$$\chi_K(D)(b_0 a_\alpha D^\alpha v) \in H^{3s-n-1-}(\mathcal{O}) \quad \text{and} \quad \chi_K(D)(a_\alpha b_0 D^\alpha v) \in H^{3s-n-1-}(\mathcal{O}),$$

by Lemma 1.7(2).

(iv) In the remaining case, both K_1 and K_2 have support near ξ_1 and ξ_2 in char(p_2) but away from K. If $\xi_1 \neq -\xi_2$, then, as in the proof of Theorem 1.12, $\chi_K(D)(b_0 a_\alpha D^\alpha v)$ and $\chi_K(D)(a_\alpha b_0 D^\alpha v)$ are smooth. And if $\xi_1 = -\xi_2$, $a_\alpha \in \tilde{H}^s(p_2)$ by Lemmas 2.3 and 1.2, v, $b_0 v \in \tilde{H}^s(p_2)$ by Lemmas 2.3 and 1.6, so by the proof of Theorem 1.12, $\chi_K(D)a_\alpha b_0 D^\alpha v, \chi_K(D)b_0 a_\alpha D^\alpha v \in H^{3s-n-1-}(\mathcal{O})$.

The rest of the argument follows as in **[5]**. Factor $p_2(x, D) = p_1(x, D)q_1(x, D)$ with q_1 microlocally elliptic near Γ, and set $W = q_1(x, D)b_0(x, D)v$. It follows from (2.8) and (i)–(iv) above that

$$p_1(x, D)w = \sum_{|\alpha|=1} a_\alpha(x, u, Du)p_{0,\alpha}(x, D)w + r(x) + \tilde{r}(x),$$

with $r(x), \tilde{r}(x) \in H_{ml}^{\nu-1+\delta}(\Gamma)$, $w \in H^{\nu-1}(\mathcal{O})$, $a_\alpha(x, u, Du) \in H^s \cap H_{ml}^\nu(\Gamma)$, and $p_{0,\alpha} \in S_{1,0}^0(\mathbb{R}^n)$. From (2.6), $w \in H_{\text{loc}}^{\nu-1+\delta}(x_0)$, and now the energy estimates in the proposition in §1 of [**5**] yield that $w \in H^{\nu-1+\delta}_{ml}(\Gamma)$, as desired. (The argument given there remains valid here because, from Lemma 1.7(1),

$$\|\chi_K(D)a_\alpha p_{0,\alpha}w\|_{H^{\nu-1+\delta}(\mathscr{O})} \leqslant C\|w\|_{H^{\nu-1+\delta}(\mathscr{O})}$$

as long as $a_\alpha \in H^s \cap H^\nu_{ml}(\Gamma)$ for any ν, with $n/2 < s \leqslant \nu$, not just for $\nu \leqslant 2s - n/2$.) ■

3. Functions satisfying quasilinear equations. In the analysis of solutions to quasilinear equations, the principal new aspect is that the space analogous to the $H^s(p_2)$ of Definition 1.1 is now defined in terms of an operator with nonsmooth coefficients. This feature introduces a few technicalities into the proofs of the algebraic properties of the space, mainly in showing that such functions may be microlocalized. But the analogues of the results of §1 do hold and may be combined with the techniques of Beals–Reed [**6**] to allow a proof of the corresponding conclusions on the propagation of smoothness. In what follows are collected the statements of the results for the quasilinear case, together with some of the ideas behind them. Complete proofs will appear elsewhere.

DEFINITION 3.1. Let $u \in H^{s+2}(\mathscr{O})$, $s > n/2$, and let $a_\alpha, f \in C^\infty$. Assume that

$$\sum_{|\alpha|=2} a_\alpha(x, u, Du)D^\alpha u = f(x, u, Du). \tag{3.1}$$

Denote $\sum_{|\alpha|=2} a_\alpha(x, u, Du)D^\alpha$ by $ap_2(x, D)$. Then $v \in \tilde{H}^s(ap_2)$ means that, for $j < s - n/2$, $(ap_2(x, D))^j v \in H^{s-j}(\mathscr{O})$, and for $s - n/2 \leqslant k < s - n/2 + 1$, $(ap_2(x, D))^k v \in D^{2s-n/2-2k-}(\mathscr{O})$.

Implicit in this definition is the assumption that the composition $(ap_2(x, D))^j$ makes sense for the range of j indicated. But it is easily seen by induction and (3.1) that

$$(ap_2(x, D))^j = \sum_{i \leqslant j} \sum_{|\beta| \leqslant 2j-i} a_\beta(x, u, \dots, D^{i+1}u)D^\beta.$$

The coefficients on the right side are in $H^{s-j+1}(\mathscr{O})$ if $u \in H^{s+2}(\mathscr{O})$, by Schauder's lemma, as long as $j < s - n/2 + 1$.

As in the semilinear case, an induction argument and Lemma 1.3 allow us to conclude that $\tilde{H}^s(ap_2)$ is an algebra and contains solutions to appropriately differentiated quasilinear equations.

LEMMA 3.2. *If $v \in \tilde{H}^s(ap_2)$, $s > n/2$, and $g \in C^\infty$, then $g(x, v) \in \tilde{H}^s(ap_2)$.*

LEMMA 3.3. *Let $u \in H^{s+2}(\mathscr{O})$, $s > n/2$, let $f \in C^\infty$, and suppose that $ap_2(x, D)u = f(x, u, Du)$ on $\mathscr{O}$. If ∂^2 is a second-order differential operator and $v = \partial^2 u$, then $v \in \tilde{H}^s(ap_2)$.*

Elements of $\tilde{H}^s(ap_2)$ may also be microlocalized, and this is the essential ingredient in showing that extra smoothness up to order $3s - n$ in characteristic directions is preserved under the action of nonlinear functions, as in §1. Using the calculus of operators with nonsmooth coefficients, developed in [**6**], or Bony's calculus (as pointed out by G. Metivier), the following result can be established.

LEMMA 3.4. *If* $v \in \tilde{H}^s(ap_2)$, $s > n/2$, *and* $b_0(x, \xi) \in S^0_{1,0}(\mathbb{R}^n)$, *then* $b_0(x, D)v \in \tilde{H}^s(ap_2)$.

The algebra lemma 1.9 remains true for strictly hyperbolic, second-order symbols ap_2 with coefficients that are no longer smooth. The analogue of Lemma 1.10 is proved by replacing properties of operators with symbols in $S^j_{1,0}(\mathbb{R}^n)$ with the corresponding properties for operators with nonsmooth symbols, as in **[6]**. Lemmas 1.7 and 3.2–3.4 then apply, as in the proof of Theorem 1.12, to yield the principal algebraic result.

DEFINITION 3.5. Let

$$ap_2(x, \xi) = \sum_{|\alpha|=2} a_\alpha(x, u, Du)\xi^\alpha$$

be the symbol of a second-order, strictly hyperbolic differential operator on $\mathcal{O} \subset \mathbb{R}^n$. Here $a_\alpha \in C^\infty$, $u \in H^{s+2}(\mathcal{O})$, $s > n/2$. If $ap_2(x_0, \xi_0) = 0$, we say that $v \in \tilde{H}^s(ap_2) \cap H^g_{ml}(x_0, \xi_0)$ if $v \in \tilde{H}^s(ap_2)$ locally near x_0 and $v \in H^g$ microlocally at (x_0, ξ_0).

THEOREM 3.6. *If* $n/2 < s \leqslant g < 3s - n$ *and* $v, u \in \tilde{H}^s(ap_2) \cap H^g_{ml}(x_0, \xi_0)$, *then* $vD^\alpha u \in H^{g-|\alpha|}(x_0, \xi_0)$ *for* $|\alpha| \leqslant 1$. *In particular,* $\tilde{H}^s(ap_2) \cap H^g_{ml}(x_0, \xi_0)$ *is an algebra for* $n/2 < s \leqslant g < 3s - n$, *and if* $f \in C^\infty$, *then* $f(x, v) \in \tilde{H}^s(ap_2) \cap H^g_{ml}(x_0, \xi_0)$.

The arguments in **[6]** may now be easily adapted to yield the theorem on propagation of smoothness. As in the semilinear case, the difference between the proof in **[6]** and the present situation is that the interaction of H^s singularities microlocally in the characteristic directions $\pm\xi_1$ results in singularities of strength $3s - n$ in other characteristic directions. But as in part (iv) of the proof of Theorem 2.5, no properties of a symbolic calculus are needed in order to estimate these terms; Theorem 3.6 may be applied directly.

THEOREM 3.7. *Let* $\mathcal{O} \subset \mathbb{R}^n$ *be an open set, and let* $u \in H^{s+2}(\mathcal{O})$, $s > n/2$. *Assume that* $ap_2(x, \xi) = \sum_{|\alpha|=1} a_\alpha(x, u, Du)\xi^\alpha$ *is strictly hyperbolic on* $\mathcal{O}$, *that* $a_\alpha, f \in C^\infty$, *and that* $\sum_{|\alpha|=2} a_\alpha(x, u, Du)D^\alpha u = f(x, u, Du)$. *Suppose that* Γ *is the unique null bicharacteristic through* (x_0, ξ_0) *for* $ap_2(x, \xi)$ (*uniqueness is automatic if* $s > n/2 + 1$) *and that* $u \in H^g_{ml}(x_0, \xi_0)$. *Then* $u \in H^g_{ml}(\Gamma)$ *as long as* $g < 3s - n + 2$.

REFERENCES

1. M. F. Atiyah, R. Bott and L. Gårding, *Lacunas for hyperbolic differential operators with constant coefficients*, Acta Math. **124** (1970), 109–189.

2. M. Beals, *Spreading of singularities for a semilinear wave equation*, Duke Math. J. **49** (1982), 275–286.

3. ______, *Self-spreading and strength of singularities for solutions to semilinear wave equations*, Ann. of Math. (2) **118** (1983), 187–214.

4. ______, *Nonlinear wave equations with data singular at one point*, Contemporary Math. **27** (1984), 83–95.

5. M. Beals and M. Reed, *Propagation of singularities for hyperbolic pseudodifferential operators with nonsmooth coefficients*, Comm. Purc Appl. Math. **35** (1982), 169–184.

6. ______, *Microlocal regularity theorems for nonsmooth pseudodifferential operators and applications to nonlinear problems*, Trans. Amer. Math. Soc. **285** (1984), 159–184.

7. J. M. Bony, *Calcul symbolique et propagation des singularités pour les équations aux derivées partielles nonlineaires*, Ann. Sci. École Norm. Sup. **14** (1981), 209–246.

8. ______, *Interaction des singularités pour les équations aux derivées partielles nonlineaires*, Sem. Goulaouic–Meyer–Schwartz, Éxpose no. 2 (1981–1982).

9. ______, *Interaction des singularités pour les équations de Klein–Gordon non lineaires*, Sem. Goulaouic–Meyer–Schwartz, Éxpose no. 10 (1983–1984).

10. L. Hörmander, *Linear differential operators*, Actes Congr. Internat. Math. (Nice, 1970), pp. 121–133.

11. ______, *Fourier integral operators*. I, Acta Math. **127** (1971), 79–183.

12. B. Lascar, *Singularités des solutions d'equations aux derivées partielles nonlinéaires*, C. R. Acad. Sci. Paris **287** (1978), 521–529.

13. R. Melrose and N. Ritter, *Interaction of nonlinear progressing waves*, Ann. of Math. (to appear).

14. Y. Meyer, *Régularité des solutions des équations aux derivées partielles non linéaires*, Sem. Bourbaki, no. 560 (1979–1980).

15. L. Nirenberg, *Lectures on linear partial differential equations*, CBMS Regional Conf. Ser. in Math., no. 17, Amer. Math. Soc., Providence, R. I., 1973.

16. J. Rauch, *Singularities of solutions to semilinear wave equations*, J. Math. Pures Appl. **58** (1979), 299–308.

17. J. Rauch and M. Reed, *Propagation of singularities for semilinear hyperbolic equations in one space variable*, Ann. of Math. (2) **111** (1980), 531–552.

18. ______, *Nonlinear microlocal analysis of semilinear hyperbolic systems in one space dimension*, Duke Math. J. **49** (1982), 397–475.

19. ______, *Striated solutions of semilinear, two speed wave equations*, Indiana Univ. Math. J. (to appear).

20. E. M. Stein, *Singular integrals and differentiability properties of functions*, Princeton Univ. Press, Princeton, N. J., 1971.

Rutgers University

Proceedings of Symposia in Pure Mathematics
Volume **43** (1985)

Multidimensional Inverse Scatterings and Nonlinear Partial Differential Equations

R. BEALS AND R. R. COIFMAN[1]

The connection between certain linear spectral problems for differential operators and corresponding nonlinear equations has motivated the study of various inverse problems [**2**, **22**, **23**, **12**, **7**].

By recasting the notion of scattering data as $\bar{\partial}$-data, it is possible to generalize the theory to higher dimensions. In §1 we give a brief exposition of this point of view in one variable and connect it to the treatment of the anti-self-dual Yang–Mills equations (for which the scattering data becomes a transition matrix for a vector bundle).

We then give in §2 a formal outline of analogous procedures for multidimensional problems, and we give an equation which formally characterizes scattering data (as giving a deformation of the $\bar{\partial}$-complex on a characteristic variety). The formal treatment leads not only to plausible results but also to various formal paradoxes. In §§3 and 4 we carry out the details of the analysis for both the direct and inverse problems associated with $\Delta - q$ in $\mathbb{R}^3$, resolving the paradoxes and characterizing the (nonphysical) scattering data.

In at least some multidimensional cases there are, as in the one-dimensional case, associated hierarchies of evolution equations. In §5 we describe in detail a hierarchy of pseudo-Lax pairs in $2 + 1$ dimensions, relating it to the $\bar{\partial}$-formalism. The system described generalizes the AKNS–ZS framework [**2**, **23**] to two space dimensions.

Some historical remarks may be in order. The suggestion that inverse scattering could be viewed as a $\bar{\partial}$-problem seems to have originated with the authors [**7**]. For most one-dimensional problems the $\bar{\partial}$-data lives on certain curves (lines) plus a discrete set, so the problem may also be thought of as a Riemann–Hilbert

---

1980 *Mathematics Subject Classification*. Primary 35R30, 35G25.
[1] Research supported by NSF grants MCS 8104234, DMS 8402637.

factorization problem with assigned singularities; but see **[7]** for an example where the Riemann–Hilbert view fails. Ablowitz, Fokas, and others applied the $\bar{\partial}$-approach in a formal way to a number of two-dimensional problems; see **[12]** and references. For the analytic theory of KP II and related equations, see **[21]**. We give a rigorous treatment of a related two-dimensional problem in §5.

In the applications just cited, the associated complex manifold is still $\mathbb{C}$. Beyond two space dimensions new features appear that are discussed formally in §2. Some of the ideas were developed independently by Nachman and Ablowitz **[17]**.

1. The one-dimensional case. For illustration we consider a first-order system characterized by a fixed invertible matrix $J \in M_n(\mathbb{C})$ and a function $q\colon \mathbb{R} \to M_n(\mathbb{C})$ assumed to be small at ∞. For $z \in \mathbb{C}$, we look for an invertible matrix-valued function $\psi(\cdot, z)$ such that

$$J^{-1}\frac{\partial\psi}{\partial x}(x, z) = z\psi(x, z) + q(x)\psi(x, z). \tag{1.1}$$

With q small at ∞ one could expect

$$\psi(x, z) \sim e^{x\xi J}a_{\pm}(z) \quad \text{as } x \to \pm\infty. \tag{1.2}$$

The "scattering matrix"

$$s(z) = a_{+}(z)a_{-}(z)^{-1} \tag{1.3}$$

is independent of the choice of solution to (1.1). A preliminary version of the *direct problem* is to determine s from q (given J), while the *inverse problem* is to determine q from s.

The (formal) connection with evolution equations can be understood as follows **[1, 22]**. If q depends also on a time parameter t, one can look for an equation of the following form, coupled with (1.1):

$$\partial\psi/\partial t = b(x, t, z)\psi(x, t, z), \tag{1.4}$$

where $b(x, t, \cdot\,)$ is a polynomial of degree m and $b(x, t, z) \sim b_0 z^m$ as $|x| \to \infty$, with $b_0 J = J b_0$. Differentiating (1.1) with respect to t and (1.4) with respect to x yield compatibility conditions that determine the coefficients of $b(x, t, \cdot\,)$ uniquely, given b_0, J, and q, and that result in a generally nonlinear evolution equation

$$\frac{\partial q}{\partial t} = F\left(q, \frac{\partial q}{\partial x}, \ldots, \frac{\partial^m q}{\partial x^m}\right). \tag{1.5}$$

The scattering matrix is also time dependent; writing it as

$$s(z, t) = \lim_{x \to +\infty} \psi(x, z, t)\psi(-x, z, t)^{-1} \tag{1.6}$$

and differentiating lead formally to the linear equation

$$\frac{\partial s}{\partial t}(z, t) = [z^m b_0, s(z, t)]. \tag{1.7}$$

Now (1.7) can be integrated trivially. Thus if the inverse problem can be solved, one has an explicit schema for solving the initial-value problem for (1.5):

$$q(\cdot,0) \to s(\cdot,0) \to s(\cdot,t) \to q(\cdot,t). \tag{1.8}$$

The asymptotics (1.2) make sense if, for example, $J + J^* = 0$ and $z \in \mathbb{R}$, but not in general; even in the case $J + J^* = 0$, one needs discrete information off $\mathbb{R}$ in order to solve the inverse problem. To handle the general case one may look for a solution of (1.1) in the form

$$\psi(x,z) = m(x,z)e^{xzJ}, \tag{1.9}$$

$$m(\cdot,z) \text{ bounded}, \qquad \lim_{x\to-\infty} m(x,z) = 1. \tag{1.10}$$

Equations (1.1) and (1.9), (1.10) determine a unique m, which for each x is a piecewise-meromorphic function of z that tends to 1 as $z \to \infty$. Therefore one can hope to recover m from its $\bar\partial$-data by

$$m(x,z) = 1 + \frac{1}{2\pi i}\int (\zeta - z)^{-1}\frac{\partial m}{\partial \bar z}(x,\zeta)\, d\zeta \wedge d\bar\zeta. \tag{1.11}$$

Equations (1.1) and (1.9) imply

$$D_z m = Qm; \qquad D_z = \partial/\partial x - z \operatorname{ad} J,\ Q = Jq. \tag{1.12}$$

Note that D_z is a derivation on the algebra of matrix-valued functions. Therefore if m_1 and m_2 are two solutions of (1.12) and m_1 is invertible, then

$$D_z(m_1^{-1}m_2) = 0. \tag{1.13}$$

Conversely, $D_z v = 0$ implies mv is a solution of (1.12) whenever m is.

Since D_z depends holomorphically on z, it commutes with $\partial/\partial\bar z$ and one has, formally, that $\partial m/\partial \bar z$ solves (1.12) whenever m does. Thus, formally,

$$\partial m/\partial\bar z = mv, \quad \text{where } D_z v = 0. \tag{1.14}$$

In case (1.12), $D_z v = 0$ implies v has the form

$$v(x,z) = \exp(xz \operatorname{ad} J)w(z) = e^{xzJ}w(z)e^{-xzJ}. \tag{1.15}$$

Combining (1.11) and (1.14) gives an integral equation for m.

Conversely, suppose one is given v with $D_z v = 0$, and suppose m solves

$$\begin{aligned} m &= 1 + \frac{1}{\pi}\int (\zeta - z)^{-1} m(x,\zeta)v(x,\zeta)\, d\zeta \wedge d\bar\zeta \\ &= 1 + C(mv) = 1 + CTm. \end{aligned} \tag{1.16}$$

Then we claim that m solves (1.12) with $Q = [D_z, C](mv)$. Indeed,

$$D_z(1) = 0, \qquad [D_z, T] = 0; \tag{1.17}$$

the second identity follows from the fact that D_z is a derivation and $D_z v = 0$. Thus

$$\begin{aligned} D_z m &= D_z(I - CT)^{-1}1 = \left[D_z, (I - CT)^{-1}\right]1 \\ &= (I - CT)^{-1}[D_z, CT]m = (I - CT)^{-1}[D_z, C](mv). \end{aligned} \tag{1.18}$$

The affine dependence of D_z on z implies that $[D_z, C]$ maps to functions that are independent of z. Therefore

$$[D_z, C]mv = Q(x) = Q(x)\cdot 1. \tag{1.19}$$

Now CT commutes with left multiplication by functions of x, so (1.18) and (1.19) give (1.12).

The formal program sketched here has been carried out for general first-order systems (1.1) in [**8**, **5**], systems with quadratic eigenvalue parameter [**14**], and higher-order scalar equations [**6**]. Like the scattering matrix s (when it exists), the "scattering data" w of (1.15) evolves linearly in conjunction with (1.4):

$$\partial w/\partial t = \left[z^m b_0, w\right]. \tag{1.20}$$

This has consequences for the evolutions (1.5) [**9**, **19**]. For the exist relationship between s (when it exists) and w, see Theorem C of [**8**] and also [**9**].

Let us discuss evolutions like (1.20) from the point of view of the inverse problem; this leads also to a natural relationship with the anti-self-dual Yang–Mills equation. Suppose that D_z^1 and D_z^2 are two commuting derivations involving differentiation with respect to two variables or sets of variables x and t, and suppose

$$D_z^j v = 0, \qquad j = 1, 2. \tag{1.21}$$

If the dependence on z is affine and m is the solution of (1.16), it follows again that

$$D_z^j m = Q^j m, \qquad j = 1, 2. \tag{1.22}$$

Cross differentiation of (1.22) leads (if m is invertible) to the simplest nonlinear compatibility equation

$$D_z^2 Q^1 - D_z^1 Q^2 + \left[Q^1, Q^2\right] = 0. \tag{1.23}$$

As an example, let J and J' be two diagonal matrices, and set

$$D_z^1 = \partial/\partial x - z \operatorname{ad} J, \qquad D_z^2 = \partial/\partial t - z \operatorname{ad} J'. \tag{1.24}$$

Then a solution v of (1.21) has the form

$$v(x, t, z) = e^{z(xJ + tJ')} w(z) e^{-z(xJ + tJ')}, \tag{1.25}$$

and $Q^j = [D_z^j, C](mv)$ has the form

$$\begin{aligned} Q^1 &= \operatorname{ad} J \int_{\mathbf{C}} m(x, t, \zeta) v(x, t, \zeta)\, d\zeta \wedge d\bar{\zeta} = \operatorname{ad} J q(x, t), \\ Q^2 &= \operatorname{ad} J' q(x, t). \end{aligned} \tag{1.26}$$

The compatibility equation (1.23) becomes

$$\operatorname{ad} J \frac{\partial q}{\partial t} - \operatorname{ad} J' \frac{\partial q}{\partial x} + [\operatorname{ad} Jq, \operatorname{ad} J'q] = 0. \tag{1.27}$$

To relate the preceding to the well-known methods of Ward, Atiyah, and Hitchin [**3**, **4**] for obtaining solutions of the anti-self-dual Yang–Mills equations, we begin with a version of these equations described by K. Pohlmeyer [**18**]. Let $x = (x_1, x_2) \in \mathbb{C}^2$ and look for $\Omega(x)$, a positive-matrix-valued function verifying

$$\partial_{\bar{x}_1}\left(\Omega^{-1}\partial_{x_1}\Omega\right) + \partial_{\bar{x}_2}\left(\Omega^{-1}\partial_{x_2}\Omega\right) = 0. \tag{1.28}$$

(In the scalar case this is Laplace's equation for $\log \Omega$.)

Let $A_j = -\Omega^{-1}\partial_{x_j}\Omega$, so

$$\partial_{x_j}\Omega = -\Omega A_j. \tag{1.29}$$

Then (1.28) becomes

$$\partial_{\bar{x}_1}A_1 + \partial_{\bar{x}_2}A_2 = 0. \tag{1.30}$$

Equations (1.29) require

$$\partial_{x_2}A_1 - \partial_{x_1}A_2 + [A_1, A_2] = 0. \tag{1.31}$$

Multiplying (1.31) by z and adding to (1.30) give

$$D_z^2 A_1 - D_z^1 A_2 + [A_1, A_2] = 0, \tag{1.32}$$

where

$$D_z^1 = \partial_{x_1} - z\partial\bar{x}_2, \qquad D_z^2 = \partial x_2 + z\partial\bar{x}_1. \tag{1.33}$$

Now the D_z^j are commuting derivations which are affine in z and (1.32) is precisely the compatibility condition for

$$D_z^j m = A_j m, \qquad j = 1, 2. \tag{1.34}$$

As we have seen, to generate solutions A_j it suffices to find V with $D_z^j V = 0$, $j = 1, 2$, and then to solve (1.16).

To find such functions V we observe that in terms of the variables

$$u_1 = zx_1 + \bar{x}_2, \qquad u_2 = zx_2 - \bar{x}_1, \tag{1.35}$$

the equations for V become

$$\partial_{\bar{u}_i}V(u_1, u_2; z) = 0. \tag{1.36}$$

Thus V is holomorphic in the u_j and we may take

$$V(x_1, x_2; z) = \sum_{k=1}^{\infty}\sum_{j=0}^{k} c_{jk}(z)(zx_1 + \bar{x}_2)^j(zx_2 - \bar{x}_1)^{k-j}; \tag{1.37}$$

where the c_{jk} are distributions on $\mathbb{C}^1$ chosen to guarantee solvability of (1.16).

Let us relate these developments to the solutions obtained by Ward and to the Penrose twistor spaces. The coordinates u_j of (1.35) arise naturally from the following fibration of $P_3(\mathbb{C})$. Let $\mathbb{H}$ denote the quaternions, identified with $\mathbb{C}^2$ and with $P_1(\mathbb{H})$ by

$$\begin{gathered}(x_1, x_2) \sim \begin{pmatrix} x_1 & x_2 \\ -\bar{x}_2 & \bar{x}_1 \end{pmatrix} \sim \xi_1 + i\eta_1 + j\xi_2 + k\eta_2 \sim x_1 + jx_2; \\ i \sim \begin{pmatrix} i & 0 \\ 0 & -i \end{pmatrix}, \quad j \sim \begin{pmatrix} 0 & 1 \\ -1 & 0 \end{pmatrix}, \quad k \sim \begin{pmatrix} 0 & i \\ i & 0 \end{pmatrix}. \end{gathered} \tag{1.38}$$

Since each element of $P_3(\mathbb{C})$ is a line in $\mathbb{C}^4$, it is imbedded in a quaternionic line in $\mathbb{H}^2$, so $P_3(\mathbb{C})$ fibers over $P_1(\mathbb{H})$. In coordinates identifying $l \in P_3(\mathbb{C})$ with $(z_1, z_2, z_3, -1) \in \mathbb{C}^4$, l imbeds in the quaternionic line with coordinates

$$\left((z_3 - j)^{-1}(z_1 + z_2 j), 1\right) \sim (x_1 + x_2 j, 1).$$

Conversely, the fiber above (x_1, x_2) is given by

$$((z_3 - j)(x_1 + jx_2), z_3, -1) = (u_1, u_2, z_3, -1),$$

where the u_i are given by (1.35). Thus if we view V as defined on $\mathbb{P}^3(C)$, it becomes holomorphic in the natural complex structure transverse to the fibers.

Suppose $\tilde{V}$ is holomorphic in the variables u_j and also in z, for z in an annulus containing the circle $|z| = 1$. Let $V = \tilde{V}\mu$, where μ is the normalized rotation-invariant measure concentrated on the circle. If m solves (1.16), then m is holomorphic off the circle and has limits satisfying $m_+ = m_-(I + V)$; moreover, m_+ (resp. m_-) is holomorphic for $|z| > 1 - \varepsilon$ (resp. $|z| < 1 + \varepsilon$). Thus m lifts to provide locally a holomorphic frame for a holomorphic vector bundle on $P_3(\mathbb{C})$ with transition matrix $I + V$. Then Ω may be viewed as defining an inner product in this frame, and the A_j are the components of the unique associated hermitian holomorphic connection. See [**3**] for details and [**20**] for a derivation of instanton solutions by a factorization problem.

2. The multidimensional case: formal theory. Consider a scalar operator $p(D) = \sum p_\alpha D^\alpha$, $D = -i\partial/\partial x$. Given $z \in \mathbb{C}^n$, we seek a solution to

$$p(D)\psi(x, z) = p(z)\psi(x, z) + q(x, D)\psi(x, z), \tag{2.1}$$

where $q \to 0$ as $x \to \infty$, with

$$\psi(x, z) \sim e^{ix\cdot z} \quad \text{as } x \to \infty, \qquad x \cdot z = \sum_{i=1}^{n} x_i z_i. \tag{2.2}$$

For simplicity we consider here a zero-order perturbation $q(x, D) = q(x)$. Again (2.2) is generally ill defined, so we look for

$$\psi(x, z) = m(x, z)e^{ix\cdot z}, \qquad m(x, z) \to 1 \quad \text{as } x \to \infty. \tag{2.3}$$

Thus

$$p_z(D)m \equiv [p(D + z) - p(z)]m = qm. \tag{2.4}$$

Suppose there is a subvariety $V \subset \mathbb{C}^n$ such that $\nabla_\xi(\operatorname{Re} p_z)$ and $\nabla_\xi(\operatorname{Im} p_z)$ are linearly independent at each point of $p_z^{-1}(0) \cap \mathbb{R}^n$. Then $1/p_z$ is locally integrable on $\mathbb{R}^n$ and defines a fundamental solution G_z for $p_z(D)$:

$$G_z f = g_z * f, \qquad \hat{g}_z = 1/p_z. \tag{2.5}$$

Thus it is natural to interpret (2.3), (2.4) as an integral equation

$$m = 1 + G_z(qm) \equiv 1 + K_z m, \qquad z \in V. \tag{2.6}$$

Since we have assumed V to be a variety, there should be a corresponding $\bar{\partial}$ equation

$$\bar{\partial} m = G_z(q\bar{\partial} m) + [\bar{\partial} G_z](qm), \tag{2.7}$$

where $\bar\partial G_z$ is the distribution derivative of the operator-valued distribution $z \mapsto G_z$. Thus, formally,

$$\bar\partial m = Tm \equiv (I - K_z)^{-1}[\bar\partial G_z](qm). \tag{2.8}$$

Under our assumptions, as distributions on $\mathbb{R}^n$,

$$1/p_z = \lim_{\varepsilon \to 0+} \left[1 - \exp\left(-|p_z|^2/\varepsilon\right)\right](p_z)^{-1}, \tag{2.9}$$

$$\begin{aligned} \bar\partial(1/p_z) &= \overline{\partial p_z} \lim_{\varepsilon \to 0+} \varepsilon^{-1} \exp\left(-|p_z|^2/\varepsilon\right) \\ &= \overline{\partial p_z}\, |\nabla_\xi \operatorname{Re} p_z|^{-1} |\nabla_\xi \operatorname{Im} p_z|^{-1} (\sin\theta)^{-1}\, d\sigma_z = \overline{\partial p_z}\, d\nu_z, \end{aligned} \tag{2.10}$$

where $d\sigma_z$ is surface measure on the surface of codimension 2, and θ is the angle between $\nabla_\xi \operatorname{Re} p_z$ and $\nabla_\xi \operatorname{Im} p_z$

$$\Sigma_z = p_z^{-1}(0) \cap \mathbb{R}^n. \tag{2.11}$$

To put (2.8) in a more perspicuous form, we need one more ingredient. Let $e_\xi = \exp(ix \cdot \xi)$, considered as a multiplication operator, for $\xi \in \mathbb{R}^n$. Then

$$e_\xi^{-1} G_z e_\xi = G_{z+\xi}, \qquad e_\xi^{-1} K_z e_\xi = K_{z+\xi}, \tag{2.12}$$

so

$$(I - K_z)^{-1}(e_\xi 1) = e_\xi (I - K_{z+\xi})^{-1} 1 = e_\xi m(\cdot, z + \xi). \tag{2.13}$$

Provided that $z + \xi \in V$ whenever $z \in V$, $\xi \in \mathbb{R}^n$ and $p_z(z + \xi) = 0$, set

$$t(z, \xi) = [q(\cdot)m(\cdot, z)]^\wedge(\xi) = \int_{\mathbb{R}^n} e^{-ix\cdot\xi} q(x) m(x, z)\, dx. \tag{2.14}$$

Then combining (2.8), (2.13), and (2.14) formally gives

$$\begin{aligned} \bar\partial m(\cdot, z) &= (I - K_z)^{-1} c_n \int e_\xi t(z, \xi)\, \overline{\partial p_z}(\xi)\, d\nu_z(\xi) \\ &= c_n \int e_\xi m(\cdot, z + \xi) t(z, \xi)\, \overline{\partial p_z}(\xi)\, d\nu_z(\xi) \\ &\equiv Tm, \qquad c_n = (2\pi)^{-n}, \end{aligned} \tag{2.15}$$

$$[Tf](x, z) = c_n \int_{\Sigma_z} e^{ix\cdot\xi} f(x, z + \xi) t(z, \xi)\, \overline{\partial p_z}(\xi)\, d\nu_z(\xi). \tag{2.16}$$

The operator T takes suitable functions on V to (0, 1)-forms. It can be extended to map (0, r)-forms to (0, $r + 1$)-forms by the same formula. Formally, it satisfies the commutation identity

$$[\bar\partial, T] = -T^2. \tag{2.17}$$

To derive (2.17) we differentiate (2.14) and use (2.15) to obtain

$$\begin{aligned} \bar\partial t(z, \xi) &= c_n \int e^{-ix\cdot\xi} q(x) \int e^{ix\eta} m(x, z + \eta) t(z, \eta)\, \overline{\partial p_z}(\eta)\, d\nu_z(\eta)\, dx \\ &= c_n \int t(z + \eta, \xi - \eta) t(z, \eta)\, \overline{\partial p_z}(\eta)\, d\nu_z(\eta). \end{aligned} \tag{2.18}$$

Now the identity $p_{z+\eta}(\xi-\eta) \equiv p_z(\xi)$ implies

$$\overline{\partial p_z}(\xi)\, d\nu_z(\xi) \equiv \overline{\partial p_{z+\eta}}(\xi-\eta)\, d\nu_{z+\eta}(\xi-\eta). \tag{2.19}$$

From (2.18) and (2.19) we obtain

$$\begin{aligned}
[\bar\partial, T]f(x,z) &= c_n \int e^{ix\cdot\xi} f(x, z+\xi)\bar\partial t(z,\xi)\overline{\partial p_z}(\xi)\, d\nu_z(\xi) \\
&= -c_n^2 \int e^{ix\cdot\eta} \int e^{ix\cdot(\xi-\eta)} f(x, z+\eta+\xi-\eta) \\
&\qquad \cdot \overline{\partial p}_{z+\eta}(\xi-\eta)\, d\nu_{z+\eta}(\xi-\eta) \wedge \overline{\partial p_z}(\eta)\, d\nu_z(\eta) \\
&= -[T^2 f](x,z).
\end{aligned} \tag{2.20}$$

We want to consider the operator T as scattering data for the potential q. This is equivalent to considering the kernel t, but it is important to note that, although (2.14) defines t on $V \times \mathbb{R}^n$ (and makes recovery of q from t trivial), the kernel is defined only on the bundle

$$E = \bigcup_{z\in V} \Sigma_z = \{(z,\xi) \colon z \in V,\ \xi \in \mathbb{R}^n, p(z+\xi) = p(z)\}. \tag{2.21}$$

In deriving (2.18) we treated z and ξ as independent variables, but in deriving (2.17) we needed only to know

$$\bar\partial t(z,\xi) \wedge \overline{\partial p}(z+\xi) \quad \text{on } E. \tag{2.22}$$

When V is an open subvariety of $p^{-1}(\lambda)$ for some $\lambda \in \mathbb{C}$, the expression (2.22) depends only on $t|E$. Indeed, suppose g is a function on $V \times \mathbb{R}^n$ vanishing on E. Then $\bar\partial g|E$ annihilates all tangential vector fields to E, so it is a linear combination of $\bar\partial p(z)$ and $\bar\partial p(z+\xi)$. The former vanishes on E, so $\bar\partial g \wedge \overline{\partial p}(z+\xi) = 0$.

We turn now to the inverse problem. Suppose T is an operator defined by (2.16), where t is a function on the bundle E of (2.21), and suppose $[\bar\partial, T] = -T^2$. Consider the equation

$$\bar\partial m = Tm \tag{2.23}$$

with boundary conditions $m \sim 1$ at ∞ in V. Let S be a corresponding homotopy operator for the Dolbeault complex on V:

$$S\bar\partial + \bar\partial S = I. \tag{2.24}$$

We convert (2.23) to

$$m = 1 + STm. \tag{2.25}$$

Suppose the operators $I \pm ST$ are invertible in $L^\infty(V \times R^n)$, so that (2.25) has a unique solution. Formally,

$$\begin{aligned}
\bar\partial m &= \bar\partial(STm) = Tm - S\bar\partial(Tm) \\
&= Tm - ST\bar\partial m - S[\bar\partial, T]m = Tm - ST\bar\partial m + ST^2 m
\end{aligned} \tag{2.26}$$

or

$$(I+ST)[\bar\partial m - Tm] = 0. \tag{2.27}$$

Thus the commutator condition $[\bar\partial, T] = -T^2$ is precisely what is needed to force (2.23) from (2.25).

It is easy to check formally that $D + z$ commutes with T; it also commutes with $\bar\partial$, so

$$[\bar\partial - T]\,p_z(D)m = p_z(D)(\bar\partial m - Tm) = 0. \tag{2.28}$$

Now

$$\begin{aligned}\bar\partial(I - ST) &= \bar\partial - T + S\bar\partial T = \bar\partial - T - ST^2 + ST\bar\partial\\ &= (I + ST)(\bar\partial - T),\end{aligned} \tag{2.29}$$

so (2.28) implies

$$\bar\partial(I - ST)p_z(D)m = 0. \tag{2.30}$$

If constants are the only bounded holomorphic functions on V, and if $p_z(D)m$ is bounded and $STp_z(D)m \to 0$ at ∞ in V, we obtain

$$p_z(D)m(x, z) = q(x) + STp_z(D)m. \tag{2.31}$$

Thus, since ST commutes with multiplication by q,

$$p_z(D)m = (I - ST)^{-1}q = qm. \tag{2.32}$$

REMARKS. This last argument illustrates the danger of purely formal manipulations, since it leads not only to the supposedly correct conclusion (2.32) but also to the incorrect conclusion that for *any* polynomial p^* there is a potential $q^*(x)$ such that $p^*(D + z)m = q^*m$. What saves matters is not algebra but analysis: the operator identities are valid only on certain domains of functions, and only $p_z(D)m$ has the requisite analytic properties. With this *caveat* concerning formal theory, we turn to careful consideration of an example in §§3, 4.

Following a suggestion of G. Zuckerman, we note that the compatibility condition (2.17) can be given a more conceptual form. Let T_1 be the operator defined as in (2.16), but with left multiplication by $\partial\overline{p_z}$; thus

$$T_1\alpha = (-1)^{\deg(\alpha)}\,T\alpha. \tag{2.33}$$

Then (2.17) becomes

$$0 = T_1^2 - \bar\partial T_1 - T_1\bar\partial = (\bar\partial - T_1)^2, \tag{2.34}$$

and $\bar\partial - T_1$ is a (nonlocal) deformation of the Dolbeault complex associated to the bundle E over V. In this connection it is interesting to note that (geometric) deformations of complex structure play an important role in the Penrose approach to the Yang–Mills and Einstein equations [**15**].

3. $\Delta - q$ in $\mathbb{R}^3$: the direct problem. Here $p(z) = z \cdot z = z_1^2 + z_2^2 + z_3^2$. We take

$$V = \{z \in \mathbb{C}^3 : z \neq 0,\ z \cdot z = 0\}. \tag{3.1}$$

Thus $p_z(\xi) = p(z + \xi)$, $z \in V$. Set

$$g(x, z) = g_z(x) = (1/p_z)^\vee(x), \tag{3.2}$$

the inverse Fourier transform. The basic analytical tool is the following lemma, which is proved at the end of this section.

LEMMA 1. *$|x|g(x, z)$ is a bounded continuous function on $(\mathbb{R}^3 \setminus 0) \times V$. The form $\bar{\partial}g$ is a bounded continuous function on $\mathbb{R}^3 \times V$.*

As above we define the operator K_z by

$$K_z u = G_z(qu) = g_z * (qu). \tag{3.3}$$

COROLLARY. *Suppose that for some ε, $0 < \varepsilon \leqslant 1/2$,*

$$q \in L^{3/2-\varepsilon}(\mathbb{R}^3) \cap L^{3/2+\varepsilon}(\mathbb{R}^3). \tag{3.4}$$

Then K_z is a compact operator in $L^\infty(\mathbb{R}^3)$ with range in $C_0(\mathbb{R}^3)$. The map $z \to K_z$ is continuous with respect to the operator norm.

PROOF. Splitting g_z into parts supported near the origin and near ∞, one obtains from Lemma 1 that

$$g_z = g_z' + g_z'', \qquad g_z' \in L^{3-\delta'}, g_z'' \in L^{3+\delta''},$$

with g_z' and g_z'' depending continuously in norm on z. Choose δ' and δ'' so that $3 - \delta'$ and $3 + \delta''$ are dual indices for $3/2 + \varepsilon$ and $3/2 - \varepsilon$, respectively. The desired conclusion is then a straightforward application of Hölder's inequality and standard approximation arguments.

DEFINITION. Y_0 is the space $C \cap L^\infty(\mathbb{R}^3 \times V)$. Y_j, $j = 1, 2$, is the space of $(0, j)$-forms that can be expressed as

$$\alpha = \sum \alpha_j(x, z)\, d\bar{z}_j \quad \text{or} \quad \beta = \sum \beta_{jk}(x, z)\, d\bar{z}_j \wedge d\bar{z}_k \tag{3.5}$$

with α_j and β_{jk} in Y_0.

Set

$$K: Y_0 \to Y_0, \qquad Kf(\cdot, z) = K_z[f(\cdot, z)]. \tag{3.6}$$

Thus the equations $m(\cdot, z) = 1 + K_z m(\cdot, z)$ are equivalent to the single equation

$$m = 1 + Km. \tag{3.7}$$

THEOREM 1. *Suppose q has sufficiently small norm in $L^{3/2 \pm \varepsilon}$, where $0 < \varepsilon \leqslant 1/2$. Then (3.7) has a unique solution $m \in Y_0$. Moreover,*

$$\lim_{|x| \to \infty} m(x, z) = 1, \quad \textit{all } z \in V; \tag{3.8}$$

$$\bar{\partial} m \in Y_1. \tag{3.9}$$

PROOF. Corollary 1 implies that K maps Y_0 to itself with norm dominated by the appropriate L^r-norms of q, so the existence and uniqueness of m is immediate. The conclusion (3.8) about the limit follows from the facts that $m(\cdot, z) - 1 = K_z m(\cdot, z)$ and K_z has range in $C_0(\mathbb{R}^3)$. The final statement follows from a term-by-term differentiation of the Neumann series for m, which gives

$$\bar{\partial} m = \sum_{0 \leqslant j < k < \infty} K^j M K^{k-j-1} 1, \tag{3.10}$$

with

$$Mf(\cdot, z) = \bar{\partial} g(\cdot, z) * f(\cdot, z).$$

For the remainder of this section we assume that the hypothesis of Theorem 1 is satisfied. As in §2 we set

$$t(z, \xi) = \int e^{-ix\cdot\xi} q(x) m(x, z)\, dx. \tag{3.11}$$

Note that $q(\cdot)m(\cdot, z)$ belongs to $L^1(\mathbb{R}^3)$ if q does, so

$$q \in L^1(\mathbb{R}^3) \quad \text{implies} \quad t \in C \cap L^\infty(V \times \mathbb{R}^3). \tag{3.12}$$

If z is in V we have

$$z = a + ib; \qquad a, b \in \mathbb{R}^3, \tag{3.13}$$

$$|a|^2 = |b|^2 = \tfrac{1}{2}|z|^2, \qquad a \cdot b = 0. \tag{3.14}$$

The set Σ_z of (2.11) is

$$\Sigma_z = \{\xi \in \mathbb{R}^3 : |\xi + a| = |a|, \xi \cdot b = 0\}. \tag{3.15}$$

Thus Σ_z is a circle of radius $|a|$ centered at $-a$. The measure $d\nu_z$ of (2.10) is

$$d\nu_z = \tfrac{1}{2}|z|^{-2}\, d\sigma_z \tag{3.16}$$

where $d\sigma_z$ is arc-length measure on Σ_z. Let T be the operator defined by (2.16).

LEMMA 2. *If q belongs to $L^1(\mathbb{R}^3)$, then T maps Y_0 to Y_1 and maps Y_1 to Y_2.*

PROOF. In the present case

$$\bar{\partial} p_z(\xi) = 2\xi \cdot d\bar{z} \tag{3.17}$$

and $|\xi| \leqslant 2|z|$ on Σ_z, so the form-valued measure that is the "kernel" of T has uniformly bounded norm. The measure is weak* continuous with respect to z; the mapping properties follow.

THEOREM 2. *Suppose q belongs to $L^1(\mathbb{R}^3)$. Then $\bar{\partial} m = Tm$.*

PROOF. Equation (3.10) gives

$$\bar{\partial} m = (I - K)^{-1} Mm, \tag{3.18}$$

and the formal calculation (2.15) is readily justified.

THEOREM 3. *Suppose q belongs to $L^1(\mathbb{R}^3)$. For any $f \in Y$ such that $\bar{\partial} f$ belongs to Y_1,*

$$[\bar{\partial}, T] f = -T^2 f. \tag{3.19}$$

PROOF. Under our hypotheses the formal calculation (2.18)–(2.20) is valid.

For the inverse problem we shall want some decay of t in the ξ-variables. This corresponds to smoothness of q.

PROPOSITION. *Suppose $D^\alpha q$ belongs to $L^1(\mathbb{R}^3)$ for each α with $|\alpha| \leqslant N$. Then*

$$|t(z, \xi)| \leqslant C(1 + |\xi|)^{-N}, \qquad z \in V, \quad \xi \in \mathbb{R}^3. \tag{3.20}$$

PROOF. In view of the definition of t and the hypothesis on q, (3.20) follows immediately from

$$D_x^\alpha m \in Y_0, \quad \text{all } |\alpha| \leqslant N. \tag{3.21}$$

As in the discussion of $\bar{\partial} m$, we may differentiate the Neumann series term by term to obtain

$$D_i m = \sum_{0 \leqslant j < k} K^j [D_i, K] K^{k-j-1} 1. \tag{3.22}$$

Now

$$[D_i, K] f(\cdot, z) = G_z((D_i q) f(\cdot, z)),$$

so $[D_i, K]$: $Y_0 \to Y_0$ and we obtain (3.21) for $|\alpha| = 1$. Iterating this argument gives (3.21).

PROOF OF LEMMA 1. A change of variables gives

$$g(\lambda z, x) = \lambda g(z, \lambda x), \qquad \lambda > 0. \tag{3.23}$$

Therefore it is enough to establish the result when, say $|z|^2 = 2$. Rotating coordinates we may assume $b = \operatorname{Im} z$ is the vector $(0, 0, -1) \in \mathbb{R}^3$. Write $x = (y, t)$, $y \in \mathbb{R}^2$, and $\xi = (\eta, \tau)$, $\eta \in \mathbb{R}^2$. Then $\operatorname{Im} z = (a, 0)$ and $|a| = 1$, so

$$\begin{aligned} g(z, x) &= (2\pi)^{-3} e^{-iy \cdot a} \iint \frac{e^{i(x \cdot \eta + t\tau)}}{|\xi|^2 - 1 - 2i\tau}\, d\eta\, d\tau \\ &= \frac{1}{2}(2\pi)^{-2} e^{-iy \cdot a} \int e^{i(y \cdot \eta)} h_t(|\eta|)\, d\eta. \end{aligned} \tag{3.24}$$

Since

$$\begin{aligned} (|\xi|^2 - 1 - 2i\tau)^{-1} &= \left[(\tau - i)^2 + |\eta|^2\right]^{-1} \\ &= \frac{1}{2i|\eta|}\left[(\tau - i - i|\eta|)^{-1} - (\tau - i + i|\eta|)^{-1}\right], \end{aligned} \tag{3.25}$$

the function h_t can be evaluated explicitly:

$$\begin{aligned} h_t(r) &= H(r-1) r^{-1} e^{t(r-1)}, \qquad t < 0; \\ &= r^{-1} e^{-t(r+1)} - H(1-r) r^{-1} e^{t(r-1)}, \qquad t > 0 \end{aligned} \tag{3.26}$$

for $r > 0$, where $H(s) = 0$, $s < 1$, and $H(s) = 1$, $s > 1$. We need to show

$$\left| \int_{\mathbb{R}^2} e^{iy \cdot \eta} h_t(|\eta|)\, d\eta \right| < C(t^2 + |y|^2)^{-1/2}. \tag{3.27}$$

Since $h_t(|\eta|)$ is radial, we can first calculate the corresponding inverse Fourier transform in $\mathbb{R}^3$. Indeed,

$$\int_{\mathbb{R}^2} e^{iy \cdot \eta} h_t(|\eta|)\, d\eta = c \int_{\mathbb{R}} k_t(y, s)\, ds = 2c \int_0^\infty k_t(y, s)\, ds, \tag{3.28}$$

where

$$k_t(x) = \int_{\mathbb{R}^3} e^{ix \cdot \eta} h_t(|\xi|)\, d\xi. \tag{3.29}$$

Since k_t is radial, we may replace x by $(0, 0, |x|)$ and compute (3.29) in spherical coordinates:

$$
\begin{aligned}
k_t(x) &= c_1 \int_0^\infty \int_0^\pi e^{i|x|r\cos\theta} r \sin\theta \, r h_t(r)\, d\theta\, dr \\
&= c_2 |x|^{-1} \int_0^\infty \left[e^{i|x|r} - e^{-i|x|r}\right] r h_t(r)\, dr \qquad (3.30)\\
&= c_3 \left[\frac{t \sin|x|}{|x|(t^2 + |x|^2)} - \frac{\cos|x|}{t^2 + |x|^2} \right].
\end{aligned}
$$

The second term on the right in (3.30), substituted into the integral (3.28), easily gives an estimate of the form (3.27). To derive the same estimate for the first term we want to show

$$
\left| \int_0^\infty \frac{t \sin\sqrt{s^2 + a^2}}{\sqrt{s^2 + a^2}\,(t^2 + s^2 + a^2)}\, ds \right| \leqslant \frac{C}{\sqrt{t^2 + a^2}}, \qquad t, a > 0. \tag{3.31}
$$

To estimate the integral in (3.31) we consider the intervals $0 \leqslant s \leqslant 1$ and $s > 1$. For the first, one has the obvious domination by $t(t^2 + a^2)^{-1}$. For the second, the integral is

$$
-\int_1^\infty \frac{d}{ds}\left[\cos\sqrt{s^2 + a^2}\right] f(s, t, a)\, ds, \quad f(s, t, a) = ts^{-1}(t^2 + s^2 + a^2)^{-1}.
$$

Since $f(1, t, a) \leqslant t(t^2 + a^2)^{-1}$ and $|(\partial f/\partial s)(s, t, a)| \leqslant 3t(t^2 + a^2)^{-1}s^{-2}$, integration by parts gives the desired estimate.

4. $\Delta - q$ in $\mathbb{R}^3$: the inverse problem. Here again $p(z) = z \cdot z$ in $\mathbb{C}^3$, and we take $V = p^{-1}(0) \setminus 0$. The associated bundle is

$$
E = \{(z, \xi) \in V \times \mathbb{R}^3 : p(z + \xi) = 0\}. \tag{4.1}
$$

This is a trivial bundle; in fact, given z as in (3.14), let c be the vector product $a \times b$ and set

$$
\xi(z, \theta) = (\cos\theta - 1)a + |a|^{-1}(\sin\theta)c. \tag{4.2}
$$

Then $\theta \to \zeta(z, \theta)$ maps S^1 onto Σ_z, and

$$
E \cong V \times S^1. \tag{4.3}
$$

Given

$$
t \in C(E), \tag{4.4}
$$

we define an operator T as in §2:

(4.5)

$$
Tf(x, z) = (2\pi)^{-3} \int_{-\pi}^{\pi} e^{ix\cdot\xi} f(x, z + \xi) \wedge t(z, \xi) |z|^{-1} \xi \cdot d\bar{z}\, d\theta, \qquad \xi = \xi(z, \theta).
$$

We make two assumptions on the kernel t of T. For some $N > 4$ we assume

$$
|t(z, \xi)| \leqslant \delta(1 + |\xi|)^{-N}, \qquad (z, \xi) \in E. \tag{4.6}
$$

In particular, this implies

$$T\colon Y_0 \to Y_1, \qquad T\colon Y_1 \to Y_2. \tag{4.7}$$

We also assume that for any $f \in Y_0$, such that $\bar{\partial} f$ is in Y_1,

$$[\bar{\partial}, T] f = -T^2 f. \tag{4.8}$$

As in §2, a sufficient condition for (4.8) is that t has an extension to $V \times \mathbb{R}^3$ with $\bar{\partial} t$ continuous and

$$\begin{aligned}\bar{\partial} t(z, \xi) \wedge \xi \cdot d\bar{z} \\ = (2\pi)^{-3} \int t(z+\eta, \xi-\eta) t(z,\eta)[\eta \cdot d\bar{z} \wedge \xi \cdot d\bar{z}]\, dv_z(\eta).\end{aligned} \tag{4.9}$$

The goal of this section is the following.

THEOREM 4. *Suppose t is such that* (4.6) *and* (4.8) *are satisfied, with δ in* (4.6) *sufficiently small. Then there is a unique $m \in Y_0$ with the properties*

$$\bar{\partial} m = Tm, \qquad \lim_{z\to\infty} m(x,z) = 1. \tag{4.10}$$

Moreover, $D_x^\alpha m$ belongs to Y_0 for $|\alpha| \leqslant 2$, and there is $q \in C \cap L^\infty(\mathbb{R}^3)$ such that

$$p(D_x + z) m(x,z) = q(x) m(x,z). \tag{4.11}$$

The proof depends on an analysis of T and a homotopy operator S.

LEMMA 3. *Suppose f is in $C(\mathbb{R}^3 \times V)$ and $g(|z|)f(x,z)$ is bounded. Then $g(|z|)(1+|z|)^2 Tf(x,z)$ is bounded.*

PROOF. Note that $|z+\xi| = |z|$ for $(z,\xi) \in E$. Therefore T commutes with multiplication by $g(|z|)$, and we may take $g \equiv 1$. Now

$$|\zeta(z,\theta)|^2 = (\cos\theta - 1)^2 |a|^2 + \sin^2\theta |b|^2 = (1-\cos\theta)|z|^2,$$

so

$$c_1|\theta z| \leqslant |\xi(z,\theta)| \leqslant c_2|\theta z|, \qquad |\theta| \leqslant \pi. \tag{4.12}$$

The necessary estimate is straightforward when $|z| \leqslant 1$. For $|z| > 1$ we split the region of integration into $|\theta z| \leqslant 1$ and $|\theta z| > 1$. The first part gives an integral dominated by

$$\int_0^{1/|z|} |\theta z|\, |z|^{-1}\, d\theta = \frac{1}{2}|z|^{-2}. \tag{4.13}$$

The second part is dominated by

$$\int_{1/|z|}^{\pi} |\theta z|\, |z|^{-1} |\theta z|^{-N}\, d\theta \leqslant C\left(|z|^{-2} + |z|^{-N}\right). \tag{4.14}$$

This proof used (4.6) only with $N > 2$. The following extension uses $N > 4$.

LEMMA 4. *Suppose $D_x^\alpha f$ is in $C(\mathbb{R}_3 \times V)$ for $|\alpha| \leqslant 2$ and $g(|z|)D_x^\alpha f$ is bounded. Then $g(|z|)(1+|z|)^2 D_x^\alpha Tf$ is bounded.*

PROOF. We need to establish that the commutators $[D_x, T]$ and the double commutators $[D_x,[D_x, T]]$ have the same mapping property in Lemma 3 as T itself. Now $[D_x, T]$ has the same form as T with the kernel replaced by $\xi t(z, \xi)$. The method of proof of Lemma 3 applies; there is an extra factor $|\theta z|^2$ and $N > 4$, for the double commutator.

In deriving a homotopy operator S, it is convenient to start in $\mathbf{C}^3 \setminus 0$. The complex dilations $\varphi_\lambda(z) = \lambda z$, $\lambda \in \mathbf{C} \setminus 0$, act on $\mathbf{C}^3 \setminus 0$ and for any form

$$(\partial/\partial\bar\lambda)\varphi_\lambda^*\alpha = \varphi_\lambda^*(L_{Z_\lambda}\alpha),$$

where

$$Z_\lambda = \bar\lambda^{-1}\sum \bar z_j \frac{\partial}{\partial \bar z_j}.$$

Therefore one should invert $\partial/\partial\bar\lambda$ at $\lambda = 1$ and take

(4.15)
$$S\alpha(z) = \int (\lambda - 1)^{-1}\varphi_\lambda^*[Z_\lambda \lrcorner \alpha](z)\, d\mu(\lambda), \qquad d\mu(\lambda) = (2\pi i)^{-1} d\lambda \wedge d\bar\lambda.$$

Explicitly, for (0, 1)-forms and (0, 2)-forms

$$\alpha = \sum \alpha_j(z)\, d\bar z_j, \tag{4.16}$$

$$\beta = \sum \beta_{jk}(z)\, d\bar z_j \wedge d\bar z_k, \tag{4.17}$$

we take

$$S\alpha(z) = \int (\lambda - 1)^{-1}\sum \alpha_j(\lambda z)\bar z_j\, d\mu(\lambda), \tag{4.18}$$

$$S\beta(z) = \sum\left[\int (\lambda - 1)^{-1}\bar\lambda\beta_{jk}(\lambda z)\, d\mu(\lambda)\right]\left[\bar z_j\, d\bar z_k \wedge d\bar z_j\right]. \tag{4.19}$$

For a formal check of the homotopy property for a (0, 1)-form α, one has, after a formal computation,

$$\begin{aligned}(\bar\partial S + S\bar\partial)\alpha(z) &= \sum\left[\int (\lambda - 1)^{-1}\frac{\partial}{\partial\bar\lambda}\{\bar\lambda\alpha_j(\lambda z)\}\, d\mu(\lambda)\right] d\bar z_j \\ &= \sum \bar\lambda\alpha_j(\lambda z)\big|_{\lambda=1}\, d\bar z_j = \alpha(z).\end{aligned} \tag{4.20}$$

Note that $\varphi_\lambda\colon V \to V$, so (4.18), (4.19) may also be used to define S on forms on V.

For the remainder of this section we fix some ε, $0 < \varepsilon < 1$.

DEFINITION. $F_0(\varepsilon)$ is the space of continuous functions on V with $|f(z)| \leqslant C(1 + |z|)^{\varepsilon-1}$. $F_1(\varepsilon)$ is the space of (0, 1)-forms on V that can be expressed in the form (4.16) with $|\alpha_j(z)| \leqslant C(1 + |z|)^{\varepsilon-2}$. $F_x(\varepsilon)$ is the space of (0, 2)-forms on V that can be expressed in the form (4.17) with $|z|\,|\beta_{jk}(z)| \leqslant C(1 + |z|)^{\varepsilon-2}$.

LEMMA 5. *S maps $F_1(\varepsilon)$ to $F_0(\varepsilon)$.*

PROOF. Suppose α belongs to $F_1(\varepsilon)$. The integral in (4.18) is dominated by

$$|\lambda - 1|^{-1}|z|(1 + |\lambda z|)^{\varepsilon-2}. \tag{4.21}$$

To estimate $S\alpha$ for $|z| \leqslant 1$ we integrate over the regions

$$|\lambda| < 2, \quad 2 \leqslant |\lambda| < |z|^{-1}, \quad |\lambda| \geqslant |z|^{-1}, \tag{4.22}$$

and replace (4.21) by, respectively,

$$|\lambda - 1|^{-1}|z|, \quad |\lambda|^{-1}|z|, \quad |z|^{\varepsilon-1}|\lambda|^{\varepsilon-3}. \tag{4.23}$$

The three integrals are dominated by 1; for example, the third is essentially

$$|z|^{\varepsilon-1}\int_{1/|z|}^{\infty} r^{\varepsilon-3} r\,dr = (1-\varepsilon)^{-1}. \tag{4.24}$$

For $|z| > 1$ we use the regions

$$|\lambda z| < 1/2, \quad 1/2 \leqslant |\lambda z| < 2|z|, \quad |\lambda| > 2, \tag{4.25}$$

and the respective kernel estimates

$$|z|, \quad |z|^{\varepsilon-1}|\lambda - 1|^{-1}, \quad |z|^{\varepsilon-1}|\lambda|^{\varepsilon-3}. \tag{4.26}$$

LEMMA 6. *Suppose β is in $F_2(\varepsilon)$. Then $\alpha = |z|S\beta$ has the form* (4.16) *with $\alpha_j \in F_0(\varepsilon)$.*

PROOF. The integral corresponding to a given α_j in (4.19), after multiplication by $|z|$, is dominated by

$$|\lambda|\,|\lambda - 1|^{-1}|z|^2(|\lambda z|)^{-1}(1 + |\lambda z|)^{\varepsilon-2}, \tag{4.27}$$

which is the same as (4.21).

LEMMA 7. *Suppose α belongs to $F_1(\varepsilon)$ and $\bar\partial\alpha$ belongs to $F_2(\varepsilon)$. Then*

$$\alpha = (S\bar\partial + \bar\partial S)\alpha. \tag{4.28}$$

PROOF. Under our assumptions the right side of (4.28) is well defined, and the formal computation (4.20) is valid. Indeed, with $g(\lambda) = \alpha_j(\lambda z)$ we have

$$\bar\lambda g(\lambda) \leqslant C(1 + |\lambda|)^{\varepsilon-1}, \quad (\partial/\partial\bar\lambda)[\bar\lambda g(\lambda)] \leqslant C(1 + |\lambda|)^{\varepsilon-2},$$

so that the inversion at the last step of (4.20) is valid.

DEFINITION. $Y(\varepsilon)$ is the space of functions $f \in C(\mathbb{R}^3 \times V)$ such that $\bar\partial f$ belongs to $C(\mathbb{R}^3 \times V)$ and

$$|f(x, z)| \leqslant C(1 + |z|)^{\varepsilon}, \tag{4.29}$$

$$|z|\,|\bar\partial f(x, z)| \leqslant C(1 + |z|)^{\varepsilon}. \tag{4.30}$$

$Z(\varepsilon)$ is the subspace of $Y(\varepsilon)$ consisting of functions satisfying

$$|f(x, z)| \leqslant C(1 + |z|)^{\varepsilon-1}. \tag{4.31}$$

LEMMA 8. *The operator ST maps $Y(\varepsilon)$ to $Z(\varepsilon)$.*

PROOF. The identity (4.8) carries over to $Y(\varepsilon)$. Thus for f in $Y(\varepsilon)$,

$$\bar\partial Tf = T\bar\partial f - T^2 f. \tag{4.32}$$

From Lemma 3 and (4.29), (4.30), (4.32) one obtains

$$|Tf| + |z|\,|\bar\partial Tf| \leqslant C(1 + |z|)^{\varepsilon-2}. \tag{4.33}$$

Therefore $Tf(x, \cdot)$ belongs to $F_1(\varepsilon)$ uniformly with respect to x, and Lemma 6 gives the inequality (4.31) for STf. Moreover, $\bar\partial Tf(x, \cdot)$ belongs to $F_2(\varepsilon)$ uniformly with respect to x, so Lemma 6 gives

$$\bar\partial STf = Tf - S\bar\partial Tf. \tag{4.34}$$

Inequality (4.30) for STf follows from (4.34), (4.33), Lemma 3, and Lemma 6.

LEMMA 9. *The following identity is valid on* $Y(\varepsilon)$:

$$(I + ST)(\bar\partial - T) = \bar\partial(I - ST). \tag{4.35}$$

PROOF. The proof of Lemma 8 showed that all terms of this identity are well defined on $Y(\varepsilon)$ and that

$$(\bar\partial S + S\bar\partial)T = T \quad \text{on } Y(\varepsilon). \tag{4.36}$$

The identity (4.35) follows from (4.8) and (4.36).

LEMMA 10. *If f belongs to $Y(\varepsilon)$ and $\bar\partial f = 0$, then f is independent of z.*

PROOF. For fixed $x \in \mathbb{R}^3$ and $z \in V$, $g(\lambda) = f(x, \lambda z)$ is holomorphic on $\mathbb{C} \setminus 0$ and dominated by $(1 + |\lambda|)^\varepsilon$. Since we assumed $\varepsilon < 1$, g is constant. Thus for fixed $x \in \mathbb{R}^3$, $f(x, z) = f(x, |z|^{-1}z)$ is holomorphic and attains a maximum on $\{z \in V: |z| = 1\}$, hence is constant.

PROOF OF THEOREM 4. Suppose that m belongs to $C \cap L^\infty(\mathbb{R}^3 \times V)$ and satisfies (4.10). Then m is in $Y(\varepsilon)$, and (4.35) implies $\bar\partial f = 0$, where $f = (I - ST)m$. By Lemma 10, f is constant. Since STm is in $Z(\varepsilon)$ and $m(x, z) \to 1$ as $|z| \to \infty$, it follows that f is identically 1. We have shown

$$m = 1 + STm. \tag{4.37}$$

If t satisfies (4.6) with sufficiently small δ, the operator ST will have norm < 1 as an operator in $Y(\varepsilon)$. Thus the solution to (4.37) is unique.

Conversely, suppose t satisfies (4.6) with δ sufficiently small. Then (4.37) has a unique solution $m \in Y(\varepsilon)$. Since $m - 1$ belongs to $Z(\varepsilon)$, we have $m(x, z) \to 1$ as $|z| \to \infty$. The identity (4.35) implies $(I + ST)(\bar\partial m - Tm) = 0$, so $\bar\partial m = Tm$.

To differentiate m we note first that $[D_x, ST] = S[D_x, T]$ has the same properties as ST; see Lemma 4. The same is true for the double commutators, so

$$D_x^\alpha f \in Y(\varepsilon) \quad \text{implies} \quad D_x^\alpha STf \in Z(\varepsilon) \qquad \text{for all } |\alpha| \leqslant 2. \tag{4.38}$$

The Neumann series for the solution to (4.37) may be differentiated term by term, therefore, to give

$$\begin{aligned} D_x m &= \sum_{j,k\geqslant 0} (ST)^j S[D_x, T](ST)^k 1 \\ &= (I - ST)^{-1} S[D_x, T]m \in Z(\varepsilon). \end{aligned} \tag{4.39}$$

Another term-by-term differentiation gives

$$D_x^\alpha m \in Z(\varepsilon), \qquad |\alpha| = 2. \tag{4.40}$$

Finally, note that $D_x + z$ commutes with $\bar\partial$ and with T. Set

$$n(x, z) = p(D_x + z)m(x, z) = D_x \cdot D_x m(x, z) + 2z \cdot D_x m(x, z).$$

Then $\bar{\partial} n = Tn$. It follows from (4.38) and (4.39) that n belongs to $Y(\varepsilon)$, so (4.35) implies $\bar{\partial}(n + STn) = 0$. Again, Lemma 10 gives

$$n(x, z) + STn(x, z) = q(x). \tag{4.41}$$

Since n is in $Y(\varepsilon)$, q belongs to $C \cap L^\infty(\mathbb{R}^3)$. Multiplication by a function of x commutes with ST, so (4.41) gives

$$n = (I - ST)^{-1} q = q(I - ST)^{-1} 1 = qm. \tag{4.42}$$

5. A two-dimensional system and the associated hierarchy of nonlinear evolutions. The two-dimensional case is similar to the one-dimensional case, since our scattering data automatically verifies the compatibility condition (2.17): both sides are (0, 2)-forms on a one-dimensional variety. As noted in the introduction, a number of examples have been described formally in the literature [**1**, **12**]. We give here an abbreviated but rigorous treatment of a spectral problem, which leads naturally to a hierarchy of nonlinear evolution equations in 2 + 1 variables, including the Davey–Stewartson equations [**10**]

$$\mathring{q} = 2i \frac{\partial^2 q}{\partial x_1 \partial x_2} - 4irq, \qquad \frac{1}{4}\Delta r = \frac{\partial^2}{\partial x_1 \partial x_2}\left(|q|^2\right). \tag{5.1}$$

The advantages of this example are the simplicity of the analysis for both the direct and inverse problems, and the inherent symmetries which lead to a simple hierarchy of pseudo-Lax pairs.

The spectral problem is associated to the system

$$\begin{gathered} D\psi \equiv \begin{pmatrix} \partial_{\bar{x}} & 0 \\ 0 & \partial_x \end{pmatrix} \psi = Q\psi; \\ Q = Q(x) = \begin{pmatrix} 0 & q_1 \\ q_2 & 0 \end{pmatrix}, \qquad x \in R^2 \approx \mathbb{C}. \end{gathered} \tag{5.2}$$

The free equation ($Q \equiv 0$) has two one-parameter families of exponential vector solutions: $(e^{xz}, 0)^t$ and $(0, e^{\bar{x}z})^t$, $z \in \mathbb{C}$. Combining, we set

$$e_z = e_z(x) = \begin{pmatrix} e^{xz} & 0 \\ 0 & e^{\bar{x}z} \end{pmatrix}, \qquad z \in \mathbb{C}, \tag{5.3}$$

and look for matrix solutions to (5.2) in the form

$$\psi(x, z) = m(x, z) e_z(x); \qquad m \to 1 \quad \text{as } |x| \to \infty. \tag{5.4}$$

Then (5.2), (5.3) become

$$D_z m \equiv \left[D(m e_z)\right] e_z^{-1} = Qm; \qquad m \to 1 \quad \text{as } |x| \to \infty. \tag{5.5}$$

The operator D has the well-known fundamental solution D^{-1}, where

$$D^{-1} f(x) = \frac{1}{\pi} \int_{\mathbb{C}} \begin{pmatrix} (x - y)^{-1} & 0 \\ 0 & (\bar{x} - \bar{y})^{-1} \end{pmatrix} f(y)\, dy. \tag{5.6}$$

Here, abusing notation, we write $dy = dy_1\,dy_2$ for Lebesgue measure. To invert D_z we note that

$$D_z = E_z^{-1} D E_z, \tag{5.7}$$

where E_z operates on matrix functions by

$$E_z a(x) = \begin{pmatrix} a_{11}(x) & e^{\bar{x}z - x\bar{z}} a_{21}(x) \\ e^{xz - \bar{x}\bar{z}} a_{12}(x) & a_{22}(x) \end{pmatrix}. \tag{5.8}$$

Since the exponentials occurring in this formulation are bounded,

$$D_z^{-1} = E_z^{-1} D^{-1} E_z. \tag{5.9}$$

Thus D_z^{-1} is given entry by entry by convolution with functions that belong to $L^{2-\varepsilon} \cap L^{2+\varepsilon}$ uniformly with respect to z.

THEOREM 5. *Suppose Q belongs to $L^{2-\varepsilon} \cap L^{2+\varepsilon}$ for some $\varepsilon > 0$ and the norms $\|Q\|_{2-\varepsilon}$, $\|Q\|_{2+\varepsilon}$ are sufficiently small. Then the equation*

$$m = 1 + D_z^{-1}(Qm) \tag{5.10}$$

has a unique bounded solution m satisfying (5.5). *If $xQ(x)$ also belongs to $L^{2-\varepsilon} \cap L^{2+\varepsilon}$, then*

$$m(x, z) = 1 + E_z^{-1} \begin{pmatrix} x^{-1} & 0 \\ 0 & \bar{x}^{-1} \end{pmatrix} S(z) + o\left(\frac{1}{|x|}\right) \tag{5.11}$$

as $|x| \to \infty$, where

$$\pi S_{jj}(z) \equiv (Qm)\widehat{_{jj}}(0), \tag{5.12}$$

$$\pi S_{12}(z) = (Qm)\widehat{_{12}}(2iz), \qquad \pi S_{21}(z) = (Qm)\widehat{_{21}}(-2i\bar{z}), \tag{5.13}$$

and we identify $z \in \mathbb{C}$ with $(z_1, z_2) \in \mathbb{R}^2$.

PROOF. The first statement follows easily from the remarks preceding the statement of the theorem: $f \mapsto D_z^{-1}Qf$ is a bounded operator in L^∞ with values in $C_0(\mathbb{R}^2)$, uniformly with respect to $z \in C$. The second statement follows from (5.10) and the form of D_z^{-1}. Indeed, writing $(x - y)^{-1} = x^{-1} + x^{-1}(x - y)^{-1}y$, one obtains, for f with $xf(x) \in L^{2-\varepsilon} \cap L^{2+\varepsilon}$, that

$$D^{-1}f(x) = \frac{1}{\pi} \begin{pmatrix} x^{-1} & 0 \\ 0 & \bar{x}^{-1} \end{pmatrix} \hat{f}(0) + o\left(|x|^{-1}\right). \tag{5.14}$$

Together with the observations

$$\begin{aligned} \bar{y}z - x\bar{z} &= i(y_1, y_2) \cdot (2z_2, -2z_1), \\ yz - \bar{y}\bar{z} &= i(y_1, y_2) \cdot (2z_2, 2z_1), \end{aligned} \tag{5.15}$$

(5.14), (5.9), (5.10) yield (5.12) and (5.13).

The matrix S in (5.11) gives scattering data in the usual sense of the term: the principal asymptotics of the solution m. As we shall see, it is also scattering data in the terminology of this paper.

Indeed, note that

$$\left[\frac{\partial}{\partial \bar{z}}, E_z\right] = -XE_z = -E_z X, \quad Xa(x) = \begin{pmatrix} 0 & xa_{12}(x) \\ \bar{x}a_{21}(x) & 0 \end{pmatrix}. \tag{5.16}$$

Therefore,

$$[\partial/\partial\bar{z}, D_z^{-1}] = E_z^{-1}[X, D^{-1}]E_z. \tag{5.17}$$

But (on functions f with $xf \in L^{2-\varepsilon} \cap L^{2+\varepsilon}$)

$$[X, D^{-1}]f(x) \equiv \frac{1}{\pi}\begin{pmatrix} 0 & \hat{f}_{12}(0) \\ \hat{f}_{21}(0) & 0 \end{pmatrix}. \tag{5.18}$$

Thus, if $(1 + |x|)Q$ belongs to $L^{2-\varepsilon} \cap L^{2+\varepsilon}$, we may differentiate the Neumann series for m term by term to obtain

$$\partial m/\partial\bar{z} = (I - D_z^{-1}Q)^{-1}[\partial/\partial\bar{z}, D_z^{-1}]Qm = (I - D_z^{-1}Q)^{-1}E_z^{-1}s, \tag{5.19}$$

where

$$s(z) = [X, D^{-1}]E_z Q_m(\cdot, z) = \begin{pmatrix} 0 & S_{12}(z) \\ S_{21}(z) & 0 \end{pmatrix}. \tag{5.20}$$

Thus

$$\frac{\partial m}{\partial\bar{z}} = n(x, z)s(z) = n(x, z)\begin{pmatrix} s_1(z) & 0 \\ 0 & s_2(z) \end{pmatrix}, \tag{5.21}$$

where $s_1 = S_{21}$, $s_2 = S_{12}$, and

$$n = (I - D_z^{-1}Q)^{-1}E_z^{-1}\begin{pmatrix} 0 & 1 \\ 1 & 0 \end{pmatrix}. \tag{5.22}$$

Let Σ_z denote the matrix operation

$$\Sigma_z a = a\begin{pmatrix} 0 & e^{x\bar{z} - \bar{x}z} \\ e^{\bar{x}\bar{z} - xz} & 0 \end{pmatrix} = a \cdot E_z^{-1}\begin{pmatrix} 0 & 1 \\ 1 & 0 \end{pmatrix}. \tag{5.23}$$

Then Σ_z commutes with Q, while an easy calculation shows

$$D_z\Sigma_z = \Sigma_z D_{\bar{z}}. \tag{5.24}$$

Then

$$n = (I - D_z^{-1}Q)^{-1}\Sigma_z 1 = \Sigma_z(I - D_{\bar{z}}^{-1}Q)^{-1}1 = \Sigma_z m(x, \bar{z}). \tag{5.25}$$

We have obtained the following result.

THEOREM 6. *Under the assumptions of Theorem* 5, *if xQ is in $L^{2-\varepsilon} \cap L^{2+\varepsilon}$, then*

$$\begin{aligned} \bar{\partial} m = Tm &\equiv (\Sigma_z m(x, \bar{z}))s(z) \\ &\equiv m(x, \bar{z})\begin{pmatrix} 0 & e^{x\bar{z} - \bar{x}z}s_2(z) \\ e^{\bar{x}\bar{z} - xz}s_1(z) & 0 \end{pmatrix}. \end{aligned} \tag{5.26}$$

Let us now consider the converse problem. As in §3, if $D_x^\alpha Q$ belongs to L^1 for every α, then s has rapid decay. Conversely, suppose s_1, s_2 are two functions of z with sufficient decay, and define T as in (5.26). We seek m satisfying (5.26) by solving

$$\begin{aligned} m(x, z) &= 1 + \frac{1}{2\pi i}\int_{\mathbb{C}} (\zeta - z)^{-1}Tm(x, \zeta)\, d\zeta \wedge d\bar{\zeta} \\ &= 1 + CTm. \end{aligned} \tag{5.27}$$

THEOREM 7. *Suppose* s *belongs to* $L^{2-\varepsilon} \cap L^{2+\varepsilon}$ *for some* $\varepsilon > 0$ *and the norms* $\|s\|_{2-\varepsilon}$, $\|s\|_{2+\varepsilon}$ *are sufficiently small. Then* (5.27) *has a unique bounded solution. If, in addition,* zs *belongs to* $L^{2-\varepsilon} \cap L^{2+\varepsilon}$, *then* m *satisfies* (5.5), *with*

$$Q = [D_z, C]Tm = \begin{pmatrix} 0 & q_1(x) \\ q_2(x) & 0 \end{pmatrix}. \tag{5.28}$$

PROOF. The first statement follows exactly as for Theorem 5. To prove the second, we argue as in §§1 and 2. An easy calculation shows that D_z commutes with T, and

$$[D_z, C]f(z) = -\frac{1}{2\pi i}\int_{\mathbf{C}} \begin{pmatrix} 0 & f_{12}(\zeta) \\ f_{21}(\zeta) & 0 \end{pmatrix} d\zeta \wedge d\bar{\zeta}. \tag{5.29}$$

Under our assumptions we may differentiate the Neumann series for m term by term to obtain

$$D_z m = (I - CT)^{-1}[D_z, C]Tm = (I - CT)^{-1}Q(x)1 = Q(x)m(x, z). \tag{5.30}$$

We turn now to a brief survey of evolution equations associated with the spectral problem (5.5). These can be approached either from the scattering side or the direct side; we begin with the scattering side. The discussion depends on several preliminary observations.

Given a matrix, we denote the diagonal and off-diagonal parts by

$$a = \Pi' a + \Pi'' a = a' + a'' = \begin{pmatrix} a_{11} & 0 \\ 0 & a_{22} \end{pmatrix} + \begin{pmatrix} 0 & a_{12} \\ a_{21} & 0 \end{pmatrix}. \tag{5.31}$$

Note that

$$D_z = D + z\Pi''. \tag{5.32}$$

It is convenient to introduce the operators

$$\overline{D} = \begin{pmatrix} \partial_x & 0 \\ 0 & \partial_{\bar{x}} \end{pmatrix} = \bar{\sigma} D \sigma, \qquad \sigma = \begin{pmatrix} 0 & 1 \\ 1 & 0 \end{pmatrix}, \tag{5.33}$$

$$\overline{D}_z = \sigma D_z \sigma = \overline{D} + z\Pi'; \tag{5.34}$$

here we abuse notation and write σ for left multiplication by the matrix σ. Since left multiplication by σ commutes with T, so does $\overline{D}_z$. Thus

$$\left[\overline{D}_z, T\right] = 0, \qquad \overline{D}_z 1 = z1. \tag{5.35}$$

Suppose the diagonal matrix s of Theorem 1 has rapid decrease. Expanding the kernel $(\zeta - z)^{-1}$, one finds from (5.27) that m has an asymptotic expansion as $|z| \to \infty$:

$$m(x, z) \sim 1 + \sum_{k \geqslant 0} m_k(x) z^{-k-1}, \qquad m_k(x) \to 0 \quad \text{as } |x| \to \infty. \tag{5.36}$$

$$m_k(x) = -\frac{1}{2\pi i}\int_{\mathbf{C}} \zeta^k Tm(x, \zeta)\, d\zeta \wedge d\bar{\zeta}. \tag{5.37}$$

The series may be differentiated with respect to x, using (5.5), to give

$$m_0'' = Q; \qquad Dm_k + m_{k+1}'' = Qm_k, \quad k \geq 0. \tag{5.38}$$

These equations may be solved recursively, first for the off-diagonal part and then the diagonal part:

$$mm_{k+1}'' = Qm_k' - Dm_k''; \qquad m_{k+1}' = D^{-1}Qm_{k+1}''. \tag{5.39}$$

In particular,

$$m_0' = D^{-1}Q^2, \qquad m_1'' = QD^{-1}Q^2 - DQ. \tag{5.40}$$

Let us now define a sequence of functions

$$n_0 = m, \qquad n_k = (1 - CT)^{-1}(z^k 1), \quad k \geq 0. \tag{5.41}$$

We may compute the n_k recursively from m and the moments m_j, using the operator $\overline{D}_z$ of (5.34). Indeed, using (5.35) we obtain

$$\begin{aligned} n_1 &= (1 - CT)^{-1}\overline{D}_z 1 = \overline{D}_z m - (1 - CT)^{-1}[\overline{D}_z, C]Tm \\ &= \overline{D}_z m - (1 - CT)^{-1}\Pi'[z, C]Tm \\ &= \overline{D}_z m - (1 - CT)^{-1}m_0' 1 = \overline{D}_z m - m_0' m. \end{aligned} \tag{5.42}$$

In general,

$$\begin{aligned} n_k &= \overline{D}_z^k m - (1 - CT)^{-1}[\overline{D}_z^k, C]Tm \\ &= \overline{D}_z^k m - (1 - CT)^{-1}\sum \binom{k}{j} \overline{D}_z^{k-j}[z^j, C]\Pi' Tm. \end{aligned} \tag{5.43}$$

Now

$$[C, z^j]f(z) = \frac{1}{2\pi i}\int \sum z^{j-i-1}\zeta^i f(\zeta)\, d\zeta \wedge d\bar{\zeta}, \tag{5.44}$$

so

$$\begin{aligned} n_k &= \overline{D}_z^k m - (1 - CT)^{-1}\sum \binom{k}{j}\left(\overline{D}^{k-j} m_i'\right) z^{j-i-1} \\ &= \overline{D}_z^k m - \sum_{i<j\leq k} \binom{k}{j}\left(\overline{D}^{k-j} m_i'\right) n_{j-i-1}. \end{aligned} \tag{5.45}$$

Suppose $a(z)$ is an entire function which is real on $\mathbb{R}$. Starting with rapidly decreasing s, set

$$s(z, t) = e^{ta(z) - ta(\bar{z})} s(z). \tag{5.46}$$

Then $|s(z, t)| \equiv |s(z)|$. The corresponding operator $T = T(t)$ of (5.26) satisfies

$$\partial T/\partial t = [a(z), T]. \tag{5.47}$$

Thus the functions m satisfy

$$\begin{aligned} \partial m/\partial t &= \left[\partial/\partial t, [I - CT]^{-1}\right]1 = (I - CT)^{-1}C[a(z), T]m \\ &= (I - CT)^{-1}\{[a(z), CT]m + [C, a(z)]Tm\} \\ &= a(z)m - (I - CT)^{-1}a(z)1 + (I - CT)^{-1}[C, a(z)]Tm. \end{aligned} \tag{5.48}$$

This evolution of m may be calculated explicitly, using the functions n_k and the moments m_k above, for any polynomial a. By taking the asymptotic expansion (5.36) one can then calculate the evolution of the potential $q = m_0''$. Suppose, for example, that $a(z) = z^k$. The corresponding evolution, using (5.48) and (5.45), becomes

$$\frac{\partial m}{\partial t} = z^k m - n_k + (I - CT)^{-1}[C, z^k]Tm \tag{5.49}$$

$$= z^k m - \bar{D}_z^k m - \sum_{i \leq k} m_i'' n_{k-i-1} + \sum_{i<j<k} \binom{k}{j} \left(\bar{D}^{k-j} m_i'\right) n_{j-i-1}.$$

If we specialize to $k = 2$ and pass to the evolution of $Q = m_0''$, we obtain

$$\partial Q/\partial t = m_2'' - \bar{D}^2 Q - Q\bar{D}m_0' - Qm_1' + (DQ)m_0' + 2\left(\bar{D}m_0'\right)Q. \tag{5.50}$$

Further reduction, using (5.39) and (5.40), gives

$$\partial Q/\partial t = (D^2 - \bar{D}^2)Q + 2\left[\bar{D}m_0', Q\right]. \tag{5.51}$$

Now $\bar{D}m_0' = \bar{D}D^{-1}Q^2 = 4\bar{D}^2\Delta^{-1}Q^2$. Therefore if Q is selfadjoint, and

$$Q = \begin{pmatrix} 0 & q \\ \bar{q} & 0 \end{pmatrix}, \tag{5.52}$$

then (5.51) gives the equations (5.1). (Although this derivation has been formal, it can be carried out rigorously to show that (5.1) with small, regular initial data can be solved for all time by the inverse scattering method.)

We conclude this section by discussing these evolutions from the direct side. Equation (5.49) and its generalization to an arbitrary polynomial $a(z)$ show that the evolution of the eigenfunctions m has the form

$$\partial m/\partial t = a(z)m - A\left(x, \bar{D}_z\right)m \equiv Am = A_Q m, \tag{5.53}$$

$$A\left(x, \bar{D}_z\right) = a\left(\bar{D}_z\right) + \sum_{0 \leq j < k} \alpha_j(x)\bar{D}_z^j; \tag{5.54}$$

$$\alpha_j(x) \in M_2(\mathbb{C}), \qquad \alpha_j(x) \to 0 \quad \text{as } |x| \to \infty.$$

The direct argument is based on the following.

PROPOSITION. *For any polynomial $a(z)$ there exist unique operators A and B of the form in* (5.53), (5.54) *such that the operator*

$$LA - BL, \tag{5.55}$$

where $L = \bar{D}_z - Q$, is (*left multiplication by*) *an off-diagonal matrix-valued function $\gamma = \gamma_Q$.*

Before proving this let us note the consequences for the direct and inverse approaches. From the form of γ_Q the equation

$$\partial Q/\partial t = \gamma_Q \quad \text{or} \quad \partial L/\partial t = BL - LA \tag{5.56}$$

makes sense. Differentiating the equation $Lm \equiv 0$ with respect to t and taking into account the asymptotics, one obtains

$$\partial m/\partial t = Am. \tag{5.57}$$

Then the asymptotics (5.11) of m as $|x| \to \infty$ give the evolution of scattering

$$\partial s/\partial t = [a(z) - a(\bar{z})]s(z, t). \tag{5.58}$$

Conversely, we know that this scattering evolution corresponds to (5.57) with the particular A given by (5.53); thus, this coincides with A in (5.55). A consequence is that the proposition gives a second method of computing the Q-evolution corresponding to the scattering evolution (5.58). In the cases when the commutator $[A, L]$ has the desired form, as in [**13**], A and L are called a *Lax pair*. Here we refer to them as a pseudo-Lax pair.

Note that undoing the conjugation (5.5) allows us to replace D_z and $\overline{D}_z$ by D and $\overline{D}$. (Then (5.53) becomes an equation for ψ of (5.4).) For notational convenience we do this here.

Proof of Proposition. The scalar polynomial $a(z)$ commutes with $D - Q$, so we ignore it and look for A and B in the form (5.54). Given operators C_1, C_2, let us write $C_1 \sim C_2$ if the difference can be written as $C_3(D - Q)$. In particular, $CD \sim CQ$. If $\alpha(x)$ is a matrix-valued function with diagonal and off-diagonal parts α', α'', the operators D, $\overline{D}$ satisfy

$$D(\alpha f) = (D\alpha)f + \alpha' Df + \alpha''\overline{D}f, \quad D(\alpha f) = (\overline{D}\alpha)f + \alpha' Df + \alpha'' Df. \tag{5.59}$$

Thus, as operators,

$$(D - Q)(\alpha\overline{D}^k) \sim [(D - Q)\alpha]\overline{D}^k + \alpha''\overline{D}^{k+1} + \alpha'\overline{D}^k Q, \tag{5.60}$$

$$\begin{aligned} \overline{D}^k Q = \sum \binom{k}{j}(\overline{D}^{k-j}Q)D^j &\sim [\overline{D}^k Q] + \sum \binom{k}{j}(\overline{D}^{k-j}Q)D^{j-1}Q \\ &\sim [\overline{D}^k Q] + \sum_{i<k} C_{ki}(Q)\overline{D}^i, \end{aligned} \tag{5.61}$$

where

$$C_{ki} = \sum_j \binom{k}{j}\binom{j-1}{i}(\overline{D}^{k-j}Q)(D^{j-1-i}Q). \tag{5.62}$$

Note that c_{ki} is diagonal. Using (5.60) and (5.61) and separating diagonal and off-diagonal parts, one sees that

$$(D - Q)\left(\sum \beta_k \overline{D}^k\right) \sim \text{multiplication by on off-diagonal function} \tag{5.63}$$

if and only if for each j,

$$\beta_j'' = Q\beta_{j+1}' - D\beta_{j+1}'', \tag{5.64}$$

$$D\beta_j' = Q\beta_j'' - \sum_{k \geqslant j} \beta_k' c_{kj}(Q). \tag{5.65}$$

Equation (5.65) determines β_j' if the right side and the asymptotic value of β_j' are known. Thus in the proposition the coefficients can be determined uniquely by recursion, starting from the maximum degree.

For a monomial $a(z) = z^k$, the leading term α_{k-1} is Q. For $a(z) = z^2$ the full operator (5.54) is

$$\overline{D}_z^2 + Q\overline{D}_z - 2D_z^{-1}\overline{D}_z(Q^2) - D_zQ. \tag{5.66}$$

This gives a second determination of (5.53).

REMARKS. Any polynomial $A_1(x, D_z, \overline{D}_z)$ reduces modulo the left ideal generated by L to one of the form $A(x, \overline{D}_z)$. Thus nothing is lost by restricting attention to the latter form.

6. Concluding remarks; open questions. The examples discussed in §§3–5 are intended to illustrate the formal program given in §2. In general, for the direct problem one has less specific information about the kernel of the fundamental solution than in these two examples. Nevertheless, it is possible by rather different techniques to obtain the necessary kinds of estimates: e.g., for Δ and $\partial/\partial t - \Delta$ in other dimensions. Thus the *local problem* (q close to zero) can be treated for many operators. So far the *global problem* (large q) is less well understood except in one dimension.

We have shown that there is a well-behaved computable family of pseudodifferential evolution equations associated with the system of §5, corresponding to terms of the operator T to

$$T(t)f = e^{ta(z)}T(e^{-ta(z)}f). \tag{6.1}$$

An operation of this form will preserve the compatibility condition $[\overline{\partial}, T] = -T^2$ in higher dimensions provided a is holomorphic on the variety V; it amounts to replacing the kernel by

$$e^{ta(z)-ta(z+\zeta)}t(z, \xi). \tag{6.2}$$

For the example in §§3–4 one cannot expect to limit the support of the kernel, and the only holomorphic functions that give rise to bounded exponentials in (6.2) are the polynomials of degree at most one, which give rise to linear evolutions of q (translations). This tends to confirm the lack of interesting nonlinear evolutions associated to the Laplacian. A major open question is to find those operators and systems that do have associated, stable computable evolutions.

REFERENCES

1. M. J. Ablowitz, D. Bar Yaacov and A. S. Fokas, *On the inverse scattering transform for the Kadomtsev-Petviashivili equation*, Stud. Appl. Math. **69** (1983), 135–143.

2. M. J. Ablowitz, D. J. Kaup, A. C. Newell and H. Segur, *The inverse scattering transform. Fourier analysis for nonlinear problems*, Stud. Appl. Math. **53** (1974), 249–315.

3. M. F. Atiyah, *Geometry of Yang–Mills fields*, Lezioni Fermiane Pisa, 1979 (Scuola Normale Superiore).

4. M. F. Atiyah and R. S. Ward, *Instantons and algebraic geometry*, Comm. Math. Phys. **55** (1977), 117–124.

5. D. Bar-Yaacov, Dissertation, Yale Univ., New Haven, Conn.

6. R. Beals, *The inverse problem for ordinary differential operators on the line*, Amer. J. Math. (to appear).

7. R. Beals and R. R. Coifman, *Scattering, transformations spectrales, et equations d'evolution non lineaires*. I, II, Séminaire Goulaouic-Meyer-Schwartz 1980–1981, Exposé 22, 1981–1982, Exposé 21, École Polytechnique, Palaiseau.

8. ______, *Scattering and inverse scattering for first order systems*, Comm. Pure Appl. Math. **37** (1984), 39–90.

9. ______, *Inverse scattering and evolution equations*, Comm. Pure Appl. Math. **38** (1985), 29–42.

10. A. Davey and K. Stewartson, *On three-dimensional packets of surface waves*, Proc. Roy. Soc. A**338** (1974), 101–110.

11. P. Deift, C. Tomei and E. Trubowitz, *Inverse scattering and the Boussinesq equation*, Comm. Pure Appl. Math. **35** (1982), 567–628.

12. A. S. Fokas and M. J. Ablowitz, *The inverse scattering transform for multidimensional* $2+1$ *problems*, Nonlinear Phenomena (Proc., Oaxtepec, Mexico, 1982, K. B. Wolf, ed.), Lecture Notes in Physics, no. 189, Springer, Berlin.

13. P. Lax, *Integrals of nonlinear equations of evolution and solitary waves*, Comm. Pure Appl. Math. **21** (1968), 467–490.

14. J.-H. Lee, *Analytic properties of Zakharov-Shabat inverse scattering problem with a polynomial spectral dependence of degree* 1 *in the potential*, Dissertation, Yale Univ., New Haven, Conn.

15. D. E. Lerner and P.D. Sommers (Editors), *Complex manifold techniques in theoretical physics*, Pitman, San Francisco, 1979.

16. S. V. Manakov, *The inverse scattering transform for the time-dependent Schroedinger equation and Kadomtsev-Petviashvili equation*, Physica **3D** (1981), 1 & 2, 420–427.

17. A. Nachman and M. J. Ablowitz, *A multidimensional inverse scattering method* (preprint).

18. K. Pohlmeyer, *On the Lagrangian theory of anti-self-dual fields in four-dimensional Euclidean space*, Comm. Math. Phys. **72** (1980), 37–47.

19. D. H. Satinger, *Inverse scattering for* $sl(2, C)$ *models* (to appear).

20. K. Ueno, Y. Nakamuma, *Transformation theory for anti-self-dual equations*, Publ. Res. Inst. Math. Sci. **19** (1983), 519–547.

21. V. Wickerhauser, dissertation, Yale University.

22. V. E. Zakharov, *The inverse scattering method*, Solitons (R. K. Bullough and P. J. Caudrey, eds.), Topics in Current Physics, no. 17, Springer, Berlin, 1980.

23. V. E. Zakharov and A. B. Shabat, *A refined theory of two-dimensional self-focussing and one-dimensional self-modulation of waves in nonlinear media*, Soviet Physics JETP **34** (1972), 62–69.

YALE UNIVERSITY

Proceedings of Symposia in Pure Mathematics
Volume **43** (1985)

Nonlinear Harmonic Analysis and Analytic Dependence

R. R. COIFMAN AND YVES MEYER

1. We would like to describe certain features of a number of nonlinear dependence problems arising naturally in analysis.[1]

Perhaps the most common occurrence of such questions involves the dependence of solutions of a differential or partial differential equation on the coefficients of the differential operator. To be specific, let $P(D)$ denote a differential operator such as

$$(1.1)\qquad \begin{gathered} a(x)\frac{d}{dx}, \quad x \in \mathbb{R}, \qquad a(z)\frac{\partial}{\partial z}, \quad z \in \mathbb{C}, \\ \sum_{i,j} \frac{\partial}{\partial x_i}\left(a_{ij}(x)\frac{\partial}{\partial x_j}\right), \quad \sum_{i,j} a_{ij}(x)\frac{\partial^2}{\partial x_i\,\partial x_j}, \qquad x \in \mathbb{R}^n. \end{gathered}$$

Let F be a function. We are interested in $F(P(D))$; for example, $\operatorname{sgn}(P(D))$, $\sqrt{P(D)}$, $\exp(t(P(D)))$, $\exp(-t\sqrt{P(D)})$.

We view $T(a) = F(P(D))$ as an operator-valued function of the coefficients of P, and we would like to obtain

(1) continuity estimates for T as a linear operator on an appropriate Banach space;

(2) obtain a precise description, say, of the kernel of T;

(3) study $T(a)$ as an operator-valued nonlinear function of a, and, in particular, show that $T(a)$ is a real or complex analytic function of a (again on an appropriate Banach space of coefficients).

A related class of problems involves the dependence on the boundary of harmonic or holomorphic solutions of a boundary value problem, say, the Dirichlet problem. The usual way to deal with such a question would be to use

1980 *Mathematics Subject Classification*. Primary 35A25.

[1] This paper is a companion to *Real analysis and operator theory* by Y. Meyer in this volume.

pseudodifferential calculus to calculate the symbol of T. This, however, only works modulo smoothing operators; it does not provide us with exact expansions or permit us to deal with coefficients having no smoothness. One of our aims is to represent $F(P(D))$ exactly by a power series and to learn to analyze such expressions.

A particularly simple example to describe is given by the Cauchy integral. Let

$$C(f) = \frac{1}{2\pi i}\int_\Gamma \frac{f(\zeta)}{\zeta - z}\,d\zeta$$

be the Cauchy integral along the unbounded curve Γ. We would like to obtain estimates, say in L^2, and analytic dependence of the operator on some parameter space for curves.

For this example we now consider curves of the form $Z(x) = x + iA(x)$ (i.e., the graph of $A(x)$) and assume that $A'(x) = a(x)$ is bounded. The question of boundedness in $L^2(\mathbb{R})$ of this operator has been resolved in the last few years by A. P. Calderón (for small L^∞ norm) and by the authors in collaboration with A. MacIntosh in [**1**, **4**], where it was shown that $C(a)(f)$, viewed as an operator-valued functional in a, is complex analytic on the domain $\|\mathrm{Re}\, a\|_\infty < 1$.

Since the main features of the theory originated, and are included in, this example, we digress for a moment to describe some basic ingredients.

We start by observing that when we suppress the i in the expression for $Z(x)$ in the Cauchy integral, we can view the corresponding operator as the conjugation of the Hilbert transform by the change of variable on the line, given as

$$UHU^{-1}, \quad \text{where } H(f) = \text{p.v.}\int_{\mathbb{R}} \frac{f(t)}{x-t}\,dt,$$

$$U(f) = f(x + A(x)).$$

If $\|a\|_\infty < 1$, this is clearly a bounded operator on L^2, and the Cauchy integral is the analytic continuation to complex a of the conjugation UHU^{-1}. This point of view is easily related to functional calculus. Since the Hilbert transform is in symbolic notation $\mathrm{sgn}(d/dx)$, the conjugation by U gives $\mathrm{sgn}((1+a)^{-1}(d/dx))$. Thus by "analytic continuation" we find

$$C(a)(f) = \mathrm{sgn}\left(\frac{1}{1+ia}\frac{d}{dx}\right). \tag{1.2}$$

It is now clear that the L^2 estimates for C are related to the analyticity in a of $\mathrm{sgn}((1+a)^{-1}(d/dx))$.

A related fundamental example is provided by Kato's problem. Here, one considers an elliptic Laplacian in divergence form given by

$$L(f) = -\sum \frac{\partial}{\partial x_i}\left(a_{ij}\frac{\partial}{\partial x_j}f\right) = -\mathrm{div}(A\,\mathrm{grad}\, f), \qquad A \in L^\infty, \tag{1.3}$$

where $\mathrm{Re}\langle A\zeta, \zeta\rangle \geq |\zeta|_c^2$. Such an operator is accretive and has a unique maximal accretive square root $L^{1/2}$. It is easy to prove (using Lax–Milgram) that when A is selfadjoint the domain of $L^{1/2}$ is $H^1(\mathbb{R}^n)$. Kato conjectured that this is the case for general accretive A. This would mean that $L^{1/2}$ extends to the complex domain as a bounded operator on H^1 and should depend complex analytically on A. This, in fact, is the case in dimension one and has been proved to be true in $\mathbb{R}^n$ for small perturbations (in L^∞) of I; see [**2**, **8**]. The proof is based on the expansion of $L^{1/2}$ in a power series in $A - I$. (The general problem is still open in $\mathbb{R}^n$, $n > 1$.)

In general, the simplest approach to prove analyticity of a functional $T(a)$, mapping complex Banach spaces B_1 to B_2, would be to prove that $T(a_1 + za_2)$ is a holomorphic (vector-valued) function of $z \in \mathbb{C}$, $|z| < 1$, for all a_1, a_2 in some ball centered at 0. This is equivalent to the existence of the Taylor expansion

$$T(a) = \sum_{k=0}^{\infty} \Lambda_k(a), \qquad \Lambda_k(a) = \Lambda_k(a_1,\ldots,a_k)|_{a_i = a}.$$

where a is in a sufficiently small ball in B_1 and Λ_k: $B_1^k \to B_2$ is a multilinear map (in a_i) satisfying the estimate

$$\|\Lambda_k(a_1,\ldots,a_k)\|_{B_2} \leqslant C^k \prod_{i=1}^{k} \|a_i\|_{B_1}. \tag{1.4}$$

We are confronted with the two basic problems involving these homogeneous polynomials on B.

Problem 1. Obtain representation theorems for multilinear operators.

Problem 2. Obtain estimates in C^k.

In the previous examples the functionals $T(a)$ have a natural invariance under translations and changes of scale. More precisely, let $f^\tau(x) = f(x - \tau)$; then

$$\Lambda_k(a_1^\tau,\ldots,a_k^\tau)(f^\tau) = [\Lambda_k(a_1,\ldots,a_k)(f)]^\tau$$

(a similar statement is often true for dilations). Under some mild continuity assumptions, such Λ_k can be represented as

(1.5)

$$\begin{aligned}
&\Lambda_k(a_1,\ldots,a_k)(f)(x)\\
&\quad = \int_{\mathbb{R}^{n(k+1)}} e^{ix\cdot(\alpha_1 + \cdots + \alpha_k + \xi)}\sigma_k(\alpha_1 \cdots \alpha_k; \xi) \prod_{i=1}^{k} \hat{a}_i(\alpha_i)\hat{f}(\xi)\, d\alpha_1 \cdots d\alpha_k\, d\xi\\
&\quad = \int_{\mathbb{R}^{n(k+1)}} \check{\sigma}_k(x - t_1, x - t_2,\ldots,x - t_k, x - y) \prod_{i=1}^{k} a_i(t_i) f(y)\, dt_1 \cdots dt_k\, dy,
\end{aligned}$$

or as limits of such expressions. σ_k is called the multilinear symbol of Λ_k. (More generally, we could consider such expressions with symbols depending also on x.)

Such expansions can be obtained explicitly for our examples by either expanding the kernel in a power series in a or by using a more symbolic operator approach. To be specific, we can write

$$C(a)(f)(x) = \frac{1}{2\pi i}\int \frac{f(y)(1+ia(y))}{x-y+i(A(x)-A(y))}\,dy$$

$$\text{(1.6a)} \qquad = \sum_0^\infty (-i)^k \int \left(\frac{A(x)-A(y)}{x-y}\right)^k \frac{f_1(y)}{x-y}\,dy$$

$$= \sum_0^\infty \int \exp\left\{ix \cdot \sum^k (\alpha_i + \xi)\right\} \sigma_k(\alpha, \xi) \prod \hat{a}(\alpha_i) \hat{f}_1(\xi)\,d\xi,$$

where $f_1(y) = f(y)(1+ia(y))$ $\|a\|_\infty < 1$ for σ_k (see [5]) and

$$\text{(1.6b)} \qquad F(P(D)) = \frac{1}{2\pi i}\int_\Gamma \frac{F(\zeta)}{\zeta - P(D)}\,d\zeta,$$

where the contour is a wedge containing the spectrum of $P(D)$, and F is holomorphic in a neighborhood of the wedge.

More generally, one could try and assign meaning to the formula

$$F(P(D)) = \frac{-1}{2\pi i}\int_{\mathbf{C}} \frac{\partial F}{\partial \bar{\zeta}} \frac{1}{\zeta - P(D)}\,d\zeta \wedge d\bar{\zeta} \quad \text{for general } F.$$

This is particularly useful when the spectrum of $P(D)$ is the whole plane: for example, $P(D) = a(z)(\partial/\partial z)$, $z \in \mathbf{C}$.

In (1.6b), if we take

$$P(D) = -\Delta + \sum_{i,j} b_{ij} \frac{\partial^2}{\partial x_i \partial x_j}$$

with $\|b_{ij}\| < \delta_0$, we can use standard identities for the resolvent to write

$$\text{(1.7)} \qquad F(P(D)) = \sum_{k=0}^\infty \int_{-\infty}^\infty F\left(\frac{1}{it}\right)(I - it\Delta)^{-1}(MR_t)^k \frac{dt}{t},$$

where F is assumed to be bounded holomorphic on the right half-plane, the integral is taken in principal value on the imaginary axis, $R_t = [tD_iD_j(I - it\Delta)^{-1}]$, and M is multiplication by the matrix $[b_{ij}]$. This kind of expression appears for most of the examples cited above.

2. Now that we have some realization for "homogeneous polynomials", we need to describe methods for analyzing and estimating these expressions.

Surprisingly, one can follow the program initiated by A. P. Calderón and A. Zygmund, in their study of (linear) singular integrals, by blending real-variable techniques on the kernel side with microlocalization methods on the Fourier transform side (for example, Littlewood–Paley decompositions). A major ingredient in these developments is the space of functions of bounded mean oscillation, B.M.O. This remarkable class of functions, discovered by F. John and L. Nirenberg in connection with a nonlinear problem in P.D.E., is an essential tool for our examples. Recall that $b \in$ B.M.O(R^n) if the mean square deviation of b is uniformly bounded on all balls B; i.e.,

$$\sup_B \frac{1}{|B|}\int_B |b(x) - m_B(b)|^2\,dx = \|b\|_*^2 < \infty, \qquad m_B(b) = \frac{1}{|B|}\int_{\mathbb{R}} b(x)\,dx.$$

The basic fact concerning this space is that Calderón–Zygmund operators map L^∞ into B.M.O.

Perhaps the most versatile tool for handling multilinear operators is provided by the following recent theorem of G. David and J. L. Journé.

THEOREM. *Let* $T(f) = \int k(x, y)f(y)\,dy$, *where* k *is locally integrable in* $x \neq y$ *such that*

$$|x-y|\,|\nabla_x k| + |x-y|\,|\nabla_y k| + |k| \leqslant |x-y|^{-n}. \tag{2.1}$$

Moreover, assume that T *has a weak cancellation property* (*see* [7] *or Y. Meyer in these proceedings*). *Then* T *is a bounded operator on* L^2 *iff* $T(1)$ *and* $T^*(1)$ *are in* B.M.O. *Moreover*,

$$c\left(1 + \|T(1)\|_* + \|T^*(1)\|_*\right) < \|T\|_{L^2} < C\left(1 + \|T(1)\|_* + \|T^*(1)\|_*\right),$$

and T *maps* L^∞ *into* B.M.O.

This remarkable theorem reduces the study of k-multilinear operators to $(k-1)$-multilinear operators and leads to estimates in C^k for the k-multilinear terms.

To understand this reduction let us consider the following corollary.

THEOREM [5]. *Let*

$$T_k(a_1, a_2, a_k : f) = \int \exp\left\{ix\left(\sum_1^k \alpha_i + \xi\right)\right\}\sigma(\alpha_1 \cdots \alpha_k, \xi) \cdot \prod_{i=1}^k \hat{a}_i(\alpha_i)\hat{f}(\xi)\,d\xi\,d\alpha_1 \cdots d\alpha_1,$$

where σ *is a classical symbol in* $\mathcal{S}^0_{1,0}$. *Then* T *is a bounded operator on* L^2 *and*

$$\|T_k(a_1, a_2, \ldots, a_k : f)\|_{L^2} \leqslant C^k \prod_{i=1}^k \|a_i\|_\infty \|f\|_{L^2}.$$

It is quite easy to verify that the kernels on T and T^* (as an operator on f) satisfy (2.1). $T(1)$ is in B.M.O. because $T_k(a_1, \ldots, a_{k-1}, f; 1)$ is a Calderón–Zygmund operator mapping L^∞ into B.M.O. and because of the corresponding theorem for $k-1$.

It is clear that this method will work whenever the kernel of the operator can be shown to satisfy the estimates of the David–Journé theorem. Originally, the various multilinear operators of the form

$$T_k(a)(f) = \int_0^\infty (I - it\Delta)^{-1}(M_a R_t)^k m(t)\,\frac{dt}{t}, \tag{2.2}$$

arising in the study of $F(P(D))$ (see (1.7)), were studied directly (without estimating the kernels) by means of quadratic Littlewood–Paley–Stein functions using Carleson measures (see [**2**, **4**, **5**]).

We conclude this section by remarking that, for all examples stated, the basic organization of the proofs and the reductions of the orders of multilinearities were guided by a careful study on the Fourier transform side of the symbols of the operators involved. This was achieved by localizing the frequencies of the various functions and studying their interactions (i.e., microlocalization). Estimates were then obtained by real-variable methods.

3. The role of parametrizations and choice of topologies on function spaces of parameters (coefficients). In the study of nonlinear functionals arising in analysis, it is often possible to expand the functional in a formal power series. The main problem is to identify the correct Banach space topology relative to which the estimates (1.4), guaranteeing convergence of the series, are valid. Since the problem can be considered on the complexification of the function spaces, it is natural to also investigate the domain of holomorphy of the functional. This, however, would make good sense only if we were to place ourselves in the largest possible Banach space for which we have analyticity. It turns out that, for many examples, this space, which we call the space of holomorphy (of the functional), can be identified.

We start by describing the space of holomorphy for the Szegö projection. We consider unbounded rectifiable Jordan curves going through the origin in the plane and splitting it into two domains, D_+ and D_-. We parametrize the curves by arc length as $z(s) = \int_0^s \exp(i\alpha(t))\,dt$ and let $H^2_+(\Gamma, ds)$ be the subspace of L^2 obtained as the closure in $L^2(ds)$ of rational functions with poles in D_-. We let $S(\alpha)$ denote the orthogonal projection in $L^2(\Gamma, ds)$ onto H^2_+. We now write

$$S(\alpha) = \sum_{k=0}^{\infty} \Lambda_i(\alpha),$$

where Λ_k are bounded operators on $L^2(\Gamma, ds)$.

We ask, What is the largest Banach space B for which

$$\|\Lambda_k(\alpha)\|_{L^2,L^2} \leqslant C^k\|\alpha\|_B^k? \tag{3.1}$$

Clearly, if we define $\|\alpha\|_* = \|\Lambda_1(\alpha)\|_{L^2,L^2}$, we must have $\|\alpha\|_* \leqslant c\|\alpha\|_B$, so that, if we can prove (3.1), relative to the norm $\|\ \|_*$, we have identified the space of holomorphy.

In this example we have

$$\Lambda_1(\alpha) = \tfrac{1}{2}[a, H] + [A(d/dx), H], \qquad a = \alpha + iH(\alpha),\ A' = \alpha, \tag{3.2}$$

and

$$\|\alpha\|_* \simeq \|\alpha\|_{\text{B.M.O.}}$$

Thus B.M.O. is the natural space of holomorphy. A slightly simpler version of (3.2) is the commutator between the Hilbert transform and multiplication by b (it can be shown that $\Lambda_1(\alpha)$ is a superposition of such operators). In this case we also have that

$$\|[b, H]\|_{L^2, L^2} \simeq \|b\|_{\text{B.M.O.}}$$

We should note that if α is in B.M.O. with small norm then the curve defined by $z' = \exp(i\alpha)$ satisfies the chord-arc condition; i.e., $\exists \delta > 0$ s.t. $\forall s, t$,

$$\delta \leqslant |z(s) - z(t)|/|s - t| \leqslant 1;$$

conversely, for any chord-arc curve there is a natural B.M.O. determination of α. We can view B.M.O. as the parameter space for the manifold of chord-arc curves. Using the preceding identification of the space of holomorphy, we can prove that the Riemann map that maps D onto the upper half-plane is a real analytic entire functional on the manifold of chord-arc curves (see [**3**]).

For most of the other examples, the space of holomorphy for the coefficients is L^∞.

There is, however, another natural example leading to an "exotic" space of holomorphy. The operator is

$$L = \frac{1}{1 + a} \frac{\partial}{\partial z}, \quad \operatorname{sgn} z = \frac{z}{|z|}, \qquad z \in C.$$

We ask for the space of holomorphy of the functional $T(a) = \operatorname{sgn}(L)$ viewed as a bounded operator on $L^2(C)$. It can be shown that the norm is given by

$$\|a\|_* \simeq \|a\|_{L^\infty} + \|a * (1/z^2)\|_{L^\infty}$$

and this is the natural space of holomorphy for other functions of L.

We conclude by observing that identifying the space of holomorphy for these examples provided the connection with geometric interpretations of the problems. It would be extremely interesting to identify the space of holomorphy for other natural, operator-valued functionals: for example, for the scattering operator viewed as a function of the potential q in $\Delta + q$. It is quite clear that this subject is in its infancy and needs to be explored.

REFERENCES

1. A. P. Calderón, *Cauchy integrals on Lipschitz curves and related operators*, Proc. Nat. Acad. Sci. U.S.A. **75** (1977), 1324–1327.

2. R. R. Coifman, D. D. Geng and Y. Meyer, *Domaine de la racine carrée de certains opérateurs différentiels accrétifs*, Ann. Inst. Fourier (Grenoble) **33** (1983), 123–134.

3. R. R. Coifman and Y. Meyer, *Lavrentiev's curves and conformal mappings*, Rep. 5, Mittag-Leffler Inst., Sweden, 1983.

4. R. R. Coifman, A. MacIntosh and Y. Meyer, *L'intégrale de Cauchy définit un opérateur borné sur L^2 pour les courbes lipschitziennes*, Ann. of Math. (2) **116** (1982), 361–387.

5. R. R. Coifman and Y. Meyer, *Au delà des opérateurs pseudodifférentiels*, Astérisque **57**, Soc. Math. France, 1978.

6. G. David, *Opérateurs intégraux singuliers sur certaines courbes du plan complexe*, Ann. Sci. École Norm. Sup. (4) **17** (1984), 157–189.

7. G. David and J. L. Journé, *A boundedness criterion for Calderón Zygmund operators*, Ann. of Math. (to appear).

8. E. Fabes, D. Jerison and C. Kenig, *Multilinear Littlewood–Paley estimates with applications to partial differential equations*, Proc. Nat. Acad. Sci. U.S.A. **79** (1982), 5746–5750.

YALE UNIVERSITY

UNIVERSITÉ PARIS - SUD

Proceedings of Symposia in Pure Mathematics
Volume **43** (1985)

On Some C^*-Algebras and Fréchet*-Algebras of Pseudodifferential Operators

H. O. CORDES

0. Introduction. It is our intention to describe results concerning certain topological algebras of pseudodifferential operators, with emphasis on two different aspects: first, on Fredholm theory, which in these algebras proves to be about as perfect as in C^*-algebras; second, we plan to discuss a possible use of some of these algebras as algebras of observables in quantum mechanics. Guided more by mathematical than physical arguments, we consider 'pseudoalgebras' of observables that seem to fit more perfectly than abstract C^*-algebras and may remove some long-standing difficulties. In particular, perhaps, our approach may be an improvement of the older work of Grossman, Loupias, and Stein [**GLS**] on pseudodifferential operators and quantum mechanics.

In §§1 and 4 we consider two different approaches, different from the conventional 'symbol-to-pseudodifferential-operator' definition. The discussion of the second (older) approach will proceed in §4. In §1 we start similarly as in [**CL**, **CE**, **CD**], but will offer a more complete and much refined theory.

§§2 and 3 provide more detail in this discussion, although complete proofs are too long for a presentation here. (For this we refer to [$\mathbf{C_2}$].) §5 continues the discussion of the ψ^*-algebras of elliptic expressions started in §4. Again, detailed proofs cannot be discussed (cf. [$\mathbf{C_2}$]).

While the general Fredholm theory of commutative and other C^*-algebras to be used was discussed in [$\mathbf{C_1}$], we also emphasize the work of B. Gramsch [**G**] on a certain special kind of Fréchet algebra with a particularly good Fredholm theory—the so-called ψ^*-algebra. The Fréchet algebras we arrive at in each of the two cases seem to be prototypes of such ψ^*-algebras.

In particular, it is proven in [**G**] that the set of Fredholm operators of ψ^*-algebras forms an analytic manifold, so that, in particular, all the Fredholm perturbation results reappear.

1980 *Mathematics Subject Classification*. Primary 47G05.

We omit looking at other Banach algebras of pseudodifferential operators, such as subalgebras of L^p, $p \neq 2$, although there are some results along similar lines (cf. Illner [**I**]). Also, we only consider C^∞-multiplication operators or C^∞-coefficients (which, hence, are continuous), while some interesting development has been the incorporation of piecewise continuous coefficients, triggered by the work of Gohberg and Krupnik [**GK**] (cf. Power [**P**]).

Although we are interested in Fredholm theory of differential operators, we only discuss results directly linked to C^*-algebras (or ψ^*-algebras). Hence, we take no position on the recent results of Melrose–Mendoza [**MM**] and Lockhart–McOwen [**LM**] on differential operators on (noncompact) manifolds with conical singularities (cf. [$\mathbf{C_2}$, Chapter 2] for some analysis).

Again the C^*- (and ψ^*-) algebras of singular elliptic differential expressions always pertain to the 'limit point case' where no boundary conditions are required. Thus we also ignore all work on algebras for expressions with boundary conditions, which are too numerous to mention.

The author is indebted to B. Gramsch and W. Kaballo for a series of helpful discussions.

1. An algebra of infinitely differentiable operators. Let us first work in Euclidean space $\mathbb{R}^n$ and the Hilbert space $\mathcal{H} = L^2(\mathbb{R}^n)$. Let $\mathcal{L}(\mathcal{H})$ be the C^*-algebra of all bounded operators $\mathcal{H} \to \mathcal{H}$ with operator norm topology.

Physical properties of an 'observable' $A \in \mathcal{L}(\mathcal{H})$ should be left invariant by a coordinate transform s: $\mathbb{R}^n \to \mathbb{R}^n$, as well as by a 'gauge transform' $u(x) \to e^{i\lambda(x)}u(x)$, with λ: $\mathbb{R}^n \to \mathbb{R}$. In particular, let us focus on this property for linear maps $s(x)$ and $\lambda(x)$. In fact, we require that $s(x)$ be a *similarity*, changing the Euclidean distance only by a constant factor. In other words, with an operator $A \in \mathcal{L}(\mathcal{H})$ we should associate all its 'elementary transforms'

$$A^{s,\lambda} = (T^{s,\lambda})^{-1} A T^{s,\lambda}, \tag{1.1}$$

where

$$(T^{s,\lambda}u)(x) = \sqrt{|\partial s/\partial x|}\, e^{i\lambda(x)} u(s(x)) \tag{1.2}$$

is a combined coordinate and gauge transform.

Explicitly, we have

$$\begin{aligned} &\lambda(x) = \zeta \cdot x + \phi, \quad s(x) = \sigma o x + z, \\ &\qquad\qquad \sigma, \phi \in \mathbb{R}, \quad z, \zeta \in \mathbb{R}^n, \quad o \in \mathcal{O}_n, \quad \sigma > 0, \end{aligned} \tag{1.3}$$

where $\mathcal{O}_n$ denotes the group of orthogonal $(n \times n)$-matrices. The operators $T^{s,\lambda}$ of (1.2) are special unitary operators of $\mathcal{H}$. We have

$$T^{s,\lambda}T^{s',\lambda'} = T^{s' \circ s, \lambda + \lambda' \circ s}, \qquad (T^{s,\lambda})^{-1} = T^{s^{-1}, -\lambda \circ s^{-1}}, \tag{1.4}$$

which shows that $\mathcal{GS} = \{T^{s,\lambda}\}$, with all s, λ of the form (1.3), forms a subgroup of the unitary group $\mathcal{U}(\mathcal{H}) \subset \mathcal{L}(\mathcal{H})$. In fact, the collection

$$\mathfrak{gs} = \{(s, \lambda) = (\sigma, o, z, \zeta, \phi) : 0 < \sigma < \infty, o \in \mathcal{O}_n, z, \zeta \in \mathbb{R}^n, \phi \in \mathbb{R}\} \tag{1.5}$$

forms a Lie group under the group operations

$$(1.6)\quad (s,\lambda)\vartriangle(s',\lambda')=(s'\circ s,\lambda+\lambda'\circ s),\qquad (s,\lambda)^{-1}=(s^{-1},-\lambda\circ s^{-1}),$$

and the manifold structure it naturally carries as a submanifold of $(\sigma, o, z, \zeta, \phi)$-space $\mathbb{R}^{n^2+2n+2}$.

Then the class $\mathcal{GS}$ appears as a representation (within the unitary group $\mathcal{U}(\mathcal{H})$ of $\mathcal{H}$) of that Lie group.

Let us observe a *deficiency* of this representation: *It is not even continuous under the norm topology of $\mathcal{U}(\mathcal{H})$, although it is in the strong topology. Moreover, it is not differentiable at all under any of the usual topologies of $\mathcal{L}(\mathcal{H})$.* In fact, the formal first partial derivatives of $T^{s,\lambda}$ for the group parameters $\sigma, \phi, z, \zeta, o$ prove to be unbounded operators of $\mathcal{H}$.

This observation might raise the question of whether, for certain special operators $A \in \mathcal{L}(\mathcal{H})$, the map $\mathfrak{gs} \to \mathcal{L}(\mathcal{H})$, defined by (1.1), is continuous, or differentiable, or even $C^\infty(\mathfrak{gs}, \mathcal{L}(\mathcal{H}))$, where $\mathcal{L}(\mathcal{H})$ is assumed to be equipped with its norm topology. Let us define

$$(1.7)\qquad \Psi\mathcal{GS}=\left\{A\in\mathcal{L}(\mathcal{H})\colon A^{s,\lambda}\in C^\infty(\mathfrak{gs},\mathcal{L}(\mathcal{H}))\right\}.$$

The class $\Psi\mathcal{GS}$ of (1.7) is the first Fréchet algebra of pseudodifferential operators we want to examine. *Our first results are that $\Psi\mathcal{GS}$ is an algebra of classical ψdo's and, moreover, a ψ^*-subalgebra of $\mathcal{L}(\mathcal{H})$, in the terminology of B. Gramsch.* (We use the abbreviation 'ψdo' for 'pseudodifferential operator'.) For more details, see Theorems 2.1 and 2.2.

In our above deduction we focused on linear maps $s(x)$ and $\lambda(x)$. In spite of this restriction, we now find that (1) the algebra $\Psi\mathcal{GS}$ is invariant under a rather general class of coordinate transforms, gauge transforms, restricted only by conditions at $|x| = \infty$. This is formulated in detail in Theorems 3.4 and 3.5. Moreover, (2) we find that for a large class of nonlinear coordinate or gauge transforms, restricted only at $|x| = \infty$ again, an operator $A \in \Psi\mathcal{GS}$ gives a differentiable dependence of $A^{s,\lambda}$ on s and λ in a variational sense.

This may be brought out more clearly if we next focus on a characterization of $\Psi\mathcal{GS}$ by the Lie algebra corresponding to $\mathfrak{gs}$. Focusing on the representation $\mathfrak{gs} \to \mathcal{GS}$ of (1.2), we find: For any one-dimensional subgroup $\{s_t, \lambda_t\colon t \in \mathbb{R}\}$ of $\mathfrak{gs}$, the infinitesimal generator of the corresponding strongly continuous group of unitary operators is an unbounded closed operator of $\mathcal{H}$, a skew-selfadjoint first-order, partial differential operator L. It turns out that L is the only skew-selfadjoint realization of its first-order expression L. All of these expressions form a Lie algebra $\mathfrak{as}$ of first-order, linear, differential expressions (folpdes): the real linear span of

$$(1.8)\quad Y_{00}=\sum_{k=1}^{n}x_k\partial_{x_k}+\frac{n}{2},\quad Y_{jl}=x_l\partial_{x_j}-x_j\partial_{x_l},\quad Y_{j0}=\partial_{x_j},\quad Y_{0j}=ix_l,$$
$$j,l=1,\ldots,n.$$

Each folpde in $\mathfrak{as}$ admits a unique skew-selfadjoint realization, obtainable as an infinitesimal generator, of some 1-dimensional subgroup of $\mathcal{GS}$.

Vice versa, for any folpde $L \in \mathfrak{as}$ we can form the 'formal exponentiation' e^{Lt}, defined by solving the Cauchy problem for a first-order partial differential equation

$$(1.9) \qquad \partial_t u - Lu = 0, \quad t \in \mathbb{R}, \qquad u(t, x) = u^0(x) \quad \text{as } t = 0,$$

and then setting $(e^{Lt}u^0)(x) = u(t, x)$. An explicit calculation shows that $\{e^{Lt}: t \in \mathbb{R}\}$ defines a 1-dimensional subgroup of $\mathcal{GS}$ with infinitesimal generator equal to (the unique skew-selfadjoint realization of) L. Also, of course, e^{Lt} proves identical to the exponential function of Lt defined by the spectral resolution of (the skew-selfadjoint) L. In other words, the relation between $\mathfrak{gs}$ and $\mathfrak{as}$ is exactly that of a Lie group and its corresponding Lie algebra.

For an $L \in \mathfrak{as}$ we now define two unbounded linear operators of $\mathcal{L}(\mathcal{H})$, denoted by ad_L and $\mathfrak{ad}_L$, by formally setting

$$(1.10) \quad \operatorname{ad}_L A = [L, A] = LA - AL, \qquad \mathfrak{ad}_L A = (d/dt)\,(e^{Lt}Ae^{-Lt})|_{t=0}.$$

More precisely, the domain of $\mathfrak{ad}_L$ is the set of all $A \in \mathcal{L}(\mathcal{H})$ for which the expression $\mathfrak{ad}_L A$ of (1.10) is well defined: The derivative exists in $\mathcal{L}(\mathcal{H})$, so that $\mathfrak{ad}_L A \in \mathcal{L}(\mathcal{H})$. On the other hand, the domain $\operatorname{dom} \operatorname{ad}_L$ consists of all $A \in \mathcal{L}(\mathcal{H})$ such that $\operatorname{dom}(L_0^{**}A) \supset \operatorname{dom} L_0^{**}$ (so that the formal commutator $[L_0^{**}, A]$ has (dense) domain $\operatorname{dom} L_0^{**}$) and $[L_0^{**}, A]$ extends to an operator in $\mathcal{L}(\mathcal{H})$. Here the closure L_0^{**} of the minimal operator L_0 (with domain $C_0^\infty(\mathbb{R}^n)$) must be the unique skew-selfadjoint realization of L, of course.

THEOREM 1.1. *We have* $\operatorname{ad}_L \supset \mathfrak{ad}_L$ *for* $L \in \mathfrak{as}$. *Moreover,*

$$(1.11)$$
$$\operatorname{dom}(\operatorname{ad}_L)^2 \subset \operatorname{dom} \mathfrak{ad}_L \subset \operatorname{dom} \operatorname{ad}_L, \qquad \operatorname{ad}_L A = \mathfrak{ad}_L A, \qquad A \in \operatorname{dom} \mathfrak{ad}_L.$$

In fact, the algebra $\Psi\mathcal{GS}$ is described as the class of all $A \in \mathcal{L}(\mathcal{H})$ allowing arbitrary finite application of ad_L (or $\mathfrak{ad}_L$), *for $L \in \mathfrak{as}$. That is, for an arbitrary choice of $L_j \in \mathfrak{as}$, $j = 1,\ldots,N$, the operator* $\operatorname{ad}_{L_1}\operatorname{ad}_{L_2}\cdots\operatorname{ad}_{L_N}A$ *(or* $\mathfrak{ad}_{L_1}\cdots\mathfrak{ad}_{L_N}A$*) is always well defined (in $\mathcal{L}(\mathcal{H})$).*

For a proof of Theorem 1.1, see [$\mathbf{C_2}$].

Let us observe that this alternative characterization of $\Psi\mathcal{GS}$ points to a simple generalization of our $\Psi\mathcal{GS}$: For a Lie algebra $\mathfrak{ax}$ of folpdes, defined on a given differentiable manifold Ω, if certain conditions are given, an algebra of ψdo's on Ω can be defined, using operators ad_L or $\mathfrak{ad}_L$, defined as above. This is investigated in [$\mathbf{C_2}$, Chapter V], but will not be discussed here.

On the other hand, for a nonlinear pair $(s(x, \kappa), \lambda(x, \kappa))$, depending formally on a parameter κ the derivative $\partial_\kappa(A^{s,\lambda})$ may be expressed in the form $(\operatorname{ad}_L A)^{s,\lambda}$, as a calculation shows, with some skew-selfadjoint realization of the expression

$$(1.12) \qquad L = L_\kappa = \sum_{j=1}^{n} \sigma_j(x)\partial_{x_j} + i\mu(x), \qquad \sigma_j = s_{j|\kappa}, \mu = \lambda_{|\kappa}.$$

This formal derivation suggests that the 'first variation' $\partial_\kappa A_{s,\lambda}$ will exist in $\mathscr{L}(\mathscr{H})$ whenever the expression $\kappa = -iL$ (L of (1.12)) defines a pseudodifferential operator $K = k(x, D)$ satisfying the assumptions of Theorem 3.3 (note formula (3.10)).

We note that the ψdo's of $\Psi\mathscr{G}\mathscr{S}$ are operators of order 0 (suggested already by the fact that they are bounded operators of $\mathscr{H}$) by definition. To obtain a class of ψdo's of arbitrary order, we now introduce the class $\Psi\mathscr{S}$ of all polynomials in $x = (x_1, \ldots, x_n)$ and $D = (D_1, \ldots, D_n)$, $D_j = -i\partial_{x_j}$, with coefficients in $\Psi\mathscr{G}\mathscr{S}$. In particular, $\Psi\mathscr{S}$ contains all linear differential operators with complex polynomials in x as coefficients. The most important observables are unbounded differential operators, not operators in $\mathscr{L}(\mathscr{H})$. Thus, the extended algebra $\Psi\mathscr{S}$ might be more appropriate as $\Psi\mathscr{G}\mathscr{S}$.

Note that the x_j, D_l and the $A \in \Psi\mathscr{G}\mathscr{S}$ do not normally commute: We have the Heisenberg commutator relations

$$(1.13) \qquad [x_j, D_l] = i\delta_{jl}.$$

Also, since ix_j, $iD_j \in \mathscr{a}\mathscr{o}$, it follows that $[x_j, A]$, $[D_j, A] \in \Psi\mathscr{G}\mathscr{S}$ whenever $A \in \Psi\mathscr{G}\mathscr{S}$, since $[x_j, A] = -i\,\mathrm{ad}_{ix_j} A$, etc. This shows that every product of x_j, D_l, and $A \in \Psi\mathscr{G}\mathscr{S}$ may be written as a sum of terms $x^\alpha A_{\alpha,\beta} D^\beta$ with multi-indices α, β. Accordingly, every $A \in \Psi\mathscr{S}$ may be written as

$$(1.14) \quad A = \sum x^\alpha A_{\alpha,\beta} D^\beta = a(x, D), \qquad a(x, \xi) = \sum a_{\alpha,\beta}(x, \xi) x^\alpha \xi^\beta,$$

with symbols $a_{\alpha,\beta}$ of $A_{\alpha,\beta}$.

For practical considerations it proves useful to use a pair of *weight functions*. We will exclusively use the pair

$$(1.15) \qquad \langle x \rangle = \left(1 + |x|^2\right)^{1/2}, \qquad \langle \xi \rangle = \left(1 + |\xi|^2\right)^{1/2},$$

although other weight functions, like those used by Beals [$\mathbf{B_1}$], may be better adapted to specific problems. With (1.15) we can make $\Psi\mathscr{S} = \bigcup\{\Psi\mathscr{S}_m\colon m \in \mathbb{R}^2\}$ a graded algebra (just like $\mathrm{Op}\,\Psi c$ in [**CE**]). Several descriptions of the symbol classes corresponding to $\Psi\mathscr{S}_m$ are given in §2 ((2.12), (2.13), and Proposition 2.3).

Our central result now is an Egorov-type conjugation invariance theorem (Theorem 3.3) for the algebra $\Psi\mathscr{S}$: *For $A = a(x, D) \in \Psi\mathscr{S}_m$ and a first-order ψdo $K = k(x, D)$ with classical symbol k satisfying certain assumptions, the operator $A_t = e^{-iKt} A e^{iKt}$ is again a ψdo in $\Psi\mathscr{S}_m$ with symbol given by a and the characteristic flow of k. Moreover, the family $\{A_t\colon t \in \mathbb{R}\}$ is $C^\infty(\mathbb{R}, \Psi\mathscr{S}_m)$ again with a suitable Fréchet topology of $\Psi\mathscr{S}_m$* (cf. Theorem 3.3). From this result we derive all coordinate and gauge invariances of $\Psi\mathscr{G}\mathscr{S}$ (and $\Psi\mathscr{S}$) mentioned (cf. Theorems 3.4 and 3.5).

Perhaps an even more interesting Fréchet algebra is obtained by examining the last type of problem in a more general setting. Suppose we consider a $(\nu \times \nu)$-matrix $A = ((A_{pq}))_{p,q=1,\dots,\nu}$ of operators in $\Psi\mathcal{S}_m$. If $K = k(x, D)$ is a first-order $(\nu \times \nu)$-matrix of classical pseudodifferential operators satisfying suitable conditions, we may examine the operator $A_t = e^{-iKt}Ae^{iKt}$ again for properties as above. In order to have a well-defined e^{iKt}, we require the system

$$(1.16) \qquad \partial_t u - ik(x, D)u = 0$$

to be 'semistrictly hyperbolic'. That is, the symbol matrix $k(x, \xi)$ has real eigenvalues of constant multiplicity for all $|x| + |\xi|$ sufficiently large, and, moreover, the eigenvalues stay apart even at infinity in accordance with a suitable weight function (details in §3). *We arrive at the following result: The condition that A_t is again a ψdo for all t with suitable existence of derivatives $\partial_t^k A_t$, describes a graded subalgebra $\mathcal{P} = \mathcal{P}_K$ of $\Psi\mathcal{S}$. In first approximation the ψdo's of $\mathcal{P}_K$ are determined by the property that the symbol matrix $a(x, \xi)$ commutes with the matrix $k(x, \xi)$, since $|x| + |\xi|$ is large* (*Theorem* 3.8).

It is this algebra $\mathcal{P}$ that may be of specific interest for quantum-mechanical applications. Actually, it may be seen that the Dirac equation is a semistrictly hyperbolic (4×4)-system of differential equations precisely satisfying our assumptions, with weight functions (1.14), if the electromagnetic potentials are subject to certain restrictions. Physically, the most severe restriction seems to be that no singularities of the potentials are permitted. We have reason, however, to assume that this restriction, if properly interpreted, might lead to interesting conclusions.

If we set $K = H$, with H the Dirac Hamiltonian, then the map $A \to A_t = e^{-iHt}Ae^{iHt}$ is called the *Heisenberg representation* of the observable A. Its physical interpretation is this: Assuming that the physical states (the functions $u \in \mathcal{H}^4$ of norm 1) are constant in time, A_t represents the observable A at time t. Here we set $\mathcal{H}^4 = \mathcal{H} \otimes \mathbb{C}^4$.

In other words, the algebra $\mathcal{P} = \mathcal{P}^H$ is just the collection of all ψdo's in some $\Psi\mathcal{S}_m$, with Heisenberg representation again in $\Psi\mathcal{S}_m$, for all times t, and such that the Heisenberg representation changes differentiably with time. Our analysis of this algebra $\mathcal{P}$ might reflect a variety of physical insights: The Dirac equation is symmetric hyperbolic but not strictly hyperbolic. The symbol matrix $h(x, \xi)$ of the Hamiltonian H has two real eigenvalues, each of multiplicity 2. Hence, the equation is semistrictly hyperbolic of type $e^1 = (1, 0)$ in our sense. The two characteristic flows corresponding to the two eigenvalues give the particle motion, of course, for particles of positive and negative energies. For an observable $A = a(x, D) \in \mathcal{P}_m$, we mentioned that the (4×4)-symbol matrix $a(x, \xi)$ 'approximately' commutes with the symbol matrix $h(x, \xi)$. To give an idea, assume that a and h commute. Then $a(x, \xi)$ is reduced by the spectral representation and may be described by giving its component (2×2)-matrix in each of the two eigenspaces of $h(x, \xi)$.

Now if these components are multiples of the (2×2)-identity, then they just flow along the characteristic flows (i.e., the particle motion), just as in the strictly hyperbolic case. However, if this is not the case, then each component also experiences a similarity transform while floating along the characteristic flow. One may recognize the classical spin propagation in the change of these (2×2)-matrices. For this the matrix, with respect to a proper base, must be written as a linear combination of the identity and the three Pauli matrices, which is always possible. In this representation the 3-vector given by the three coefficients of the Pauli matrices propagates as the magnetic moment of the electron. Of course, all these discussions are approximate—i.e., they hold only with certain lower-order perturbation symbols. Thus, for example, the 'Stern–Gerlach' effect—i.e., the influence of the electron spin on the particle orbit—is reflected by a symbol of lower order and cannot be captured with this approach. Let us also notice that all of our asymptotic expansions of symbols hold only for large $|x| + |\xi|$—that is, if either $|x|$ or $|\xi|$ (or both) are large. Thus our theory can only make statements at $x = \infty$ or for very large momenta (or both)—an attitiude which seems to fit with the scattering approach.

We might have offered enough motivation now to examine the usefulness of the algebra $\mathcal{P}^H$ as an algebra of observables for the Dirac equation. In this respect we make the following comments (cf. **[CD]**): $\mathcal{P}^H = \mathcal{P}$ contains almost none of the standard dynamical observables, such as space or momentum coordinates, angular momentum, particle velocity, etc. However, one may offer slightly changed observables within $\mathcal{P}$ instead. Accepting these new observables would complicate matters because they are normally ψdo's, not the simple differential operators used so far. On the other hand, a variety of inconsistencies of Dirac theory, such as 'Zitterbewegung', noncommutativity of velocity components, Klein paradox, etc., would disappear.

The above facts have been examined in detail for a slightly different algebra $\mathcal{P}$ of ψdo's **[CD, CF]**. The change here to $\mathcal{P}$ of (3.22) was made in the interest of a more natural definition. For proofs in the new setting we refer to **[C$_2$]**.

2. Details on the algebras $\Psi\mathcal{G}\mathcal{S}$ and $\Psi\mathcal{S}$ and the classes $\Psi\mathcal{S}_m$. Let us return to the class $\Psi\mathcal{G}\mathcal{S}$ of (1.7). Clearly,

$$(AB)^{s,\lambda} = A^{s,\lambda}B^{s,\lambda}, \qquad A, B \in \mathcal{L}(\mathcal{H}). \tag{2.1}$$

From Leibnitz's formula for the differentiation of a product, it follows at once that $(AB)^{s,\lambda}$ is C^∞ whenever $A^{s,\lambda}$ and $B^{s,\lambda}$ are C^∞. This implies that $\Psi\mathcal{G}\mathcal{S}$ is a subalgebra of $\mathcal{L}(\mathcal{H})$. Since the $T^{s,\lambda}$ are unitary, we get $(A^*)^{s,\lambda} = (A^{s,\lambda})^*$, where the right side is differentiable in the norm topology if and only if $A^{s,\lambda}$ is differentiable. This shows that $\Psi\mathcal{G}\mathcal{S}$ is adjoint invariant; i.e., it is a $*$-subalgebra of $\mathcal{L}(\mathcal{H})$.

To note another important property, let $A \in \Psi\mathcal{G}\mathcal{S}$ be invertible in $\mathcal{L}(\mathcal{H})$. Then by the quotient rule of differentiation it follows that $(A^{-1})^{s,\lambda} = (A^{s,\lambda})^{-1}$ is C^∞. In other words, if the inverse A^{-1} of A exists in $\mathcal{L}(\mathcal{H})$, then it is contained in $\Psi\mathcal{GS}$. Thus the algebra $\Psi\mathcal{GS}$ shares the most important property of a C^*-subalgebra of $\mathcal{L}(\mathcal{H})$: it contains all its inverses with respect to $\mathcal{L}(\mathcal{H})$.

On the other hand, $\Psi\mathcal{GS}$ may be provided with a convenient Fréchet topology generated by the seminorms

$$(2.2) \qquad \| \|A\| \| = \|\nabla^\alpha A\|, \qquad \alpha \in \mathbb{Z}_+^{(n+1)(n+2)/2}.$$

Here we refer the function $A^{s,\lambda}$ to a local coordinate base near the unit element $(\sigma, o, z, \zeta, \phi) = (1,1,0,0,0) = e$ of $\mathcal{gs}$. For example, we might choose $o = e^h$ with a skew-symmetric real $(n \times n)$-matrix $h = ((h_{jl}))$ and then use the $(n+1)(n+2)/2$ coordinates $(\sigma, h_{jl}, z_j, \zeta_j)$, $j, l = 1, \ldots, n, j < l$, noting that $A^{s,\lambda}$ is independent of ϕ. In (2.2) $\nabla^\alpha A$ just denotes the αth partial derivative at e for these coordinates in conventional multi-index notation. Also $\|\cdot\|$ denotes the operator norm of $\mathcal{L}(\mathcal{H})$.

In [**C$_2$**] we prove that $\Psi\mathcal{GS}$ is a Fréchet space with respect to this topology. Generally, it is well known, of course, that differentiability of a function over a group is a matter of differentiability at the unit element. In particular, there are transition formulas giving the derivatives at any point in terms of the derivatives at e with (locally) bounded coefficients. Hence, for a convergent sequence A_j, under (2.2), one obtains locally uniform convergence of all derivatives in operator-norm convergence, giving differentiability of the limit $A^{s,\lambda}$ by elementary theorems. This, in essence, gives the completeness, hence the Fréchet property. It is again a matter of using Leibnitz's formula to prove that the algebra operations in $\Psi\mathcal{GS}$ are continuous under the topology introduced.

We have obtained the following result.

THEOREM 2.1. *The set $\Psi\mathcal{GS}$ is a ψ^*-subalgebra of the C^*-algebra $\mathcal{L}(\mathcal{H})$.*

Note that the concept of ψ^*-subalgebra of a C^*-algebra $\mathcal{A}$ was introduced by B. Gramsch [**G**]: *This is just a $*$-subalgebra of $\mathcal{A}$ with Fréchet topology containing all its inverses in $\mathcal{A}$*. Hence, after above remarks, there is nothing to prove.

The importance of Theorem 2.1 lies in the fact that Gramsch [**G**] proved that the Fredholm operators of a ψ^*-algebra possess most of the important properties of Fredholm operators on Banach spaces or C^*-algebras. In particular, there is an analytic operator calculus and a perturbation theory. In fact, Gramsch shows that the set of Fredholm operators in a ψ^*-algebra forms an analytic manifold.

The next theorem is similar in nature and proof to results discussed in [**CL**]. There we consider two similar Lie groups and corresponding representations in $\mathcal{U}(\mathcal{H})$: The groups $\mathcal{gt}$ and $\mathcal{gl}$ are obtained from $\mathcal{gs}$ by restricting $s(x)$ to translations only (i.e., $\sigma = 1$, $o = 1$) or admitting all invertible linear maps $s(x)$, respectively. Corresponding classes $\Psi\mathcal{GT}$ and $\Psi\mathcal{GL}$ were proved to be algebras of ψdo's in [**CL**], and it is trivial that

$$(2.3) \qquad \Psi\mathcal{GL} \subset \Psi\mathcal{GS} \subset \Psi\mathcal{GT}.$$

(In [**CL**] we used the notations $\Psi\mathcal{GL} = nclc^\infty$ and $\Psi\mathcal{GT} = nctc^\infty$.) While $\Psi\mathcal{GL}$ is perhaps slightly more restricted than necessary, $\Psi\mathcal{GT}$ appears too large: Its operators are ψdo's, but the conventional calculus of ψdo's does not apply. The result on $\Psi\mathcal{G}\mathcal{T}$ is similar in nature to theorems of Beals [**B**$_2$] and Dunau [**D**]; cf. also [**C**$_1$, Chapter 4]. In all of these publications, suitable weighted L^2-Sobolev spaces are used instead of $\mathcal{H}$, resulting in classes which allow ψdo-calculus again. We make a special point of the fact that Sobolev spaces are not required for our more general Lie groups $\mathfrak{gl}$ and $\mathfrak{gs}$, and we still obtain a calculus of ψdo's in $\Psi\mathcal{G}\mathcal{X}$, $\mathcal{X} = \mathcal{S}, \mathcal{L}$.

For our theorem we introduce the differential expressions in phase space

$$\text{(2.4)}\qquad \begin{aligned} \eta_{00} &= \sum_{j=1}^{n} \left(\xi_j \partial_{\xi_j} - x_j \partial_{x_j}\right), \\ \eta_{jl} &= \left(\xi_j \partial_{\xi_l} - \xi_l \partial_{\xi_j}\right) + \left(x_j \partial_{x_l} - x_l \partial_{x_j}\right), \qquad j, l \geqslant 1. \end{aligned}$$

Let

$$\text{(2.5)}\qquad \begin{aligned} \psi\ell_0 &= CB^\infty(\mathbb{R}^{2n}) \\ &= \left\{ a(x, \xi) \in C^\infty(\mathbb{R}^{2n}) \colon a_{(\beta)}^{(\alpha)}(x, \xi) = O(1) \text{ for all } \alpha, \beta \right\}, \end{aligned}$$

with $a_{(\beta)}^{(\alpha)} = \partial_\xi^\alpha \partial_x^\beta a$, and let ψs_0 denote the class of all $a \in \psi\ell_0$ such that an arbitrary finite application

$$\text{(2.6)}\qquad \eta_{j_0 l_0} \eta_{j_1 l_1} \cdots \eta_{j_N l_N} a$$

of the expressions (2.4) is always in $\psi\ell_0$.

It is easily checked that ψs_0 is an algebra under pointwise multiplication of functions, and that ψs_0 is invariant under η_{jl} of (2.4), as well as

$$\text{(2.7)}\qquad \eta_{j0} = \partial_{x_j}, \quad \eta_{0j} = \eta_{\xi_j}, \qquad j = 1, \ldots, n,$$

which is introduced for convenience.

THEOREM 2.2. *We have*

$$\text{(2.8)}\qquad \Psi GS = \operatorname{Op} \psi s_0 = \left\{ a(x, D) \colon a \in \psi s_0 \right\}.$$

The proof is very similar in structure to the proof of Theorem 1.2 of [**CL**]. It is discussed in detail in [**C**$_2$]. Recall that

$$\text{(2.9)}\qquad a(x, D) u(x) = (2\pi)^{-n/2} \int d\xi\, dy\, e^{i\xi(x-y)} a(x, \xi) u(y), \qquad u \in \mathcal{S},$$

which one proves to provide a map $\Psi s_0 \to \operatorname{Op} \psi s_0 = \Psi\mathcal{G}\mathcal{S}$. An essential part of the proof of Theorem 2.2 is the construction of an inverse of the map $a \to A$ of (2.9). Such an inverse is provided by the formula

$$\text{(2.10)}\qquad a(z, \zeta) = (2\pi)^n \operatorname{trace}\left(Q_-^* P(\partial_z, \partial_\zeta) A^{x+z, \zeta\xi}\right).$$

Here

$$P(x, \xi) = \prod_{j=1}^{n} \left((1 + x_j)^2 (1 + \xi_j)^2\right)$$

and $Q_- = q(-x, -D)$, where $q(x, \xi)$ is a suitable fundamental solution of the differential operator $P(\partial_x, \partial_\xi)$—in essence the product of Green's functions of the ordinary differential expressions $(1 + d/dt)^2$, with $t = x_j$ and $t = \xi_j$. It is shown that Q_- is of trace class, and $P(\partial_z, \partial_\zeta)A^{x-z,\zeta x}$ is bounded with all its derivatives for z and ζ if only $A \in \Psi\mathscr{G}\mathscr{S}$ or, at least, $A \in \Psi\mathscr{G}\mathscr{T}$. This makes the right side of (2.10) well defined, since the trace class is an ideal of $\mathscr{L}(\mathscr{H})$. For more detail, see [**CL** or $\mathbf{C_2}$].

Note that (2.9) and (2.10) provide a 1-1-correspondence $\psi c_0 \leftrightarrow \Psi\mathscr{G}\mathscr{S}$. This map is continuous in either direction if we equip the symbol algebra ψc_0 with the Fréchet topology of the seminorms

$$\left\| \prod_{j=1}^{n} \eta_{p_j q_j} a \right\|_{L^\infty}, \tag{2.11}$$

where we must take every finite product of the η_{jl} of (2.4) and (2.7). In other words, *the Fréchet topologies of* (2.2) *and* (2.11) *are equivalent under the above* 1-1-*correspondence of operators and symbols*. For details of the proof see [$\mathbf{C_2}$].

Next we focus on the algebra $\Psi\mathscr{S}$ of all operators (1.14) and its symbol class ψc. Using the two weight functions (1.15), we define $\Psi\mathscr{S}_m = \operatorname{Op} \psi c_m$, where ψc_m is from (2.12). We easily see that

$$\begin{gathered} \psi c = \bigcup_{m=(m_1, m_2) \in \mathbf{R}^2} \psi c_m, \\ \psi c_m = \left\{ a \in C^\infty(\mathbb{R}^{2n}) \colon b = \langle \xi \rangle^{-m_1} \langle x \rangle^{-m_2} a \in \psi c_0 \right\}. \end{gathered} \tag{2.12}$$

On ψc_m define a Fréchet topology by the map $a \leftrightarrow a\pi_m$, $\psi c_0 \leftrightarrow \psi c_m$. We obtain

$$\psi c_m = \left\{ a \in C^\infty(\mathbb{R}^{2n}) \colon \prod_{j=1}^{N} \eta_{p_j q_j} a = O(\pi_m(x, \xi)) \text{ for all such products} \right\}, \tag{2.13}$$

as is easily seen, with the more general weight function

$$\pi_m(x, \xi) = \langle \xi \rangle^{m_1} \langle x \rangle^{m_2}. \tag{2.14}$$

Another feature, bringing out the classical nature of the symbols in ψc, is

PROPOSITION 2.3. *For $a \in \psi c_m$ we have the estimates*

$$a_{(\beta)}^{(\alpha)}(x, \xi) = O\left(\pi_{m + r(e^1 - e^2)}(x, \xi)\right), \qquad -|\alpha| \leqslant r \leqslant |\beta|, \tag{2.15}$$

where $e^1 = (1, 0)$, $e^2 = (0, 1)$.

(See [$\mathbf{C_2}$] for the proof.)

If x is restricted to a compact set, then (2.15) implies

$$a_{(\beta)}^{(\alpha)}(x, \xi) = O(\langle \xi \rangle^{m_2 - |\alpha|}), \tag{2.16}$$

which is the set of inequalities describing $S^{m_1} = S^{m_1, 1, 0}$ of Hörmander's symbol classes [**Ho**]. Moreover, for $m = 0 = (0, 0)$, estimates (2.15) describe the symbol class Z_n of Weinstein [**W**] and Zelditch [**Z**]. In other words, $\psi c_0 \subset Z_n$. From [**W**] and Proposition 2.3 it follows that the operators in $\Psi\mathcal{S}$ do not enlarge the wave front sets of distributions with compact support.

Finally, we mention the order concept introduced in **[CE]**, under the present conditions. Clearly,

$$(2.17)\qquad \psi s_m \subset \psi t_m = \left\{ a \in C^\infty(\mathbb{R}^{2n}) \colon a^{(\beta)}_{(\beta)}(x, \xi) = O\left(\pi_m(x, \xi)\right)\right\}.$$

For $m = (m_1, m_2)$ let us introduce the weighted L^2-Sobolev spaces and norms

$$(2.18)\quad \mathcal{H}_m = \left\{ u \in \mathcal{S}' \colon \pi_{-m}(x, D)u \in \mathcal{H} \right\}, \qquad \|u\|_m = \|\pi_{-m}(x, D)u\|_{L^2}.$$

Clearly,

$$(2.19)\qquad \mathcal{S} = \bigcap\left\{ \mathcal{H}_m \colon m \in \mathbb{R}^2 \right\}, \qquad \mathcal{S}' = \bigcap\left\{ \mathcal{H}_m \colon m \in \mathbb{R}^2 \right\}.$$

An operator $A \colon \mathcal{S} \to \mathcal{S}$ is said to be of order $m = (m_1, m_2)$ if it has continuous extensions $A_m \in \mathcal{L}(\mathcal{H}_s, \mathcal{H}_{s-m})$ for all $s \in \mathbb{R}^2$. The class of all such operators is denoted by $\mathcal{O}(m)$, and we set

$$(2.20)\quad \mathcal{O}(\infty) = \bigcup\left\{ \mathcal{O}(m) \colon m \in \mathbb{R}^2 \right\}, \qquad \mathcal{O}(-\infty) = \bigcap\left\{ \mathcal{O}(m) \colon m \in \mathbb{R}^2 \right\}.$$

It is easily seen that $\mathcal{O}(-\infty)$ coincides with the class of all integral operators $Ku(x) = \int k(x, y)u(y)\, dy$ with integral over $\mathbb{R}^n$ and kernel $k \in \mathcal{S}(\mathbb{R}^{2n})$. In **[CE]** we found that $\mathrm{Op}\, \psi t_m \subset \mathcal{O}(m)$; hence, we get

$$(2.21)\qquad \Psi\mathcal{S}_m = \mathrm{Op}\, \psi s_m \subset \mathrm{Op}\, \psi t_m \subset \mathcal{O}(m)$$

as a consequence of (2.17).

As the most important consequence of the symbol estimates of Proposition 2.3, we note that a global calculus of ψdo's still holds (outlined in Proposition 2.4). This will be an important element of proof for our Egorov-type discussion in §3.

Let us define the functions

$$(2.22)\qquad \eta^{00} = -x \cdot \xi, \quad \eta^{j0} = \xi_j, \quad \eta^{0l} = -x_l, \quad \eta^{jl} = x_j\xi_l - \xi_j x_l, \qquad j, l = 1, \ldots, n,$$

noting that

$$(2.23)\qquad \eta_{pq}a = \langle \eta^{pq}, a\rangle, \qquad p, q = 0, 1, \ldots, n,$$

with the Poisson bracket (2.26). Let $\psi\sigma_e$ denote the class of all symbols a of the form (with $e^1 = (1, 0)$, $e^2 = (0, 1)$)

$$(2.24)\quad a = \sum_{p,q=0}^{n} \gamma_{pq}\eta^{pq} + a_d + a_m, \qquad a_d \in \psi c_{e^1}, a_m \in \psi c_{e^2}, \gamma_{pq} \in \mathbb{C},$$

and the classes ψc_m of classical symbols (cf. **[CP, CE]**), where

$$(2.25)\quad \psi c_m = \left\{ a \in C^\infty(\mathbb{R}^{2n}) \colon a^{(\alpha)}_{(\beta)}(x, \xi) = O\left(\pi_{m-|\alpha|e^1-|\beta|e^2}(x, \xi)\right)\right\}.$$

For the symbol c of a commutator $[a(x, D), b(x, D)]$ of two ψdo's, one expects an asymptotic expansion of the form

$$(2.26)\qquad c = -i\langle a, b\rangle + \sum_{j=2}^{\infty} \frac{(-i)^j}{j!}\langle a, b\rangle_j \quad \left(\mathrm{mod}\, \mathcal{S}(\mathbb{R}^{2n})\right)$$

with the (iterated) Poisson brackets

$$(2.27)\quad \langle a, b\rangle = a_{|\xi}\cdot b_{|x} - b_{|\xi}\cdot a_{|x}, \quad \langle a, b\rangle_j = \sum_{|\theta|=j} \theta!\big(a^{(\theta)}b_{(\theta)} - b^{(\theta)}a_{(\theta)}\big).$$

If the symbols a and b are $(\nu\times\nu)$-matrix-valued, one expects the corresponding formula (2.26) with an additional additive term $[a, b]$ (vanishing for $\nu = 1$), where now the order of terms in all matrix products of (2.27) must be kept exactly as listed.

PROPOSITION 2.4. *For $k \in \psi\sigma_e$ and $a \in \psi\mathscr{o}_m$ we have*

$$(2.28)\qquad \langle k, a\rangle_j \in \psi\mathscr{o}_{m-re^1-r'e^2} \quad \textit{for all } r + r' = j - 1, j = 1, 2, \ldots.$$

Moreover, the asymptotic expansion (2.26) *of the commutator symbol $c(x, \xi)$, such that $c(x, D) = [k(x, D), a(x, D)]$, holds in the sense that*

$$(2.29)\quad c - [k, a] - \sum_{j=1}^{N} \frac{(-i)^j}{j!}\langle k, a\rangle_j \in \psi\mathscr{o}_{m-re^1-r'e^2}, \qquad r + r' = N.$$

Here (2.29) *has been written such that it remains true for $(\nu\times\nu)$-matrices k and a of symbols with entries in $\psi\sigma_e$ and $\psi\mathscr{o}_m$, respectively.*

For the proof we again refer to [$\mathbf{C_2}$].

3. Conjugation of $\Psi\mathscr{G}\mathscr{S}$ with a ψdo; coordinate and gauge invariance. First we look at the characteristic flow of a single hyperbolic pseudodifferential equation

$$(3.1)\qquad \partial_t u - ik(x, D)u = 0,$$

where k is a real-valued symbol in $\psi\mathscr{o}_e$. That is, we consider the flow $\nu_t = (x_t(x, \xi), \xi_t(x, \xi))$ given by the $(2n \times 2n)$-system of ordinary differential equations

$$(3.2)\qquad \frac{dx}{dt} = k_{|\xi}(x, \xi), \qquad \frac{d\xi}{dt} = -k_{|x}(x, \xi).$$

For fixed $x = x^0$ and $\xi = \xi^0$, the functions $f(t) = x_t(x^0, \xi^0)$ and $\phi(t) = \xi_t(x^0, \xi^0)$ are the unique solutions of the Cauchy problem

$$(3.3)\quad \frac{df}{dt} = k_{|\xi}(f, \phi), \quad \frac{d\phi}{dt} = -k_{|x}(x, \xi), \qquad f(0) = x^0, \phi(0) = \xi^0.$$

A priori estimates can easily be given to verify that the flow ν_t is globally well defined: For each $t \in \mathbb{R}^n$ the map ν_t: $\mathbb{R}^{2n} \to \mathbb{R}^{2n}$ defines a diffeomorphism $\mathbb{R}^{2n} \leftrightarrow \mathbb{R}^{2n}$ (cf. [**CE**]). An important question is this: For a symbol $a \in \psi\mathscr{o}_m$ will the composed function

$$(3.4)\qquad a_t = a \circ \nu_t$$

again be a symbol in $\psi\mathscr{o}_m$? This question is answered as follows: Let ψs_e denote the class of all $C^\infty(\mathbb{R}^{2n})$-functions $b(x, \xi)$ with the property that, for every finite application $c = \prod_{j=1}^N \eta_{p_j q_j} b$ of the differential expressions η_{jl} of (2.4) onto b, the first-order differential expression defined by the map $a \to \langle c, a\rangle$ is a finite linear combination of the expressions η_{pq} with coefficients in ψc_0:

$$(3.5)\qquad \langle c, \cdot\rangle = \sum_{p,q=0}^{n} \gamma_{pq}\eta_{pq}, \qquad \gamma_{pq} \in \psi c_0.$$

It is easily seen that $\psi\sigma_e \subset \psi s_e$, where $\psi\sigma_e$ is from (2.24). For application to hyperbolic systems the following is still important.

PROPOSITION 3.1. *With the functions η^{pq} of* (2.22) *let $\zeta_{pq} = \eta^{pq}\pi_{-e}$, and let*

$$(3.6)\qquad a = \pi_e P(\zeta) + a_d + a_m, \qquad a_d \in \psi c_{e^1},\ a_m \in \psi c_{e^2},$$

where $P(\zeta)$ denotes any polynomial in the $(n+1)^2$ variables ζ_{pq}. Then a is a symbol in ψs_e. Moreover, let the polynomial

$$(3.7)\qquad \Theta(\lambda) = \lambda N + \theta_1\lambda N - 1 + \cdots + \theta_N$$

have the properties: (i) $\kappa_j = \theta_j\pi_i$ *are all functions of x, ξ and symbols of the form* (3.6) *for $j = 1,\dots,N$*; (ii) *for all sufficiently large $|x| + |\xi|$ the polynomial has ρ' roots μ_j of constant multiplicity ν_j*; *and* (iii) *the differences $\mu_j - \mu_{j-1}$ are bounded and bounded away from zero. Then each of the roots $\mu_j(x,\xi)$, $|x| + |\xi| \geq \eta$, extends into $\mathbb{R}^{2n}$ such that $\lambda_j = \pi_e\mu_j \in \psi s_e$.*

This proposition is proved in [$\mathbf{C_2}$].

THEOREM 3.2. *Let $k \in \psi s_e$ be real-valued. Then for every $a \in \psi\mathscr{s}_m$, $m \in \mathbb{R}^2$, and $t \in \mathbb{R}$ the function a_t of* (3.4) *is a symbol in $\psi\mathscr{s}_m$. Moreover, the function $t \to a_t$ is $C^\infty(\mathbb{R}, \psi\mathscr{s}_m)$ in the Fréchet topology of $\psi\mathscr{s}_m$ (defined by the map $a \leftrightarrow \pi_{-m}a$ of $\psi\mathscr{s}_m \leftrightarrow \psi\mathscr{s}_0$). Each t-derivative is bounded on compact sets of $\mathbb{R}$.*

The proof uses the well-known fact that $a_t(x,\xi)$ satisfies the linear partial differential equation

$$(3.8)\qquad \partial_t a_t + \langle k, a_t\rangle = 0.$$

One may apply the operators η_{pq} onto (3.8), using the equality $\eta_{pq} = \langle \eta^{pq}, \cdot\rangle$. Using the basic property of symbols in ψs_e—that $\langle \eta_{pq}k, \cdot\rangle$ is a linear combination of the η_{rs}—one may express the differentiated equations (3.8) as a system of linear differential equations for the functions $a_{pq} = \eta_{pq}a$. All equations turn out to have the same principal part, so that we may express them as a system of ODEs along the flow of (3.2) with bounded coefficients (bounded by $\pi_m(x,\xi)$, to be precise, also in the following). Since the initial values for $t = 0$ are bounded by π_m as well, in view of $a \in \psi\mathscr{s}_m$ we get boundedness of the $\eta_{pq}a_t$ over $\mathbb{R}^{2n} \times I$ for every bounded interval $I \subset \mathbb{R}$. The procedure may be iterated for higher applications of the η_{pq}. This gives $a_t \in \psi\mathscr{s}_m$ as stated. Also the systems obtained may be differentiated for t and treated similarly to get results on the derivatives $\partial_t^n a_t$. Details are again found in [$\mathbf{C_2}$].

Theorem 3.2 may be used to prove the following result on conjugation of $\Psi\mathscr{G}\mathscr{S}$ or $\Psi\mathscr{S}_m$ with e^{iKt} for $K = k(x, D)$ with complex-valued symbol.

THEOREM 3.3. *Let $K = k(x, D)$, where $k = k_0 + \kappa$, with a real-valued $k_0 \in \psi\sigma_e$ and a general complex-valued $\kappa \in \psi c_0$. Then for every $A = a(x, D) \in \mathrm{Op}\,\psi s_m$, $m \in \mathbb{R}^2$, we have*

$$A_t = e^{-iKt}Ae^{iKt} = a(t, x, D) \in \mathrm{Op}\,\psi s_m. \tag{3.9}$$

Moreover,

$$a(t, x, \xi) - (a \circ \nu_t)(x, \xi) \in \psi s_{m-e^1} \cap \psi s_{m-e^2}. \tag{3.10}$$

Also, the family $\{A_t: t \in \mathbb{R}\}$ is infinitely differentiable in $\mathrm{Op}\,\psi s_m$, and

$$ad_{iK}A = -dA_t/dt|_{t=0} = i[A, K] \in \mathrm{Op}\,\psi s_m. \tag{3.11}$$

The proof is modeled after the proof of Theorem 8.1 in [**CG**], where the necessary changes are given by Proposition 3.1 and Theorem 3.2 above. For details see [$\mathbf{C_2}$].

As a consequence of Theorem 3.3, we now state a theorem on coordinate invariance of the algebras $\Psi\mathcal{GS}$ and $\Psi\mathcal{S}$.

THEOREM 3.4. *Let the diffeomorphism $\phi: \mathbb{R}^n \to \mathbb{R}^n$ and its inverse $\chi: \mathbb{R}^n \to \mathbb{R}^n$ both be of the form*

$$\phi(x) = \sigma o x + \omega(x), \qquad \chi(x) = \sigma^{-1}o^{-1}x + \psi(x), \tag{3.12}$$

with positive real σ, an orthogonal matrix o, and bounded C^∞-functions ω, ψ. Let $\omega, \psi \in \psi c_0$. That is, let

$$\omega^{(\alpha)}(x) = O(\langle x\rangle^{-|\alpha|}), \qquad \psi^{(\alpha)}(x) = O(\langle x\rangle^{-|\alpha|}) \tag{3.13}$$

for all multi-indices α. Then the coordinate transforms

$$u \to u \circ \phi, \qquad u \to u \circ \chi, \tag{3.14}$$

leave the algebra $\Psi\mathcal{GS}$ (and also each $\Psi\mathcal{S}_m$) invariant.

OUTLINE OF THE PROOF. Under such a coordinate transform, the Lie algebra $\mathfrak{as}$ of (1.8) will transform into a Lie algebra $\mathfrak{as}^\sim$ of folpde. Under our assumptions each expression $L \in \mathfrak{as}^\sim$ is a ψdo satisfying the assumptions of Theorem 3.3. From this it follows that the condition of Theorem 1.1, after the coordinate transform, is precisely equivalent to differentiability for e^{Lt}, $L \in \mathfrak{as}^\sim$. This indeed gives coordinate invariance of the algebras.

Similarly, for a 'gauge invariance' we have

THEOREM 3.5. *Let the 'gauge function' $\mu(x)$ be real-valued and a symbol in ψc_{e^2} independent of ξ:*

$$\mu^{(\alpha)}(x) = O(\langle x\rangle^{1-|\alpha|}). \tag{3.15}$$

Then the corresponding gauge transform $A \to e^{-i\mu(x)}Ae^{i\mu(x)}$ leaves $\Psi\mathcal{GS}$ and $\Psi\mathcal{S}_m$ invariant.

For the proof we refer to [$\mathbf{C_2}$]. (Actually there we impose a somewhat weaker condition for $\mu(x)$.)

Finally, let us discuss a result corresponding to Theorem 3.3, but for a $(\nu \times \nu)$-matrix $K = k(x, D)$ of ψdo's. We focus on a $(\nu \times \nu)$-matrix

$$k(x,\xi) = ((k_{jl}(x,\xi)))_{j,l=1,\dots,\nu}, \qquad k_{jl} \in \psi\sigma_e, \tag{3.16}$$

with the class $\psi\sigma_e$ of (2.24) where complex-valued symbols are permitted. In order to have the operator e^{iKt} well defined, we need a condition of hyperbolicity: $k(x, \xi)$ (or rather the system (3.1) with symbol k) is called *semistrictly hyperbolic* (*of type* f, where $f = e = (1, 1), f = e^{-1}$, or $f = e^2$) if

$$k(x,\xi) = k_f(x,\xi) + \kappa(x,\xi), \qquad \kappa \in \psi c_{f-e},\ k_f = \pi_f k_0 \in \psi c_f, \tag{3.17}$$

where k_f is of the form (2.24) (in particular, $k_f \in \psi c_f$ reduces to the second or third term of (2.24) only if $f \neq e$). In addition, we assume that the matrix function $k_f(x, \xi)$ (hence $k_0(x, \xi)$) is diagonalizable for all $|x| + |\xi| \geqslant \eta$, η sufficiently large. We also require (ii) all eigenvalues μ_j of k_0 to be real and of constant multiplicity ν_j (independent of x and ξ), $|x| + |\xi| \geqslant \eta$. (In particular , this implies that we may arrange for

$$\begin{gathered} \mu_1(x,\xi) < \mu_2(x,\xi) < \cdots < \mu_\rho(x,\xi), \\ \nu_1 + \cdots + \nu_\rho = \nu, \qquad |x| + |\xi| \geqslant \eta, \end{gathered} \tag{3.18}$$

where the μ_j are C^∞, by well-known perturbation arguments for matrices.) (iii) We also assume that

$$\mu_{j+1}(x,\xi) - \mu_j(x,\xi) \geqslant \eta_0 > 0, \qquad |x| + |\xi| \geqslant \eta, \tag{3.19}$$

with some positive η_0 independent of x and ξ.

PROPOSITION 3.6. *For a semistrictly hyperbolic $(\nu \times \nu)$-symbol $k(x, \xi)$ of type f, the ρ eigenvalues $\lambda_j(x, \xi) = \pi_f \mu_j(x, \xi)$, $|x| + |\xi| \geqslant \eta$, of $k_f(x, \xi)$ extend to symbols in ψs_e, so that the flows v_t^j of the λ_j (i.e., of (3.2) with $k(x, \xi) = \lambda_j(x, \xi)$) leave $\psi\mathfrak{s}_m$ invariant in the sense of Theorem 3.2.*

The proof essentially follows from Proposition 3.1.

We also note that a semistrictly hyperbolic system (3.1) allows the symmetrizer symbol

$$r(x,\xi) = \sum_{j=1}^{\rho} \bar{p}_j^T(x,\xi) p_j(x,\xi) \in \psi c_0, \tag{3.20}$$

where $p_j(x, \xi)$ is the spectral projection matrix onto the eigenspace corresponding to $\lambda_j(x, \xi)$, $|x| + |\xi| \geqslant \eta$. (Indeed, the p_j, hence r, extend to symbols in ψc_0 as shown in [$\mathbf{C_2}$].)

PROPOSITION 3.7. *If k is semistrictly hyperbolic of type f, then the solution operator e^{iKt} of system* (3.1) *exists as an operator in $\mathcal{O}(0)$, and*

$$\partial_t^j(e^{iKt}) = K^j e^{iKt} \in \mathcal{O}(jf). \tag{3.21}$$

This result is discussed in detail in [**CE**] and [$\mathbf{C_2}$].

Note that one cannot expect the algebra of all $(\nu \times \nu)$-matrices

$$(3.22)\qquad A = ((A_{jl})) = ((a_{jl}(x, D))), \qquad a_{jl} \in \psi\sigma_m,$$

to be invariant under conjugation with e^{iKt}. One might expect this to be true, modulo a term of lower order, if only $((a_{jl}(x, \xi)))$ and $k_f(x, \xi)$ commute for large $|x| + |\xi|$. We shall obtain a more precise answer.

For a semistrictly hyperbolic k of type $f = e$, let $\mathcal{P}_m = \mathcal{P}_m^K$, $m \in \mathbb{R}^2$, denote the class of all $A = a(x, D)$ of the form (3.22) such that

$$(3.23)\qquad A_t = e^{-iKt}Ae^{iKt} \in C^\infty(\mathbb{R}, \Psi\mathcal{S}_m).$$

On the other hand, for $f = e^1$ or $f = e^2$ we require, in addition to (3.23), that

$$(3.24)\qquad \partial_t^j A_t \in C^\infty(\mathbb{R}, \Psi\mathcal{S}_{m-j(e-f)}), \qquad j = 1, 2, \ldots.$$

Also, let $\mathcal{P} = \mathcal{P}^K = \bigcup\{\mathcal{P}_m^K\colon m \in \mathbb{R}^2\}$, and let the set of all $(\nu \times \nu)$-matrices of symbols in $\psi\sigma_m$ be denoted by $\psi\sigma_m^\nu$, etc.

The next theorem clarifies the structure of the algebra $\mathcal{P}$.

THEOREM 3.8. *Let the* $(\nu \times \nu)$*-system* (3.1) *be semistrictly hyperbolic of type* f.

(1) *For each* $(\nu \times \nu)$*-matrix-valued symbol* q *with*

$$(3.25)\qquad q \in \psi\sigma_m^\nu, \quad [k_0(x, \xi), q(x, \xi)] = 0, \quad |x| + |\xi| \geq \eta,$$

there exists a 'correction symbol' $z \in \psi\sigma_{m-e}$ *such that* $A = a(x, D)$, *with* $a = q + z$, *is an operator in* $\mathcal{P}_m^K$.

(2) *Vice versa, if* $A = a(x, D) \in \mathcal{P}_m^K$ *is given, then there exists a decomposition of* a_t, $a_t(x, D) = A_t = e^{-iKt}Ae^{iKt}$, *for all* t:

$$(3.26)\quad a_t = q_t + z_t, \quad z_t \in \psi\sigma_{m-e}, \quad [k_0(x, \xi), a_t(x, \xi)] = 0, \quad |x| + |\xi| \geq \eta.$$

Moreover, the decomposition may be differentiated for t *for the corresponding decomposition* $\partial_t^j a_t = \partial_t^j q_t + \partial_t^j z_t$ *of the symbol* $\partial_t^j a_t$ *of* $((\mathrm{ad}_{iK})^j A)_t$. *In particular,* $\partial_t^j z_t \in \psi\sigma_{m-j(e-f)-e}^\nu$.

(3) *If, for any* q *satisfying* (3.25), *the symbols* z_1, z_2 *are both proper correction symbols satisfying* (1), *then* $z_1 - z_2$ *allows a decomposition* (3.26) *for* $m - e$.

The proof of Theorem 3.8 is given in [$\mathbf{C_2}$]. It closely follows the model of Theorem 5.1 in [**CD**]. The principal changes are again indicated by Proposition 2.4 (which gives the calculus of ψdo's), Theorem 3.2, and Proposition 3.6, which insure the usefulness of the characteristic flows corresponding to the eigenvalues λ_j. A recursion will again have to be set up, after the model of [**CD**], where a set of commutator equations $[k_0(x, \xi), z_j(x, \xi)] = f_j(x, \xi)$ must be successively solved. At each step the solution z_{j-1} must be selected among the infinity of solutions such that the jth equation can be solved. The correction symbol z is then the asymptotic sum of the symbols z_j.

In particular we may ask for the 'propagation of the commuting part q_t of a_t'. If all multiplicities ν_j are 1, then, clearly, $q_t = \sum_{j=1}^{\nu} \kappa_{jt} p_j$, with complex-valued κ_{jt}.

Then we can simply set $\kappa_t = q \circ \nu_t^j$ for a valid decomposition (3.26). In the general case

$$(3.27) \qquad q_t = \sum_{j=1}^{\rho} q_{jt}, \qquad p_j q_{jt} = q_{jt} p_j = q_{jt}.$$

Then the components q_{jt} of q_t within the eigenspaces im p_j, while floating along the flow of λ_j, are also 'twisted' within the eigenspaces of $\lambda_j(x, \xi)$ according to some well-defined rules. For details we refer to [**CD**, Theorem 5.1(4)], where a slightly different algebra is discussed (a precise proof for the present algebra is discussed in [$\mathbf{C_2}$]).

Finally, let us mention that the algebra $\mathscr{P}_0$ is a ψ^*-algebra again if we only assume that $K = K^*$ (in addition to semistrict hyperbolicity of K, of course). This may be proven similarly to Theorem 2.1.

4. The C^*-algebra of an elliptic differential expression. In this section we sketch a somewhat older approach, worked out by the author and his associates, E. Herman, R. McOwen, and H. Sohrab [$\mathbf{C_1}$, **H**, $\mathbf{CM_1}$, $\mathbf{CM_2}$, $\mathbf{S_1}$, $\mathbf{S_2}$], which results in a different type of C^*-algebra (or ψ^*-algebra) of (generalized) pseudo-differential operators.

Consider a paracompact differentiable manifold Ω equipped with a positive C^∞-measure $d\mu$. Let $\mathscr{H} = L^2(\Omega, d\mu)$ be the Hilbert space of square-integrable functions over Ω. Also on Ω let there be defined a selfadjoint, positive-definite, second-order, elliptic differential expression H. In local coordinates $x = (x^1, \ldots, x^n)$,

$$(4.1) \qquad d\mu = \kappa\, dx, \quad H = -\kappa^{-1}\partial_{x^j}\kappa h^{jk}\partial_{x^k} + q,$$

with C^∞-functions κ, h^{jk}, q (and summation convention), where (h^{pq}) is the (contravariant) (positive-definite) principal part tensor. We assume that all coefficients are real-valued and, moreover,

$$(4.2) \qquad \int_\Omega d\mu\, \bar{u}Hu = \int_\Omega d\mu\left(h^{jk}\partial_{x^j}\bar{u}\partial_{x^k}u + q|u|^2\right) \geqslant \int_\Omega d\mu\, |u|^2.$$

It is then convenient to introduce a corresponding Riemannian metric on Ω, setting

$$(4.3) \qquad ds^2 = h_{jk}\, dx^j\, dx^k, \qquad ((h_{jk})) = ((h^{jk}))^{-1},$$

so that Ω becomes a Riemannian space. If desired, a transformation of the dependent variable $u = \gamma v$ with a (global) positive C^∞-function c may be carried out, resulting in a new measure $d\mu^\sim = \gamma^2\, d\mu$ and an expression $H^\sim = \gamma^{-1}H\gamma$ such that

$$(4.4) \qquad d\mu^\sim = d\sigma = \text{surface measure of } ds, \quad H = -\Delta + q,$$

with the Beltrami–Laplace operator Δ of ds. Or else, by another such transform, using (4.2), we may achieve $q = 1$ for all $x \in \Omega$, assuming that Ω is noncompact or, at least, that $q = $ const. if Ω is compact.

It is clear that the semibounded selfadjoint expression H admits selfadjoint realizations—i.e., selfadjoint extensions (allowing a spectral decomposition) of the 'minimal operator' H_0, with domain dom $H_0 = C_0^\infty(\Omega)$. The following construction will work only if there is precisely one such realization (that is, if H_0 has a selfadjoint closure). The other case, where boundary conditions are required, has been studied in the special case of a half-space Ω by the author and P. Colella [**CC**, $\mathbf{C_1}$], but it will be omitted here.

In fact, the theory will become more perfect if not only H_0 but also higher powers H_0^k, $k = 1, 2, \ldots$, of H_0 have selfadjoint closures. It was shown by the author [**CWS**], and later by Chernoff [**Ch**], that all powers of H_0^k have selfadjoint closure if the Riemannian space Ω (under the metric (4.3)) is complete. On the other hand, in [**CWS**], we also have results for the case where Ω is the interior of a complete C^∞-Riemannian space $\Omega^{\sim}$ with boundary. In that case we must require H to be singular at the boundary. For example, in the case of a compact $\partial\Omega = \Omega^{\sim} \setminus \Omega$, Theorem 4 of [**CWS**] implies the following: *There exist positive constants* α_k, $k = 1, 2, \ldots$, *such that*

$$q(x) \geq \alpha_k / \mathrm{dist}(x, \partial\Omega) \quad \text{for all } x \text{ sufficiently close to } \partial\Omega, \tag{4.5}$$

implies that, for the expression

$$H = -\Delta + q, \tag{4.6}$$

all powers of H_0 *up to* H_0^k *inclusive are essentially selfadjoint* (*i.e. have a selfadjoint closure*).

We shall speak of condition (s_k) if this last property holds for k, and, specifically, of conditions (s) and (w) when $k = \infty$ and $k = 1$, respectively.

Assuming at least (w), we introduce a certain algebra $\mathscr{A}^\#$ of C^∞-functions over Ω and also a class $\mathscr{D}^\#$ of first-order linear partial differential expressions (folpdes) with C^∞-coefficients over Ω. $\mathscr{A}^\#$ and $\mathscr{D}^\#$ contain all functions (folpdes) with compact support. Let

$$\Lambda = H^{-1/2}, \tag{4.7}$$

which is a well-defined, bounded, selfadjoint operator, and a ψdo of order -1 on Ω, assuming (w). Assuming that all functions $a \in \mathscr{A}^\#$ are bounded and, for all $D \in \mathscr{D}^\#$,

$$D = b^j \partial_{x^j} + p, \qquad (b^j) = O(1),\ p = O(\sqrt{q}), \tag{4.8}$$

where the principal part tensor (b^j) is measured with the metric tensor (h^{jk}), consider the L^2-bounded pseudodifferential operators

$$a(M), \quad D\Lambda, \qquad a \in \mathscr{A}^\#, D \in \mathscr{D}^\#, \tag{4.9}$$

and the norm-closed Banach subalgebra of $\mathscr{L}(\mathscr{H})$ generated by the operators (4.9). Here $a(M)$ denotes the multiplication operator $u(x) \to a(x)u(x)$ (often written briefly as $a(M) = a$). The algebra $\mathscr{A}$ is generated as the norm closure of the finitely generated algebra $\mathscr{A}^0$ in $\mathscr{L}(\mathscr{H})$, $\mathscr{H} = L^2(\mathbb{R}^n)$.

In [**CS**, $\mathbf{CM_1}$, $\mathbf{CM_2}$] further conditions on $\mathscr{A}^\#$ and $\mathscr{D}^\#$ are imposed to insure that the algebra $\mathscr{A}$ (c_1) is adjoint invariant, (c_2) contains the compact ideal $\mathscr{K}(\mathscr{H})$ of $\mathscr{H}$, (c_3) has compact commutator, and (c_4) contains the identity operator of $\mathscr{L}(\mathscr{H})$. For example, the following set of conditions is sufficient (in addition to (w)):

As $\mathscr{A}^\#$ choose the class of all bounded C^∞-functions with

$$\nabla a = o^{\sim}(\sqrt{q}); \tag{4.10}$$

as $\mathscr{D}^\#$ choose the class of C^∞-coefficient folpdes with (4.8) *and*

$$(b^j_{|k}) = o^{\sim}(\beta), \quad \nabla p = o^{\sim}(\sqrt{q}), \quad \nabla(b^j_{|j}) = o^{\sim}(\sqrt{q}), \quad b^j q_{|j} = o(\sqrt{q}), \tag{4.11}$$

with a positive C^∞-function β on Ω satisfying

$$\beta = o^{\sim}(\sqrt{q}), \qquad \nabla\beta = o(\sqrt{q}). \tag{4.12}$$

In (4.11) and (4.12), "$_{|j}$" denotes the covariant derivative with respect to the Riemannian connection of the metric (4.3). Also, "$z = o^{\sim}(\gamma)$" stands for "$z = O(1)$ and $z = o(\gamma)$".

It is clear then that $\mathscr{A}$ has all the properties required for a theory such as that developed in [**CI**] for algebras of singular integral operators over $\mathbb{R}^n$ (cf. also [**C**$_1$, **BC**$_1$, **BC**$_2$]). In particular, we introduce the 'symbol space' $\mathscr{M} = \mathscr{M}(\mathscr{A})$ of $\mathscr{A}$ as the maximal ideal space of the quotient algebra $\mathscr{A}/\mathscr{K}(\mathscr{H})$ (which is a commutative C^*-algebra and, hence, isometrically isomorphic to a function algebra $C(\mathscr{M})$ with a compact Hausdorff space $\mathscr{M}$). Then the symbol σ_A of an operator $A \in \mathscr{A}$ is defined as the continuous complex-valued function over $\mathscr{M}$ corresponding to the coset of A mod $\mathscr{K}(\mathscr{H})$.

THEOREM 4.1. *Let* (c_1)–(c_4) *hold. Then a necessary and sufficient condition for $A \in \mathscr{A}$ to be a Fredholm operator is that $\sigma_A \neq 0$ on all of $\mathscr{M}$. Also, an 'index homomorphism' ι: $\pi^*(\mathscr{M}) \to \mathbb{Z}$ is well defined, giving the Fredholm index* ind A *of $A \in \mathscr{A}$ from the first cohomotopy class $(\sigma_A)^\wedge$ of its symbol*

$$\operatorname{ind} A = \iota((\sigma_A)^\wedge). \tag{4.13}$$

Moreover, for a $(\nu \times \nu)$-matrix $A = ((A_{jk}))$, $A_{jk} \in \mathscr{A}$, with symbol defined as $\sigma_A = ((\sigma_{A_{jk}}))$, the matrix A, considered as an operator $\mathscr{H}^\nu \to \mathscr{H}^\nu$, with $\mathscr{H}^\nu = \mathscr{H} \otimes \mathbb{C}^\nu$ the Hilbert space of all ν-component vectors of $L^2(\Omega, d\mu)$-functions, is Fredholm if and only if the $(\nu \times \nu)$-matrix σ_A is invertible at each point of $\mathscr{M}$. Again, an index homomorphism ι can be defined: The Fredholm index is given by a homomorphism from the group of homotopy classes of maps, $\mathscr{M} \to \mathscr{U}_\nu$, to $\mathbb{Z}$, with the unitary group $\mathscr{U}_\nu$ of dimension ν (cf. [**BC**$_2$]).

A similar set of statements also holds for operators between L^2-spaces of sections of vector bundles over Ω with suitable behaviour at infinity.

For a first-order differential expression $L \in \mathscr{D}^\#$, or a $(\nu \times \nu)$-matrix L of such expressions (or an expression L acting between the sections of a vector bundle F with suitable behaviour at infinity, with symbol a homomorphism of the lifted bundle F^* over $T^*\Omega$:

$$a(x, \xi) = b^j(x)\xi_j + p(x) \in \operatorname{Hom} F^*), \tag{4.14}$$

a necessary and sufficient condition for the Fredholm property and an index formula are immediate: Define the first L^2-Sobolev space $\mathscr{H}_1 \subset \mathscr{H} = \mathscr{H}_0$ as the completion of $C_0^\infty(\Omega)$ under the norm

$$(4.15)\quad \|u\|_1 = \left\{\int_\Omega d\mu\left(h^{jk}\bar{u}_{|x^j}u_{|x^k} + q|u|^2\right)\right\}^{1/2} \geqslant \left\{\int_\Omega d\mu|u|^2\right\}^{1/2} = \|u\|$$

(using (4.2)). For each expression L as above (also for a system or an expression on vector bundles, correspondingly), a differential operator L (a realization) may be defined by setting

$$(4.16)\qquad \operatorname{dom} L = \mathscr{H}_1, \qquad L = L_0^{**}|\mathscr{H}_1.$$

Also, for such an expression the 'algebra symbol' is defined as

$$(4.17)\qquad \sigma_L = \sigma_{L\Lambda}.$$

Then it follows that the operator L is Fredholm if and only if its algebra symbol σ_L is nonzero (or an invertible matrix, or an invertible bundle homomorphism). A corresponding index formula also results.

Note that these abstractly formulated results can be useful only if (i) a sufficient insight into the structure of the compact Hausdorff space $\mathscr{M}$ can be obtained, and (ii) explicit construction of the algebra symbol σ_L, say, from the ψdo symbol of L can be achieved. Also, (iii) it is desirable to obtain the index homomorphism ι explicitly for a given space Ω and dimension ν or vector bundle. Furthermore, (iv) while (i)–(iii) promise to be useful for first-order elliptic systems and their Fredholm theory on compact or noncompact manifolds, it would be interesting to also formulate and prove similar statements for higher elliptic systems, perhaps under additional assumptions.

In [**C**$_1$] we answered these questions for $\Omega = \mathbb{R}^n$, with $H = 1 - \Delta$, $\Delta = \Sigma\partial_{x^j}{}^2$, using various classes $\mathscr{A}^\#$ and $\mathscr{D}^\#$. Generally, the algebras $\mathscr{A}$ considered all contain a dense 'core' of classical ψdo's, such as all operators of the form

$$(4.18)\qquad A = \sum a_\alpha(M)s^\alpha(D),$$

with $s_0(x) = (1 + |x|^2)^{-1/2}$, $s_j(x) = x_j s_0(x)$, and bounded $a_\alpha \in C^\infty(\mathbb{R}^n)$, having derivatives vanishing at $|x| = \infty$. The continuous functions

$$(4.19)\qquad a(x,\xi) = \sum a_\alpha(x)s^\alpha(\xi)$$

define a certain unique compactification $\mathscr{W}^n$ of $\mathbb{R}^{2n} = \mathbb{R}^n \times \mathbb{R}^n$ as the smallest compactification onto which all of these functions can be continuously extended (cf. [**CH**$_1$, **C**$_1$]). ($\mathscr{W}^n$ also may be defined as the maximal ideal space of the commutative function algebra obtained by taking the closure of the algebra of all functions (4.19) under the sup norm.) Then the symbol space $\mathscr{M}$ proves to be the boundary of $\mathbb{R}^{2n}$ in this compactification:

$$(4.20)\qquad \mathscr{M} = \mathscr{W}^n \setminus \mathbb{R}^{2n}.$$

Also, for a ψdo of the form (4.18) the symbol is just the restriction to $\mathscr{M}$ of the (continuous extension to $\mathscr{W}^n$ of the) ψdo symbol $a(x,\xi)$ of A. The index homomorphism ι may be obtained by a topological construction similar to that of Seeley [$\mathbf{S_2}$]. Also, for an elliptic differential expression $L = \sum a_\alpha D^\alpha$ of order N, the ψdo

$$(4.21) \qquad A = L\Lambda^N = L\left(1 + |D|^2\right)^{-N/2} = \sum a_\alpha s_0^{N-|\alpha|}(D) s^\alpha(D)$$

is of the form (4.18) if the coefficients satisfy the conditions mentioned. This gives a simple statement regarding the Fredholm property of the operator L in the Sobolev space $\mathscr{H}_N = \operatorname{dom} L$.

Here we formulate an extension of these results for general manifolds Ω and "comparison operators" H, as discussed in detail in [**CS**, $\mathbf{CM_1}$, $\mathbf{CM_2}$]. In the general case, without adding further assumptions, we do not have a complete description of the symbol space $\mathscr{M}$ as in $\mathbb{R}^n$, although we can show that $\mathscr{M}$ is a compact subset of the corresponding space (often a proper subset).

In more detail, the function class $\mathscr{A}^\#$ defines a unique compactification $\mathscr{M}^0$ of the manifold Ω (again either as the maximal ideal space of the function algebra closure, or else as the smallest compactification onto which all functions in $\mathscr{A}^\#$ may be continuously extended). Moreover, for a $D \in \mathscr{D}^\#$ consider the 'ψdo symbol' $d(x, \xi) = b^j(x)\xi_j + p(x)$, which defines a continuous function over the cotangent space $T^*\Omega$. Note that the class of all functions

$$(4.22) \quad a(x), \quad a \in \mathscr{A}^\#, \qquad a(x, \xi) = \frac{d(x, \xi)}{\left(h^{jk}(x)\xi_j\xi_k + q(x)\right)^{1/2}}, \qquad D \in \mathscr{D}^\#,$$

again defines a compactification $\mathscr{W} = \mathscr{W}(\Omega, H)$ of $T^*\Omega$. Moreover, since the C^*-function algebra generated by $\mathscr{A}^\#$ is contained in the algebra generated by (4.22), the associated dual map of the injection map defines a projection π: $\mathscr{W} \to \mathscr{M}_0$.

In [**CS**, $\mathbf{C_2}$] under the assumption of (4.10)–(4.12), it was proven that (i) *the symbol space $\mathscr{M}$ of $\mathscr{A}$ is a compact subset of $\partial T^*\Omega = \mathscr{W} \setminus T^*\Omega$. Moreover,* (ii) *the 'wave front space' $\mathbf{W} = (\pi^{-1}\Omega) \cap \partial T^*\Omega$ is entirely contained in $\mathscr{M}$. Also,* (iii) *the space $\mathbf{W}$ is homeomorphic to the bundle $S^*\Omega$ of unit spheres*

$$(4.23) \qquad h^{jk}(x)\xi_j\xi_k = 1$$

over Ω. Under this homeomorphism the algebra symbol of an operator $D \in \mathscr{D}^\#$ is given as the restriction of the principal symbol $b^j(x)\xi_j$ to $S^*\Omega$ on the subset $\mathbf{W} \subset \mathscr{M}$. On the other hand, on the entire space $\mathscr{M}$ the symbol $\sigma_{D\Lambda}$ equals the function $a(x, \xi)$ of (4.22).

This was first proven in [**CS**] under slightly stronger assumptions (but the proof will apply to the general case). A simpler proof was given by McOwen [$\mathbf{CM_1}$] in the special case $H = 1 - \Delta$ on a general manifold Ω. Again this proof also works in the general case.

Now the more precise control of the 'secondary symbol space' $\mathscr{M}_s = \mathscr{M} \setminus \mathbf{W}^{\text{clos}}$ is important, because it may be seen that for an elliptic system the values of the symbol on $\mathscr{M}_s$ determine the essential spectrum of the operator [$\mathbf{C_1}$]. Simple examples, like the torus or the infinite cylinder, show that $\mathcal{M}_s$ may be a proper subset of $\partial T^*\Omega \setminus \mathbf{W}^{\text{clos}}$, or it may even be empty.

While the case of $\mathbb{R}^n$ with $H = 1 - \Delta$, Δ = ordinary Laplacian, was discussed in [$\mathbf{C_1}$], H. Sohrab [$\mathbf{S_1}$] has completely discussed the case of $\mathbb{R}^n$ with $H = -\Delta + |x|^2$ (i.e., the case of the harmonic oscillator in quantum mechanics). In [$\mathbf{S_2}$] and [$\mathbf{S_3}$] Sohrab has generalized his results to certain other singular Sturm–Liouville operators on $\mathbb{R}^1$ and a class of rotationally symmetric operators on $\mathbb{R}^n$ containing the anharmonic oscillator.

Finally, an answer to (iv) will be given in §5 in combination with the construction of higher-order Sobolev spaces and a ψ^*-algebra under stronger assumptions on the expression H.

5. The ψ^*-algebra of a singular elliptic expression. Let us now require condition (s) of §4 to hold for the expression H of (4.1). Then we may define Sobolev spaces $\mathcal{H}_s$ of arbitrary real order s by setting

$$\mathcal{H}_s = \text{dom}\, \Lambda^s = \text{dom}\, H^{s/2}, \quad \|u\|_s = \|\Lambda^{-s}u\|, \qquad s > 0, \tag{5.1}$$

with the unique selfadjoint realization of H (denoted H again), which we know to be positive (by (4.2) and, since $H = H_0^{**}$, by (s)), so that the powers $H^{s/2}$ are well defined. In fact, since all operators H_0^{m**} are selfadjoint (and $H_0^{m**} = H^m$) for $m = 1, 2, \ldots$, it follows easily that $\mathcal{H}_s$ is the completion of $C_0^\infty(\Omega)$ under the norm $\|\cdot\|_s$; i.e., C_0^∞ is dense in $\mathcal{H}_s$ under its norm. On the other hand, for $s < 0$ we define the norm $\|\cdot\|_s$ as in (5.1) and then define $\mathcal{H}_s$ as the completion of $\mathcal{H}$ (or C_0^∞) under that norm.

We will now choose a class of bounded $C^\infty(\Omega)$-functions and a class of first-order, linear, differential operators on Ω satisfying (4.8) and again obtain a finitely generated algebra $\mathcal{A}^0 \subset \mathcal{L}(\mathcal{H}_0)$. However, instead of (4.10)–(4.12) we will now impose a stronger set of conditions, which, in addition to (c_1)–(c_4) for the algebra $\mathcal{A}$ generated by (4.9), not only implies that $L\Lambda^N \in \mathcal{A}$, for a suitable class of Nth order differential operators L, but also gives that

$$A - \Lambda^{-s}A\Lambda^s \in \mathcal{K}(\mathcal{H}) \quad \text{as } A \in \mathcal{A}^0, \qquad s \in \mathbb{R}. \tag{5.2}$$

Note that

$$\Lambda^s : \mathcal{H}_s \to \mathcal{H}, \qquad s > 0, \tag{5.3}$$

defines an isometry between $\mathcal{H}_s$ and $\mathcal{H}_0$, and that, moreover, for each $s, t \in \mathbb{R}$, Λ^s extends to an isometry $\mathcal{H}_t \leftrightarrow \mathcal{H}_{t-s}$. From (5.2) one first concludes that

$$A - \Lambda^{-s}A\Lambda^s \in \bigcap \{\mathcal{K}(\mathcal{H}_t) : t \in \mathbb{R}\}, \qquad s \in \mathbb{R}, A \in \mathcal{A}^0, \tag{5.4}$$

where the intersection stands for all operators $C_0^\infty \to \bigcap \mathcal{H}_s$ extending to operators in $\mathcal{K}(\mathcal{H}_t)$ for all t. These reflections lead to the following abstract result.

THEOREM 5.1. *Assume that* (c_1)–(c_4), *and* (5.2) (*and* (s)) *hold. Then*

$$\mathcal{A}^0 \subset \mathcal{L}(\mathcal{H}_s), \qquad s \in \mathbb{R}. \tag{5.5}$$

(*That is, more precisely, for every $A \in \mathcal{A}^0$ the restriction $\mathcal{A}|C_0^\infty$ extends continuously to an operator in $\mathcal{L}(\mathcal{H}_s)$.*) *Moreover, the norm closure $\mathcal{A}_s$ of $\mathcal{A}^0$ in $\mathcal{L}(\mathcal{H}_s)$ is a*

C^*-subalgebra of $\mathscr{L}(\mathscr{H}_s)$ containing $\mathscr{K}(\mathscr{H}_s)$ and having commutator in $\mathscr{K}(\mathscr{H}_s)$. Also, the symbol space of $\mathscr{A}_s$ is (homeomorphic to) $\mathscr{M}$, the symbol space of $\mathscr{A} = \mathscr{A}_0$, and (under that homeomorphism) the symbol σ_A of an operator $A \in \mathscr{A}^0 \subset A_s$ is independent of s for $s \subset \mathbb{R}$.

Note that similarly, as discussed in [**C$_1$**], for $\mathbb{R}^n$ the topology generated on $\mathscr{A}^0$ by all operator norms

$$(5.6) \qquad \left\{ \|A\|_s = \sup\{ \|Au\|_s\colon \|u\|_s = 1\} : s \in \mathbb{R} \right\}$$

is again a Fréchet topology under the conditions of Theorem 5.1. Indeed, the proof of [**C$_1$**, Theorem III, 3.4], is easily amended to this case: Just replace the holomorphic function $f(z)$ of [**C$_1$**, III, (3.21)], by

$$(5.7) \qquad f(z) = (A\Lambda^{s(\bar{z})}u, \Lambda^{-s(z)}v), \qquad u, v \in C_0^\infty(\Omega), u, v \text{ fixed}.$$

Using (c_1)–(c_4) and (s), we again find that the set of norms (5.6) is equivalent to the set

$$(5.8) \qquad \left\{ \|\cdot\|_j : j \in \mathbb{Z} \right\}.$$

We again define $\mathscr{A}_\infty$ as the closure of $\mathscr{A}^0$ under this Fréchet topology. (For compact manifolds see Seeley [**Se$_1$**] or Calderón [**Ca**].)

The next theorem holds under the same abstract assumptions as Theorem 5.1.

THEOREM 5.2. *The algebra $\mathscr{A}_\infty$ is a ψ^*-subalgebra of every $\mathscr{L}(\mathscr{H}_s)$, $s \in \mathbb{R}$. Moreover, for $A \in \mathscr{A}_\infty \subset \mathscr{A}_s$ the symbol σ_A is independent of s. Thus A is Fredholm in $\mathscr{L}(\mathscr{H}_s)$ if and only if it is Fredholm in every $\mathscr{L}(\mathscr{H}_t)$, $t \in \mathbb{R}$. Also, the Fredholm index as well as nullity and deficiency of (the continuous extension to $\mathscr{H}_s$ of) A are independent of s.*

Again the proof is identical to the proof of [**C$_1$**, IV, Theorem 4.1].

In [**C$_1$**] we investigated the question of *surjectivity of the symbol map* σ: $\mathscr{A}_\infty \to C(\mathscr{M})$. While the map $\sigma\colon \mathscr{A}_s \to C(\mathscr{M})$ is surjective for every finite s, as an immediate consequence of the Gelfand homomorphism's surjectivity for C^*-algebras, this question seems open for the ψ^*-algebra $\mathscr{A}_\infty$. For the special case of $\Omega = \mathbb{R}^n$, $H = 1 - \Delta$, $\Delta = \sum \partial_{x_j}^2$, we have shown in [**C$_1$**, IV] that all $a \in C(\mathscr{M})$ with all x-derivatives existing in sup-norm convergence over the wave front space **W** (but under no further condition on the secondary symbol space) are symbols of some operator in $\mathscr{A}_\infty$. The condition appears too strong: We believe that it suffices to require at each $(x, \xi) \in \mathbf{W}$ only the existence of the directional derivatives in the directions of the characteristic flow of the hyperbolic pseudodifferential equation $\partial_t u = i\Lambda^{-1}u$, $\Lambda = (1 - \Delta)^{-1/2}$. In fact, the main ingredient of the proof of Theorem 0.1 of [**CW**] is just the existence of all commutators $(\mathrm{ad}_{\Lambda^{-1}})^j A$, $j = 1, 2, \ldots$, in $\mathscr{H}$. That, on the other hand, can be reduced to Theorem 3.3 (added in proof; cf. [**CSch**]).

Even this condition appears much too strong.

In order to make the above abstract consideration useful, we now must specify explicit conditions on the classes $\mathscr{A}^\#$ and $\mathscr{D}^\#$ that imply (5.2) (apart from (c_1)–(c_4), of course).

In **[CS]** we have worked with the classes $\mathscr{A}^\infty$ and $\mathscr{D}^\infty$, defined as follows.

(e_1) *$\mathscr{A}^\infty$ is the algebra of all bounded $C^\infty(\Omega)$-functions such that all covariant derivatives (of arbitrary order) are $o(\sqrt{q})$ as $x \to \infty$ (which gives $\mathscr{A}^\infty = C^\infty(\Omega) = C_0^\infty(\Omega)$ if Ω is compact).*

(e_2) *$\mathscr{D}^\infty$ is the Lie algebra of all folpdes satisfying* (3.8) *such that the covariant derivatives of the tensor (b^j) and the function p are $o(1)$ and $o(\sqrt{q})$, respectively, as $x \to \infty$.*

Here the covariant derivatives are taken with respect to the Riemannian connection of the metric h_{jk}.

With these classes $\mathscr{A}^\infty$ and $\mathscr{D}^\infty$ we must also impose the following additional conditions in order to satisfy (c_1)–(c_4) and (5.2).

(e_3) The differential expression H allows a representation

$$H = \sum a_j D_j F_j + \sum b_j G_j + c, \qquad a_j, b_j, c \in \mathscr{A}^\infty, \; D_j, F_j, G_j \in \mathscr{D}^\infty. \tag{5.9}$$

(e_4) The classes $\mathscr{A}^\infty$ and $\mathscr{D}^\infty$ are adjoint invariant.

(Note that (e_4) is satisfied, for example, if $q = 1$.)

Unfortunately, (e_3) is quite restrictive. It implies that the Riemann curvature tensor is $o(1)$ as $x \to \infty$ together with all its covariant derivatives (cf. **[CM$_1$]**). On the other hand, if we replace $\mathscr{D}^\infty$ by the subset $\mathscr{D}^0$ of all $D \in \mathscr{D}^\infty$ with $(b^j) = o(1)$ and $p = o(\sqrt{q})$, then (e_3) is not required.

It is possible to choose covariant derivatives with respect to a slightly different affine connection. However, it may differ from the Riemannian connection at most by a bounded tensor (cf. **[CM$_1$]**).

In **[CM$_1$]** and **[CM$_2$]** the case of $q = 1$ was worked out in more detail and proofs were simplified. In **[M$_1$]**, again for the case of $q = 1$, McOwen works with slightly modified conditions, requiring only boundedness of the Riemann tensor R_{ijkl}. However, since the second covariant derivatives of the (b^j) are required to vanish at ∞, it still follows that $R_{ijkl} b^i = o(1)$ as $x \to \infty$ (from the commutator rules for covariant differentiation). Thus the tensor R_{ijkl} must be $o(1)$ in a subspace, or else we get $\mathscr{D}^\infty = \mathscr{D}^0$.

Let $\mathscr{L}_N$ denote the class of all partial differential expressions that are finite, linear combinations of products of at most N folpdes in $\mathscr{D}^\infty$ with coefficients in $\mathscr{A}^\infty$.

THEOREM 5.3. *For $L \in \mathscr{L}_N$ we have $A = L\Lambda^{-N} \in \mathscr{A}_\infty$. Thus, for $L \in \mathscr{L}_N$, the differential operator L with domain* dom $L = \mathscr{H}_N$ *is a closed unbounded Fredholm operator if and only if $\sigma_A \neq 0$ on $\mathscr{M}$. Also, we again get $L \in \mathscr{L}(\mathscr{H}_s, \mathscr{H}_{s-N})$ for all $s \in \mathbb{R}$, and this operator is Fredholm if and only if $\sigma_A \neq 0$ on $\mathscr{M}$.*

Corresponding statements hold for systems of differential operators, etc. We also again obtain a formula for the Fredholm index of L.

REFERENCES

[AS] M. Atiah and I. Singer, *The index of elliptic operators*, Ann. of Math. (2) **66** (1968), 484–530.

[B_1] R. Beals, *A general calculus of pseudo-differential operators*, Duke Math. J. **42** (1975), 1–42.

[B_2] ______, *Characterization of pseudodifferential operators and applications*, Duke Math. J. **44** (1977), 45–57; correction, Duke Math. J. **46** (1979), 215.

[B_3] ______, *On the boundedness of pseudodifferential operators*, Comm. Partial Differential Equations **2** (10) (1977), 1063–1070.

[BC_1] M. Breuer and H. O. Cordes, *On Banach algebras with σ-symbol*. I, J. Math. Mech. **13** (1964), 313–324.

[BC_2] ______, *On Banach algebras with σ-symbol*. II, J. Math. Mech. **14** (1965), 299–314.

[Ca] A. Calderón, *Intermediate spaces and interpolation, the complex method*, Studia Math. **24** (1964), 113–190.

[Ch] P. Chernoff, *Essential selfadjointness of powers of generators of hyperbolic equations*, J. Funct. Anal. **12** (1973), 402–414.

[CC] P. Colella and H. O. Cordes, *The C^*-algebra of the elliptic boundary problem*, Rocky Mountain J. Math. **10** (1980), 217–238.

[C_1] H. O. Cordes, *Elliptic pseudo-differential operators, an abstract theory*, Lecture Notes in Math., vol. 756, Springer-Verlag, 1979.

[C_2] ______, *Techniques in pseudodifferential operators* (to appear).

[CD] ______, *A pseudo-algebra of observables for the Dirac equation*, Manuscripta Math. **45** (1983), 77–105.

[CE] ______, *A version of Egorov's theorem for systems of hyperbolic pseudo-differential equations*, J. Funct. Anal. **48** (1982), 285–300.

[CF] ______, *A pseudodifferential Foldy–Wouthuysen transform*, Comm. Partial Differential Equations **8** (13) (1983), 1475–1485.

[CG] ______, *On geometrical optics*, Lecture notes, Berkeley, 1982.

[CI] ______, *The algebra of singular integral operators in $\mathbb{R}^n$*, J. Math. Mech. **14** (1965), 1007–1032.

[CL] ______, *On pseudodifferential operators and smoothness of special Lie-group representations*, Manuscripta Math. **28** (1979), 51–69.

[CP] ______, *A global parametrix for pseudodifferential operators over $\mathbb{R}^n$*, Preprint SFB 72, Bonn, 1976 (available as preprint from the author).

[CS] ______, *Banach algebras, singular integral operators and partial differential equations*, Lecture notes, Lund, 1971 (available from the author).

[CT] ______, *An algebra of singular integral operators with two symbol homomorphisms*, Bull. Amer. Math. Soc. **75** (1969), 37–42.

[CWS] ______, *Selfadjointness of powers of elliptic operators on noncompact manifolds*, Math. Ann. **195** (1972), 257–272.

[CEr] H. O. Cordes and A. Erkip, *The N-th order elliptic boundary problem for non-compact boundaries*, Rocky Mountain J. Math. **10** (1980), 7–24.

[CH_1] H. O. Cordes and E. Herman, *Gelfand theory of pseudo-differential operators*, Amer. J. Math. **90** (1968), 681–717.

[CH_2] ______, *Singular integral operators on a half-line*, Proc. Nat. Acad. Sci. U.S.A. **56** (1966), 1668–1673.

[CM_2] H. O. Cordes and R. McOwen, *The C^*-algebra of a singular elliptic problem on a non-compact Riemannian manifold*, Math. Z. **153** (1977), 101–116.

[CM_2] ______, *Remarks on singular elliptic theory for complete Riemannian manifolds*, Pacific J. Math. **70** (1977), 133–141.

[CSch] H. O. Cordes and E. Schrohe, *On the symbol homomorphism of a certain Fréchet algebra of singular integral generators* (to appear).

[CW] H. O. Cordes and D. Williams, *An algebra of pseudo-differential operators with non-smooth symbol*, Pacific J. Math. **78** (1978), 279–290.

[D] J. Dunau, *Fonctions d'un operateur elliptique sur une varietée compacte*, J. Math. Pures Appl. **56** (1977), 367–391.

[F] J. Frehse, *Essential selfadjointness of singular elliptic operators*, Bol. Soc. Brasil Mat. **8** (2) (1977), 87–107.

[G] B. Gramsch, *Relative Inversion in der Stoerungstheorie von Operatoren und ψ-Algebren* (to appear).

[GK] I. Gohberg and N. Krupnik, *Einfuehrung in die Theorie der eindimensionalen singulaeren Integrale*, Birkhauser, Basel, Boston, 1979.

[GLS] A. Grossman, G. Loupias and E. Stein, *An algebra of pseudodifferential operators and quantum mechanics in phase space*, Ann. Inst. Fourier (Grenoble) **18** (1968), 343–368.

[H] E. Herman, *The symbol of the algebra of singular integral operators*, J. Math. Mech. **15** (1966), 147–156.

[Ho] L. Hörmander, *Pseudo-differential operators and hypoelliptic equations*, Proc. Sympos. Pure Appl. Math., Vol. 10, 1966, pp. 138–183.

[I] R. Illner, *On algebras of pseudodifferential operators in $L^p(\mathbb{R}^n)$*, Comm. Partial Differential Equations **2** (1977), 133–141.

[K] J. L. Kelley, *General topology*, Van Nostrand, Princeton, 1955.

[Ku] H. Komano-go, *Pseudodifferential operators*, MIT Press, Boston, 1982.

[LM] R. Lockhart and R. McOwen, *Elliptic differential operators on noncompact manifolds* (to appear).

[M_1] R. McOwen, *Fredholm theory of partial differential equations on complete Riemannian manifolds*, Pacific J. Math. **87** (1980), 169–185.

[M_2] ______, *On elliptic operators in $\mathbb{R}^n$*, Comm. Partial Differential Equations **5** (9) (1980), 913–933.

[MM] R. Melrose and G. Mendoza, *Elliptic boundary problems on spaces with conic points* (to appear).

[P] S. C. Power, *Fredholm theory of piecewise continuous Fourier integral operators on Hilbert space*, J. Operator Theory **7** (1982), 52–60.

[Se_1] R. T. Seeley, *Topics in pseudo-differential operators*, C.I.M.E. Conf. (Stresa, 1968).

[Se_2] ______, *Integro-differential operators on vector bundles*, Trans. Amer. Math. Soc. **117** (1965), 167–204.

[S_1] H. Sohrab, *The C*-algebra of the n-dimensional harmonic oscillator*, Manuscripta Math. **34** (1981), 45–70.

[S_2] ______, *C*-algebras of singular integral operators on the line related to singular Sturm–Liouville problems*, Manuscripta Math. **34** (1981), 45–70.

[S_3] ______, *Pseudodifferential C*-algebras related to Schroedinger operators with radially symmetric potentials* (to appear).

[So] A. Sommerfeld, *Atombau und Spektrallinien*, Vieweg, Braunschweig, 1957.

[Ta] M. Taylor, *Pseudodifferential operators*, Princeton Univ. Press, Princeton, 1981.

[T] F. Treves, *Introduction to pseudodifferential operators and Fourier integral operators*, Plenum, New York, 1980.

[W] A. Weinstein, *A symbol class for some Schroedinger equations on $\mathbb{R}^n$*, Amer. J. Math. **107** (1985), 1–21.

[Z] S. Zelditch, *Reconstruction of singularities for solutions of Schroedinger equations*, Ph.D. Thesis, Univ. of Calif., Berkeley, 1981.

UNIVERSITY OF CALIFORNIA, BERKELEY

Proceedings of Symposia in Pure Mathematics
Volume **43** (1985)

Boundary-Value Problems for Second-Order Elliptic Equations in Domains with Corners

G. ESKIN

1. Introduction. Let Ω be a bounded domain in $\mathbf{R}^n$ $(n \geqslant 3)$ with boundary $\Sigma = \partial\Omega$ such that $\Sigma = \overline{\Sigma}_1 \cup \overline{\Sigma}_2$, $\overline{\Sigma}_1 \cap \overline{\Sigma}_2 = \Sigma_0$, where Σ_1 and Σ_2 are smooth $(n-1)$-dimensional surfaces in $\mathbf{R}^n$ having a common smooth $(n-2)$-dimensional boundary Σ_0. Here $\overline{\Sigma}_k$ means the closure of Σ_k, $k = 1, 2$. Consider in Ω a second-order elliptic equation

$$A(x, D)u(x) = f(x), \qquad x \in \Omega, \tag{1.1}$$

with boundary conditions

$$B_1(x, D)u(x)|_{\Sigma_1} = g_1(x'), \qquad x' \in \Sigma_1, \tag{1.2}$$

$$B_2(x, D)u(x)|_{\Sigma_2} = g_2(x'), \qquad x' \in \Sigma_2, \tag{1.3}$$

where $B_k(x, D)$ are differential operators of degree m_k, $k = 1, 2$.

It is natural to assume that $B_k(x, D)$ satisfies the Shapiro–Lopatinskii condition on $\overline{\Sigma}_k$, $k = 1, 2$ (see (3.23), (3.23′)). Our main goal will be to show that in the case of the second-order elliptic equation, one can choose functional spaces for $u(x)$ and the right sides of (1.1)–(1.3) such that the boundary-value problem (1.1)–(1.3) will be Fredholm (or normally solvable).

We shall use Sobolev spaces with weights as in §23 of [**1**]. These spaces are different from weighted Sobolev spaces usually used in such problems (see, for example, [**5**]). The boundary-value problem (1.1)–(1.3) was studied in many papers (see, for example, the recent survey [**5**]). The main results for $n \geqslant 3$ belong to Mazya and Plamenevskii [**6**], Komech [**4**], and Reisman [**7**]. Our result is more explicit and complete, and our approach may be used to study initial-boundary value problems for hyperbolic equations in domains with corners. We shall consider these problems in another paper. As usual we start with the study of a model problem.

1980 *Mathematics Subject Classification*. Primary 35J25.

2. Model problem in the wedge. Let G be the wedge $y_2 > 0$, $y_2 = x_1 \sin\alpha - x_2\cos\alpha$, $x_2 > 0$, $x'' = (x_3, \ldots, x_n) \in \mathbf{R}^{n-2}$. We consider at first the case when the angle α satisfies $0 < \alpha \leqslant \pi$. The case $\pi < \alpha < 2\pi$ will be treated in Remark 3.2. Denote by $H_{s,N}(\mathbf{R}^n)$ the space of distribution in $\mathbf{R}^n$ with finite norm

$$\|[f]\|_{s,N}^2 = \sum_{k_1+k_2=0} \|x_1^{k_1}x_2^{k_2}f\|_{s+k_1+k_2}^2, \tag{2.1}$$

where $\|h\|_s$ is a Sobolev norm in $\mathbf{R}^n$, $s \in \mathbf{R}$. Let p be the restriction operator to G. Set $H_{s,N}(G) = pH_{s,N}(\mathbf{R}^n)$ with norm

$$|[u]|_{s,N} = \inf_{lu} \|[lu]\|_{s,N}, \tag{2.2}$$

where lu is an arbitrary extension of lu to $\mathbf{R}^n$. Denote by Γ_1 and Γ_2 the planes $x_2 = 0$ and $y_2 = x_1 \sin\alpha - x_2 \cos\alpha = 0$, respectively, and by $\Gamma_1^+ \subset \Gamma_1 (\Gamma_1^- \subset \Gamma_1)$ the half-plane $x_2 = 0, x_1 > 0$ $(x_2 = 0, x_1 < 0)$. Also denote by $\Gamma_2^+ \subset \Gamma_2 (\Gamma_2^- \subset \Gamma_2)$ the half-plane $y_2 = 0, y_1 > 0$ $(y_2 = 0, y_1 < 0)$, where

$$y_1 = -x_1\cos\alpha - x_2 \sin\alpha, \qquad y_2 = x_1 \sin\alpha - x_2\cos\alpha.$$

Let $p_1^\pm, p_2^\pm$ be the restriction operators to $\Gamma_1^\pm, \Gamma_2^\pm$. Denote by $H_{s,N}(\Gamma_1)$ the space of distributions on Γ_1 with finite norm

$$[g]_{s,N}^2 = \sum_{k=0}^{N} \|x_1^k g(x_1, x'')\|_{s+k}^2, \tag{2.3}$$

and let $H_{s,N}(\Gamma_1^+) = p_1^+ H_{s,N}(\Gamma_1)$. The norm in $H_{s,N}(\Gamma_1^+)$ is defined analogously to (2.2). Similarly, denote by $H_{s,N}(\Gamma_2)$ the space of distributions on Γ_2 with norm of the form (2.3), and set $H_{s,N}(\Gamma_2^-) = p_2^- H_{s,N}(\Gamma_2)$.

The following lemma was proved in [**1**] (see Lemma 24.5 in [**1**]).

LEMMA 2.1. *The following estimate holds for any* $u \in H_{s,N}(\mathbf{R}^n)$:

$$[p_1^+u]_{s-1/2,N} \leqslant C|[u]|_{s,N}, \qquad [p_2^-u]_{s-1/2,N} \leqslant C|[u]|_{s,N}, \tag{2.4}$$

assuming that $s + N > 1/2$ *and* s *is noninteger if* $s \leqslant 1/2$.

Consider the following, model boundary-value problem in G:

$$(-\Delta + 1)u = f(x), \qquad x \in G, \tag{2.5}$$

$$p_1^+\hat{B}_1(D)u = g_1(x'), \qquad x' \in \Gamma_1^+, \tag{2.6}$$

$$p_2^-\hat{B}_2(D)u = g_2(x'), \qquad x' \in \Gamma_2^-, \tag{2.7}$$

where

$\Delta = \Sigma_{k=1}^n \partial^2/\partial x_k^2$,

$D = (i\,\partial/\partial x_1, \ldots, i\,\partial/\partial x_n)$,

$B_k(\xi)$ are homogeneous polynomials of degree m_k, $k = 1, 2$,

$\hat{B}_k(\xi_1, \xi_2, \xi'') = B_k(\xi_1, \xi_2, (1 + |\xi''|^2)^{1/2}\omega)$,

$\omega = \xi''/|\xi''|$.

Here

$f \in H_{s-2,N}(G)$,

$g_1 \in H_{s-m_1-1/2,N}(\Gamma_1^+)$,

$g_2 \in H_{s-m_2-1/2,N}(\Gamma_2^-)$,

$s + N > \max(m_1, m_2) + 1/2$,

s is noninteger if $s \leqslant \max(m_1, m_2) + 1/2$,

and we are looking for a solution $u \in H_{s,N}(G)$. Let $u_0 = (-\Delta + 1)^{-1} lf$, where lf is an arbitrary extension of $f(x)$ such that $lf \in H_{s-2,N}(\mathbf{R}^n)$. Note that $u_0 \in H_{s,N}(\mathbf{R}^n)$ (see, for example, Lemma 24.3 in [**1**]). The boundary-value problem (2.5)–(2.7) can be reduced to

$$(2.5') \qquad (-\Delta + 1)v = 0, \qquad x \in G,$$

$$(2.6') \qquad p_1^+ \hat{B}_1(D)v = h_1(x'), \qquad x' \in \Gamma_1^+,$$

$$(2.7') \qquad p_2^- \hat{B}_2(D)v = h_2(x'), \qquad x' \in \Gamma_2^-,$$

where $v = u - u_0$, $v \in H_{s,N}(G)$, $h_k(x') = g_k(x') - p_k^{\pm} \hat{B}_k(D) u_0$.

It follows from Lemma 2.1 that $h_1(x') \in H_{s-m_1-1/2,N}(\Gamma_1^+)$, $h_2(x') \in H_{s-m_2-1/2,N}(\Gamma_2^-)$. Denote by $(\Lambda_1^-)^s$ the pseudodifferential operator (ψdo) with symbol

$$(\Lambda_1^-)^s = \left(\xi_1 - i|\hat{\xi}''|\right)^s, \qquad \xi'' = (\xi_3, \ldots, \xi_n),\ |\hat{\xi}''| = \left(1 + |\xi''|^2\right)^{1/2},$$

$$\left(\xi_1 - i|\hat{\xi}''|\right)^s = \exp s \ln\left(\xi_1 - i|\hat{\xi}''|\right),$$

and we consider the branch of ln z that is real for real positive z. We have

$$(2.8) \quad (\Lambda_1^-)^s v = \frac{1}{(2\pi)^n} \int_{\mathbf{R}^n} \left(\xi_1 - i|\hat{\xi}''|\right)^s e^{-ix_1\xi_1 - i(x'',\xi'')} \tilde{v}(\xi_1, \xi'')\, d\xi_1\, d\xi''.$$

Denote the kernel of the ψdo $(\Lambda_1^-)^s$ by $\lambda_{1,s}^-$; i.e. $a\lambda_{1,s}^- = F^{-1}(\xi_1 - i|\hat{\xi}''|)^s$, where F^{-1} is the inverse Fourier transform of the distribution a. We have (see, for example, [**1**, §2])

$$\lambda_{1,s}^- = C\delta(x_2) x_{1,-}^{-s-1} F_{\xi''}^{-1} e^{x_1|\hat{\xi}''|},$$

where C is a constant and $x_{1,-}^{-s-1} = 0$ for $x_1 > 0$. Note that $(\Lambda_1^-)^s$ commutes with the restriction operator p. Indeed, if $v_- = 0$ in G, then $(\Lambda_1^-)^s v_- = \lambda_{1,s}^- * v_- = 0$ in G.

LEMMA 2.2. *The operator* $(\Lambda_1^-)^{s-1/2}$ *gives an isomorphism between* $H_{s,N}(G) \cap$ ker$(-\Delta + 1)$ *and* $H_{1/2,N}(G) \cap$ ker$(-\Delta + 1)$, *where* ker$(-\Delta + 1)$ *means the space of distribution solutions of* $(-\Delta + 1)u = 0$ *in* G.

PROOF. Since $(\Lambda_1^-)^s$ commutes with p,

$$(2.9) \qquad v_0 = p(\Lambda_1^-)^{s-1/2} lv$$

will be a solution of $(-\Delta + 1)u = 0$ in G, where $v \in H_{s,N}(G)$, $(-\Delta + 1)v = 0$ in G, and lv is an arbitrary extension of v belonging to $H_{s,N}(\mathbf{R}^n)$. If $s - 1/2 > 0$, then it is obvious that $v_0 \in H_{1/2,N}(G)$. Consider the case when $s - 1/2 < 0$. It follows from equation $(-\Delta + 1)v_0 = 0$ in G that

$$\partial^2 v_0 / \partial x_2^2 \in p(\Lambda_1^-)^{s-1/2+2} H_{s,N}(\mathbf{R}^n).$$

Analogously,

$$\partial^{2k} v_0/\partial x_2^{2k} \in p(\Lambda_1^-)^{s-1/2+2k} H_{s,N}(\mathbf{R}^n) \quad \text{for any } k > 0.$$

Take $k > 1/2|s - 1/2|$. Then $\partial^{2k} v_0/\partial x_2^{2k} \in H_{1/2-2k,N}(G)$. Since also $\partial^{2k} v_0/\partial x_1^{2k} \in H_{1/2-2k,N}(G)$ and $\partial^{2k} v_0/\partial x_j^{2k} \in H_{1/2-2k,N}(G)$, $3 \leqslant j \leqslant n$, we obtain that $v_0 \in H_{1/2,N}(G)$ (see **[8]**). Denote by $G_1 v_1$ the solution of the Dirichlet problem in the half-space $x_2 > 0$:

$$(-\Delta + 1) G_1 v_1 = 0, \qquad x_2 > 0, \tag{2.10}$$

$$G_1 v_1|_{x_2=0} = v_1(x_1, x''). \tag{2.11}$$

Taking the Fourier transform in x_1 and x'', we obtain

$$G_1 v_1 = \frac{1}{(2\pi)^{n-1}} \int_{\mathbf{R}^{n-1}} \exp\left\{-x_2\sqrt{\xi_1^2 + |\hat{\xi}''|^2} - ix_1\xi_1 - i(x'', \xi'')\right\} \cdot \tilde{v}_1(\xi_1, \xi'')\, d\xi_1\, d\xi'', \tag{2.12}$$

where

$$\tilde{v}_1(\xi_1, \xi'') = \int_{\mathbf{R}^{n-1}} v_1(x_1, x'') e^{ix_1\xi_1 + i(x'', \xi'')}\, d\xi_1\, d\xi''.$$

Analogously, the solution of the Dirichlet problem in the half-space $y_2 = x_1 \sin\alpha - x_2 \cos\alpha > 0$,

$$(-\Delta + 1) G_2 v_2 = 0, \qquad y_2 = x_1 \sin\alpha - x_2\cos\alpha > 0, \tag{2.13}$$

$$G_2 v_2|_{y_2=0} = v_2(y_1, x''), \qquad y_1 = -(\cos\alpha x_1 + \sin\alpha x_2), \tag{2.13'}$$

has the form

$$G_2 v_2 = \frac{1}{(2\pi)^{n-1}} \int_{\mathbf{R}^{n-1}} \exp\left\{-(x_1 \sin\alpha - x_2\cos\alpha)\sqrt{\xi_2^2 + |\hat{\xi}|^2} + i(\cos\alpha x_1 + \sin\alpha x_2)\xi_2 - i(x'', \xi'')\right\} \tilde{v}_2(\xi_2, \xi'')\, d\xi_2\, d\xi''. \tag{2.14}$$

LEMMA 2.3. *Any $v_0 \in H_{1/2,N}(G)$, such that $(-\Delta + 1)v_0 = 0$ in G, can be uniquely represented in the form*

$$v_0 = G_1 v_1^+ + G_2 v_2^-, \tag{2.15}$$

where $v_1^+ = H_{0,N}(\Gamma_1)$, $v_1^+ = 0$ for $x_1 < 0$, $v_2^- \in H_{0,N}(\Gamma_2)$, $v_2^-(y_1, x'') = 0$ for $y_1 = (\cos ax_1 + \sin ax_2) > 0$, $N > 0$.

PROOF. Let $f_1 = p_1^+ v_0$, $f_2 = p_2^- v_0$. It follows from Lemma 2.1 that $f_1 \in H_{0,N}(\Gamma_1^+)$, $f_2 \in H_{0,N}(\Gamma_1^-)$, $N > 0$. Taking the restriction of (2.15) to Γ_1^+ and Γ_2^- and taking the Fourier transform in x'', we obtain

$$\begin{aligned} \tilde{v}_1^+(x_1, \xi'') + K_2 \tilde{v}_2^- &= \tilde{f}_1(x_1, \xi''), \quad x_1 > 0, \\ \tilde{v}_2^-(y_1, \xi'') + K_1 \tilde{v}_1^+ &= \tilde{f}_2(y_1, \xi''), \quad y_1 < 0, \end{aligned} \tag{2.16}$$

where $y_1 = -(x_1 \cos\alpha + x_2 \sin\alpha)$ for $\sin\alpha x_1 - \cos\alpha x_2 = 0$, so that $x_2 = -y_1 \sin\alpha$, $x_1 = -y_1 \cos\alpha$,

$$(2.17)\qquad \begin{aligned} K_2\tilde{\nu}_2^- &= \frac{1}{2\pi}\int_{-\infty}^{\infty} \exp\left\{-x_1 \sin\alpha\sqrt{\xi_2^2 + |\xi''|^2} + ix_1\cos\alpha\xi_2\right\}\tilde{\nu}_2^-(\xi_2, \xi'')\,d\xi_2, \\ &\qquad x_1 > 0, \\ K_1\tilde{\nu}_1^+ &= \frac{1}{2\pi}\int_{-\infty}^{\infty} \exp\left\{y_1 \sin\alpha\sqrt{\xi_1^2 + |\xi''|^2} + iy_1\cos\alpha\xi_1\right\}\tilde{\nu}_1^+(\xi_1, \xi'')\,d\xi_1, \\ &\qquad y_1 < 0. \end{aligned}$$

Denote

$$(2.18)\qquad \begin{aligned} K_0\nu_+ &= \frac{1}{2\pi}\int_{-\infty}^{\infty} e^{-x_1\sin\alpha|\xi_2| - ix_1\cos\alpha\xi_2}\tilde{\nu}_+(\xi_2)\,d\xi_2 \\ &= \frac{1}{2\pi}\int_{-\infty}^{\infty}\int_0^{\infty} e^{-x_1\sin\alpha|\xi_2| - ix\cos\alpha\xi_2 + iy_2\xi_2}\nu_+(y_2)\,dy_2\,d\xi_2, \qquad x_1 > 0. \end{aligned}$$

Computing the integral in ξ_2, we obtain

$$(2.18')\qquad K_0\nu_+ = \frac{1}{\pi}\int_0^{\infty} \frac{x_1\sin\alpha}{x_1^2\sin\alpha + (x_1\cos\alpha - y_2)^2}\nu_+(y_2)\,dy_2.$$

We shall show that K_0 is a bounded operator in $L_2(\mathbf{R}^1_+)$ with norm satisfying

$$(2.19)\qquad \|K_0\| \leqslant \sin(|\pi - \alpha|/2).$$

Applying the Mellin transform

$$\hat{f}(z) = \int_0^{\infty} t^{z-1}f(t)\,dt,$$

we obtain

$$\hat{K}_0\nu_+ = \frac{1}{\pi}\int_0^{\infty}\int_0^{\infty} \frac{x_1^{z-1}x_1\sin\alpha}{x_1^2\sin^2\alpha + (x_1\cos\alpha - y_2)^2}\nu + (y_2)\,dy_2\,dx_1.$$

Changing variables $x_1 = ty_2$, we have (cf. §15 of [**1**])

$$(2.20)\qquad \hat{K}_0\nu_+ = K_0(z)\hat{\nu}_+(z),$$

where $\hat{\nu}_+(z)$ is the Mellin transform of $\nu_+(y_2)$ and

$$K_0(z) = \frac{1}{\pi}\int_0^{\infty} \frac{t^{z-1}t\sin\alpha\,dt}{t^2\sin^2\alpha + (t\cos\alpha - 1)^2}.$$

Note that

$$\frac{t\sin\alpha}{t^2\sin^2\alpha + (t\cos\alpha - 1)^2} = \frac{1}{2i}\left(\frac{e^{i\alpha}}{t - e^{i\alpha}} - \frac{e^{-i\alpha}}{t - e^{-i\alpha}}\right).$$

Therefore, computing the Mellin transform (see Example 2.1 in [**1**]), we obtain

$$(2.21)\qquad \begin{aligned} K_0(z) &= \frac{1}{1-e^{2\pi iz}}\left(e^{i\alpha}e^{(z-1)i\alpha} - e^{-i\alpha}e^{(z-1)i(2\pi-\alpha)}\right) \\ &= \frac{e^{i\alpha z} - e^{i(2\pi-\alpha)z}}{1-e^{2\pi iz}} = \frac{e^{i(\alpha-\pi)z} - e^{i(\pi-\alpha)z}}{e^{-i\pi z} - e^{i\pi z}} = \frac{\sin(\pi-\alpha)z}{\sin \pi z}. \end{aligned}$$

Note that

$$\begin{aligned} K_0(1/2+i\tau) &= \frac{\sin(\pi-\alpha)(1/2+i\tau)}{\sin(1/2+i\tau)\pi} \\ &= \frac{\sin\frac{1}{2}(\pi-\alpha)\cos h(\pi-\alpha)\tau + i\cos\frac{1}{2}(\pi-\alpha)\sin h(\pi-\alpha)\tau}{\cos h\pi\tau}. \end{aligned}$$

Therefore

$$(2.22)\qquad \begin{aligned} |K_0(1/2+i\tau)|^2 &= \frac{\cos h^2(\pi-\alpha)\tau - \cos^2\frac{1}{2}(\pi-\alpha)}{\cos h^2\pi\tau} \\ &\leqslant 1 - \cos^2\frac{\pi-\alpha}{2} = \sin^2\frac{\pi-\alpha}{2}, \end{aligned}$$

since $\cos h^2\pi\tau \geqslant 1$, $\cos h^2(\pi-\alpha)\tau \leqslant \cos h^2\pi\tau$. Using Parseval's formula for the Mellin transform, we obtain

$$(2.23)\qquad \begin{aligned} \int_0^\infty |K_0 v_+|^2\,dx_1 &= \frac{1}{2\pi}\int_{1/2-i\infty}^{1/2+i\infty} |K_0(z)\hat{v}_+(z)|^2\,d\tau \\ &\leqslant \sin\frac{|\pi-\alpha|}{2}\int_0^\infty |v_+(x_1)|^2\,dx_1, \end{aligned}$$

so that (2.19) is proved.

REMARK 2.1. Consider the system

$$(2.24)\qquad \begin{aligned} v_1(x_1) + K_0v_2 = f_1(x_1), &\qquad x_1>0,\ v_2(x_1)+K_0v_1 \\ = f_2(x_1), &\qquad x_1 > 0, \end{aligned}$$

where $f_k(x_1)\in L_2(\mathbf{R}^1_+)$. This system is uniquely solvable in $L_2(\mathbf{R}^1_+) = H_0(\mathbf{R}^1_+)$ since $\|K_0\| < 1$. Applying the Mellin transform gives

$$(2.25)\qquad \hat{v}_1(z) + K_0(z)\hat{v}_2(z) = \hat{f}_1(z), \qquad \hat{v}_2(z_1) + K_0(z)\hat{v}_1(z) = \hat{f}_2(z).$$

Therefore

$$(2.26)\qquad \hat{v}_1(z) = \frac{\hat{f}_1 - K_0(z)\hat{f}_2}{1-K_0^2(z)}, \qquad \hat{v}_2(z) = \frac{\hat{f}_2 - K_0(z)\hat{f}_1}{1-K_0^2(z)},$$

$$(2.26')\qquad v_k(x_1) = \frac{1}{2\pi i}\int_{1/2-i\infty}^{1/2+i\infty}\hat{v}_k(z)x_1^{-z}\,dz, \qquad z = 1/2+i\tau.$$

Formulas (2.26), (2.26′) give an explicit solution of (2.24).

Now we show that

$$(2.27)\qquad \|K_p\| \leqslant \|K_0\| \leqslant \sin\tfrac{1}{2}|\pi-\alpha|, \qquad p = 1,2,$$

where $\|K_p\|$ is the norm of the operator K_p (see (2.17)) acting in the spaces $L_2(\mathbf{R}^1_\pm)$.

Using the formula (see, for example, (4.65) in [**1**])

$$\frac{1}{(2\pi)^2}\int_{\mathbf{R}^3}\exp\left\{-x_1\sqrt{\xi_2^2+\xi_3^2}-ix_2\xi_2-ix_3\xi_3\right\}d\xi_2\,d\xi_3 = \frac{1}{2\pi}\frac{x_1}{\left(x_1^2+x_2^2+x_3^2\right)^{3/2}},\qquad x_1>0, \tag{2.28}$$

we obtain

$$\begin{aligned} K_2(x_1,y_2) &= \frac{1}{2\pi}\int_{-\infty}^{\infty}\exp\left\{-x_1\sin\alpha\sqrt{\xi_2^2+|\xi''|^2}+ix_1\cos\alpha\xi_2+iy_2\xi_2\right\}d\xi_2 \\ &= \frac{1}{2\pi}\int_{-\infty}^{\infty}\frac{x_1\sin\alpha\, e^{ix_3|\xi''|}\,dx_3}{\left(x_1^2\sin^2\alpha+(x_1\cos\alpha+y_2)^2+x_3^2\right)^{3/2}}. \end{aligned} \tag{2.29}$$

Therefore, using the change of variables $x_3=(x_1^2\sin^2\alpha+(x_1\cos\alpha+y_2)^2)^{1/2}y_3$, we obtain

$$\begin{aligned} |K_2(x_1,y_2)| &\leqslant \frac{1}{2\pi}\int_{-\infty}^{\infty}\frac{x_1\sin\alpha\,dx_3}{\left(x_1^2\sin^2\alpha+(x_1\cos\alpha+y_2)^2+x_3^2\right)^{3/2}} \\ &= \frac{1}{2\pi}\frac{x_1\sin\alpha}{x_1^2\sin^2\alpha+(x_1\cos\alpha+y_2)^2}\int_{-\infty}^{\infty}\frac{dy_3}{\left(1+y_3^2\right)^{3/2}} \\ &= \frac{1}{\pi}\frac{x_1\sin\alpha}{x_1^2\sin^2\alpha+(x_1\cos\alpha+y_2)^2} \end{aligned}$$

so that

$$|K_2(x_1,x_2)|\leqslant K_0(x_1,-y_2), \tag{2.30}$$

where $K_0(x_1,y_2)$ is the kernel of K_0. The same estimate holds for $K_1(y_1,y_2)$, since $K_1(y_1,y_2)=K_2(-y_1,-y_2)$. Inequality (2.27) shows that the system (2.16) has a unique solution $v_1^+\in H_{0,0}(\Gamma_1^+)$, $v_2^-\in H_{0,0}(\Gamma_2^-)$ for any $f\in H_{0,0}(\Gamma_1^+)$, $f_2\in H_{0,0}(\Gamma_2^-)$. Suppose $f_1\in H_{0,N}(\Gamma_1^+)$, $f_2\in H_{0,N}(\Gamma_2^-)$, $N>0$. Note the following property of the operators $K_p(p=1,2)$: The composition $(d/dx_1\,x_1)^rK_p$ is a bounded operator in $H_{0,0}$ for any $r\geqslant 0$, so that K_p is bounded from $H_{0,0}$ to $H_{0,N}$ for any $N>0$. Therefore, if $f_1\in H_{0,N}(\Gamma_1^+)$, $f_2\in H_{0,N}(\Gamma_1^-)$, then $v_1^+\in H_{0,N}(\Gamma_1^+)$, $v_1^-\in H_{0,N}(\Gamma_2^-)$ for any N.

We already proved that for any $v_0\in H_{1/2,N}(G)$, $(-\Delta+1)v_0=0$ in G there exist a unique pair $v_1^+\in H_{0,N}(\Gamma_1)$, $v_2^-\in H_{0,N}(\Gamma_2)$ such that (2.16) holds. To complete the proof of Lemma 2.3, it remains to prove that if $v_0\in H_{1/2,N}(G)$, $(-\Delta+1)v_0=0$ in G, $p_1^+v_0=0$, $p_2^-v_0=0$, then $v_0=0$. Taking the Fourier transform in x'', we obtain

$$\left(\frac{\partial^2}{\partial x_1^2}+\frac{\partial^2}{\partial x_2}\right)\tilde{v}_0(x_1,x_2,\xi'')=\left(|\xi''|^2+1\right)\tilde{v}_0(x_1,x_2,\xi''), \tag{2.31}$$
$$p_1^+\tilde{v}_0=0,\qquad p_2^-\tilde{v}_0=0.$$

We consider ξ'' as a parameter and denote also by G the two-dimensional sector $x_2>0$, $x_1\sin\alpha-x_2\cos\alpha>0$. Since $(|\xi''|^2+1)\tilde{v}_0(x_1,x_2,\xi'')\in H_{1/2,N}(G)$ in (x_1,x_2) and $N>0$, we have $\tilde{v}_0\in H_{3/2}$ for $x_1^2+x_2^2>\varepsilon$, $\forall\varepsilon>0$. Let $w_0(x_1,x_2)$ be the solution of the problem

$$\left(\frac{\partial^2}{\partial x_1^2}+\frac{\partial^2}{\partial x_2^2}\right)w_0=\left(|\xi''|^2+1\right)\tilde{v}_0(x_1,x_2,\xi''),\qquad (x_1,x_2)\in G_1, \tag{2.32}$$

$$p_1^+w_0=0,\quad p_2^-w_0=0,\quad w_0=\tilde{v}_0(x_1,x_2,\xi'')\quad \text{for } x_1^2+x_2^2=1,\quad (x_1,x_2)\in G.$$

Here G_1 is the intersection of G with the disc $x_1^2+x_2^2\leqslant 1$. Since problem (2.32) is strongly elliptic, it has a solution w_0 belonging at least to $H_1(G_1)$. Therefore, $w_1=\tilde{v}_0-w_0\in H_{1/2}(G_1)$, and

$$\left(\frac{\partial^2}{\partial x_1^2}+\frac{\partial^2}{\partial x_2^2}\right)w_1=0\quad \text{in } G_1,\qquad w_1|_{\partial G_1}=0. \tag{2.32$'$}$$

We map G_1 conformally onto the half-disc, using the change of variables $\zeta=z^{\pi/\alpha}$. Then we take the odd extension of w_1 to the whole unit disc C_1. Denote the function w_1 in the new coordinates by $w_2(y)$. Then

$$\left(\frac{\partial^2}{\partial y_1^2}+\frac{\partial^2}{\partial y_2^2}\right)w_2(y)=0\quad \text{for } (y,y_2)\neq(0,0) \text{ and } (y_1,y_2)\in C_1, \tag{2.33}$$
$$w_2(y)=0\quad \text{for } y_1^2+y_2^2=1.$$

Since $(\partial^2/\partial y_1^2+\partial^2/\partial y_2^2)w_2$ is a distribution with support at the origin, we have

$$w_2(y)=P(\partial/\partial y)\ln|y|+w_3(y), \tag{2.34}$$

where $P(\partial/\partial y)$ is a differential operator and $w_3(y)$ is a smooth function in C_1. Returning to the old coordinates in G_1 and using the fact that $w_1\in H_{1/2}(G_1)$ and $p_1^+w_1=p_2^+w_2=0$, we conclude that $P(\partial/\partial y)\equiv 0$. Indeed, $|x|=|y|^{\alpha/\pi}$ and, therefore, $(\partial^k/\partial y^k)\ln|y|$, $k\geqslant 1$, corresponds to a function in x-coordinates that does not belong to $H_{1/2}(G_1)$. Also $c_0\ln|y|$ satisfies in x-coordinates the conditions $p_1^+w=0$, $p_2^-w=0$ iff $c_0=0$. Therefore, $P(\partial/\partial y)\equiv 0$ and $w_2(y)=w_3(y)$. Therefore, $w_1(x_1,x_2)\in H_1(G_1)$, so that $\tilde{v}_0\in H_1(G)$. Now the uniqueness of the solution of problem (2.31) follows by integration by parts.

3. The index of the boundary-value problem depending on a parameter. Consider the boundary-value problem (2.5′)–(2.7′), assuming $h_1\in H_{s-m_1\text{-}1/2,N}(\Gamma_1^+)$, $h_2\in H_{s-m_2-1/2,N}(\Gamma_2^-)$, and $v\in H_{s,N}(G)$. Using Lemma 2.2 we can replace v by $v_0=p(\Lambda_1^-)^{s-1/2}lv\in H_{1/2,N}(G)$:

$$(-\Delta+1)v_0=0,\qquad x\in G, \tag{3.1}$$

$$p_1^+\hat{B}_1(D)(\Lambda_1^-)^{-(s-1/2)}lv_0 = h_1(x'), \qquad x' \in \Gamma_1^+, \tag{3.2}$$

$$p_2^-\hat{B}_2(D)(\Lambda_1^-)^{-(s-1/2)}lv_0 = h_2(x'), \qquad x' \in \Gamma_2^-, \tag{3.3}$$

where lv_0 is the arbitrary extension of v_0 to $\mathbf{R}^n$. Note that $\psi\mathrm{do}(\Lambda_1^-)^s$ maps $H_{s,N}(\Gamma_1^+)$ isomorphically onto $H_{0,N}(\Gamma_1^+)$ (see, for example, Lemma 4.6 in [**1**]).

Analogously, $\psi\mathrm{do}(\Lambda_3^+)^s$, with symbol

$$\left(\zeta_1 + i|\hat{\xi}''|\right)^s = \left(-\xi_1\cos\alpha - \xi_2\sin\alpha + i|\hat{\xi}''|\right)^s,$$

maps $H_{s,N}(\Gamma_2^-)$ isomorphically onto $H_{0,N}(\Gamma_2^-)$. As in the case of $(\Lambda_1^-)^s$, we take in $(\Lambda_3^+)^s$ the branch of z^s that is positive for positive z. Since $(\Lambda_1^-)^s$ commutes with p_1^+, (3.2) is equivalent to

$$p_1^+\hat{B}_1(D)(\Lambda_1^-)^{-m_1}lv_0 = h_{10}(x'), \tag{3.4}$$

where $h_{10} = p_1^+(\Lambda_1^-)^{s-m_1-1/2}lh_1 \in H_{0,N}(\Gamma_1^+)$. Analogously, from (3.3),

$$p_2^-\hat{B}_2(D)(\Lambda_3^+)^{s-m_2-1/2}(\Lambda_1^-)^{1/2-s}lv_0 = h_{20}(x'), \tag{3.5}$$

where $h_{20} = p_2^-(\Lambda_3^+)^{s-m_2-1/2}lh_2 \in H_{0,N}(\Gamma_2^-)$. Therefore, the boundary-value problem (1.1)–(1.3) is equivalent to the boundary-value problem (3.1), (3.4), (3.5). Taking the Fourier transform in x'', we obtain a family of two-dimensional boundary problems depending on the parameter $|\hat{\xi}''|$:

$$\left(-\frac{\partial^2}{\partial x_1^2} - \frac{\partial^2}{\partial x_2^2} + |\hat{\xi}''|^2 + 1\right)\tilde{v}_0(x_1, x_2, \xi'') = 0, \qquad (x_1, x_2) \in G, \tag{3.6}$$

$$p_1^+\hat{B}_1(D_1, D_2, \xi'')(\Lambda_1^-)^{-m_1}l\tilde{v}_0(x_1, x_2, \xi'') = \tilde{h}_{10}(x_1, \xi''), \qquad x_1 > 0, \tag{3.7}$$

$$\begin{aligned} p_2^-\hat{B}_2(D_1, D_2, \xi'')(\Lambda_3^+)^{s-m_2-1/2}(\Lambda_1^-)^{1/2-s}l\tilde{v}_0(x_1, x_2, \xi'') \\ = \tilde{h}_{20}(y_1, \xi''), \qquad y_1 < 0. \end{aligned} \tag{3.8}$$

Now G means the two-dimensional sector $x_2 > 0$, $y_2 = x_1\sin\alpha - x_2\cos\alpha > 0$, Γ_1^+ is the semiaxis $x_2 = 0$, $x_1 > 0$, and Γ_2^- is the semiaxis $y_2 = x_1\sin\alpha - x_2\cos\alpha = 0$, $y_1 = -(x_1\cos\alpha + x_2\sin\alpha) < 0$. In order to prove the existence and the uniqueness of the solution of problem (3.1), (3.4), (3.5), one should prove the invertibility of the operator corresponding to (3.6)–(3.8) with the norm of the inverse operator uniformly bounded in ξ''. According to Lemma 2.3, $\tilde{v}_0(x_1, x_2, \xi'')$ can be uniquely represented as

$$\begin{aligned} \tilde{v}_0(x_1, x_2, \xi'') &= G_1v_1^+ + G_2v_2^- \\ &= \frac{1}{2\pi}\int_{-\infty}^{\infty}\exp\left\{-x_2\sqrt{\xi_1^2 + |\hat{\xi}''|^2} - ix_1\xi\right\}\tilde{v}_1^+(\xi_1, \xi'')\,d\xi_1 \\ &\quad + \frac{1}{2\pi}\int_{-\infty}^{\infty}\exp\left\{-(x_1\sin\alpha - x_2\cos\alpha)\sqrt{\xi_2^2 + |\hat{\xi}'|^2}\right. \\ &\qquad \left. + i(\cos\alpha x_1 + \sin\alpha x_2)\xi_2\right\}\tilde{v}_2^-(\xi_1, \xi'')\,d\xi_1, \end{aligned} \tag{3.9}$$

where

$$\begin{aligned} &\nu_1^+(x_1,\xi'') \in H_{0,N}(\Gamma_1), \qquad \nu_1^+ = 0 \quad \text{for } x_1 < 0, \\ &\nu_2^-(y_1,\xi'') \in H_{0,N}(\Gamma_2), \qquad \nu_2^- = 0 \quad \text{for } y_1 > 0. \end{aligned} \tag{3.10}$$

We fix the extension $l\tilde{\nu}_0$ of $\tilde{\nu}_0(x_1, x_2, \xi'')$: $G_1\nu_1^+$ is defined by the integral for $x_2 > 0$. Define $(G_1\nu_1^+)^+ = G_1\nu_1^+$ for $x_2 > 0$, $(G_1\nu_1^+)^+ = 0$ for $x_2 < 0$. Analogously, $G_2\nu_2^-$ is defined by the second integral in (3.9) in the half-plane $x_1 \sin\alpha - x_2\cos\alpha > 0$. Define $(G_2\nu_2^-)^+ = 0$ for $y_2 = x_1\sin\alpha - x_2\cos\alpha < 0$ and $(G_2\nu_2^-)^+ = G_2\nu_2^-$ for $y_2 > 0$. Therefore, $l\tilde{\nu}_0 = (G_1\nu_1^+)^+ + (G_2\nu_2^-)^+$ is the extension of $\tilde{\nu}_0$ to $\mathbf{R}^2$. We shall substitute $l\tilde{\nu}_0$ in (3.7) and in (3.8). We have

$$\begin{aligned} p_1^+\hat{B}_1(\Lambda_1^-)^{-m_1}(G_1\nu_1^+)^+ = \frac{1}{2\pi}\int_{-\infty}^{\infty} &\hat{B}_1\left(\xi_1, -i\sqrt{\xi_1^2 + |\hat{\xi}''|^2}, \xi''\right) \\ &\cdot\left(\xi_1 - i|\hat{\xi}''|\right)^{-m_1} e^{-ix_1\xi_1}\tilde{\nu}_1^+(\xi_1,\xi'')\, d\xi_1. \end{aligned} \tag{3.11}$$

We show that

$$\begin{aligned} p(\Lambda_1^-)^{1/2-s}(G_2\nu_2^-)^+ = \frac{1}{2\pi}\int_{-\infty}^{\infty} &\left(-i\sin\alpha\sqrt{\xi_2^2 + |\hat{\xi}''|^2} - \cos\alpha\xi_2 - i|\hat{\xi}''|\right)^{1/2-s} \\ &\cdot\exp\left\{-(x_1\sin\alpha - x_2\cos\alpha)\sqrt{\xi_2^2 + |\hat{\xi}''|^2}\right\} \\ &\cdot e^{+i(x_1\cos\alpha + x_2\sin\alpha)\xi_2}\tilde{\nu}_2^-(\xi_2,\xi'')\, d\xi_2. \end{aligned} \tag{3.12}$$

Indeed,

$$\begin{aligned} F_{x_1}(G_2\nu_2^-)^+ &= \frac{1}{2\pi}\int_{x_2\cot\alpha}^{\infty}\int_{-\infty}^{\infty} \exp\Big\{-(x_1\sin\alpha - x_2\cos\alpha)\sqrt{\xi_2^2 + |\hat{\xi}''|^2} \\ &\qquad + i(x_1\cos\alpha + x_2\sin\alpha)\xi_2\Big\} \\ &\qquad \cdot\tilde{\nu}_2^-(\xi_2,\xi'')e^{ix_1\zeta_1}\, d\xi_2\, dx_1 \\ &= \frac{1}{2\pi}\int_{-\infty}^{\infty} \exp\left\{x_2\cos\alpha\sqrt{\xi_2^2 + |\hat{\xi}''|^2} + ix_2\sin\alpha\xi_2\right\}\tilde{\nu}_2^-(\xi_2,\xi'') \\ &\qquad \cdot\frac{\exp\left\{x_2\cot\alpha\left(-\sin\alpha\sqrt{\xi_2^2 + |\xi''|^2} + i\xi_2\cos\alpha + i\zeta_1\right)\right\}}{\sin\alpha\sqrt{\xi_2^2 + |\xi''|^2} - i\xi_2\cos\alpha - i\zeta_1}\, d\xi_2, \end{aligned} \tag{3.13}$$

where

$$F_{x_1}w(x_1) = \int_{-\infty}^{\infty} w(x_1)e^{ix_1\zeta_1}\, dx_1.$$

We have

$$(\Lambda_1^-)^{1/2-s}(G_2\nu_2^-)^+ = \frac{1}{2\pi}\int_{-\infty}^{\infty}\left(\zeta_1 - i|\hat{\xi}''|\right)^{1/2-s}\left(F_{x_1}(G_2\nu_2^-)^+\right)e^{-ix_1\zeta_1}\, d\zeta_1. \tag{3.14}$$

Substitute (3.14) in (3.13) and compute the integral in ζ_1 using the Jordan lemma. Note that

$$x_2 \cot \alpha - x_1 = (1/\sin \alpha)(x_2 \cos \alpha - x_1 \sin \alpha) < 0 \quad \text{in } G$$

and the only pole in the half-plane $\operatorname{Im} \zeta_1 < 0$ is at the point

$$\zeta_1 = -i \sin \alpha \sqrt{\xi_2^2 + |\hat{\xi}''|^2} - \xi_2 \cos \alpha.$$

Therefore (3.12) holds. Therefore substituting (3.9) into the boundary condition (3.7), we obtain

$$p_1^+(b_1 v_1^+ + M_2 v_2^-) = h_{10}(x_1, \xi''), \qquad x_1 > 0, \tag{3.15}$$

where b_1 is a ψdo on $\mathbf{R}^1$ with symbol

$$b_1(\xi_1, \xi'') = B_1\left(\xi_1, -i\sqrt{\xi_1^2 + |\hat{\xi}''|^2}, \xi''\right)\left(\xi_1 - i|\hat{\xi}''|\right)^{-m_1}; \tag{3.16}$$

i.e.,

$$b_1 v_1^+ = \frac{1}{2\pi} \int_{-\infty}^{\infty} b_1(\xi_1, \xi'') e^{-ix_1\xi_1} \tilde{v}_1^+(\xi_1, \xi'')\, d\xi_1,$$

and

$$\begin{aligned} M_2 v_2^- = \frac{1}{2\pi} \int_{-\infty}^{\infty} M_2(\xi_1, \xi'') \exp\left\{-x_1 \sin \alpha \sqrt{\xi_1^2 + |\hat{\xi}''|^2|} + ix_1\xi_1 \cos \alpha\right\} \\ \cdot \tilde{v}_2^-(\xi_1, \xi'')\, d\xi_1, \end{aligned} \tag{3.17}$$

where

$$\begin{aligned} M_2(\xi_1, \xi'') = B_1\left(-i \sin \alpha \sqrt{\xi_1^2 + |\hat{\xi}''|^2} - \xi_1 \cos \alpha, i \cos \alpha \sqrt{\xi_1^2 + |\hat{\xi}''|^2} - \xi_1 \sin \alpha, \xi''\right) \\ \cdot \left(-i \sin \alpha \sqrt{\xi_1^2 + |\xi''|^2} - \xi_1 \cos \alpha - i|\hat{\xi}''|\right)^{-m_1}. \end{aligned} \tag{3.18}$$

Analogously substituting (3.9) into (3.8), we obtain

$$p_2^-(b_2 v_2^- + M_1 v_1^+) = h_{20}(y_1, \xi''), \qquad y_1 < 0, \tag{3.19}$$

where b_2 is a ψdo on $\mathbf{R}^1$ with symbol

$$\begin{aligned} b_2(\xi_1, \xi'') = B_2\left(-i \sin \alpha \sqrt{\xi_1^2 + |\hat{\xi}''|^2} - \xi_1 \cos \alpha, i \cos \alpha \sqrt{\xi_1^2 + |\hat{\xi}''|^2} - \xi_1 \sin \alpha, \xi''\right) \\ \cdot \left(\xi_1 + i|\hat{\xi}''|\right)^{s - m_2 - 1/2}\left(-i \sin \alpha \sqrt{\xi_1^2 + |\hat{\xi}''|^2} - \xi_1 \cos \alpha - i|\hat{\xi}''|\right)^{1/2 - s} \end{aligned} \tag{3.20}$$

and

$$\begin{aligned} M_1 v_1^+ = \frac{1}{2\pi} \int_{-\infty}^{\infty} M_1(\xi_1, \xi'') \exp\left\{y_1 \sin \alpha \sqrt{\xi_1^2 + |\hat{\xi}''|^2} + iy_1\xi_1 \cos \alpha\right\} \\ \cdot \tilde{v}_1^+(\xi_1, \xi'')\, d\xi_1, \end{aligned} \tag{3.21}$$

$$\begin{aligned} M_1(\xi_1, \xi'') = B_2\left(\xi_1, -i\sqrt{\xi_1^2 + |\hat{\xi}''|^2}, \xi''\right)\left(\xi_1 - i|\hat{\xi}''|\right)^{1/2 - s} \\ \cdot \left(-\xi_1 \cos \alpha + i \sin \alpha \sqrt{\xi_1^2 + |\hat{\xi}''|^2} + i|\hat{\xi}''|\right)^{s - m_2 - 1/2}. \end{aligned} \tag{3.22}$$

Therefore, we have

LEMMA 3.1. *For every* ξ'' *the solution of the boundary-value problem* (3.6)–(3.8) *is equivalent to the solution of the system* (3.15), (3.19), *where* $v_1^+ \in H_{0,N}(\mathbf{R}^1)$, $v_2^- \in H_{0,N}(\mathbf{R}^1)$, $v_1^+ = 0$ *for* $x_1 < 0$, $v_2^- = 0$ *for* $y_1 > 0$, $h_{10} \in H_{0,N}(\mathbf{R}^1_+)$, $h_{20} \in H_{0,N}(\mathbf{R}^1_-)$.

Note that if (3.15), (3.19) is uniquely solvable for $N = 0$, then the same fact holds for any $N > 0$. Indeed, M_k are bounded from H_0 to $H_{0,N}$ for any N, and the commutators of x_1^k and $b(D_1, \xi'')$ have the form

$$x_1^k b = b x_1^k + \sum_{r=1}^{k} b^{(r)} x_1^{k-r},$$

where $\operatorname{ord} b^{(r)} \leqslant -r$. Therefore, if the right sides of (3.15), (3.19) belong to $H_{0,N}$ and the solution belongs to $H_{0,0}$, we obtain that the solution also belongs to $H_{0,N}$. We assume that $B_1(D)$ and $B_2(D)$ satisfy the Shapiro–Lopatinskii condition on $\overline{\Gamma}_1^+$ and $\overline{\Gamma}_2^-$; i.e., the corresponding boundary-value problem in the half-spaces $x_2 > 0$ and $y_2 = x_1 \sin\alpha - x_2 \cos\alpha > 0$ is uniquely solvable. The Shapiro–Lopatinskii condition has the form

$$B_1\left(\xi_1, -i\sqrt{\xi_1^2 + |\xi''|^2}, \xi''\right) \neq 0 \quad \text{for } \xi_1^2 + |\xi''|^2 > 0, \tag{3.23}$$

$$B_2\left(-i\sin\alpha\sqrt{\xi_1^2 + |\xi''|^2} - \xi_1\cos\alpha, i\cos\alpha\sqrt{\xi_1^2 + |\xi''|^2} - \xi_1 \sin\alpha, \xi''\right) \neq 0 \quad \text{for } \xi_1^2 + |\xi''|^2 > 0. \tag{3.23'}$$

We prove in this section that the operator $A(\alpha)$ defined by the left sides of (3.15), (3.19) is a Fredholm operator, and we find a formula for the index of $A(\alpha)$. It is convenient to set $y_1 = -x_1$, $0 < x_1 < +\infty$, in (3.15), (3.19). Then the system (3.15), (3.19) has the form

$$p_1^+\left(b_1 v_1^+ + M_2^1 v_2^+\right) = h_{10}\left(x_1, \xi''\right), \qquad x_1 > 0, \tag{3.15'}$$

$$p_1^+\left(b_2' v_2^+ + M_1^1 v_1^+\right) = h_{20}\left(-x_1, \xi''\right), \qquad x_1 > 0, \tag{3.19'}$$

where $v_2^+(x_1, \xi'') = v_2^-(-x_1, \xi'')$, $\tilde{v}_2^+(\xi_1, \xi'') = \tilde{v}_2^-(-\xi_1, \xi'')$, $b_2'(\xi_1, \xi'') = b_2(-\xi_1, \xi'')$, $M_2^1 v_2^+$ is obtained from (3.17) by changing ξ_1 to $-\xi_1$, and $M_1^1 v_1^+$ is obtained from (3.21) by changing y_1 to $-x_1$.

In addition to (3.15′), (3.19′), we consider the system of equations

$$\begin{aligned} p_1^+\left(b_1^{(0)} v_1^+ + M_2^{(0)} v_2^+\right) &= h_{10}(x_1), \qquad x_1 > 0, \\ p_1^+\left(b_2^{(0)} v_2^+ + M_1^{(0)} v_1^+\right) &= h_{20}(x_1), \qquad x_1 > 0, \end{aligned} \tag{3.24}$$

where $b_k^{(0)}$, $k = 1, 2$, are ψdo's on $\mathbf{R}^1$ with symbols

$$\begin{aligned} b_1^{(0)}(\xi_1) &= B_1\left(\xi_1, -i|\xi_1|, 0\right)\left(\xi_1 - i0\right)^{-m_1}, \\ b_2^{(0)}(\xi_2) &= B_2\left(-i\sin\alpha|\xi_1| + \xi_1\cos\alpha, i|\xi_1|\cos\alpha + \xi_1\sin\alpha, 0\right) \\ &\quad \cdot\left(-\xi_1 + i0\right)^{s-m_2-1/2}\left(-i\sin\alpha|\xi_1| + \xi_1\cos\alpha - i0\right)^{1/2-s}, \end{aligned} \tag{3.25}$$

and

$$M_k^{(0)} v_k^+ = \frac{1}{2\pi} \int_{-\infty}^{\infty} M_k^{(0)}(\xi_1) e^{-x_1 \sin\alpha|\xi_1| - i x_1 \xi_1 \cos\alpha} \tilde{v}_k^+(\xi_1)\, d\xi_1, \tag{3.26}$$

where

(3.27)

$$M_1^{(0)}(\xi_1) = B_2(\xi_1, -i|\xi_1|, 0)(\xi_1 - i0)^{1/2-s}(-\xi_1 \cos\alpha + i \sin\alpha|\xi_1| + i0)^{s-m_1-1/2},$$

$$M_2^{(0)}(\xi_1) = B_1(-i\sin\alpha|\xi_1| + \xi_1\cos\alpha, i\cos\alpha|\xi_1| + \xi_1\sin\alpha, 0)$$
$$\cdot(-i\sin\alpha|\xi_1| + \xi_1\cos\alpha - i0)^{-m_1}.$$

For the definition of $(\xi_1 \pm i0)^s$ see, for example, [**1**, §2].

Note that system (3.24) corresponds to system (3.15′), (3.19′) when the B_k are independent of ξ'' and (3.1) is replaced by the two-dimensional Laplacian $-(\partial^2/\partial x_1^2 + \partial^2/\partial x_2^2)$. System (3.24) can be solved explicitly using the Mellin transform (cf. [**1**, **2**]). We have

$$\begin{aligned} b_k^{(0)}(\xi_1) &= b_k^+ \theta(\xi_1) + b_k^-(1 - \theta(\xi_1)), \\ M_k^{(0)}(\xi_1) &= M_k^+ \theta(\xi_1) + M_k^-(1 - \theta(\xi_1)), \end{aligned} \qquad k = 1,2, \tag{3.28}$$

where

$$\theta(\xi_1) = 0 \quad \text{for } \xi_1 < 0, \qquad \theta(\xi_1) = 1 \quad \text{for } \xi_1 > 0,$$

$$\begin{aligned}
b_1^+ &= B_1(1, -i, 0), \qquad b_1^- = B_1(1, i, 0), \\
b_2^+ &= B_2(e^{-i\alpha}, ie^{-i\alpha}, 0) e^{i\pi(s-m_2-1/2)} e^{-i\alpha(1/2-s)} \\
&= B_2(1, i, 0) e^{i(s-m_2-1/2)(\alpha+\pi)}, \\
b_2^- &= B_2(-e^{i\alpha}, ie^{i\alpha}, 0) e^{-i(\pi-\alpha)(1/2-s)} = B_2(1, -i, 0) e^{i(\pi-\alpha)(s-m_2-1/2)}, \\
M_1^+ &= B_2(1, -i, 0) e^{i(\pi-\alpha)(s-m_2-1/2)} = b_2^-, \\
M_1^- &= B_2(-1, -i, 0) e^{-i\pi(1/2-s)} e^{i\alpha(s-m_2-1/2)} \\
&= B_2(1, i, 0) e^{i(\pi+\alpha)(s-m_2-1/2)} = b_2^+, \\
M_2^+ &= B_1(e^{-i\alpha}, ie^{-i\alpha}, 0) e^{i\alpha m_1} = B_1(1, i, 0) = b_1^-, \\
M_2^- &= B_1(-e^{i\alpha}, ie^{i\alpha}, 0) e^{im_1(\pi-\alpha)} = B_1(1, -i, 0) = b_1^+.
\end{aligned} \tag{3.29}$$

Applying the Mellin transform we obtain, analogously to (2.25) (cf. [**1**, §15; **2**]),

$$\begin{aligned}
\frac{b_1^+ - b_1^- e^{2\pi i z}}{1 - e^{2\pi i z}} \hat{v}_1^+(z) + \frac{b_1^- e^{i\alpha z} - b_1^+ e^{i(2\pi-\alpha)z}}{1 - e^{2\pi i z}} \hat{v}_2^+(z) &= \hat{h}_{10}(z), \\
\frac{b_2^+ - b_2^- e^{2\pi i z}}{1 - e^{2\pi i z}} \hat{v}_2^+(z) + \frac{b_2^- e^{i\alpha z} - b_2^+ e^{i(2\pi-\alpha)z}}{1 - e^{2\pi i z}} \hat{v}_1(z) &= \hat{h}_{20}(z).
\end{aligned} \tag{3.30}$$

We assume that

$$\det M(z) \neq 0 \quad \text{for } z = 1/2 + i\tau, \ -\infty < \tau < +\infty, \tag{3.31}$$

where $M(z)$ is the matrix

$$(3.32)\quad M(z)=(1-e^{2\pi iz})^{-1}\begin{Vmatrix} b_1^+-b_1^-e^{2\pi iz}, & b_1^-e^{i\alpha z}-b_1^+e^{i(2\pi-\alpha)z}\\ b_2^-e^{i\alpha z}-b_2^+e^{i(2\pi-\alpha)z}, & b_2^+-b_2^-e^{2\pi iz}\end{Vmatrix}.$$

Without loss of generality we can assume that $B_1(1,-i,0)=1$, $B_2(1,-i,0)=e^{i(\pi-\alpha)m_2}$. Therefore,

$$(3.33)\quad \begin{aligned} b_1^+&=1,\\ b_1^-&=B_1(1,i,0),\qquad b_2^+=B_2(1,i,0)e^{i(s-m_2-1/2)(\pi+\alpha)},\\ b_2^-&=e^{i(\pi-\alpha)(s-1/2)}.\end{aligned}$$

Condition (3.31) is equivalent to the condition

$$(3.34)\qquad \mathscr{D}(z,\alpha,s)\neq 0\quad \text{for } z=1/2+i\tau,\ -\infty<\tau<+\infty,$$

where

$$(3.34')\qquad \mathscr{D}(z,\alpha,s)=(1-e^{2\pi iz})^2\det M(z).$$

Note that $\mathscr{D}(z,\alpha,s)$ is an entire analytic function of (z,α,s):

$$\begin{aligned}\mathscr{D}(z,\alpha,s)&=b_2^+(1-e^{2(2\pi-\alpha)zi})-B_1(1,i,0)e^{i(\pi-\alpha)(s-1/2)}(e^{2i\alpha z}-e^{4\pi iz})\\ &=b_2^+e^{-i(s-1/2)(\pi+\alpha)}\big(e^{i(s-1/2)(\pi+\alpha)}(1-e^{(4\pi-2\alpha)zi})\\ &\qquad\qquad -b_3e^{i(\pi-\alpha)(s-1/2)}(e^{2i\alpha z}-e^{4\pi iz})\big),\end{aligned}$$

where $b_3=B_1(1,i,0)B_2^{-1}(1,i,0)e^{im_2(\pi+\alpha)}$. Set

$$(3.35)\qquad \gamma=\frac{1}{2\pi i}\ln\left(\frac{B_1(1,i,0)}{B_2(1,i,0)}e^{im_2(\pi+\alpha)}\right),\quad \text{i.e.}\quad b_3=e^{2\pi\gamma i},$$

where we take an arbitrary branch of $\ln z$. We have

$$(3.36)\quad \det M(z,\alpha,s)=b_2^+e^{-i(s-1/2)\alpha+i\pi\gamma}\frac{\sin(2\pi-\alpha)z\sin((z-s+1/2)\alpha+\pi\gamma)}{\sin^2\pi z}.$$

Denote by $S(\alpha)$ the set of all $s\in\mathbf{R}$ such that (3.31) fails. It follows from (3.36) that $S(\alpha)$ consists of the points

$$(3.37)\qquad s_k=\operatorname{Re}(\pi\gamma/\alpha)+1+\pi k/\alpha,\qquad -\infty<k<+\infty.$$

Note that the only root of $\mathscr{D}(z,\alpha,s_k)$ on the line $\operatorname{Re} z=1/2$ is simple and has the form

$$(3.38)\qquad z_k=1/2-i(\pi/\alpha)\operatorname{Im}\gamma.$$

Now consider the system (3.15′), (3.19′). Note that M_1^1, M_2^1 can be represented as

$$(3.39)\qquad M_k^1v_k^+=\theta_1M_k^{(0)}v_k^++T_kv_k^+,\qquad k=1,2,$$

where $\theta_1=1$ for $0\leqslant x_1\leqslant 1$, $\theta_1=0$ for $x_1>0$, $M_k^{(0)}$ are the same as in (3.26), and T_k are Hilbert-Schmidt operators in $L_2(\mathbf{R}^1_+)$; i.e.,

$$T_kv_k^+=\int_0^\infty T_k(x_1,y_1)v_k^+(y_1)\,dy_1,$$

where

$$\int_0^\infty \int_0^\infty |T_k(x_1, y_1)|^2 \, dx_1 \, dy_1 < +\infty.$$

Therefore system (3.15′), (3.19′) is of the form considered in §15 of [**1**]. It is proved in [**1**] that operators of the form $b + \theta_1 M^{(0)} + T$ form an algebra. One can assign a symbol to the operator $A(\alpha)$ defined by the left sides of (3.15′), (3.19′). It will be a pair of matrices

$$\left\| \begin{matrix} b_1(\xi_1, \xi'') & 0 \\ 0 & b_2'(\xi_1, \zeta'') \end{matrix} \right\|, \quad M(z), \tag{3.40}$$

where $-\infty < -\xi_1 < +\infty$, $z = 1/2 + i\tau$, $-\infty < \tau < +\infty$. Here b_1, b_2' are the same as in (3.15′), (3.19′), and $M(z)$ is the same as in (3.32). The symbol of the composition of two such operators is the product of symbols. The necessary and sufficient condition for $A(\alpha)$ to be a Fredholm operator is that the symbol (3.40) is invertible. This is equivalent to the Shapiro–Lopatinskii conditions (3.23), (3.23′) together with condition (3.31). Moreover, when $A(\alpha)$ is a Fredholm operator the following formula for the index of $A(\alpha)$ holds:

$$\begin{aligned} \operatorname{ind} A(\alpha) = {} & (1/2\pi)\Delta \arg\big(b_1(\xi_1, \xi'')b_2'(\zeta_1, \xi'')\big)\big|_{\xi_1 = +\infty}^{-\infty} \\ & + (1/2\pi)\Delta \arg \det M(z)\big|_{z=1/2-i\infty}^{1/2+i\infty}, \end{aligned} \tag{3.41}$$

where $\Delta \arg c|_{\xi_1 = +\infty}^{-\infty}$ means the increment of the argument of $c(\xi_1)$ as ξ_1 changes from $+\infty$ to $-\infty$. Since $b_2'(\xi_1, \xi'') = b_2(-\xi_1, \xi'')$, we have, taking into account (3.36),

$$\begin{aligned} \operatorname{ind} A(\alpha) = {} & \frac{1}{2\pi}\Delta \arg \left.\frac{b_1(\xi_1, \xi'')}{b_2(\zeta_1, \xi'')}\right|_{\xi_1 = +\infty}^{-\infty} \\ & + \frac{1}{2\pi}\Delta \arg \left.\frac{\sin(2\pi - \alpha)z \sin((z - s + 1/2)\alpha + \pi\gamma)}{\sin^2 \pi z}\right|_{z=1/2-i\infty}^{1/2+i\infty}, \end{aligned} \tag{3.42}$$

where γ is the same as in (3.35), $b_1(\xi_1, \xi'')$ is given by (3.16), $b_2(\xi_1, \xi'')$ is given by (3.20), and $\xi'' \neq 0$ is fixed. Set

$$\begin{aligned} & b_3(\xi_1, \xi'') \\ & = \frac{B_1\left(\xi_1, -i\sqrt{\xi_1^2 + |\hat{\xi}''|^2}, \xi''\right)(\xi_1 - i|\hat{\xi}''|)^{-m_1}(\xi_1 + i|\hat{\xi}''|)^{m_2}}{B_2\left(-i\sin\alpha\sqrt{\xi_1^2 + |\hat{\xi}''|^2} - \xi_1\cos\alpha, i\cos\alpha\sqrt{\xi_1^2 + |\hat{\xi}''|^2} - \xi_1 \sin\alpha, \xi''\right)}. \end{aligned} \tag{3.43}$$

Since

$$\begin{aligned} & \Delta \arg \left(-i\sin\alpha\sqrt{\xi_1^2 + |\hat{\xi}''|^2} - \xi_1\cos\alpha - i|\hat{\xi}''|\right)^{1/2-s}(\xi_1 + i|\hat{\xi}''|)^{s-1/2}\Big|_{\xi_1 = +\infty}^{-\infty} \\ & = \pi(s - 1/2) + (\pi - 2\alpha)(1/2 - s) = 2(s - 1/2)\alpha, \end{aligned}$$

we finally obtain the simple formula

(3.44)

$$\operatorname{ind} A(\alpha) = \left| \frac{1}{2\pi} \Delta \arg b_3(\xi_1, \xi'') \right|_{\xi_1 = +\infty}^{-\infty} + \left(s - \frac{1}{2} \right) \frac{\alpha}{\pi}$$

$$+ \frac{1}{2\pi} \Delta \arg \left. \frac{\sin(2\pi - \alpha) z \sin((z - s + 1/2)\alpha + \pi\gamma)}{\sin^2 \pi z} \right|_{z=1/2-i\infty}^{1/2+i\infty},$$

where b_3 is given by (3.43) and γ is given by (3.35). Therefore we have proved

THEOREM 3.1. *Consider the boundary-value problem*

$$(3.45) \qquad \left(-\partial^2/\partial x_1^2 - \partial^2/\partial x_2^2 + |\xi''|^2 + 1\right) u(x_1, x_2) = f(x_1, x_2),$$

$$(3.46) \qquad p_1^+ \hat{B}_1 \left(D_1, D_2, \xi''\right) u(x_1, x_2)|_{\Gamma_1^+} = g_1(x_1),$$

$$(3.47) \qquad p_2^- \hat{B}_2 \left(D_1, D_2, \xi''\right) u(x_1, x_2)|_{\Gamma_2^-} = g_2(y_1),$$

where ξ'' is a parameter, $f \in H_{s-2,N}(G)$, $g_1 \in H_{s-m_1-1/2,N}(\Gamma_1^+)$, $g_2 \in H_{s-m_2-1/2,N}(\Gamma_2^-)$, and we are looking for $u(x_1, x_2) \in H_{s,N}(G)$. Assume that $s + N > \max_k(m_k + 1/2)$, $N \geqslant 1$, s is a noninteger if $s \leqslant \max_k(m_k + 1/2)$. Assume that the Shapiro–Lopatinskii conditions (3.23), (3.23′) *are satisfied and $s \notin S(\alpha)$. Then the left sides of* (3.45)–(3.47) *define a Fredholm operator from $H_{s,N}(G)$ to $H_{s-2,N}(G) \times H_{s-m_1-1/2,N}(\Gamma_1^+) \times H_{s-m_2-1/2,N}(\Gamma_2^-)$, and the index of this operator is given by* (3.42) *or* (3.44).

Note that the set $\mathbf{R}^1 - S(\alpha)$, where $A(\alpha)$ is a Fredholm operator, consists of intervals (s_k, s_{k+1}), $-\infty < k < +\infty$, where s_k are the same as in (3.37). The index of $A(\alpha)$ is a constant on each interval (s_k, s_{k+1}) and increases by 1 as s decreases from $s_k + \varepsilon \in (s_k, s_{k+1})$ to $s_k - \varepsilon \in (s_{k-1}, s_k)$. The reason is that $\mathscr{D}(z, d, s)$ has a simple zero in z on the line $\operatorname{Re} z = 1/2$ for $s = s_k$. Therefore, the range of ind $A(\alpha)$ is the set of all integers. Let $m_+(s)$ ($m_-(s)$) be the dimension of the kernel (cokernel) of $A(\alpha)$:

$$m_+(s) = \dim \ker A(\alpha), \quad m_-(s) = \dim \operatorname{coker} A(\alpha).$$

Note that $m_\pm(s)$ are independent of $s \in (s_k, s_{k+1})$ in each interval (s_k, s_{k+1}). Indeed, $m_+(s) \geqslant m_+(s_0)$, $m_-(s) \leqslant m_-(s_0)$ for $s \leqslant s_0$, since $H_{s_0,N} \subset H_{s,N}$ for $s \leqslant s_0$. Therefore, $m_+(s) = m_+(s_0)$, $m_-(s) = m_-(s_0)$ for $s \in (s_k, s_{k+1})$, $s_0 \in (s_k, s_{k+1})$, since $m_+(s) - m_-(s) = m_+(s_0) - m_-(s_0) = \operatorname{ind} A(\alpha)$. Note that we can consider s to be a complex parameter in all formulas of this section. Than $A(\alpha)$ will depend analytically on s in the strip $s_k < \operatorname{Re} s < s_{k+1}$. As before, for any fixed $\tau = \operatorname{Im} s$, $m_+(\operatorname{Re} s + i\tau)$ and $m_-(\operatorname{Re} s + i\tau)$ are independent of $\operatorname{Re} s$ for $s_k < \operatorname{Re} s < s_{k+1}$. Since a change of $m_+(s)$ ($m_-(s)$) may occur only on a discrete set (because of the analyticity of $A(\alpha)$ in s), $m_+(s)$ ($m_-(s)$) are constant in each strip $s_k < \operatorname{Re} s < s_{k+1}$. Denote by (s_-, s_+) the interval where ind $A(\alpha) = 0$. In the next section we show that $m_+(s) = m_-(s) = 0$ when $s \in (s_-, s_+)$. Set

$$(3.48) \qquad s_{k0} = \frac{\pi\gamma}{\alpha} + \frac{1}{2} + \frac{\pi(k+1)}{\alpha}, \qquad -\infty < k < +\infty.$$

Note that $\operatorname{Re} s_{k0} \in (s_k, s_{k+1}), \forall_k$. Then

$$\frac{1}{2\pi}\Delta \arg \left.\frac{\sin(2\pi - \alpha)z\sin((z - s_{k0} + 1/2)\alpha + \pi\gamma)}{\sin^2 \pi z}\right|_{1/2-i\infty}^{1/2+i\infty}$$

$$= \frac{1}{2\pi}\Delta \arg \left|\frac{\sin(2\pi - \alpha)z\sin z\alpha}{\sin^2 \pi z}\right|_{1/2-i\infty}^{1/2+i\infty} = 0.$$

Therefore,

$$\operatorname{ind} A(\alpha) = \frac{1}{2\pi}\Delta \arg \left.\frac{b_1(\xi_1, \xi'')}{b_2(\xi_1, \xi'')}\right|_{+\infty}^{-\infty} \quad \text{for } s = s_{k0}. \tag{3.49}$$

In particular,

$$\Delta \arg \left.\frac{b_1(\xi_1, \xi'')}{b_2(\xi_1, \xi'')}\right|_{+\infty}^{-\infty} = 0 \quad \text{when } s^{(0)} = s_+ - 1/2 - \operatorname{Im}\frac{\pi\gamma}{\alpha}.$$

REMARK 3.1. It was shown in [**1**, §13], by the Wiener–Hopf, or factorization, method, that when $\alpha = \pi$ the boundary-value problem (3.45)–(3.47) has a unique solution when $s \in (s_-, s_+)$. Since system (3.15′), (3.19′) is equivalent to the boundary-value problem (3.45)–(3.47), this system is uniquely solvable when $\alpha = \pi$. It is easy to show that the operator $A(\alpha)$, defined by the left sides of (3.15′), (3.19′), is a bounded operator in $L_2(\mathbf{R}^1_+)$ with norm continuously (and even analytically) dependent on α. Therefore, for $|\alpha - \pi|$ small system (3.15′), (3.19′) is also uniquely solvable, and, therefore, we have proved the main result for α closed to π. In the next section we prove that (3.15), (3.19) is uniquely solvable for any α by explicitly solving this system.

REMARK 3.2. We show that all results of this section also remain valid in the case when $\pi < \alpha < 2\pi$. We note a few changes that are needed to treat this case. In Lemma 2.2 one should replace $(\Lambda_1^-)^{s-1/2}$ by $(\Lambda_1^+)^{s-1/2}$, where $(\Lambda_1^+)^{s-1/2}$ is the ψdo with symbol $(\xi_1 + i|\hat{\xi}''|)^{s-1/2}$. Since $(\Lambda_1^+)^{s-1/2}$ commutes with the restriction operator p when $\pi < \alpha < 2\pi$, the proof of Lemma 2.2 can be repeated without additional changes. So instead of (2.9) we have $v_0 = p(\Lambda_1^+)^{s-1/2}lv$. Formulas (2.12), (2.14) define operators G_1 and G_2 only in the half-spaces $x_2 > 0$ and $y_2 = x_1 \sin\alpha - x_2\cos\alpha > 0$, respectively. Since $G_1 v_1^+ = 0$ for $x_2 = 0, x_1 < 0$, the odd extension of $G_1 v_1^+$ in x_2 will be a solution of $(-\Delta + 1)v = 0$ in the wedge G. Analogously, the odd extension of $G_2 v_2^-$ in $y_2 = x_1\sin\alpha - x_2\cos\alpha$ will be a solution of $(-\Delta + 1)v = 0$ in G satisfying $G_2 v_2^- = v_2^-$ for $y_2 = x_1\sin\alpha - x_2\cos\alpha = 0$ and $y_1 = -(x_1\cos\alpha + x_2\sin\alpha) < 0$. Therefore, we have the following formulas for $G_1 v_1^+$ when $x_2 < 0$ and for $G_2 v_2^-$ when $y_2 = x_1\sin\alpha - x_2\cos\alpha < 0$:

$$\begin{aligned} G_1 v_1^+ = -\frac{1}{(2\pi)^{n-1}}\int_{-\infty}^{\infty} &\exp\left\{-|x_2|\sqrt{\xi_1^2 + |\hat{\xi}''|^2} - ix_1\xi_1 - i(x'', \xi'')\right\} \\ &\cdot \tilde{v}_1^+(\xi_1, \xi'')\, d\xi_1\, d\xi'', \end{aligned} \tag{3.50}$$

when $x_2 < 0$,

$$G_2 \nu_2^- = -\frac{1}{(2\pi)^{n-1}} \int_{-\infty}^{\infty} \exp\Big\{ |x_1 \sin\alpha - x_2 \cos\alpha| \sqrt{\xi_2^2 + |\xi''|^2} + i(x_1 \cos\alpha + x_2 \sin\alpha)\xi_2 - i(x'', \xi'') \Big\} \tilde{\nu}_2^-(\xi_2, \xi'')\, d\xi_2\, d\xi''. \tag{3.51}$$

Now, using (2.12), (2.14) when $x_2 > 0$ or $x_1 \sin\alpha - x_2 \cos\alpha > 0$ and (3.50), (3.51) when $x_2 < 0$ or $x_1 \sin\alpha - x_2\cos\alpha < 0$, we can repeat all proofs in §2 and, therefore, Lemma 2.3 remains true. In §3 we should change $(\Lambda_1^-)^{s-1/2}$ to $(\Lambda_1^+)^{s-1/2}$, as mentioned before, and also use (3.50), (3.51). Then we can repeat all constructions of §3 with only minor changes.

4. Solution of homogeneous system. In this section we show that the homogeneous system (3.15), (3.19)—i.e., the system

$$p_1^+(b_1 \nu_1^+ + M_2 \nu_2^-) = 0, \qquad x_1 > 0, \tag{4.1}$$

$$p_2^-(b_2 \nu_2^- + M_1 \nu_1^+) = 0, \qquad y_1 < 0, \tag{4.2}$$

—has only a trivial solution in $H_{0,N}$ when $s \in (s_-, s_+)$. For simplicity we take $s^{(0)} = s_+ - 1/2 - \operatorname{Im}(\pi\gamma/\alpha)$. Then (see (3.48), (3.49))

$$\frac{1}{2\pi} \Delta \arg \frac{b_1(\xi_1, \xi'')}{b_2(\xi_1, \xi'')} \Bigg|_{\xi_1 = +\infty}^{-\infty} = 0. \tag{4.3}$$

Without loss of generality we assume that $\xi'' = \omega$, $|\omega| = 1$, and $|\xi''|$ is replaced by 1. This can be achieved by changing variables x_1, y_1 to $x_1/|\xi''|$, $y_1/|\xi''|$. Extend (4.1) to zero for $x_1 < 0$ and (4.2) to zero for $y_1 > 0$, and take the Fourier transform. We obtain

$$\Pi^+ b_1(\xi_1, \omega)\tilde{\nu}_1^+(\xi_1, \omega) + \frac{1}{2\pi} \int_{-\infty}^{\infty} \frac{M_2(\xi_1, \omega)\tilde{\nu}_2^-(\zeta_1, \omega)\, d\xi_1}{\sin\alpha\sqrt{\xi_1^2 + 1} - i\xi_1 \cos\alpha - i\zeta_1} = 0, \tag{4.4}$$

$$\Pi^- b_2(\xi_1, \omega)\tilde{\nu}_2^-(\xi_1, \omega) + \frac{1}{2\pi} \int_{-\infty}^{\infty} \frac{M_1(\xi_1, \omega)\tilde{\nu}_1^+(\xi_1, \omega)\, d\xi_1}{\sin\alpha\sqrt{\xi_1^2 + 1} + i\xi_1 \cos\alpha + i\zeta_1} = 0, \tag{4.5}$$

where

$$\Pi^{\pm} f(\xi_1) = \pm \frac{i}{2\pi} \int_{-\infty}^{\infty} \frac{f(\xi_1)\, d\xi_1}{\zeta_1 - \xi_1 \pm i0}. \tag{4.6}$$

Denote the hyperbola

$$z = \xi_1 \cos\beta + i \sin\beta \sqrt{\xi_1^2 + 1}, \qquad -\infty < \xi_1 < +\infty, \tag{4.7}$$

by H_β. Denote the region in the complex plane $\mathbf{C}$ above H_β by $\mathcal{D}_\beta^+$; the region below H_β, by $\mathcal{D}_\beta^-$; $-\pi/2 < \beta < \pi/2$. We first consider the case $\pi/2 < \alpha \leqslant \pi$. The case $0 < \alpha \leqslant \pi/2$ is considered in Remark 4.2. We show that $\tilde{\nu}_1^-(z_1, \omega)$ can be extended from the half-plane $\operatorname{Im} z < 0$ to the domain $\mathcal{D}_{\pi-\alpha}^-$ as a meromorphic function. The second integral in (4.5) is obviously analytic when $\zeta_1 \in \mathscr{D}^-_{\pi-\alpha}$. Therefore, $\Pi^- b_2(\xi_1, \omega)\tilde{v}_2^-(\xi_1, \omega)$ has analytic extension from $\operatorname{Im}\zeta_1 < 0$ to $\mathscr{D}^-_{\pi-\alpha}$. We have

$$(4.8) \qquad b_2(\xi_1, \omega)\tilde{v}_2^-(\xi_1, \omega) = \Pi^+ b_2\tilde{v}_2^- + \Pi^- b_2\tilde{v}_2^- .$$

Since $\Pi^+ b_2\tilde{v}_2^-$ is analytic for $\operatorname{Im}\zeta_1 > 0$, then $b_2(\xi_1, \omega)\tilde{v}_2^-(\xi_2, \omega)$ can be extended analytically to $\mathscr{D}^-_{\pi-\alpha} \cap \{\operatorname{Im} z > 0\}$. Since $\tilde{v}_2^-(z, \omega)$ is analytic for $\operatorname{Im} z < 0$ and $b_2(\xi_1, \omega)$ can be extneded analytically to $\mathscr{D}^-_{\pi-\alpha} \cap \{\operatorname{Im} z \geqslant 0\}$, we obtain that $\tilde{v}_2^-(z, \omega)$ is analytic in $\mathscr{D}^-_{\pi-\alpha}$ except at a finite number of poles in $\overline{\mathscr{D}^-_{\pi-\alpha}}$ that are zeros of $b_z(z, \omega)$. Consider first the case when there are no zeros of $b_2(z, \omega)$ in $\overline{\mathscr{D}^-_{\pi-\alpha}} \cap \{\operatorname{Im} z > 0\}$. Then $\tilde{v}_2^-(z, \omega)$ is analytic in $\mathscr{D}^-_{\pi-\alpha}$. Set

$$(4.7') \qquad \phi_\beta(z) = z\cos\beta + i\sin\beta\sqrt{z^2+1}\,,$$

where $\sqrt{z^2+1}$ is the positive branch for real z. Note that $\phi_\beta(z)$ is a single-valued analytic function in $\mathscr{D}^-_{\pi/2} \cap \mathscr{D}^+_{-\pi/2}$—i.e., in the complex plane $\mathbf{C}$ with cuts along the imaginary axis from $-i$ to $-i\infty$ and from $+i$ to $+i\infty$. Make the change of variable

$$(4.9) \qquad z = \tfrac{1}{2}(w - 1/w).$$

Note that (4.9) maps the right half-plane $\operatorname{Re} w > 0$ onto $\mathscr{D}_{\pi/2} \cap \mathscr{D}_{-\pi/2}$. We have

$$(4.10) \qquad \begin{aligned} \phi_\beta\left(\frac{1}{2}\left(w - \frac{1}{w}\right)\right) &= \frac{1}{2}\left(\left(w - \frac{1}{w}\right)\cos\beta + i\sin\beta\left(w + \frac{1}{w}\right)\right) \\ &= \frac{1}{2}\left(we^{i\beta} - \frac{1}{we^{i\beta}}\right), \end{aligned}$$

so ϕ_β corresponds to a rotation by the angle β in the half-plane $\operatorname{Re} w > 0$. Note that $\phi_{\pi-\alpha}(\frac{1}{2}(w - 1/w))$ maps conformally the sector $-(\pi-\alpha) \leqslant \arg w \leqslant 0$ onto the domain $\overline{\mathscr{D}_{\pi-\alpha}} \cap \{\operatorname{Im} z \geqslant 0\}$. Therefore, $\phi_{\pi-\alpha}(z)$ maps conformally $\overline{\mathscr{D}^+_{-(\pi-\alpha)}} \cap \{\operatorname{Im} z \leqslant 0\}$ onto $\overline{\mathscr{D}^-_{\pi-\alpha}} \cap \{\operatorname{Im} z \geqslant 0\}$, and the inverse function is equal to $\phi_{-(\pi-\alpha)}(z)$: $\phi^{-1}_{\pi-\alpha}(z) = \phi_{-(\pi-\alpha)}(z)$ for $z \in \overline{\mathscr{D}^-_{\pi-\alpha}} \cap \{\operatorname{Im} z \geqslant 0\}$. Make the change of variable $z = \phi_{-(\pi-\alpha)}(\xi_1)$ in the second integral in (4.4). We obtain

$$(4.11) \qquad \begin{aligned} &\Pi^+ b_1(\xi_1, \omega)\tilde{v}_1^+(\xi_1, \omega) \\ &\quad + \frac{1}{2\pi}\int_{H_{-(\pi-\alpha)}} \frac{M_2(\phi_{\pi-\alpha}(z), \omega)\tilde{v}_2^-(\phi_{\pi-\alpha}(z), \omega)a(z)\,dz}{iz - i\zeta_1} = 0, \end{aligned}$$

where

$$(4.12) \qquad a(z) = \frac{d\phi_{\pi-\alpha}(z)}{dz} = -\cos\alpha + i\sin\alpha\frac{z}{\sqrt{z^2+1}}.$$

Using the Cauchy theorem we can deform $H_{-(\pi-\alpha)}$ into the real axis:

$$(4.13) \qquad \begin{aligned} &\Pi^+ b_1(\xi_1, \omega)\tilde{v}_1^+(\xi_1, \omega) \\ &\quad + \frac{i}{2\pi}\int_{-\infty}^{\infty} \frac{M_2(\phi_{\pi-\alpha}(\xi_1), \omega)\tilde{v}_2^-(\phi_{\pi-\alpha}(\xi_1), \omega)a(\xi_1)\,d\xi_1}{\zeta_1 - \xi_1 + i0} = 0. \end{aligned}$$

In (4.13) we used the fact that $\tilde{v}_2^-(z)$ has no poles in $\overline{\mathscr{D}_{\pi-\alpha}^-} \cap \{\operatorname{Im} z > 0\}$. We have

$$M_2(\phi_{\pi-\alpha}(\xi_1), \omega) = b_1(\xi_1, \omega). \tag{4.14}$$

Indeed,

$$M_2(\xi_1, \omega) = B_1\left(\phi_{-(\pi-\alpha)}(\xi_1), -i\sqrt{(\phi_{-(\pi-\alpha)}(\xi_1))^2 + 1}, \omega\right)(\phi_{\pi-\alpha}(\xi_1) - i)^{-m_1},$$

and this equality can be extended analytically to $\mathscr{D}_{\pi-\alpha}^- \cap \{\operatorname{Im} z > 0\}$. Therefore, (4.13) can be rewritten as

$$\Pi^+(b_1(\xi_1, \omega))(\tilde{v}_1^+(\xi_1, \omega) + a(\xi_1)\tilde{v}_2^-(\phi_{\pi-\alpha}(\xi_1), \omega)) = 0. \tag{4.15}$$

It follows from (4.15) that

$$b_1(\xi_1, \omega)(\tilde{v}_1^+(\xi_1, \omega) + a(\xi_1)\tilde{v}_2^-(\phi_{\pi-\alpha}(\xi_1), \omega)) = w_-(\xi_1), \tag{4.16}$$

where $w_-(\xi_1)$ is analytic for $\operatorname{Im} \xi_1 < 0$. Now consider equation (4.5). Changing the variable $z = \phi_{-(\pi-\alpha)}(\xi_1)$ in the first integral in (4.5), we obtain

$$\begin{aligned} -\frac{i}{2\pi} \int_{H_{-(\pi-\alpha)}} & \frac{b_2(\phi_{\pi-\alpha}(z), \omega)\tilde{v}_2^-(\phi_{\pi-\alpha}(z), \omega)a(z)\, dz}{\zeta_1 - \phi_{\pi-\alpha}(z) - i0} \\ & - \frac{i}{2\pi} \int_{-\infty}^{\infty} \frac{M_1(\xi_1, \omega)\tilde{v}_1^+(\xi_1, \omega)\, d\xi_1}{\zeta_1 - \phi_{\pi-\alpha}(\xi_1)} = 0. \end{aligned} \tag{4.17}$$

Using the Cauchy theorem and the fact that

$$b_2(\phi_{\pi-\alpha}(\xi_1), \omega) = M_1(\xi_1, \omega), \tag{4.18}$$

we obtain

$$-\frac{i}{2\pi} \int_{-\infty}^{\infty} \frac{b_2(\phi_{\pi-\alpha}(\xi_1), \omega)(\tilde{v}_1^+(\xi_1, \omega) + a(\xi_1)\tilde{v}_2^-(\phi_{\pi-\alpha}(\xi_1), w))}{\zeta_1 - \phi_{\pi-\alpha}(\xi_1)} = 0. \tag{4.19}$$

Set

$$w_+(z) = -\frac{i}{2\pi} \int_{-\infty}^{\infty} \frac{C(\xi_1)\, d\xi_1}{z - \phi_{\pi-\alpha}(\xi_1)}, \tag{4.20}$$

where

$$C(\xi_1) = b_2(\phi_{\pi-\alpha}(\xi_1), \omega)(\tilde{v}_1^+(\xi_1, \omega) + a(\xi_1)\tilde{v}_2^-(\phi_{\pi-\alpha}(\xi_1), \omega)).$$

The function $w_+(z)$ is analytic in $\mathscr{D}_{\pi-\alpha}^+$, and (4.19) implies that $w_+(z) = 0$ in $\mathscr{D}_{\pi-\alpha}^-$. The formula for the jump of the Cauchy-type integral gives

$$w_+(\phi_{\pi-\alpha}(\zeta_1) + i0) - w_+(\phi_{\pi-\alpha}(\zeta_1) - i0) = C(\zeta_1)(d\phi_{\pi-\alpha}(\zeta_1)/d\zeta_1)^{-1}. \tag{4.21}$$

Therefore,

$$\begin{aligned} b_2(\phi_{\pi-\alpha}(\xi_1), \omega)(\tilde{v}_1^+(\xi_1, \omega) + a(\xi_1)\tilde{v}_2^-(\phi_{\pi-\alpha}(\xi_1), \omega)) & \\ = a(\xi_1)w_+(\phi_{\pi-\alpha}(\xi_1)). & \end{aligned} \tag{4.22}$$

Substituting (4.16) into (4.22), we obtain

$$(4.23)\qquad b_1^{-1}(\xi_1,\omega)b_2(\phi_{\pi-\alpha}(\xi_1),\omega)a^{-1}(\xi_1)w_-(\xi_1) = w_+(\phi_{\pi-a}(\xi_1)).$$

Since we assumed that $b_2(z,\omega)$ has no zeros in $\overline{\mathscr{D}_{\pi-\alpha}^-}\cap\{\operatorname{Im} z>0\}$, we have

$$(1/2\pi)\Delta\arg b_2(\xi_1,\omega)\big|_{+\infty}^{-\infty} = (1/2\pi)\Delta\arg b_2(\phi_{\pi-\alpha}(\xi_1),\omega)\big|_{+\infty}^{-\infty}.$$

Therefore, (4.3) implies

$$(4.24)\qquad (1/2\pi)\Delta\arg b_1^{-1}(\xi_1,\omega)b_2(\phi_{\pi-\alpha}(\xi_1),\omega)\}\big|_{\xi_1=+\infty}^{-\infty} = 0.$$

Equation (4.23) is an example of a Riemann–Hilbert problem with a shift (see [**3**]). The solution of such a problem can be reduced to the solution of a usual Riemann–Hilbert problem without shift. We take a conformal mapping $\delta^+(z)$ of the upper half-plane $\operatorname{Im} z>0$ onto $\mathscr{D}_{\pi-\alpha}^+$ such that

$$(4.25)\qquad w^+(\phi_{\pi-\alpha}(\xi_1)) = w^+(\delta^+(\xi_1)).$$

It is proven in [**3**] that such a $\delta^+(z)$ exists. Then (4.23) becomes the Riemann–Hilbert problem without shift:

$$(4.25')\qquad b_1^{-1}(\xi_1,\omega)b_2(\phi_{\pi-\alpha}(\xi_1),\omega)a^{-1}(\xi_1)w_-(\xi_1) = w_+(\delta^+(\xi_1)),$$

where $w_+(\delta^+(z))$ is analytic for $\operatorname{Im} z>0$ and $w_-(z)$ is analytic for $\operatorname{Im} z<0$. Note that

$$(4.26)\qquad \frac{1}{2\pi}\Delta\arg a(\xi_1)\Big|_{\xi_1=-\infty}^{+\infty} = \frac{\pi-\alpha}{\pi}<\frac{1}{2}.$$

Therefore (4.26) and (4.24) imply that (4.25′) has only a trivial solution $w_+=0$, $w_-=0$ in the class of w_+, w_- vanishing at infinity. When $w_-(\xi_1)=0$, we obtain from (4.16) that

$$(4.27)\qquad \tilde{v}_1^+(\xi_1,\omega)+a(\xi_1)\tilde{v}_2^-(\phi_{\pi-\alpha}(\xi_1),\omega)=0.$$

It is also a Riemann–Hilbert problem with a shift. Analogously to (4.23), we can reduce this to a Riemann–Hilbert problem without shift by taking the conformal map $\delta^-(z)$ of $\operatorname{Im} z<0$ onto $\mathscr{D}_{\pi-\alpha}^-$. It follows from (4.26) that (4.27) has only a trivial solution. Therefore, we proved that the system (4.1), (4.2) has only a trivial solution, assuming that $b_2(z,w)$ has no zeros in $\overline{\mathscr{D}_{\pi-\alpha}^-}\cap\{\operatorname{Im} z>0\}$. Now consider the case when $b_2(z,\omega)$ has $m>0$ zeros $z_1,\dots,z_m$ in $\overline{\mathscr{D}_{\pi-\alpha}^-}\cap\{\operatorname{Im} z>0\}$. We assume, only for notational simplicity, that all zeros are simple. It follows from (4.7) that $\tilde{v}_2^-(z,\omega)$ is a meromorphic function in $\mathscr{D}_{\pi-\alpha}^-$ that may have poles at $z_1,\dots,z_m$ such that $b_2(z,\omega)\tilde{v}_2^-(z,\omega)$ is analytic in $\mathscr{D}_{\pi-\alpha}^-$. Therefore (4.22) holds without change. Instead of (4.15) we obtain

$$(4.28)\qquad \begin{aligned}\Pi^+ b_1(\xi_1,\omega)(\tilde{v}_1^+(\xi_1,\omega)+a(\xi_1)\tilde{v}_2^-(\phi_{\pi-\alpha}(\xi_1),\omega))&\\ +\sum_{k=1}^m\frac{b_1(z_k,\omega)a(z_k)c_k}{\xi_1-z_k}&=0,\end{aligned}$$

where c_k are the residues of $\tilde{v}_1^+(z,\omega)$ at z_k, $k=1,\dots,m$. Therefore

$$(4.29)\qquad \begin{aligned}b_1(\xi_1,\omega)(\tilde{v}_1^+(\xi_1,\omega)+a(\xi_1)\tilde{v}_2^-(\phi_{\pi-\alpha}(\xi_1),\omega))&\\ =w_-(\xi_1)-\sum_{k=1}^m\frac{b_1(z_k,\omega)}{\xi_1-z_k}a(z_k)c_k&,\end{aligned}$$

where $w_-(z)$ is analytic for $\operatorname{Im} z < 0$. Substituting (4.22) into (4.29), we obtain

$$\begin{aligned}&b_1(\xi_1,\omega)b_2^{-1}(\phi_{\pi-\alpha}(\xi_1),\omega)a(\xi_1)w_+(\phi_{\pi-\alpha}(\xi_1))\\ &\qquad = w_-(\xi_1) - \sum_{k=1}^{m}\frac{b_1(z_k,\omega)}{\xi_1 - z_k}a(z_k)c_k.\end{aligned}\tag{4.30}$$

Since $b_2(z,\omega)$ has m zeros in $\mathscr{D}^-_{\pi-\alpha}\cap\{\operatorname{Im} z > 0\}$, we obtain, taking into account (4.3),

$$(1/2\pi)\Delta\arg b_1(\xi_1,\omega)b_2^{-1}(\phi_{\pi-\alpha}(\xi_1),\omega)\Big|_{+\infty}^{-\infty} = -m.\tag{4.31}$$

Therefore the solution of equation (4.30) exists only if the right side $\sum_{k=1}^m b_1(z_k,\omega)(\xi_1 - z_k)^{-1}a(z_k)c_k$ satisfies m orthogonality conditions (see [**3**]). Therefore c_k must be zeros, $k = 1,\dots,m$, and there is only a trivial solution $w_+ = 0$, $w_- = 0$ of (4.30). When $w_- = 0$ and $c_k = 0,\dots,c_m = 0$, there is only a trivial solution of (4.28). So that in the case $\pi/2 < \alpha \leqslant \pi$, we have proved

THEOREM 4.1. *Under the same assumptions as in Theorem* 3.1 *and the additional assumption that* $s\in(s_-,s_+)$, *there exists a unique solution* $u\in H_{s,N}(G)$ *of the boundary-value problem* (3.45)–(3.47).

REMARK 4.1. We have proved that the homogeneous system (4.4), (4.5) has only a trivial solution in $H_{0,N}$. Indeed, the same method allows us to solve the nonhomogeneous system

$$\begin{aligned}&\Pi^+ b_1(\xi_1,\omega)\tilde v_1^+(\xi_1,\omega)\\ &\qquad + \frac{1}{2\pi}\int_{-\infty}^{\infty}\frac{M_2(\xi_1,\omega)\tilde v_2^-(\xi_1,\omega)\,d\xi_1}{\sin\alpha\sqrt{\xi_1^2+1} - i\xi_1\cos\alpha - i\zeta_1} = \tilde h_1^+(\xi_1),\end{aligned}\tag{4.32}$$

$$\begin{aligned}&\Pi^- b_2(\xi_1,\omega)\tilde v_2^-(\xi_1,\omega)\\ &\qquad + \frac{1}{2\pi}\int_{-\infty}^{\infty}\frac{M_1(\xi_1,\omega)\tilde v_1^+(\xi_1,\omega)\,d\xi_1}{\sin\alpha\sqrt{\xi_1^2+1} + i\xi_1\cos\alpha + i\zeta_1} = \tilde h_2^-(\xi_1).\end{aligned}\tag{4.33}$$

We consider separately two cases: (a) $h_2^-(y_1) = 0$, $h_1^+(x_1)\in H_{0,N}(\mathbf{R}^1_+)$ is arbitrary, $h_1^+ = 0$ for $x_1 < 0$; and (b) $h_1^+ = 0$, $h_2^-(y_1)\in H_{0,N}(\mathbf{R}^1_-)$ is arbitrary, $h_2^-(y_1) = 0$ for $y_1 > 0$. Then the solution for the general case will be the sum of the solutions in cases (a) and (b). Consider the solution in case (a). Case (b) can be treated analogously. Since $\tilde h_2^-(\xi_1) = 0$, we have, analogously to (4.5), that $b_2(z,\omega)\tilde v_2^-(z,\omega)$ can be extended as an analytic function to $\mathscr{D}^-_{\pi-\alpha}\cap\{\operatorname{Im} z\geqslant 0\}$. Consider, for simplicity, the case when $b_2(z,\omega)$ has no zeros in $\mathscr{D}^-_{\pi-\alpha}\cap\{\operatorname{Im} z > 0\}$. Then $\tilde v_2^-(z,\omega)$ is analytic in $\mathscr{D}^-_{\pi-\alpha}\cap\{\operatorname{Im} z\leqslant 0\}$. Analogously to (4.16), we have

$$b_1(\xi_1,\omega)(\tilde v_1^+(\xi_1,\omega) + a(\xi_1)\tilde v_2^-(\phi_{\pi-\alpha}(\xi_1),\omega)) = w_-(\xi_1) + \tilde h_1^+(\xi_1),\tag{4.34}$$

where $w_-(z)$ is analytic for $\operatorname{Im} z < 0$. As in (4.22) we have

$$\begin{aligned}&b_2(\phi_{\pi-\alpha}(\xi_1),\omega)(\tilde v_1^+(\xi_1,\omega) + a(\xi_1)\tilde v_2^-(\phi_{\pi-\alpha}(\xi_1),\omega)\\ &\qquad = a(\xi_1)w^+(\phi_{\pi-\alpha}(\xi_1))).\end{aligned}\tag{4.35}$$

Substituting (4.35) into (4.34), we obtain

$$b_1(\xi_1,\omega)b_2^{-1}(\phi_{\pi-\alpha}(\xi_1),\omega)a(\xi_1)w^+(\phi_{\pi-a}(\xi_1)) = w_-(\xi_1)+\tilde{h}_1^+(\xi_1). \tag{4.36}$$

Equation (4.36) is a nonhomogeneous Riemann–Hilbert problem with a shift. Solving this problem (see **[3]**), we find $w_+(z)$, $w_-(z)$. When $w_-(\xi_1)$ is known, then (4.34) is also a nonhomogeneous Riemann–Hilbert problem with a shift. Solving this problem, we find $\tilde{\nu}_1^+(\xi_1)$, $\tilde{\nu}_2^-(\xi_1)$.

REMARK 4.2. In this remark we prove that (4.4), (4.5) has only a trivial solution for angles $0<\alpha\leqslant\pi/2$. Take γ such that

$$0<\gamma<\pi/2,\quad 0<\pi-\alpha-\gamma<\pi/2,\quad \text{i.e.}\quad \pi/2-\alpha<\gamma<\pi/2. \tag{4.37}$$

The function $\phi_\beta(z)$ is a well-defined analytic function on the Riemann surface of $\sqrt{z^2+1}$. Since $\pi-\alpha>\pi/2$, $\phi_{\pi-\alpha}(z)$ maps the real line $-\infty<\xi_1<+\infty$ on the second sheet of the Riemann surface of $\sqrt{z^2+1}$. If γ satisfies (4.37), then

$$\phi_{\pi-\alpha-\gamma}(z)=\phi_{\pi-\alpha}(\phi_{-\gamma}(z))\quad \text{for } z\in H_\gamma \tag{4.38}$$

and

$$\phi_{-(\pi-\alpha)}(\phi_{\pi-\alpha-\gamma}(z))=\phi_{-\gamma}(z)\quad \text{for } z\in H_{-(\pi-\alpha-\gamma)}. \tag{4.39}$$

To prove (4.38), (4.39) one should use (4.9), (4.10). Therefore, rotating the map $\phi_{\pi-\alpha}$ back by the angle $-\gamma$, we obtain that the maps $\phi_{\pi-\alpha-\gamma}$ and $\phi_{-\gamma}$ act on the same sheet of the Riemann surface of $\sqrt{z^2+1}$. We assume for simplicity that $b_2(z,\omega)$ has no zeros in $\overline{\mathscr{D}_{\pi-\alpha-\gamma}^-}\cap\{\operatorname{Im} z>0\}$ and $b_1(z,\omega)$ has no zeros in $\overline{\mathscr{D}_{-\gamma}^+}\cap\{\operatorname{Im} z<0\}$. The modifications of the proof for the case when there are zeros are the same as in (4.28)–(4.31). It follows from (4.5) that $\Pi^-b_2(\xi_1,\omega)\tilde{\nu}_2^-(\xi_1,\omega)$ can be extended analytically to $\mathscr{D}_\alpha^-\cap\{\operatorname{Im} z>0\}$ since $\phi_{\pi-\alpha}(\xi_1)=\phi_\alpha(-\xi_1)$ and the second integral in (4.5) is analytic in $\mathscr{D}_\alpha^-$. Therefore (4.8) implies that $b_2(z,\omega)\tilde{\nu}_2^-(z,\omega)$ is analytic in $\mathscr{D}_\alpha^-\cap\{\operatorname{Im} z>0\}$. Similarly, it follows from (4.4) that $b_1(z,\omega)\tilde{\nu}_1^+(z,\omega)$ can be extended analytically to $\mathscr{D}_{-\alpha}^+\cap\{\operatorname{Im} z<0\}$. Suppose $\pi/3<\alpha\leqslant\pi/2$. Then one can choose γ such that $\pi-\alpha-\gamma\leqslant\alpha$, $\gamma\leqslant\alpha$. Therefore, $\mathscr{D}_{-\gamma}^+\subset\mathscr{D}_{-\alpha}^+$ and $\mathscr{D}_{\pi-\alpha-\gamma}^-\subset\mathscr{D}_\alpha^-$. Analogously to (4.11), (4.13) we make the change of variable $z=\phi_{-(\pi-\alpha-\gamma)}(\xi_1)$ in the second integral in (4.4). Then using the Cauchy theorem and (4.39), we obtain

$$\begin{aligned}&\Pi^+b_1(\xi_1,\omega)\tilde{\nu}_1^+(\xi_1,\omega)\\ &\quad+\frac{i}{2\pi}\int_{-\infty}^{\infty}\frac{M_2(\phi_{\pi-\alpha-\gamma}(\xi_1),\omega)\tilde{\nu}_2^-(\phi_{\pi-\alpha-\gamma}(\xi_1))\,(d\phi_{\pi-\alpha-\gamma}/d\xi_1)\,d\xi_1}{\zeta_1-\phi_{-\gamma}(\xi_1)}\\ &=0.\end{aligned} \tag{4.40}$$

Analogously making the change of variable $z=\phi_\gamma(\xi_1)$ in the second integral in (4.5) and using the Cauchy theorem and (4.38), we obtain

$$\begin{aligned}&\Pi^-b_2(\xi_1,\omega)\tilde{\nu}_2^-(\xi_1,\omega)\\ &\quad-\frac{i}{2\pi}\int_{-\infty}^{\infty}\frac{M_1(\phi_{-\gamma}(\xi_1),\omega)\tilde{\nu}_1^+(\phi_{-\gamma}(\xi_1),\omega)\,(d\phi_{-\gamma}/d\xi_1)\,d\xi_1}{\zeta_1-\phi_{\pi-\alpha-\gamma}(\xi_1)}=0.\end{aligned} \tag{4.41}$$

Now change the variable $\xi_1 = \phi_{-\gamma}(z)$ in the first integral in (4.40). Using the Cauchy theorem, we obtain, analogously to (4.17), (4.19),

(4.42)

$$\frac{i}{2\pi}\int_{-\infty}^{\infty}\frac{b_1(\phi_{-\gamma}(\xi_1),\omega)\tilde{v}_1^+(\phi_{-\gamma}(\xi_1),\omega)\,(d\phi_{-\gamma}/d\xi_1)\,d\xi_1}{\zeta_1-\phi_{-\gamma}(\xi_1)}$$
$$+\frac{i}{2\pi}\int_{-\infty}^{\infty}\frac{M_2(\phi_{\pi-\alpha-\gamma}(\xi_1),\omega)\tilde{v}_2^-(\phi_{\pi-\alpha-\gamma}(\xi_1),\omega)\,(d\phi_{\pi-\alpha-\gamma}/d\xi_1)\,d\xi_1}{\zeta_1-\phi_{-\gamma}(\xi_1)}$$
$$=0.$$

Note that (cf. (4.14))

$$b_1(\phi_{-\gamma}(\xi_1),\omega)=M_2(\phi_{\pi-\alpha-\gamma}(\xi_1),\omega). \tag{4.43}$$

Therefore (4.42) can be written as

(4.44)

$$\frac{i}{2\pi}\int_{-\infty}^{\infty}\frac{b_1(\phi_{-\gamma}(\xi_1),\omega)\big(\tilde{v}_1^+(\phi_{-\gamma}(\xi_1),\omega)\,d\phi_{-\gamma}/d\xi_1+\tilde{v}_2^-(\phi_{\pi-\alpha-\gamma}(\xi_1),\omega)\,d\phi_{\pi-\alpha-\gamma}/d\xi_1\big)\,d\xi_1}{\zeta_1-\phi_{-\gamma}(\xi_1)}$$
$$=0.$$

Analogously, we make the change of variable $\xi_1 = \phi_{\pi-\alpha-\gamma}(z)$ in the first integral in (4.41). Using the Cauchy theorem, we obtain, analogously to (4.42),

(4.45)

$$-\frac{i}{2\pi}\int_{-\infty}^{\infty}\frac{b_2(\phi_{\pi-\alpha-\gamma}(\xi_1),\omega)\tilde{v}_2^-(\phi_{\pi-\alpha-\gamma}(\xi_1),\omega)\,((d\phi_{\pi-\alpha-\gamma})/d\xi_1)\,d\xi_1}{\zeta_1-\phi_{\pi-\alpha-\gamma}(\xi_1)}$$
$$-\frac{i}{2\pi}\int_{-\infty}^{\infty}\frac{M_1(\phi_{-\gamma}(\xi_1),\omega)\tilde{v}_1^+(\phi_{-\gamma}(\xi_1),\omega)\,(d\phi_{-\gamma}/d\xi_1)\,d\xi_1}{\zeta_1-\phi_{\pi-\alpha-\gamma}(\xi_1)}=0.$$

Since (cf. (4.18))

$$b_2(\phi_{\pi-\alpha-\gamma}(\xi_1),\omega)=M_1(\phi_{-\gamma}(\xi_1),\omega), \tag{4.46}$$

we obtain

(4.47)

$$-\frac{i}{2\pi}\int_{-\infty}^{\infty}b_2(\phi_{\pi-\alpha-\gamma}(\xi_1),\omega)$$
$$\cdot\big(\tilde{v}_2^-(\phi_{\pi-\alpha-\gamma}(\xi_1),\omega)\big)\frac{d\phi_{\pi-\alpha-\gamma}}{d\xi_1}+\tilde{v}_1^+\left(\frac{\phi_{-\gamma}(\xi_1),\omega\, d\phi_{-\gamma}}{d\xi_1}\right)\frac{1}{\zeta_1-\phi_{\pi-\alpha-\gamma}(\xi_1)}\,d\xi_1$$
$$=0.$$

Analogously to (4.19)–(4.22), we obtain from (4.44) and (4.47), using the formula for the jump of the Cauchy-type integral (cf. (4.21)),

(4.48)

$$\frac{d\phi_{-\gamma}(\xi_1)}{d\xi_1}b_1(\phi_{-\gamma}(\xi_1),\omega)\left(\tilde{v}_1^+(\phi_{-\gamma}(\xi_1),\omega)\frac{d\phi_{-\gamma}}{d\xi_1}+\tilde{v}_2^-(\phi_{\pi-\alpha-\gamma}(\xi_1),\omega)\frac{d\phi_{\pi-\alpha-\gamma}}{d\xi_1}\right)$$
$$=w_-(\phi_{-\gamma}(\xi_1)),$$

$$\begin{aligned}
&\frac{d\phi_{\pi-\alpha-\gamma}}{d\xi_1} b_2(\phi_{\pi-\alpha-\gamma}(\xi_1), \omega) \\
&\quad\cdot\left(\tilde{\nu}_1^+(\phi_{-\gamma}(\xi_1), \omega)\frac{d\phi_{-\gamma}}{d\xi_1} + \tilde{\nu}_2^-(\phi_{\pi-\alpha-\gamma}(\xi_1), \omega)\frac{d\phi_{\pi-\alpha-\gamma}}{d\xi_1}\right) \\
&= w_+(\phi_{\pi-\alpha-\gamma}(\xi_1)),
\end{aligned} \tag{4.49}$$

where $w_+(z)$ is analytic in $\mathscr{D}^+_{\pi-\alpha-\gamma}$ and $w_-(z)$ is analytic in $\mathscr{D}^-_{-\gamma}$. It follows from (4.48), (4.49) that

$$\begin{aligned}
&b_1(\phi_{-\gamma}(\xi_1), \omega) b_2^{-1}(\phi_{\pi-\alpha-\gamma}(\xi_1), \omega)\frac{d\phi_{-\gamma}}{d\xi_1}\left(\frac{d\phi_{\pi-\alpha-\gamma}}{d\xi_1}\right)^{-1} w_+(\phi_{\pi-\alpha-\gamma}(\xi_1)) \\
&= w_-(\phi_{-\gamma}(\xi_1)).
\end{aligned} \tag{4.50}$$

Since we have assumed that $b_1(z, \omega)$ has no zeros in $\overline{\mathscr{D}^+_{-\gamma}} \cap \{\operatorname{Im} z < 0\}$ and $b_2(z, \omega)$ has no zeros in $\overline{\mathscr{D}^-_{\pi-\alpha-\gamma}} \cap \{\operatorname{Im} z > 0\}$ we have

$$(1/2\pi)\Delta \arg b_1(\phi_{-\gamma}(\xi_1), \omega) b_2^{-1}(\phi_{\pi-\alpha-\gamma}(\xi_1), \omega)\Big|_{\xi_1=+\infty}^{-\infty} = 0. \tag{4.51}$$

Note that (cf. (4.26))

$$\frac{1}{2\pi}\Delta \arg \frac{d\phi_{-\gamma}}{d\xi_1}\left(\frac{d\phi_{\pi-\alpha-\gamma}}{d\xi_1}\right)^{-1}\Bigg|_{\xi_1=+\infty}^{-\infty} = -\frac{\pi-\alpha}{\pi}. \tag{4.52}$$

Therefore problem (4.50), which is the Riemann–Hilbert problem with a shift, has only a trivial solution $w_+ = 0$, $w_- = 0$ (cf. (4.23)), so that, from (4.49),

$$\tilde{\nu}_1^+(\phi_{-\gamma}(\xi_1, \omega))\frac{d\phi_{-\gamma}}{d\xi_1} + \tilde{\nu}_2^-(\phi_{\pi-\alpha-\gamma}(\xi_1), \omega)\frac{d\phi_{\pi-\alpha-\gamma}}{d\xi_1} = 0. \tag{4.53}$$

This Riemann–Hilbert problem also has only a trivial solution (cf. (4.27)). Therefore, $\tilde{\nu}_1^+ = 0$ and $\tilde{\nu}_2^- = 0$. It remains to consider the case when $\gamma > \alpha$ or $\pi - \alpha - \gamma > \alpha$. Let $\alpha_1 = \alpha$ if $\alpha < \pi/4$ and $\alpha_1 = \pi/2 - \alpha$ if $\alpha \geqslant \pi/4$. Since $b_2(z, w)\tilde{\nu}_2^-(z, w)$ is analytic in $\mathscr{D}^-_\alpha \cap \{\operatorname{Im} z > 0\}$, we can make a change of variable $z = \phi_{-\alpha_1}(\xi_1)$ in the second integral in (4.4) and use the Cauchy theorem. Taking into account that

$$\phi_{-(\pi-\alpha)}(\phi_{\alpha_1}(z)) = \phi_{-(\pi-\alpha-\alpha_1)}(z), \qquad z \in H_{-\alpha_1},$$

we obtain (cf. (4.40))

$$\begin{aligned}
&\Pi^+ b_1(\xi_1, \omega)\tilde{\nu}_1^+(\xi_1, \omega) \\
&\quad + \frac{i}{2\pi}\int_{-\infty}^{\infty}\frac{M_2(\phi_{\alpha_1}(\xi_1), \omega)\tilde{\nu}_2^-(\phi_{\alpha_1}(\xi_1))(d\phi_{\alpha_1}/d\xi_1)\,d\xi_1}{\zeta_1 - \phi_{-(\pi-\alpha-\alpha_1)}(\xi_1)} = 0.
\end{aligned} \tag{4.54}$$

Now the second integral in (4.54) is analytic for $\zeta_1 \in \mathscr{D}^+_{-(\alpha+\alpha_1)}$. As before, this implies that $b_1(z, \omega)\tilde{\nu}_1^+(z, \omega)$ has an analytic extension to $\mathscr{D}^+_{-(\alpha_1+\alpha)} \cap \{\operatorname{Im} z < 0\}$. Analogously making the change of variable $z = \phi_{\alpha_1}(\xi_1)$ in the second integral in

(4.5) and using the Cauchy theorem and the relation $\phi_{\pi-\alpha}(\phi_{-\alpha_1}(z)) = \phi_{\pi-\alpha-\alpha_1}(z)$ for $z \in H_{\alpha_1}$, we obtain (cf. (4.11))

$$\Pi^- b_2(\xi_1, \omega)\tilde{v}_2^-(\xi_1, \omega) \tag{4.55}$$
$$-\frac{i}{2\pi}\int_{-\infty}^{\infty}\frac{M_1(\phi_{-\alpha_1}(\xi_1), \omega)\tilde{v}_1(\phi_{-\alpha_1}(\xi_1), \omega)\,(d\phi_{-\alpha_1}/d\xi_1)\,d\xi_1}{\zeta_1 - \phi_{\pi-\alpha-\alpha_1}(\xi_1)} = 0.$$

It follows from (4.55) that $\Pi^- b_2(\xi_1, \omega)\tilde{v}_2^-(\xi_1, \omega)$ has an analytic extension to $\mathscr{D}^-_{\alpha+\alpha_1} \cap \{\operatorname{Im} z > 0\}$, and, therefore, $b_2(z, \omega)\tilde{v}_2^-(z, \omega)$ is analytic there. If $\mathscr{D}^+_{-(\alpha+\alpha_1)} \supset \mathscr{D}_{-\gamma}$ and $\mathscr{D}^-_{\alpha+\alpha_1} \supset \mathscr{D}^-_{\pi-\alpha-\gamma}$, then $\tilde{v}_1^+(z, \omega)$ is analytic in $\mathscr{D}^+_{-\gamma}$ and $\tilde{v}_2^-(a, \omega)$ is analytic in $\mathscr{D}^-_{\pi-\alpha-\gamma}$, and the previous proof applies. If $\alpha_1 + \alpha < \gamma$ or $\alpha + \alpha_1 < \pi - \alpha - \gamma$, we make the changes of variable $z = \phi_{-\alpha_2}(\xi_1)$ in the second integral in (4.54) and $z = \phi_{\alpha_2}(\xi_1)$ in the second integral in (4.55), where $\alpha_2 = \alpha + \alpha_1$ if $\alpha + \alpha_1 < \pi/4$ and $\alpha_2 = \pi/2 - \alpha - \alpha_1$ if $\alpha + \alpha_1 \geqslant \pi/4$. After a finite number of steps we obtain that $\tilde{v}_2^-(z, \omega)$ is analytic in $\mathscr{D}^-_{\pi-\alpha-\gamma}$ and $\tilde{v}_1^+(z, \omega)$ is analytic in $\mathscr{D}^+_{-\gamma}$. Therefore, we have proved the uniqueness of the solution of (4.3), (4.4) for the angles $0 < \alpha \leqslant \pi/2$. Note that $\pi \leqslant \alpha < 3\pi/2$ and $3\pi/2 \leqslant \alpha < 2\pi$ can be treated analogously to $\pi/2 < \alpha \leqslant \pi$ and $0 < \alpha \leqslant \pi/2$.

5. General boundary-value problem in a domain with corners. We briefly show how to treat the boundary-value problem (1.1)–(1.3), using the results of §§2–4. Take any $x_0'' \in \Sigma_0 = \overline{\Sigma}_1 \cap \overline{\Sigma}_2$ and introduce local coordinates in a neighborhood $U(x_0'') \subset \mathbf{R}^n$ of x_0'' such that the equations of Σ_1 and Σ_2 will be $x_1 = 0$ and $x_2 = 0$, respectively. Write $A(x, D)$, $B_k(x, D)$, $k = 1, 2$, in these local coordinates and "freeze" the coefficients of the principal parts of A, B_k at the point x_0''. Making an additional linear transformation of variables to obtain the Laplacian and adding the lower-order terms, we obtain a boundary-value problem of the form (2.5)–(2.7). Assuming that the conditions of Theorem 4.1 are satisfied, we obtain that there exists an operator $\mathbf{R}_0$ acting from $H_{s-2,N}(G) \times H_{s-m_1-1/2}(\Gamma_1^+) \times H_{s-m_2-1/2}(\Gamma_2^-)$ to $H_{s,N}(G)$ that is the inverse of the operator $\mathscr{A}_0$ defined by the left sides of (2.5)–(2.7). Here $s \in (s_-(x_0''), s_+(x_0''))$. Note that $s_-(x_0'') < s_+(x_0'')$ are two C^∞-functions on $\Sigma_0 = \overline{\Sigma}_1 \cap \overline{\Sigma}_2$. We denote by $s(x')$ any C^∞-function such that $s_-(x'') < s(x'') < s_+(x'')$ for all $x'' \in \Sigma_0$, and we consider the operator $\mathscr{A}$ defined by (1.1)–(1.3) as acting from $H_{s(x''),N}(\Omega)$ to

$$\mathscr{H}^{(1)}_{s(x''),N} = H_{s(x'')-2,N}(\Omega) \times H_{s(x'')-m_1-1/2}(\Sigma_1) \times H_{s(x'')} - m_{2-1/2}(\Sigma_2).$$

Here we used the Sobolev spaces of variable order. Analogously, one can consider the Sobolev spaces of piecewise-constant order $H_{(s_j),N}$ (see §25 of [**1**]), where $s_-(x'') < s_j < s_+(x'')$ for all $x'' \in \Sigma_0 \cap \overline{U}_j$, $|s_i - s_j| < 1$ when $U_i \cap U_j \neq \varnothing$. Note that the definition of Sobolev spaces with a weight in Ω and on Σ_1 and Σ_2 is the same as in §24 of [**1**]. For any point $x_0 \in \Sigma_k \setminus \Sigma_0$, $k = 1$ or 2, we have a neighborhood where the boundary problem (1.1)–(1.3) reduces to a boundary problem for the second-order, elliptic equation in a smooth domain, and for $x_0 \in G$, $x_0 \notin \partial G$ there is a neighborhood $U(x_0)$ such that $U(x_0) \subset G$ and

$U(x_0) \cap \partial G = \varnothing$, so that the problem reduces to the inversion of the second-order elliptic operator in the whole space. Therefore, we are in the same situation as in the proof of Theorem 22.1 of [**1**, pp. 261–264] (cf. also Theorem 24.3 and §25). Therefore, we have the following result.

THEOREM 5.1. *Assume that* $B_1(x, D)$ $(B_2(x, D))$ *satisfies the Shapiro–Lopatinskii condition on* $\overline{\Sigma}_1$ $(\overline{\Sigma}_2)$. *Let* $s_-(x'') < s_+(x'')$ *be the* C^∞*-functions on* $\Sigma_0 = \overline{\Sigma}_1 \cap \overline{\Sigma}_2$ *defined as in* §3 *for any* $x'' \in \Sigma_0$. *Then the boundary-value problem* (1.1)–(1.3) *defines a Fredholm operator from* $H_{(s_j),N}(\Omega)$ *to*

$$H_{(s_j)-2,N}(\Omega) \times H_{(s_j)-m_1-1/2,N}(\Sigma_1) \times H_{(s_j)-m_2-1/2,N}(\Sigma_2),$$

where s_j *are nonintegers if* $s_j - m_1 - 1/2 < 0$ *or* $s_j - m_2 - 1/2 < 0$, $N \geqslant 1$. *An analogous theorem holds also in the Sobolev spaces* $H_{s(x''),N}$ *of variable order. Note that if there is* s_0 *such that* $s_-(x'') < s_0 < s_+(x'')$ *for all* $x'' \in \Sigma_0$, *then we can avoid the use of Sobolev spaces of variable or piecewise-constant order.*

REFERENCES

1. G. I. Eskin, *Boundary value problems for elliptic pseudodifferential equations*, Trans. Math. Mono., vol. 52, Amer. Math. Soc., Providence, RI, 1981.
2. ______, *The conjugacy problem for equations of principal type with two independent variables*, Trans. Moscow Math. Soc. **21** (1970), 263–316.
3. F. D. Gahov, *Boundary value problems*, Addison-Wesley, Reading, Mass., 1966.
4. A. I. Komech, *Elliptic boundary value problems on manifolds with a piecewise smooth boundary*, Math. USSR-Sb. **21** (1973), 91–135.
5. V. A. Kondrat′ev and O. A. Oleinik, *Boundary value problems for partial differential equations in non-smooth domains*, Russian Math. Surveys **38**. (1983), 1–86.
6. V. G. Maz′ya and B. A. Plamenevskiĭ, *On boundary value problems for a second order elliptic equation in a domain with edges*, Vestnik Leningrad Univ. Math. **8** (1980), 99–106.
7. H. Reisman, *Second order elliptic boundary value problems in a domain with edges*, Comm. Partial Differential Equations **6** (1981), 1023–1042.
8. K. T. Smith, *Formulas to represent functions by their derivatives*, Math. Ann. **188** (1970), 53–77.

UNIVERSITY OF CALIFORNIA AT LOS ANGELES

Proceedings of Symposia in Pure Mathematics
Volume **43** (1985)

Imbedding $\mathbb{C}^n$ in H_n

PETER C. GREINER[1]

1. Introduction. Let H_n denote the nth Heisenberg group with underlying manifold

$$(1.1) \qquad \mathbb{R} \times \mathbb{C}^n = \{(t, z)\} = \{(t, z_1, \ldots, z_n)\}$$

and group law

$$(1.2) \qquad uv = (t, z)(s, w) = \left(t + s - 2\,\mathrm{Im} \sum_{j=1}^{n} z_j \bar{w}_j, z + w\right).$$

Given functions ϕ, $\psi \in C_0^\infty(H_n)$, the left-invariant Heisenberg convolution (H-convolution) is given by

$$(1.3) \qquad \phi *_{\mathrm{H}} \psi(u) = \int_{H_n} \phi(v^{-1}u)\psi(v)\, dv,$$

where $v = (s, w)$, $w = (w_1, \ldots, w_n)$, $w_j = y_j + iy_{j+n}$, $j = 1, \ldots, n$, and

$$(1.4) \qquad dv = dt\, dy$$

is the Lebesgue measure on $\mathbb{R}^{2n+1}$. We are interested in principal value convolution operators on H_n. Let

$$(1.5) \qquad r(t, z) = (r^2 t, rz), \qquad r > 0,$$

define the Heisenberg dilation. f is said to be Heisenberg homogeneous (H-homogeneous) of degree m on H_n if

$$(1.6) \qquad f(r^2 t, rz) = r^m f(t, z), \qquad r > 0.$$

We also define the norm on H_n by

$$(1.7) \qquad |u|_{\mathrm{H}} = |(t, z)|_{\mathrm{H}} = \left(t^2 + |z|^4\right)^{1/4},$$

1980 *Mathematics Subject Classification*. Primary 43A80.

[1] This research was partially supported by the Natural Sciences and Engineering Research Council of Canada under Grant No. A-3017.

which is H-homogeneous of degree one. This yields a distance function: The distance $d(u, v)$ between points $u, v \in H_n$ is

$$d(u, v) = d(v^{-1}u, 0) = |v^{-1}u|_{\mathrm{H}}. \tag{1.8}$$

d is left-invariant and satisfies the triangle inequality.

Suppose $f \in C^\infty(H_n \setminus 0)$ is H-homogeneous of degree γ. Then f is integrable near the origin if $\gamma > -2n - 2$.

We are interested in convolution operators on H_n that are induced by H-homogeneous functions of the critical degree $-2n - 2$.

1.9. DEFINITION. *Let $F \in C^\infty(H_n \setminus 0)$ be* H-*homogeneous of degree $-2n - 2$. F is said to have vanishing principal value if*

$$\int_{|u|_{\mathrm{H}}=1} F(u)\, d\sigma(u) = 0, \tag{1.10}$$

where $d\sigma(u)$ is the induced measure on the unit Heisenberg sphere.

The following result justifies the notion of vanishing principal value.

1.11. PROPOSITION. *Let $F \in C^\infty(H_n \setminus 0)$ be* H-*homogeneous of degree $-2n - 2$ with zero principal value. Then F induces a principal value (PV) convolution operator on functions $\phi \in C_0^\infty(H_n)$ as follows*:

$$F *_{\mathrm{H}} \phi(u) = \lim_{\varepsilon \to 0} \int_{d(u,v) > \varepsilon} F(v^{-1}u)\phi(v)\, dv. \tag{1.12}$$

The operator $F *_{\mathrm{H}}$ defined by (1.12) *can be extended to a bounded operator*:

$$F *_{\mathrm{H}} : L^2(H_n) \to L^2(H_n). \tag{1.13}$$

The best-known examples of left-invariant PV convolution operators on H_n are induced by the "Cauchy–Szegö" kernels $S_\pm(v^{-1}u)$, where

$$S_\pm(t, z) = (2^{n-1}n!/\pi^{n+1})\left(|z|^2 \mp it\right)^{-n-1}. \tag{1.14}$$

$S_\pm$ are projections:

$$S_\pm *_{\mathrm{H}} S_\pm = S_\pm \quad \text{or} \quad S_\pm^2 = S_\pm. \tag{1.15}$$

2. The Laguerre expansion. We denote the partial Fourier transform of $\phi(t, z) \in \mathbf{C}_0^\infty(\mathbf{R} \times \mathbf{C}^n)$ by

$$\tilde{\phi}(t, z) = \int_{-\infty}^{\infty} e^{-it\tau}\phi(t, z)\, dt, \tag{2.1}$$

and we use its inverse to define functions on $\mathbf{R} \times \mathbf{C}^n$ via their partial Fourier transform.

2.2. DEFINITION. *We define the functions $\tilde{\mathscr{L}}_k^{(p)}(\tau, z)$, $\tau \in \mathbf{R}$, $z \in \mathbf{C}$, $\pm p, k = 0, 1, 2, \ldots,$ by*

$$\tilde{\mathscr{L}}_k^{(p)}(\tau, z) = 2|\tau|\pi^{-1}(\operatorname{sgn} p)^p l_k^{(|p|)}\left(\left|\sqrt{2|\tau|}\, z\right|^2\right)e^{ip\phi}, \tag{2.3}$$

where $(\operatorname{sgn} 0)^0 = 1$,

$$z = |z|e^{i\phi}, \tag{2.4}$$

and

$$l_k^{(p)}(x) = [\Gamma(k+1)/\Gamma(p+k+1)]^{1/2} x^{p/2} L_k^{(p)}(x) e^{-x/2}, \tag{2.5}$$

with $L_k^{(p)}(x)$, $p, k = 0, 1, 2, \ldots$, *the generalized Laguerre polynomials given via the classical generating function formula*

$$\sum_{k=0}^{\infty} L_k^{(p)}(x) z^k = \frac{1}{(1-z)^{p+1}} e^{-xz/(1-z)}. \tag{2.6}$$

We note that the exponential Laguerre functions

$$\{\tilde{\mathscr{L}}_k^{(p)}(\tfrac{1}{2}, z), z \in \mathbb{C}, \pm p, k = 0, 1, 2, \ldots\} \tag{2.7}$$

form a complete orthonormal set of functions in $L^2(\mathbb{C})$.

We define the n-dimensional version of the exponential Laguerre functions by the n-fold product

$$\tilde{\mathscr{L}}_{k_1, \ldots, k_n}^{(p_1, \ldots, p_n)}(\tau, z) = \prod_{j=1}^{n} \tilde{\mathscr{L}}_{k_j}^{(p_j)}(\tau, z_j), \tag{2.8}$$

$\pm p_j, k_j = 0, 1, 2, \ldots, j = 1, \ldots, n$, *where* $\tilde{\mathscr{L}}_{k_j}^{(p_j)}$ *are given by* (2.3).

2.9. DEFINITION. *Let* $F(t, z)$ *induce a PV convolution operator on* H_n. *We define* $F_{\pm}(t, z)$ *via the partial Fourier transform* $\tilde{F}_{\pm}(\tau, z)$:

$$\tilde{F}_{\pm}(\tau, z) = \begin{cases} \tilde{F}(\tau, z), & \tau \gtrless 0, \\ 0, & \tau \lessgtr 0. \end{cases} \tag{2.10}$$

It is interesting to note that

$$S_{\pm}(t, z) = \mathscr{L}_{0, \ldots, 0; \pm}^{(0, \ldots, 0)}(t, z). \tag{2.11}$$

Consequently, we refer to $\mathscr{L}_{k_1, \ldots, k_n}^{(p_1, \ldots, p_n)}(t, z)$ as the generalized Cauchy–Szegö kernels.

The most important property of the generalized Cauchy–Szegö kernels is that the H-convolution of two of them produces a third one. To make this statement precise we need the partial Fourier transform of the Heisenberg convolution.

The twisted convolution $*_\tau$ *on* H_n *is defined by*

$$\tilde{\phi} *_\tau \tilde{\psi}(\tau, z) = \int_{\mathbb{R}^{2n}} \exp\left(-\tau \sum_{j=1}^{n} (z_j \bar{w}_j - \bar{z}_j w_j)\right) \tilde{\phi}(\tau, z - w) \tilde{\psi}(\tau, w)\, dy, \tag{2.12}$$

where $w_j = y_j + i y_{j+n}$, $j = 1, \ldots, n$. *The twisted convolution is the partial Fourier transform of the Heisenberg convolution*:

$$\phi *_{\mathrm{H}} \psi(u) = \int_{H_n} \phi(v^{-1}u) \psi(v)\, dv = \frac{1}{2\pi} \int_{-\infty}^{\infty} e^{it\tau} (\tilde{\phi} *_\tau \tilde{\psi})(\tau, z)\, d\tau. \tag{2.13}$$

2.14. PROPOSITION. *Let* $p_j, k_j, q_j, m_j = 1, 2, \ldots$ *for* $j = 1, \ldots, n$. *Then*

$$\begin{aligned} &\left(\prod_{j=1}^{n} \tilde{\mathscr{L}}_{p_j \wedge k_j - 1}^{(p_j - k_j)}(\tau, z_j)\right) *_{|\tau|} \left(\prod_{j=1}^{n} \tilde{\mathscr{L}}_{q_j \wedge m_j - 1}^{(q_j - m_j)}(\tau, z_j)\right) \\ &\qquad = \prod_{j=1}^{n} \delta_{k_j}^{(q_j)} \tilde{\mathscr{L}}_{p_j \wedge m_j - 1}^{(p_j - m_j)}(\tau, z_j), \end{aligned} \tag{2.15}$$

where $\delta_k^{(p)}$ *is the Kronecker delta*: $= 1$ *if* $p = k$ *and* 0 *otherwise, and* $m \wedge n = \min(m, n)$.

Thus the behaviour of the generalized Cauchy–Szegö kernels with respect to the Heisenberg convolution is analogous to the behaviour of the spherical harmonics on $\mathbb{C}^n$ with respect to the Euclidean convolution; see [**7**, **16**], and [**17**]. Proposition 2.14 can be found in [**1**, **2**], and [**8**].

Let $F(t, z)$ induce a principal value convolution operator on H_n; i.e., $F \in C^\infty(H_n \setminus 0)$ is H-homogeneous of degree $-2n-2$, and has zero principal value. According to the n-dimensional version of (2.7), the n-fold exponential Laguerre functions yield a complete orthonormal system in $L^2(\mathbb{C}^n)$. Therefore we can expand $\tilde{F}$ in a Laguerre series

$$\tilde{F}(\tau, z) = \sum_{\pm p_j,\, k_j=0}^{\infty} \tilde{F}_{k_1,\dots,k_n}^{(p_1,\dots,p_n)} \prod_{j=1}^{n} \tilde{\mathcal{L}}_{k_j}^{(p_j)}(\tau, z_j) \tag{2.16}$$

as long as $\tau \neq 0$. In [**1**, **2**], and [**8**] we studied the behaviour of F as a PV convolution operator on H_n via its Laguerre series. Here we shall discuss

$$\lim_{\tau\to 0} \sum_{\pm p_j,\, k_j=0}^{\infty} \tilde{F}_{k_1,\dots,k_n}^{(p_1,\dots,p_n)} \prod_{j=1}^{n} \tilde{\mathcal{L}}_{k_j}^{(p_j)}(\tau, z_j). \tag{2.17}$$

In other words, we want to find the extension of the right side of (2.17) to $\tau = 0$.

It turns out to be more natural to evaluate the limit as $\tau \to 0$ of the full Fourier transform of (2.16). *Given $\phi \in C_0^\infty(H_n)$, its full Euclidean Fourier transform is*

$$\hat{\phi}(\tau, \zeta) = \int_{\mathbb{R}^{2n}} e^{-i\langle \xi, x\rangle} \tilde{\phi}(\tau, x)\, dx, \tag{2.18}$$

where, with a slight abuse of notation, we set

$$\phi(t, z) = \phi(t, x), \qquad z_j = x_j + i x_{j+n}, \quad j = 1,\dots,n, \tag{2.19}$$

and

$$\zeta = (\zeta_1,\dots,\zeta_n), \qquad \zeta_j = \xi_j + i\xi_{j+n}, \quad j = 1,\dots,n. \tag{2.20}$$

Let F induce a PV convolution operator on H_n. Then $\hat{F}(\tau, \zeta) \in C^\infty(\mathbb{R} \times \mathbb{C}^n \setminus 0)$ and is H-*homogeneous of degree zero*; *i.e.*,

$$\hat{F}(r^2\tau, r\zeta) = \hat{F}(\tau, \zeta), \qquad r > 0. \tag{2.21}$$

Therefore its restriction to $\tau = 0$, $\hat{F}(0, \zeta)$, is also homogeneous of degree zero and, as such, induces a PV convolution operator on $\mathbb{C}^n$ whose kernel is $\tilde{F}(0, z - w)$.

Let $\tilde{F}(\tau, z)$ be given by (2.16). Then

$$\hat{F}(\tau, \zeta) = \sum_{\pm p_j,\, k_j=0}^{\infty} \tilde{F}_{k_1,\dots,k_n}^{(p_1,\dots,p_n)} \prod_{j=1}^{n} \hat{\mathcal{L}}_{k_j}^{(p_j)}(\tau, \zeta_j), \tag{2.22}$$

where

$$\hat{\mathcal{L}}_{k_j}^{(p_j)}(\tau, \zeta_j) = \hat{\mathcal{L}}_{k_j}^{(p_j)}\left(1/2, \zeta_j/\sqrt{2|\tau|}\right), \tag{2.23}$$

with

$$\hat{\mathcal{L}}_{k_j}^{(p_j)}\left(\tfrac{1}{2}, \zeta_j\right) = 2(-1)^{p_j}(-1)^{k_j} l_{k_j}^{(|p_j|)}\left(|\zeta_j|^2\right) e^{ip_j\phi_j} \tag{2.24}$$

and

$$\zeta_j = |\zeta_j| e^{i\phi_j}, \qquad j = 1, \ldots, n. \tag{2.25}$$

The purpose of this article is to find

$$\lim_{\tau \to 0} \sum_{\pm p_j, k_j = 0}^{\infty} \tilde{F}_{k_1, \ldots, k_n}^{(p_1, \ldots, p_n)} \prod_{j=1}^{n} \hat{\mathscr{L}}_{k_j}^{(p_j)}(\tau, \zeta_j). \tag{2.26}$$

When $n = 1$, (2.26) was found in [**8**].

2.27. THEOREM. *Let $\hat{F}(\tau, \zeta)$, $\tau \in \mathbb{R} \setminus 0$, $\zeta \in \mathbb{C}$ be given by*

$$\hat{F}(\tau, \zeta) = \sum_{\pm p, k = 0}^{\infty} \tilde{F}_k^{(p)} \hat{\mathscr{L}}_k^{(p)}(\tau, \zeta). \tag{2.28}$$

Assume that

$$\lim_{\tau \to 0} \hat{F}(\tau, \zeta) = \hat{F}(0, \zeta) \tag{2.29}$$

exists and thus defines $\hat{F} \in \mathbb{C}(\mathbb{R} \times C \setminus 0)$. Then $\hat{F}(0, \zeta)$ can be represented as a Fourier series

$$\hat{F}(0, \zeta) = \sum_{p = -\infty}^{\infty} \tilde{F}^{(p)} e^{ip(\phi - \pi/2)}, \tag{2.30}$$

where $\zeta = |\zeta| e^{i\phi}$ and

$$\tilde{F}^{(p)} = \lim_{k \to \infty} \tilde{F}_k^{(p)}, \qquad p = 0, \pm 1, \pm 2, \ldots. \tag{2.31}$$

REMARK. We note that

$$\lim_{\tau \to 0} \hat{\mathscr{L}}_k^{(p)}(\tau, \zeta) = 0, \qquad \pm p, k = 0, 1, 2, \ldots. \tag{2.32}$$

This gives a more intuitive meaning to (2.31).

In §3 we construct the limit (2.26) for $n = 2$. Then in §4 we state the corresponding result for H_n.

REMARK. In [**5**] Geller used group contractions to study restrictions from H_n to $\mathbb{C}^n$.

3. The limit (2.26) in H_n. We introduce spherical coordinates in $\mathbb{C}^2$,

$$\zeta_1 = |\zeta_1| e^{i\phi_1} = r \cos\theta \, e^{i\phi_1}, \tag{3.1}$$

$$\zeta_2 = |\zeta_2| e^{i\phi_2} = r \sin\theta \, e^{i\phi_2}, \tag{3.2}$$

where

$$r^2 = |\zeta|^2 = |\zeta_1|^2 + |\zeta_2|^2 \tag{3.3}$$

and

$$0 \leqslant \phi_1, \phi_2 \leqslant 2\pi, \qquad 0 \leqslant \theta \leqslant \pi/2. \tag{3.4}$$

(2.32) suggests that to obtain (2.26) we need only the limiting value of the coefficients $\tilde{F}_{k_1, k_2}^{(p_1, p_2)}$ as $|k| = k_1 + k_2 \to \infty$. More precisely, we need the notion of the Laguerre limit of the coefficients.

3.5. DEFINITION. *Let* $a(k_1, k_2) \in \mathbb{C}$ *be defined on* $\mathbb{Z}_+ \times \mathbb{Z}_+$, *where* $\mathbb{Z}_+ = \{0, 1, 2, \dots\}$. *For a given* $\theta \in [0, \pi/2]$ *we define*

$$a(\theta) = \lim_{k \to \infty} a(k_1, k_2), \tag{3.6}$$

where $k = (k_1, k_2) \to \infty$, *so that*

$$k_1/|k| = k_1/(k_1 + k_2) \to \cos^2\theta. \tag{3.7}$$

Assuming the limit (3.6) *exists, we call* $a(\theta)$ *the Laguerre limit of* $a(k_1, k_2)$ *and denote it by*

$$a(\theta) = \mathscr{L}\text{-}\lim_{k \to \infty} a(k_1, k_2). \tag{3.8}$$

REMARK. (3.7) is equivalent to $k_2/|k| \to \sin^2\theta$ and also to

$$k_2/k_1 \to \tan^2\theta. \tag{3.9}$$

3.10. THEOREM ((2.26) *in* H_2). *Let* $\hat{F}(\tau, \zeta)$, $\tau \in \mathbb{R} \setminus 0$, $\zeta \in \mathbb{C}^2$, *be defined by*

$$\hat{F}(\tau, \zeta) = \sum_{\pm p_j,\, k_j = 0}^{\infty} \tilde{F}_{k_1,k_2}^{(p_1,p_2)} \prod_{j=1}^{2} \hat{\mathscr{L}}_{k_j}^{(p_j)}(\tau, \zeta_j). \tag{3.11}$$

Assume that

$$\lim_{\tau \to 0} \hat{F}(\tau, \zeta) = \hat{F}(0, \zeta) \tag{3.12}$$

exists and thus defines $\hat{F} \in C^\infty(\mathbb{R} \times \mathbb{C}^2 \setminus 0)$. *Then*

$$\hat{F}(0, \zeta) = \sum_{p_j = -\infty}^{\infty} \tilde{F}^{(p_1,p_2)}(\theta) \prod_{j=1}^{2} e^{ip_j(\phi_j - \pi/2)}, \tag{3.13}$$

where

$$\tilde{F}^{(p_1,p_2)}(\theta) = \mathscr{L}\text{-}\lim_{k \to \infty} \tilde{F}_{k_1,k_2}^{(p_1,p_2)}. \tag{3.14}$$

(3.13) *defines* $\hat{F}(0, \zeta)$ *on* $\mathbb{C}^2$ *as a homogeneous function of degree zero.*

In the rest of §3 we prove Theorem 3.10. Here is the idea behind the proof.

(i) Given $g(\zeta) \in C^\infty(\mathbb{C}^2 \setminus 0)$, homogeneous of degree zero, we construct a function $g(\tau, \zeta) \in C^\infty(\mathbb{R} \times \mathbb{C}^2 \setminus 0)$, H-homogeneous of degree zero, so that

$$g(0, \zeta) = g(\zeta). \tag{3.15}$$

(ii) We write $g(\tau, \zeta)$ of (i) in a Laguerre series:

$$g(\tau, \zeta) = \sum_{\pm p_j,\, k_j = 0}^{\infty} g_{k_1,k_2}^{(p_1,p_2)} \prod_{j=1}^{2} \hat{\mathscr{L}}_{k_j}^{(p_j)}(\tau, \zeta_j). \tag{3.16}$$

(iii) We show that Theorem 3.10 holds for $g(\tau, \zeta)$ of (i); i.e.,

$$g(0, \zeta) = \sum_{p_j = -\infty}^{\infty} g^{(p_1,p_2)}(\theta) \prod_{j=1}^{2} e^{ip_j(\phi_j - \pi/2)}, \tag{3.17}$$

where

$$g^{(p_1,p_2)}(\theta) = \mathscr{L}\text{-}\lim_{k\to\infty} g_{k_1,k_2}^{(p_1,p_2)}. \tag{3.18}$$

(iv) It remains to show that

$$\mathscr{L}\text{-}\lim_{k\to\infty} g_{k_1,k_2}^{(p_1,p_2)} = 0, \qquad p_1, p_2 = 0, \pm 1, \pm 2, \dots, \tag{3.19}$$

implies

$$\lim_{\tau\to 0} \sum_{\pm p_j,\, k_j=0}^{\infty} g_{k_1,k_2}^{(p_1,p_2)} \prod_{j=1}^{2} \hat{\mathscr{L}}_{k_j}^{(p_j)}(\tau, \zeta_j) = 0. \tag{3.20}$$

Next we carry out the details.

(i) Let $g(\zeta) \in C^\infty(\mathbb{C}^2 \setminus 0)$ be homogeneous of degree zero. Our first task is to construct $g(\tau, \zeta) \in C^\infty(\mathbb{R} \times \mathbb{C}^2 \setminus 0)$, H-homogeneous of degree zero, such that

$$g(0, \zeta) = g(\zeta). \tag{3.21}$$

We start by writing $g(\zeta)$ as a sum of spherical harmonics. Let $\zeta = (\zeta_1, \zeta_2)$ be given in the spherical coordinates (3.1)–(3.4).

3.22. PROPOSITION. *A basis for the solid spherical harmonics of degree n, $n = 0, 1, 2, \dots,$ on $\mathbb{C}^2$ is given by*

$$r^n S_n^{(p_1,p_2)}(\theta, \phi_1, \phi_2) = r^n(\cos\theta)^{|p_1|}(\sin\theta)^{|p_2|} \cdot P_{(n-|p_1|-|p_2|)/2}^{(|p_1|,|p_2|)}(\cos 2\theta)\, e^{i(p_1\phi_1 + p_2\phi_2)}, \tag{3.23}$$

$$-\infty < p_1, p_2 < \infty, \tag{3.24}$$

$$\tfrac{1}{2}(n - |p_1| - |p_2|) = 0, 1, 2, \dots, \tag{3.25}$$

where $P_m^{(\alpha,\beta)}(x)$, $-1 < x < 1$, are the Jacobi polynomials

$$2^m m! P_m^{(\alpha,\beta)}(x) = (-1)^m (1-x)^\alpha (1+x)^\beta \left(\frac{d}{dx}\right)^m \left((1-x)^{\alpha+m}(1+x)^{\beta+m}\right). \tag{3.26}$$

$g(\zeta)$ is homogeneous of degree zero; hence, we may assume that $\zeta \in \Omega_4$, the unit sphere in $\mathbb{C}^2$. Since the spherical harmonics are orthogonal on Ω_4 with respect to the induced surface measure, we may write

$$g(\zeta) = \sum_{n=0}^{\infty} \sum_{(n-|p_1|-|p_2|)/2=0}^{\infty} g_n^{(p_1,p_2)} S_n^{(p_1,p_2)}(\theta, \phi_1, \phi_2) \tag{3.27}$$

in a unique fashion. Consequently, to extend $g(\zeta)$ to $g(\tau, \zeta)$ we need only extend $S_n^{(p_1,p_2)}(\theta, \phi_1, \phi_2)$ for each $\pm p_j$, $n = 0, 1, 2, \dots,$ that satisfies the condition $n - |p_1| - |p_2| \in 2\mathbb{Z}_+$. $S_n^{(p_1,p_2)}$ is a homogeneous polynomial of degree n in ζ_j, $\bar{\zeta}_j$, $j = 1, 2$, divided by $r^n = |\zeta|^n$; i.e., $S_n^{(p_1,p_2)}$ is a finite sum of "monomials" of the form

$$\zeta_1^{k_1}\bar{\zeta}_1^{l_1}\zeta_2^{k_2}\bar{\zeta}_2^{l_2}/|\zeta|^n, \tag{3.28}$$

where

$$\sum_{j=1}^{2}(k_j+l_j)=n. \tag{3.29}$$

(3.28) can be simplified to

$$\zeta_1^{p_1}|\zeta_1|^{2q_1}\zeta_2^{p_2}|\zeta_2|^{2q_2}/|\zeta|^n \tag{3.30}$$

or to similar monomials with ζ_1 or ζ_2 or both of them barred. In (3.30)

$$\sum_{j=1}^{2}(p_j+2q_j)=n. \tag{3.31}$$

What is important for the calculations is that in the monomial, like (3.30), for each j, $j=1,2$, only powers of ζ_j or $\bar{\zeta}_j$ occur, but not both. This simplifies the work a great deal.

This discussion shows that to extend $S_n^{(p_1,p_2)}$ it suffices to extend monomials like (3.30). We shall construct the extension for (3.30); the other possible types can be extended similarly. Let

$$Z_j=\partial/\partial z_j-i\bar{z}_j(\partial/\partial t), \tag{3.32}$$

$$\bar{Z}_j=\partial/\partial \bar{z}_j+iz_j(\partial/\partial t), \tag{3.33}$$

$j=1,2$, and

$$T=\partial/\partial t \tag{3.34}$$

denote a complex basis for the Lie algebra of left-invariant vector fields on H_n. The symbols of these "pseudodifferential" operators are given by

$$-2i\sigma(Z_j)=\bar{\zeta}_j-2i\bar{z}_j\tau, \tag{3.35}$$

$$-2i\sigma(\bar{Z}_j)=\zeta_j+2iz_j\tau, \tag{3.36}$$

$j=1,2$. Thus

$$-2i\sigma(Z_j)(0,\zeta)=\bar{\zeta}_j, \tag{3.37}$$

$$-2i\sigma(\bar{Z}_j)(0,\zeta)=\zeta_j, \tag{3.38}$$

$j=1,2$. We also need

$$\Box=\sum_{j=1}^{2}\Box_j, \tag{3.39}$$

where

$$\begin{aligned}\Box_j&=-\frac{1}{2}(Z_j\bar{Z}_j+\bar{Z}_jZ_j)\\&=-\frac{\partial^2}{\partial z_j\partial\bar{z}_j}-i\frac{\partial}{\partial t}\left(z_j\frac{\partial}{\partial z_j}-\bar{z}_j\frac{\partial}{\partial\bar{z}_j}\right)-|z_j|^2\frac{\partial^2}{\partial t^2},\end{aligned} \tag{3.40}$$

$j=1,2$. Then

$$\sigma(4\Box)(0,\zeta)=|\zeta|^2. \tag{3.41}$$

We note that the symbols in (3.37), (3.38), and (3.41) are not extensions of the right sides, since they depend on the base variables z. On the other hand, they suggest what the extensions should be. In particular, the required extension of $|\zeta|^{-n}$ turns out to be the "Fourier transform" of the left-invariant convolution operator $(4\Box)^{-n/2}$.

3.42. DEFINITION. *Let P denote a left-invariant pseudodifferential operator on H_2. Its symbol, $\sigma(P)$, has the form*

$$\sigma(P) = \hat{P}(\tau, \sigma(-2iZ), \sigma(-2i\bar{Z})) \tag{3.43}$$

with some function $\hat{P}(\tau, \zeta, \bar{\zeta})$. We shall refer to $\hat{P}(\tau, \zeta, \bar{\zeta}) = \hat{P}(\tau, \zeta)$ as the Fourier transform of P.

According to [2],

$$4\hat{\Box}(\tau, \zeta) = \sum_{k_j=0}^{\infty} \left(\sum_{j=1}^{2} 4|\tau|(2k_j + 1) \right) \prod_{j=1}^{2} \hat{\mathscr{L}}_{k_j}^{(0)}(\tau, \zeta_j). \tag{3.44}$$

Since the $\hat{\mathscr{L}}_k^{(0)}$'s induce projections on H_n, we set

$$\left[(4\Box)^{-n/2}\right]^{\wedge}(\tau, \zeta) = \sum_{k_j=0}^{\infty} \left(\sum_{j=1}^{2} 4|\tau|(2k_j + 1) \right)^{-n/2} \prod_{j=1}^{2} \hat{\mathscr{L}}_{k_j}^{(0)}(\tau, \zeta_j). \tag{3.45}$$

Using

$$t^{-\alpha} = \frac{1}{\Gamma(\alpha)} \int_0^{\infty} s^{\alpha} e^{-ts} \frac{ds}{s}, \tag{3.46}$$

we can rewrite (3.45) in the following form:

$$\begin{aligned} &\left[(4\Box)^{-n/2}\right]^{\wedge}(\tau, \zeta) \\ &= \frac{1}{\Gamma(n/2)} \int_0^{\infty} s^{n/2} \sum_{k_j=0}^{\infty} \exp\left(- \sum_{j=1}^{2} 4|\tau|(2k_j + 1)s \right) \frac{ds}{s} \prod_{j=1}^{2} \hat{\mathscr{L}}_{k_j}^{(0)}(\tau, \zeta_j) \\ &= \frac{1}{\Gamma(n/2)} \int_0^{\infty} s^{n/2-1} \left\{ [\cosh(4s|\tau|)]^{-2} \exp\left(\frac{-|\zeta|^2 \tanh 4s|\tau|}{4|\tau|} \right) \right\} ds. \end{aligned} \tag{3.47}$$

A simple calculation shows that, letting $\tau \to 0$, we obtain

$$\left[(4\Box)^{-n/2}\right]^{\wedge}(\tau, \zeta) \to \frac{1}{\Gamma(n/2)} \int_0^{\infty} s^{n/2} e^{-|\zeta|^2 s} \frac{ds}{s} = |\zeta|^{-n}. \tag{3.48}$$

Furthermore, (3.47) shows that $[(4\Box)^{-n/2}]^{\wedge}(\tau, \zeta) \in C^{\infty}(\mathbb{R} \times \mathbb{C}^2 \setminus 0)$ and is H-homogeneous of degree $-n$. Thus $[(4\Box)^{-n/2}]^{\wedge}(\tau, \zeta)$ is the required extension of $|\zeta|^{-n}$. Finally, the required extension of (3.30) is given in

3.49. LEMMA. *Let* (3.43) *and* (3.47) *define* $(4\Box)^{-n/2}$. *Then*

$$\left[\prod_{j=1}^{2} \left((-2i\bar{Z}_j)^{p_j} (4\Box_j)^{q_j} \right) (4\Box)^{-n/2} \right]^{\wedge} (0, \zeta) = \prod_{j=1}^{2} \left(\zeta_j^{p_j} |\zeta_j|^{2q_j} \right) |\zeta|^{-n}, \tag{3.50}$$

where

$$\sum_{j=1}^{2} (p_j + 2q_j) = n. \tag{3.51}$$

PROOF. Under the Fourier integral

$$-2i\bar{Z}_j \quad \text{becomes} \quad \zeta_j - 4\tau\left(\partial/\partial\bar{\zeta}_j\right), \qquad j = 1,2. \tag{3.52}$$

Also

$$4\square_j f\left(t, |z_j|^2\right) = -4\left(\frac{\partial^2}{\partial z_j \partial \bar{z}_j} + |z_j|^2 \frac{\partial^2}{\partial t^2}\right) f\left(t, |z_j|^2\right), \tag{3.53}$$

which, under the Fourier integral, becomes

$$|\zeta_j|^2 - 16\tau^2\left(\partial^2/\partial\zeta_j \partial\bar{\zeta}_j\right), \qquad j = 1,2. \tag{3.54}$$

Therefore

$$\begin{aligned} &\left[\prod_{j=1}^{2}\left((-2i\bar{Z}_j)^{p_j}(4\square_j)^{q_j}\right)(4\square)^{-n/2}\right]^{\wedge}(\tau,\zeta) \\ &= \prod_{j=1}^{2}\left[\left(\zeta_j - 4\tau\frac{\partial}{\partial\bar{\zeta}_j}\right)^{p_j}\left(|\zeta_j|^2 - 16\tau^2\frac{\partial^2}{\partial\zeta_j\partial\bar{\zeta}_j}\right)^{q_j}\right]\left[(4\square)^{-n/2}\right]^{\wedge}(\tau,\zeta) \\ &\overset{\tau\to 0}{\to} \prod_{j=1}^{2}\left(\zeta_j^{p_j}|\zeta_j|^{2q_j}\right)|\zeta|^{-n} \end{aligned} \tag{3.55}$$

in view of (3.48). This proves Lemma 3.49.

Thus the required extension of $\prod_{j=1}^{2}(\zeta_j^{p_j}|\zeta_j|^{2q_j})|\zeta|^{-n}$ *with* $\Sigma_{j=1}^{2}(p_j + 2q_j) = n$ *is*

$$\left[\prod_{j=1}^{2}\left((-2i\bar{Z}_j)^{p_j}(4\square_j)^{q_j}\right)(4\square)^{-n/2}\right]^{\wedge}(\tau,\zeta). \tag{3.56}$$

(ii) Next we obtain the Laguerre series expansion of (3.56). Starting with the partial Fourier transform of (3.45),

$$\left[(4\square)^{-n/2}\right]^{\sim}(\tau, z) = \sum_{k_j=0}^{\infty}\left(\sum_{j=1}^{2} 4|\tau|(2k_j + 1)\right)^{-n/2}\prod_{j=1}^{2}\tilde{\mathscr{L}}_{k_j}^{(0)}(\tau, z_j), \tag{3.57}$$

we use Lemma 4.37 of [**8**] to obtain

$$\begin{aligned} &\left[\prod_{j=1}^{2}(4\square_j)^{q_j}(4\square)^{-n/2}\right]^{\sim}(\tau, z) \\ &= \sum_{k_j=0}^{\infty}\frac{\Pi_{j=1}^{2}\left(4|\tau|(2k_j+1)\right)^{q_j}}{\left(\Sigma_{j=1}^{2}4|\tau|(2k_j+1)\right)^{n/2}}\prod_{j=1}^{2}\tilde{\mathscr{L}}_{k_j}^{(0)}(\tau, z_j). \end{aligned} \tag{3.58}$$

To apply Z_j or $\bar{Z}_j$, $j = 1,2$, to (3.58), we need the following result.

3.59. LEMMA. *Let* $p, k = 0, 1, 2, \ldots$. *Then*

$$\tilde{Z}\tilde{\mathscr{L}}_k^{(-p)} = \begin{cases} (2|\tau|)^{1/2}k^{1/2}\tilde{\mathscr{L}}_{k-1}^{(-p-1)}, & \tau > 0, \\ (2|\tau|)^{1/2}(k+p+1)^{1/2}\tilde{\mathscr{L}}_k^{(-p-1)}, & \tau < 0, \end{cases} \tag{3.60}$$

and

$$(3.61)\qquad \tilde{\bar{Z}}\tilde{\mathscr{L}}_k^{(p)} = \begin{cases} -(2|\tau|)^{1/2}(k+p+1)^{1/2}\tilde{\mathscr{L}}_k^{(p+1)}, & \tau > 0, \\ -(2|\tau|)^{1/2}k^{1/2}\tilde{\mathscr{L}}_{k-1}^{(p+1)}, & \tau < 0, \end{cases}$$

where we suppressed the subscript j, $j = 1, 2$.

We are interested in the limit as $\tau \to 0$. Thus, we may as well assume that $\tau > 0$. Iterating (3.61) we have

$$(3.62)\qquad \begin{aligned} (-2i\tilde{\bar{Z}})^p\tilde{\mathscr{L}}_k^{(0)} &= (-2i)^p(-\sqrt{2|\tau|})^p[(k+1)\cdots(k+p)]^{1/2}\tilde{\mathscr{L}}_k^{(p)} \\ &= (2i\sqrt{2|\tau|})^p\left[\frac{\Gamma(k+p+1)}{\Gamma(k+1)}\right]^{1/2}\tilde{\mathscr{L}}_k^{(p)}, \qquad \tau > 0. \end{aligned}$$

Using (3.58), (3.62), and calculating the full Fourier transform, we obtain the Laguerre series expansion of (3.56):

$$(3.63)\qquad \begin{aligned} &\left[\prod_{j=1}^{2}\left((-2i\bar{Z}_j)^{p_j}(4\square_j)^{q_j}\right)(4\square)^{-n/2}\right]^{\wedge}(\tau,\zeta) \\ &= (\sqrt{2}\,i)^{p_1+p_2}\sum_{k_j=0}^{\infty}\frac{\prod_{j=1}^{2}\left[(2k_j+1)^{q_j}\left(\Gamma(k_j+p_j+1)/\Gamma(k_j+1)\right)^{1/2}\right]}{\left[\sum_{j=1}^{2}(2k_j+1)\right]^{n/2}} \\ &\qquad\cdot\prod_{j=1}^{2}\hat{\mathscr{L}}_{k_j}^{(p_j)}(\tau,\zeta_j), \qquad \tau > 0. \end{aligned}$$

(iii) We start with

3.64. LEMMA. *Theorem* 3.10 *holds for*

$$(3.56)\qquad \left[\prod_{j=1}^{2}\left((-2i\bar{Z}_j)^{p_j}(4\square_j)^{q_j}\right)(4\square)^{-n/2}\right]^{\wedge}(\tau,\zeta).$$

PROOF. We need only calculate the Laguerre limit of the coefficients in (3.63). Thus

$$(3.65)$$

$$\begin{aligned} &\mathscr{L}\text{-}\lim_{k\to\infty}(\sqrt{2}\,i)^{p_1+p_2}\sum_{k_j=0}^{\infty}\frac{\prod_{j=1}^{2}\left[(2k_j+1)^{q_j}\left(\Gamma(k_j+p_j+1)/\Gamma(k_j+1)\right)^{1/2}\right]}{\left[\sum_{j=1}^{2}(2k_j+1)\right]^{n/2}} \\ &= \mathscr{L}\text{-}\lim_{k\to\infty} i^{p_1+p_2}\sum_{k_j=0}^{\infty}\prod_{j=1}^{2}\left(\frac{k_j}{k_1+k_2}\right)^{p_j/2+q_j} \\ &= i^{p_1+p_2}(\cos\theta)^{p_1+2q_1}(\sin\theta)^{p_2+2q_2}. \end{aligned}$$

This proves Lemma 3.64, since Lemma 3.49 shows that

$$\begin{aligned}(3.66)\qquad &\left[\prod_{j=1}^{2}\left(\left(-2i\bar{Z}_j\right)^{p_j}(4\Box_j)^{q_j}\right)(4\Box)^{-n/2}\right]^{\wedge}(0,\zeta) = \prod_{j=1}^{2}\left(\zeta_j^{p_j}|\zeta_j|^{2q_j}\right)|\zeta|^{-n}\\ &\qquad = (\cos\theta)^{p_1+2q_1}(\sin\theta)^{p_2+2q_2}\prod_{j=1}^{2}e^{ip_j\phi_j}.\end{aligned}$$

Consequently, we can extend arbitrary spherical harmonics from $\mathbb{C}^2\setminus 0$ to $\mathbb{R}\times\mathbb{C}^2\setminus 0$ so that Theorem 3.10 applies. Furthermore, this construction works for sums of spherical harmonics on $\mathbb{C}^2$. Thus we have derived

3.67. PROPOSITION. *Let $g(\zeta)\in C^\infty(\mathbb{C}^2\setminus 0)$ be homogeneous of degree zero. There exists a function $g(\tau,\zeta)\in C^\infty(\mathbb{R}\times\mathbb{C}^2\setminus 0)$, H-homogeneous of degree zero, such that*

$$(3.68)\qquad g(0,\zeta)=g(\zeta).$$

Furthermore, writing

$$(3.69)\qquad g(\tau,\zeta)=\sum_{\pm p_j,\,k_j=0}^{\infty} g_{k_1,k_2}^{(p_1,p_2)}\prod_{j=1}^{2}\hat{\mathscr{L}}_{k_j}^{(p_j)}(\tau,\zeta_j),$$

we have

$$(3.70)\qquad g(\zeta)=\sum_{p_j=-\infty}^{\infty} g^{(p_1,p_2)}(\theta)\prod_{j=1}^{2}e^{ip_j(\phi_j-\pi/2)},$$

where

$$(3.71)\qquad g^{(p_1,p_2)}(\theta)=\mathscr{L}\text{-}\lim_{k\to\infty} g_{k_1,k_2}^{(p_1,p_2)},\qquad 0\leqslant\theta\leqslant\pi/2.$$

(iv) To finish the proof of Theorem 3.10 we need only prove that

$$(3.72)\qquad \mathscr{L}\text{-}\lim_{k\to 0} g_{k_1,k_2}^{(p_1,p_2)}=0$$

implies

$$(3.73)\qquad \lim_{\tau\to 0}\sum_{k_j=0}^{\infty} g_{k_1,k_2}^{(p_1,p_2)}\prod_{j=1}^{2}\hat{\mathscr{L}}_{k_j}^{(p_j)}(\tau,\zeta_j)=0,$$

which is equivalent to

$$(3.74)\qquad \lim_{\tau\to 0}\sum_{k_j=0}^{\infty}(-1)^{k_1+k_2}g_{k_1,k_2}^{(p_1,p_2)}\prod_{j=1}^{2}l_{k_j}^{(|p_j|)}\left(\frac{|\zeta_j|^2}{2|\tau|}\right)=0,$$

for every (p_1,p_2), $p_j=0,\pm1,\pm2,\dots$, $j=1,2$. Suppose

$$(3.75)\qquad \sum_{k_j=0}^{\infty}\left(g_{k_1,k_2}^{(p_1,p_2)}\right)^2<\infty.$$

Clearly,

$$(3.76)\qquad (3.75)\Rightarrow(3.72).$$

Actually, (3.75) is stronger than necessary for (3.72) to hold. On the other hand, (3.75) has the advantage that it makes (3.74) self-evident. To see this we set $x_j = |\zeta_j|^2/2|\tau|$, $j = 1, 2$, and note that

$$\left\{\prod_{j=1}^{2} l_{k_j}^{(p_j)}(x_j);\ x_j \geq 0,\ k_j = 0, 1, 2, \ldots, j = 1, 2\right\} \tag{3.77}$$

is an orthonormal set of functions in $L^2(\{(x_1, x_2);\ x_1, x_2 \geq 0\})$. Then (3.75) $\Rightarrow$ (3.74) by

3.78. THEOREM. *Let*

$$g(x_1, x_2) = \sum_{k_j=0}^{\infty} g_{k_1,k_2}^{(p_1,p_2)} \prod_{j=1}^{2} l_{k_j}^{(p_j)}(x_j) \tag{3.79}$$

represent an L^2-*function on* $\{(x_1, x_2);\ x_j \geq 0,\ j = 1, 2\}$. *Suppose*

$$\lim_{|x|\to\infty} g(x_1, x_2) = g(\theta) \tag{3.80}$$

exists, $\theta = \tan^{-1}(x_2/x_1)$, *and is a* C^∞ *function on* $[0, \pi/2]$. *Then*

$$g(\theta) = 0. \tag{3.81}$$

4. The limit (2.26) in H_n. We shall only state the result in H_n. The proof is analogous to the proof given in §3 for H_2. First we introduce spherical coordinates in $\mathbf{C}^n$.

$$\zeta_1 = |\zeta_1|e^{i\phi_1} = r\cos\theta_1\, e^{i\phi_1}, \tag{4.1}$$

$$\zeta_2 = |\zeta_2|e^{i\phi_2} = r\sin\theta_1\cos\theta_2\, e^{i\phi_2}, \tag{4.2}$$

$$\cdots$$

$$\zeta_{n-1} = |\zeta_{n-1}|e^{i\phi_{n-1}} = r\sin\theta_1 \cdots \sin\theta_{n-2}\cos\theta_{n-1}\, e^{i\phi_{n-1}}, \tag{4.3}$$

$$\zeta_n = |\zeta_n|e^{i\phi_n} = r\sin\theta_1 \cdots \sin\theta_{n-2}\sin\theta_{n-1}\, e^{i\phi_n}, \tag{4.4}$$

where

$$r^2 = |\zeta|^2 = |\zeta_1|^2 + \cdots + |\zeta_n|^2, \tag{4.5}$$

$$0 \leq \phi_j \leq 2\pi, \qquad j = 1, \ldots, n, \tag{4.6}$$

and

$$0 \leq \theta_j \leq \pi/2, \qquad j = 1, \ldots, n-1. \tag{4.7}$$

4.8. DEFINITION. *Let* $a(k_1, \ldots, k_n) \in \mathbf{C}$ *denote a function on* $\mathbf{Z}_+ \times \cdots \times \mathbf{Z}_+$, n *times, where* $\mathbf{Z}_+ = \{0, 1, 2, \ldots\}$. *For a given* $(\theta_1, \ldots, \theta_{n-1})$, $\theta_j \in [0, \pi/2]$, *we define*

$$a(\theta_1, \ldots, \theta_{n-1}) = \lim_{k\to\infty} a(k_1, \ldots, k_n), \tag{4.9}$$

where $k = (k_1, \ldots, k_n) \to \infty$, *so that*

$$k_1/|k| \to \cos^2\theta_1, \tag{4.10}$$

$$k_2/|k| \to \sin^2\theta_1\cos^2\theta_2, \tag{4.11}$$

$$\cdots$$

$$k_{n-1}/|k| \to \sin^2\theta_1 \cdots \sin^2\theta_{n-2}\cos^2\theta_{n-1}, \tag{4.12}$$

$|k| = k_1 + \cdots + k_n$. *Assuming the limit* (4.9) *exists, we call* $a(\theta_1,\ldots,\theta_{n-1})$ *the Laguerre limit of* $a(k_1,\ldots,k_n)$ *and denote it by*

$$a(\theta_1,\ldots,\theta_{n-1}) = \mathscr{L}\text{-}\lim_{k\to\infty} a(k_1,\ldots,k_n). \tag{4.13}$$

4.14. THEOREM. *Define* $\hat{F}(\tau,\zeta)$ *by*

$$\hat{F}(\tau,\zeta) = \sum_{\pm p_j,\, k_j=0}^{\infty} \tilde{F}^{(p_1,\ldots,p_n)}_{k_1,\ldots,k_n} \prod_{j=1}^{n} \hat{\mathscr{L}}^{(p_j)}_{k_j}(\tau,\zeta_j) \tag{4.15}$$

for $\tau \neq 0$. *Assume that*

$$\lim_{\tau\to 0} \hat{F}(\tau,\zeta) = \hat{F}(0,\zeta) \tag{4.16}$$

exists and thus defines $\hat{F} \in C^\infty(\mathbb{R}\times\mathbb{C}^n\setminus 0)$. *Then* $\hat{F}(0,\zeta) \in C^\infty(\mathbb{C}^n\setminus 0)$ *is homogeneous of degree zero and has the following representation*:

$$\hat{F}(0,\zeta) = \sum_{p_j=-\infty}^{\infty} \tilde{F}^{(p_1,\ldots,p_n)}(\theta_1,\ldots,\theta_{n-1}) \prod_{j=1}^{n} e^{ip_j(\phi_j-\pi/2)}, \tag{4.17}$$

where

$$\tilde{F}^{(p_1,\ldots,p_n)}(\theta_1,\ldots,\theta_{n-1}) = \mathscr{L}\text{-}\lim_{k\to\infty} \tilde{F}^{(p_1,\ldots,p_n)}_{k_1,\ldots,k_n}. \tag{4.18}$$

REFERENCES

1. R. W. Beals and P. C. Greiner, *Non-elliptic differential operators of type* $\Box_b$ (in preparation).

2. R. W. Beals, P. C. Greiner and J. Vauthier, *The Laguerre calculus on the Heisenberg group*, Reider, Amsterdam, 1984, pp. 1–28.

3. G. B. Folland, *A fundamental solution for a subelliptic operator*, Bull. Amer. Math. Soc. **79** (1973), 373–376.

4. G. B. Folland and E. M. Stein, *Estimates for the* $\bar{\partial}_b$*-complex and analysis on the Heisenberg group*, Comm. Pure Appl. Math. **27** (1974), 429–522.

5. D. Geller, *Fourier analysis on the Heisenberg group*, Proc. Nat. Acad. Sci. U.S.A. **74** (1977), 1328–1331.

6. ______, *Fourier analysis on the Heisenberg group*. I: *Schwartz space*, J. Funct. Anal. **36** (1980), 205–254.

7. G. Giraud, *Sur certaines opérations du type elliptique*, C. R. Acad. Sci. Paris **200** (1935), 1651–1653.

8. P. C. Greiner, *On the Laguerre calculus of left-invariant convolution (pseudo-differential) operators on the Heisenberg group*, Seminaire Goulaouic–Meyer–Schwartz, 1980–1981, Exp. XI, pp. 1–39.

9. P. C. Greiner, J. J. Kohn and E. M. Stein, *Necessary and sufficient conditions for the solvability of the Lewy equation*, Proc. Nat. Acad. Sci. U.S.A. **72** (1975), 3287–3289.

10. P. C. Greiner and E. M. Stein, *Estimates for the* $\bar{\partial}$*-Neumann problem*, Math. Notes, no. 19, Princeton Univ. Press, Princeton, N. J., 1977.

11. ______, *On the solvability of some differential operators of type* $\Box_b$, Proc. Seminar Several Complex Variables (Cortona, Italy, 1976–1977), pp. 106–165.

12. T. H. Koornwinder, *The addition formula for Jacobi polynomials*. II, Report TW 133, Math. Centrum, Amsterdam, 1972.

13. A. Koranyi and S. Vagi, *Singular integrals in homogeneous spaces and some problems of classical analysis*, Ann. Scuola Norm. Sup. Pisa **25** (1971), 575–648.

14. H. Lewy, *An example of a smooth linear partial differential equation without solution*, Ann. of Math. (2) **66** (1957), 155–158.

15. G. Mauceri, *The Weyl transform and bounded operators on* $L^p(\mathbb{R}^n)$, Report no. 54, Math. Inst., Univ. of Genova, 1980.

16. S. G. Mikhlin, *Compounding of double singular integrals*, Dokl. Akad. Nauk SSSR **2** (**11**), No. 1(87) (1936), 3–6.

17. ______, *Multidimensional singular integrals and integral equations*, Pergamon, 1965.

18. E. M. Stein and G. Weiss, *Introduction to Fourier analysis on Euclidean spaces*, Princeton Univ. Press, Princeton, N. J., 1971.

19. G. Szegö, *Orthogonal polynomials*, Amer. Math. Soc. Colloq. Publ., vol. 23, Amer. Math. Soc., Providence, R. I., 1939.

UNIVERSITY OF TORONTO

Proceedings of Symposia in Pure Mathematics
Volume **43** (1985)

On Some Results of Gelfand in Integral Geometry[1]

VICTOR GUILLEMIN

1. The purpose of this talk is to examine some very beautiful results of Gelfand–Graev–Shapiro on the Plancherel formula for $\mathrm{SL}(n, \mathbf{C})$ and to speculate how these results might be extended to nongroup theoretic situations. For simplicity I will confine myself to $\mathrm{SL}(2, \mathbf{C})$. First let me recall that the "horocycles" in $\mathrm{SL}(2, \mathbf{C})$ are the subsets $A\mathcal{N}B$, where A and B are fixed matrices of modulus one, and $\mathcal{N}$ is the maximal unipotent subgroup

$$\mathcal{N} = \left\{ \begin{pmatrix} 1 & c \\ 0 & 1 \end{pmatrix}, c \in \mathbf{C} \right\} \tag{1.1}$$

equipped with the invariant measure $(dc \wedge d\bar{c})/2i$. It turns out that the problem of finding the explicit Plancherel formula for $\mathrm{SL}(2, \mathbf{C})$ is more or less equivalent to the following problem: Given a function f on $\mathrm{SL}(2, \mathbf{C})$, recover f from its integrals over the horocycles. Let us examine this problem in a little more detail. We think of $\mathrm{SL}(2, \mathbf{C})$ as the submanifold of $\mathbf{C}^4$ consisting of the 4-tuples (a, b, c, d) with $ad - bc = 1$. The subset where $a \neq 0$ admits a, b, and c as global coordinates, with

$$d = (bc + 1)/a. \tag{1.2}$$

Let us determine what the horocycles are in the a, b, c coordinates. Since each horocycle l is a line in $\mathbf{C}^4$, it can be parametrized by equations $a = t$, $b = \beta_1 t + \beta_2$, $c = \gamma_1 t + \gamma_2$, $d = \delta_1 t + \delta_2$. Substituting these equations into (1.2), we see that $\beta_2\gamma_2 + 1$ must be zero in order for l to lie in $\mathrm{SL}(2, \mathbf{C})$. If we set $t = 0$ in the above equations, we see that l contains the point $(0, \beta_2, \gamma_2)$, with $\beta_2\gamma_2 = -1$. In other words, for l to be a horocycle it must intersect the hyperbola

$$\{(0, s, -1/s), s \in \mathbf{C}\}. \tag{1.3}$$

1980 *Mathematics Subject Classification*. Primary 58G99.

[1] This article is based on some talks that I gave at the Symposium on Global Analysis, held in Durham, England, July 1976. My talk at the Conference on Microlocal Analysis was an abbreviated and slightly updated presentation of this material.

The converse is also not hard to prove. Hence the Plancherel formula for SL(2, **C**) can be translated into the following problem: Given a function $f \in C_0^\infty(\mathbf{C}^3)$, recover, f from its line integrals over the lines intersecting (1.3).

Lets first consider a simpler problem: Recover $f \in C_0^\infty(\mathbf{C}^3)$ from its line integrals over *all* lines l in $\mathbf{C}^3$. This problem is not hard to solve. In fact, to determine f at $x \in \mathbf{C}^3$ let H be any 2-plane containing x. The Radon inversion formula for $H = \mathbf{C}^2$ says that $f\,|H$ is determined by its integrals over all lines in H.

Let G^4 be the set of all lines in $\mathbf{C}^3$. (G^4 is clearly a 4-dimensional manifold and is, in fact, an open subset of the complex Grassmannian Gr(4, 2, **C**).) Let R: $C_0^\infty(C^3) \to C^\infty(G^4)$ be the map that assigns to each $l \in G^4$ the integral of f over l. Since $\dim G = \dim C^3 + 1$, we do not expect R to be onto. Let us try to determine its range. To do so let $l \in G^4$. In an appropriate coordinate system l can be parametrized by the equations $x = \alpha_1 t + \beta_1, y = \alpha_2 t + \beta_2, z = t$, and

$$\hat{f}(l) = \hat{f}(\alpha_1, \alpha_2, \beta_1, \beta_1, \beta_2) = \int f(\alpha_1 t + \beta_1, \alpha_2 t + \beta_2, t)\, dt \wedge d\bar{t}$$

(here $\hat{f} = Rf$ for short). Differentiating under the integral sign, one sees immediately that

$$\text{(1.4)} \qquad \left(\frac{\partial}{\partial \alpha_1}\frac{\partial}{\partial \beta_2} - \frac{\partial}{\partial \beta_1}\frac{\partial}{\partial \alpha_2}\right)\hat{f} = 0.$$

Conversely, one can show that if $\hat{f}$ satisfies (1.4), then $\hat{f}$ is in the range of R.

Consider now the 3-dimensional submanifold S of G^4 consisting of all lines intersecting the curve (1.3). It turns out that S has a remarkable property: It is *characteristic* for equation (1.4). In some sense this is the key to the Plancherel formula. In fact, consider any 3-dimensional submanifold S of G^4, and let us ask ourselves if $f \in C_0^\infty(\mathbf{C}^3)$ is determined by its integrals over the lines $l \in S$. Equivalently, we can ask if Rf is determined by its restriction to S. Since Rf satisfies (1.4), we are essentially asking a question about uniqueness of the Cauchy problem for the operator

$$\text{(1.5)} \qquad \frac{\partial}{\partial \alpha_2}\frac{\partial}{\partial \beta_2} - \frac{\partial}{\partial \beta_1}\frac{\partial}{\partial \alpha_2}$$

with Dirichlet data on the surface S. If S were noncharacteristic, the value of $Rf\,|S$ would not suffice to determine it. We would need to know the normal derivative as well. Thus the problem we have just proposed is a reasonable problem only if S is a *characteristic* surface of (1.5).

In [7] Gelfand–Graev–Vilenkin show that, conversely, if S is characteristic, $f \in C_0^\infty(\mathbf{C}^3)$ *is* determined by its integrals over the lines belonging to S; and, in fact, they obtain an explicit formula for f in terms of $\hat{f}\,|S$. They also obtain a remarkable characterization of surfaces S in G^4 that are characteristic for (1.5): With a few degenerate exceptions, *every such S consists either of all $l \in G^4$ passing through a fixed curve in $\mathbf{C}^3$ or of all $l \in G^4$ tangent to a fixed surface in $\mathbf{C}^3$.* We give a proof of this in the appendix.

2. W now try to describe these results from a slightly more general point of view. Consider a diagram of manifolds and maps

$$\begin{array}{ccc} & Z & \\ {}^{\pi}\swarrow & & \searrow{}^{\rho} \\ X & & Y \end{array} \tag{2.1}$$

where π and ρ are fiber maps, π is proper, and $\dim X \leqslant \dim Y$. We assume that $\pi \times \rho$: $Z \to X \times Y$ imbeds Z as a submanifold of $X \times Y$. This means that the fibers F_y of ρ: $Z \to Y$ can be regarded as submanifolds at X, and the fibers G_x of π: $Z \to X$ can a be regarded as submanifolds of Y. Note that the G_x's are compact since π is proper. Suppose now that we have a nowhere-vanishing, smooth density μ on Z. Then we can define a generalized Radon transform

$$R\colon C_0^\infty(X) \to C_0^\infty(|\Lambda|Y) \tag{2.2}$$

by the formula

$$Rf = \rho_*(\pi^* f\mu). \tag{2.3}$$

(A few words about notation: $|\Lambda|Y$ is the density bundle of Y; its sections are smooth measures. π^* is the usual "pull-back" operation for functions, and ρ_* is the dual "push-forward" operation for measures. Intuitively, one can think of $Rf(y)$ as the integral of the function f over the submanifold F_y of X.) The transpose of R,

$$R^{\mathrm{t}}\colon C^\infty(Y) \to C^\infty(|\Lambda|X), \tag{2.4}$$

is also a Radon transform, since $R^{\mathrm{t}}g = \rho_*(\rho^* g u)$. The problem we propose to examine is this: When is $f \in C_0^\infty(X)$ determined by its Radon transform? We show that, for f with sufficiently small support, Rf determines f provided that the G_x's (the fibers of π: $Z \to X$) have nice intersection properties in Y. Let us first see why the intersection properties of the G_x's arise in this problem. We consider a simple finite model, of our problem due to Ethan Bolker. Let X, Y, and Z be finite sets, and let π and ρ in diagram (2.1) be surjective mappings such that $\pi \times \rho$: $Z \to X \times Y$ is injective. If f is a function on X, we let

$$(Rf)(y) = \sum_{x \in F_y} f(x),$$

and we want to know when f is determined by its "Radon transform" Rf. Let us make the following assumptions: (a) The number of points in G_x is the same for all $x \in X$. (b) The number of points in $G_{x_1} \cap G_{x_2}$ is the same for all $x_1 \neq x_2$. Given $x_0 \in X$ let δ_{x_0} be the function which is one at x_0 and zero elsewhere. Then $(R\delta_{x_0})(y)$ is one or zero depending on whether or not x_0 is in F_y. In other words,

$$R\delta_{x_0} = \delta_{G_{x_0}}.$$

Next consider $(R^{\mathrm{t}}R)\delta_{x_0}(x)$. This is the sum over the set G_x of the function $\delta_{G_{x_0}}$ on Y; so

$$(R^{\mathrm{t}}R)\delta_{x_0}(x) = \# G_{x_0} \cap G_x. \tag{2.5}$$

With our assumptions about the intersection properties of the G_x's, this can be rewritten as

$$R^{t}R\delta_{x_0} = c_1\delta_{x_0} + c_2$$

for appropriate c_1 and c_2. From this formula it is easy to see that $R^{t}R$ is invertible. (Bolker's result, incidentally, has an amusing corollary. Let X be a projective n-space over a finite field, Y the dual space of hyperplanes, and $Z \subset X \times Y$ the incidence relation. Assumptions (a) and (b) are obviously satisfied, so we get a "Radon inversion formula" reminiscent of the classical Radon inversion formulas for the projective spaces over **R** and **C**.) Let us now go back to differentiable manifolds. Choose positive, nowhere-vanising densities ν and ω on X and Y. Then we get identifications $C^\infty(X) \cong C^\infty(|\Lambda|X)$ and $C^\infty(Y) \cong C^\infty(|\Lambda|Y)$ given by $f \to f\nu$ and $g \to g\omega$. With these identifications $R^{t}R$ is well defined and maps $C_0^\infty(X)$ into $C^\infty(X)$. Let us now make the assumption

(A) $$G_{x_1} \pitchfork G_{x_2} \quad \text{for } x_1 \neq x_2.$$

This is, in some sense, the analogue of Bolker's assumption that $\# G_{x_1} \cap G_{x_2}$ is independent of x_1 and x_2. It is not hard to show that, with (A), the densities μ, ν, and ω on Z, X, and Y give rise to densities on G_x and $G_{x_1} \cap G_{x_2}$. A slight modification of Bolker's argument then shows that

(2.6) $$\left(R^{t}R\delta_{x_0}\right)(x) = \operatorname{vol}\left(G_{x_0} \cap G_x\right) \quad \text{for } x_0 \neq x.$$

In particular, since the right side of (2.6) is a smooth function of x_0 and x for $x_0 \neq x$, $R^{t}R$ is a pseudolocal operator: Its Schwartz kernel has singular support along the diagonal. We now show that if hypothesis (A) is slightly strengthened, $R^{t}R$ is in fact an elliptic pseudodifferential operator, and from that we get the local invertibility of R. To do this we must first examine the microlocal aspects of diagram (2.1). The density μ on $Z \subset X \times Y$ defines a delta function δ_Z with support on Z, and it is clear that δ_Z is the Schwartz kernel of R. The wave front set of δ_Z is the conormal bundle of Z in $T^*(X \times Y)\setminus 0$, which we denote by Γ. Γ is a Lagrangian submanifold of $T^*(X \times Y)\setminus 0$, and the fact that π and ρ are fibrations implies $\Gamma \subset (T^*X\setminus 0) \times (T^*Y\setminus 0)$. Now consider the microlocal version of diagram (2.1):

$$\begin{array}{ccc} & \Gamma & \\ \pi \swarrow & & \rho \searrow \\ T^*X\setminus 0 & & T^*\setminus 0. \end{array}$$

One can show that hypothesis (A) is equivalent to the assertion that ρ is injective. We replace this by the slightly stronger assertion that

(A′) ρ is an injective immersion.

Let Σ be its image. By assumption, Σ is a one-one immersed submanifold of $T^*Y\setminus 0$. We call Σ the *characteristic variety* associated with diagram (2.1). One can easily see that Σ is a coisotropic submanifold of $T^*Y\setminus 0$ of codimension

$k = \dim Y - \dim X$. Hypothesis (A′) also implies that $\pi\colon \Gamma \to T^*X \setminus 0$ is a fiber mapping. We now prove

THEOREM 1. *If (A′) holds, R^tR is an elliptic pseudodifferential operator of order s, where $-s = \dim Z - \dim Y$.*

PROOF. The Schwartz kernel of R is δ_Z, which is a Fourier integral distribution, in the sense of Hörmander, associated with the canonical relation Γ; so R itself is a Fourier integral operator associated with Γ. Similarly, R^t is a Fourier integral operator associated with the inverse canonical relation Γ^{-1}. Using results of Duistermaat and me [**1**] on "clean" compositions of Fourier integrals operators, one can show that R^tR is a Fourier integral operator associated with the identity canonical relation on X; in other words, it is a pseudodifferential operator. Moreover, by means of the symbol calculus for "clean" compositions developed in [**1**], one can show that it is elliptic.

COROLLARY. *For $f \in C_0^\infty(X)$ of sufficiently small support, Rf determines f.*

Let us now see what the significance is of the characteristic variety Σ. Let I_R be the left ideal of pseudodifferential operators on Y such that

$$P \in I_R \Leftrightarrow PR \text{ is a smoothing operator.}$$

THEOREM 2. *The characteristic variety of I_R is Σ. Moreover, for each $(x, \xi) \in \Sigma$ there exist k commuting pseudodifferential operators $P_1, P_2, \ldots, P_k \in I_R$ whose symbols are defining functions of Σ near (x, ξ) such that I_R is generated, near (x, ξ), by the P_i's. If f is a compactly supported distribution on Y such that Pf is smooth for all $P \in I_R$, then $f = Rg + h$, where g is a compactly supported distribution on X, and h is smooth.*

Let S be a submanifold of Y, and consider the restricted Radon transform

$$f \in C_0^\infty(X) \to Rf|S. \tag{2.7}$$

With the techniques above, one can reduce the study of the inversion problem for (2.7) to a question of uniqueness for the Cauchy problem for I_R, just as Gelfand has done for the example discussed in §1. We carry out the details elsewhere.

Appendix. In [**3**] Gelfand and Graev show that if a codimension-one submanifold S of G^4 is characteristic for equation (1.4), then (with a few degenerate exceptions) S consists either of all lines $l \in G^4$ passing through a fixed curve in $\mathbf{C}^3$, or of all lines $l \in G^4$ tangent to a fixed surface in $\mathbf{C}^3$. In this appendix we sketch a proof of Gelfand's result and indicate a generalization of it to the situation discussed in §2. We recall from (1) that a diagram of maps

$$\begin{array}{ccc} X & \xrightarrow{f} & Z \\ & & \uparrow g \\ & & Y \end{array}$$

is called *clean* if the fiber product

$$F = \{(x, y) \in X \times Y, f(x) = g(y)\}$$

is a submanifold of $X \times Y$ and, in addition, for each $p = (x, y) \in \Gamma$, T_pF is the fiber product in the diagram

$$\begin{array}{ccc} T_xX & \overset{df}{\rightarrow} & T_z \\ & & \uparrow dg \\ & & T_yY \end{array}$$

with $z = f(x) = g(y)$.

Let M and N be symplectic manifolds, let Γ be a Lagrangian submanifold of $M \times N$, and let Λ be a Lagrangian submanifold of M. We denote by $\Gamma \circ \Lambda$ the set of points $\{n \in N, \exists m \in \Lambda, (m, n) \in \Gamma\}$. In other words, if we think of Γ as a "relation" between M and N, then $\Gamma \circ \Lambda$ is the range of Λ under Γ in the usual sense of relations. In **[1]** we prove

LEMMA. *Suppose*

$$\begin{array}{ccc} \Gamma & \overset{\text{proj.}}{\rightarrow} & M \\ & & \uparrow \text{incl.} \\ & & \Lambda \end{array}$$

is a clean diagram. Then $\Gamma \circ \Lambda$ is an (immersed) Lagrangian submanifold of N.

In particular, let X, Y, and Z be as in diagram (2.1), and let Γ be the conormal bundle of Z in $(T^*X \setminus 0) \times (T^*Y \setminus 0)$. Assume Axiom (A′) holds so that the image Σ of Γ in $T^*Y \setminus 0$ is a one-one immersed submanifold. We prove

THEOREM. *If $\Lambda \subset \Sigma$ is a Lagrangian submanifold of $T^* \setminus 0$ then*

$$\Lambda = \Gamma \cdot \Lambda_1, \tag{A1}$$

where Λ_1 is a Lagrangian submanifold of $T^*X \setminus 0$.

PROOF. We apply the lemma to Γ^{-1} to conclude that $\Lambda_1 = \Gamma^{-1} \cdot \Lambda$ is a Lagrangian submanifold of $T^*X \setminus 0$, and then apply the lemma again to conclude that $\Gamma \cdot \Lambda_1$ is Lagrangian and, hence, identical with Λ. We must verify that the diagrams

$$\begin{array}{ccc} \Gamma^{-1} & \overset{\text{proj.}}{\rightarrow} & T^*Y \\ & & \uparrow \text{incl.} \\ & & \Lambda \end{array}$$

and

$$\begin{array}{ccc} \Gamma & \overset{\text{proj.}}{\rightarrow} & T^*X \\ & & \uparrow \text{incl.} \\ & & \Lambda_1 \end{array}$$

are clean. The first is clean since $\Lambda \subset \Sigma$ and $\rho \colon \Gamma \to \Sigma$ is a diffeomorphism. The second is clean since $\Gamma \to T^*X \setminus 0$ is a fiber mapping. Q.E.D.

Now let $X = \mathbf{C}^3$, $Y = G^4$; let S be a hypersurface in G^4, and let Λ be the conormal bundle of S in $T^*G^4 \setminus 0$. That S is characteristic for equation (1.4) means, by definition, that $\Lambda \subset \Sigma$; so by the theorem, $\Lambda = \Gamma \cdot \Lambda_1$, where Λ_1 is a Lagrangian submanifold of $T^*\mathbf{C}^3 \setminus 0$. Since Λ and Γ are homogeneous Lagrangian manifolds, so is Λ_1. We say that a point $(x, \xi) \in \Lambda_1$ is *generic* if the projection map $\Lambda_1 \to X$ attains its maximal rank at (x, ξ). It is a well-known elementary fact (see [**8**]) that if (x, ξ) is generic, then, near (x, ξ) in the cotangent bundle, Λ is of the form N^*W, where W is a submanifold of X. In our case, X is 3-dimensional, so W is either a curve or a surface. Suppose first that W is a curve. Let $l \in G^4$ belong to S. Then by (A1) there exists $(p, \eta) \in N^*W$ and $\xi \in N_l^*S$ such that $(p, \eta, l, \xi) \in N^*Z$, where Z is the point-line incidence relation. In particular, since p is incident to l and $p \in W$, l intersects W. Thus S consists of all lines passing through W. Next, suppose that W is a surface. Then, with the same notation, ξ is normal to l at p and is also normal to W, so l is tangent to W at p, and S consists of all lines tangent to W. This concludes the proof of Gelfand's result.

REMARK. In a certain sense the theorem is stronger than Gelfand's result, since it applies to the degenerate cases when Λ_1 is *not* a conormal bundle.

REFERENCES

1. J. J. Duistermaat and V. Guillemin, *The spectrum of positive elliptic operators and periodic geodesics*, Invent. Math. **29** (1975).

2. I. M. Gelfand and M. I. Graev, *Complexes of k planes in C^n and the Plancherel formula for the group* GL(n, **C**), Dokl. Akad. Nauk SSSR **179** (3) (1968).

3. ______, *Line complexes in the space* $\mathbf{C}^n$, Funksional. Anal. i Prilozhen **2** (1968), no. 3, 39–52. (Russian)

4. I. M. Gelfand, M. I. Graev and Z. Ya. Shapiro, *Integral geometry in k-dimensional spaces*, Funktsional Anal. i Prilozhen **1** (1967), no. 1, 15–31. (Russian)

5. ______, *Differential forms and integral geometry*, Funktsional Anal. i Prilozhen **3** (1969), no. 2, 24–40. (Russian)

6. ______, *Integral geometry in projective space*, Funktsional Anal. i Prilozhen **4** (1970), no. 1, 14–32. (Russian)

7. I. M. Gelfand, M. I. Graev and N. Ya. Vilenkin, *Generalized functions*, Vol. 5, Academic Press, New York, 1968.

8. V. Guillemin and S. Sternberg, *Geometric asymptotics*, Math. Surveys, no. 14, Amer. Math. Soc., Providence, R. I., 1977.

9. L. Hörmander, *Fourier integral operators*. II, Acta Math. **127** (1971), 79–183.

MASSACHUSETTS INSTITUTE OF TECHNOLOGY

Proceedings of Symposia in Pure Mathematics
Volume **43** (1985)

The Propagation of Singularities for Solutions of the Dirichlet Problem

LARS HÖRMANDER

1. Introduction. We discuss the propagation of singularities of solutions of a second-order differential equation satisfying the Dirichlet boundary condition on a noncharacteristic part of the boundary. For a detailed exposition we refer to [**3**]; references to the literature will be given in §5 below.

By X we denote a C^∞ manifold with boundary ∂X and interior $X^\circ = X \setminus \partial X$. Let P be a second-order differential operator in X with C^∞ coefficients and real principal symbol p such that ∂X is noncharacteristic. Our purpose is to discuss the singularities of $u \in \bar{\mathscr{D}}'(X^\circ)$ (the space of extendible distributions), assuming that

$$Pu = f \in C^\infty(X); \qquad u|_{\partial X} = u_0 \in C^\infty(\partial X). \tag{1.1}$$

In X° it is well known that $\mathrm{WF}(u) \subset \mathrm{Char}\, p = p^{-1}(0)$ and that $\mathrm{WF}(u)$ is invariant under the flow defined by the Hamilton vector field in $T^*(X^\circ)\setminus 0$. Locally these are the only conditions required for a set to be the wave front set of a distribution with $Pu \in C^\infty$ (at least if H_p is not radial). Our task is to give a similar result at the boundary. In §2 we define the analogue of the wave front set at ∂X, and in §3 we give analogues of the inclusion $\mathrm{WF}(u) \subset \mathrm{Char}\, p$. Results on the propagation of singularities are then discussed in §§4–7. The paper ends with some brief comments on the existence of solutions with prescribed wave front set compatible with the propagation theorem.

2. The wave front set at the boundary. Let us choose local coordinates at a point in ∂X so that X is defined by $x_1 \geqslant 0$ and ∂X is defined by $x_1 > 0$. Thus we consider X now as an open set in $\overline{\mathbf{R}}^n_+ = \{x;\, x_1 \geqslant 0\}$. We may assume that the planes $x_1 =$ constant are all noncharacteristic in X. If $a \in S^m(X \times \mathbf{R}^{n-1})$ is a symbol of order m and type 1, 0 and the tangential pseudodifferential operator $a(x, D')$, $D' = (D_2, \ldots, D_n)$, is properly supported in X, then

$$\mathrm{WF}\big(a(x, D')u\big) \subset \mathrm{WF}(u) \quad \text{in } X^\circ$$

1980 *Mathematics Subject Classification*. Primary 35G15.

if $u \in \bar{\mathscr{D}}'(X^\circ)$ and $Pu \in C^\infty(X)$. This follows because $(X^\circ \times e_1) \cap \mathrm{WF}(u) = \varnothing$ if e_1 is the unit vector along the ξ_1 axis. This suggests the following definition of

$$\mathrm{WF}_b(u) \subset (T^*(X^\circ)\setminus 0) \cup (T^*(\partial X)\setminus 0).$$

DEFINITION 2.1. $(y', \eta') \in T^*(\partial X)\setminus 0$ is not in $\mathrm{WF}_b(u)$ if and only if, for some $a \in S^m(X \times \mathbf{R}^{n-1})$ that is noncharacteristic at $(0, y', \eta')$, the operator $a(x, D')$ is properly supported in X and $a(x, D')u \in C^\infty$. Over X° we define $\mathrm{WF}_b(u)$ equal to $\mathrm{WF}(u)$.

That this definition is invariant follows from the fact that it agrees with the general definition in terms of the totally characteristic calculus established by Melrose [**6**]. Note that $T^*(\partial X)\setminus 0$ is the quotient of $T^*(X)_{|\partial X}\setminus N^*(\partial X)$ by $N^*(\partial X)$, where $N^*(\partial X)$ is the conormal bundle of ∂X. Thus

$$(T^*(X^\circ)\setminus 0) \cup (T^*(\partial X)\setminus 0) = (T^*(X)\setminus(0 \cup N^*(\partial X)))/N^*(\partial X),$$

which gives the appropriate topology. The characteristic set

$$\Sigma = \operatorname{Char} p = \{(x,\xi) \in T^*(X)\setminus 0;\, p(x,\xi) = 0\}$$

should accordingly be reduced to $\tilde{\Sigma}$ by this equivalence relation, which means that

$$\tilde{\Sigma}_{|\partial X} = \{(x', \xi') \in T^*(\partial X)\setminus 0;\, p(0, x', \xi', \xi_n) = 0 \text{ for some } \xi_n \in \mathbf{R}\}.$$

3. Ellipticity and hypoellipticity. The microlocal version of the basic regularity theorem for solutions of elliptic equations gives $\mathrm{WF}(u) \subset \Sigma$ in X°. Similarly, the microlocal version of the basic regularity theorem for the Dirichlet boundary problem for elliptic operators gives

(3.1) $$\mathrm{WF}_b(u) \subset \tilde{\Sigma} \quad \text{if } u \text{ satisfies (1.1)}.$$

However, there may be a large subset of $\tilde{\Sigma}_{|\partial X}$ that cannot carry singularities. An example is the Tricomi operator

$$D_1^2 + x_1 \sum_2^n D_j^2$$

in the elliptic half-plane $x_1 \geqslant 0$. To single out points with similar properties, we choose our coordinates geodesic with respect to ∂X, which means that, possibly after a change of sign,

(3.2) $$p(x,\xi) = \xi_1^2 - r(x,\xi'), \qquad \xi' = (\xi_2, \ldots, \xi_n).$$

Then we have

$$\tilde{\Sigma}_{|\partial X} = \{(x', \xi');\, r_0(x', \xi') \geqslant 0\},$$

where $r_0(x', \xi') = r(0, x', \xi')$. Set

$$HE(p) = \{(x', \xi') \in T^*(\partial X)\setminus 0;\, r_0 \leqslant 0 \text{ in a neighborhood and } r_1(x', \xi') < 0\}$$

where $r_1(x', \xi') = \partial r(0, x', \xi')/\partial x_1$. It is easily seen that this is an invariant definition. Then, just as for the Tricomi equation,

(3.3) $$\mathrm{WF}_b(u) \cap HE(p) = \varnothing.$$

In fact, if we only know that $Pu = f \in \overline{H}^{\mathrm{loc}}_{(s)}(X^\circ)$, $u_{|\partial X} \in H^{\mathrm{loc}}_{(s+5/3)}(\partial X)$, then it is easy to show that $u \in \overline{H}^{\mathrm{loc}}_{(s+4/3)}(X^\circ)$. (By $\overline{H}^{\mathrm{loc}}_{(s)}(X^\circ)$ we denote the distributions in X° which are restrictions of elements in the Sobolev space $\overline{H}^{\mathrm{loc}}_{(s)}(Y)$ for some open neighborhood Y of X in $\mathbf{R}^n$.)

4. The hyperbolic set. The hyperbolic set is defined by

$$H = \{(x', \xi') \in T^*(\partial X)\setminus 0;\ r_0(x', \xi') > 0\}$$

when p is of the form (3.2), which we assume throughout. Thus p has two different zeros in the fiber of $T^*(X)$ over (x', ξ') if and only if $(x', \xi') \in H$. Microlocally the boundary problem (1.1) differs from the Cauchy problem for a strictly hyperbolic operator only in the lack of one boundary condition. Factoring P into first order factors, which are pseudodifferential along the boundary, one proves that $(x', \xi') \in H \cap \mathrm{WF}_b(u)$ implies that $\mathrm{WF}(u)$ contains both the bicharacteristic of p coming in at $(0, x', -r_0(x', \xi')^{1/2}, \xi')$ and the one going out from $(0, x', r_0(x', \xi')^{1/2}, \xi')$. (Conversely, if $\mathrm{WF}_b(u)$ contains some point on these, it follows from the interior propagation theorem that $(x', \xi') \in \mathrm{WF}_b(u)$, for this is a closed set.) Note that the Hamilton equations

$$\text{(4.1)} \qquad \frac{dx_1}{dt} = 2\xi_1, \quad \frac{d\xi_1}{dt} = \frac{\partial r}{\partial x_1}, \quad \frac{dx'}{dt} = -\frac{\partial r}{\partial \xi'}, \quad \frac{d\xi'}{\partial t} = \frac{\partial r}{\partial x'}$$

show that x_1 is increasing on a bicharacteristic when $\xi_1 > 0$. The two bicharacteristics form a continuous curve in $\tilde{\Sigma}$. Together they make up a *broken bicharacteristic*. Note that ξ_1 has a positive jump at the reflection point, while all other coordinates are continuous.

5. The glancing set. What is left to discuss now is the singularities in the *glancing set* $G \subset \tilde{\Sigma}_{|\partial X}$ defined in local coordinates by $r_0 = 0$. It can be identified with its inverse image in Σ defined invariantly by $x_1 = p = H_p x_1 = 0$, where x_1 is any function vanishing simply on ∂X. The *diffractive set* $G_d \subset G$ is the subset where $\partial r/\partial x_1 > 0$. By (4.1) this means that $H_p^2 x_1 > 0$. From this convexity it follows that there are three kinds of bicharacteristics nearby: those which do not reach ∂X because x_1 has a positive minimum; those which are tangent to ∂X because the minimum of x_1 is 0; and those which arrive and are reflected at the hyperbolic set. Since ξ_1 has a positive jump, it follows from (4.1) that x_1 is also a strictly convex function of t on such broken bicharacteristics, so they have to stay away from the boundary for a fixed time before and after the reflection (if $|\xi'| = 1$ at the reflection point). We shall mainly be concerned with the proof of the following theorem.

THEOREM 5.1. *If* $(x', \xi') \in \mathrm{WF}_b(u) \cap G_d$ *and* u *satisfies* (1.1), *then* $\mathrm{WF}_b(u)$ *contains the bicharacteristic of* p *through* $(0, x', 0, \xi')$.

Theorem 5.1 is due to Melrose [**5**] and Taylor [**8**] when $dr_0(x', \xi')$ is not a multiple of the one-form $\langle x', d\xi'\rangle$. Their proofs depend on parametrix constructions, later supplemented by reduction to model operators. (See also Eskin [**2**].)

The propagation of singularities in the *gliding set* $G_g \subset G$, defined by $\partial r/\partial x_1 < 0$ or, equivalently, $H_p^2 x_1 < 0$, was investigated by Andersson and Melrose [**1**] who showed that the singularities propagate along the gliding vector field $-H_{r_0}$, which is described invariantly as the only vector $H_p + \lambda H_{x_1}$ tangential to the glancing set. Their proof was based on energy integral arguments and a very precise control of the symplectic geometry of the pair of surfaces $p = 0$ and ∂X. A much simpler argument was found by Melrose and Sjöstrand [**7**], who also handled the general glancing set, introducing the notion of generalized bicharacteristic there. This is a (discontinuous) curve in Σ that becomes continuous when mapped into $\tilde{\Sigma}$, and it is equal to a (broken) bicharacteristic except at $G \setminus G_d$, where it is differentiable with derivative equal to the gliding vector. The result of Melrose and Sjöstrand is that every point in $\mathrm{WF}_b(u) \cap (G \setminus G_d)$ where the gliding vector is not zero is in the interior of a generalized bicharacteristic contained in $\mathrm{WF}_b(u)$ after projection into $\tilde{\Sigma}$.

The first study of the diffractive case by means of energy estimates was made by Ivrii [**4**] (who considered symmetric hyperbolic systems). Our purpose here is to present his methods and apply them to sketch a proof of Theorem 5.1.

6. An energy identity and estimates for some quadratic forms. We assume throughout that $X = \overline{\mathbf{R}}^n_+$ and that the principal symbol p of P is of the form (3.2). By conjugation of P with a multiplication operator, we can remove the D_1 term also and have

$$P = D_1^2 - R(x, D'),$$

where R has the principal symbol r. It is technically convenient to assume that the coefficients of R are in $C_0^\infty(\mathbf{R}^n)$, and this does not affect any local arguments. Let

$$Q(x, D) = Q_1(x, D')D_1 + Q_0(x, D'),$$

where $Q_j \in S^{-j}(\overline{\mathbf{R}}^n_+ \times \mathbf{R}^{n-1})$ has homogeneous principal symbol q_j; we write

$$q(x, \xi) = q_1(x, \xi')\xi_1 + q_0(x, \xi')$$

for the principal symbol of Q. We assume that Q is selfadjoint; that is,

$$Q_1^* = Q_1, \quad Q_0^* = Q_0 - [D_1, Q_1],$$

which implies that q_1 and q_0 are real valued. Let $u \in C_0^\infty(X)$ and set $u_j = D_1^j u$, $j = 0, 1$. With $(\ ,\ )_X$ and $(\ ,\ )_{\partial X}$ denoting the usual sequilinear scalar products in X and in ∂X, we have the following energy identity.

LEMMA 6.1. *If* $u \in C_0^\infty(\overline{\mathbf{R}}^n_+)$ *then*

$$2\,\mathrm{Im}(Pu, Qu)_X = \sum_0^1 \big(B_{jk}(x', D')u_k, u_j\big)_{\partial X} + \sum_0^1 \big(C_{jk}(x, D')u_k, u_j\big)_X, \tag{6.1}$$

where $B_{11} = Q_1$, $B_{01}^* = B_{10} = Q_0$, $B_{00} = Q_1(R + R^*)/2$ *for* $x_1 = 0$, *the principal symbol* c_{jk} (*of order* $1 - j - k$) *of* C_{jk} *is real*, $c_{01} = c_{10}$, *and*

$$\sum c_{jk}(x, \xi')\xi_1^{j+k} = \{p, q\} + 2q \operatorname{Im} p^s, \tag{6.2}$$

$$p^s(x, \xi') = -\frac{i}{2}\sum_2^n r_{(j)}^{(j)}(x, \xi') - \sum_2^n R_j(x)\xi_j$$

is the subprincipal symbol of p.

The proof is immediate. Note that (6.2) is just the principal symbol that one would obtain if there were no boundary. To derive an estimate from (6.1) we estimate the left side from below and the right side from above by the negative of the square of the norm of u that we must control. To do so we first observe that if $A_{jk} \in S^{1-j-k}(\overline{\mathbf{R}}^n_+ \times \mathbf{R}^{n-1})$ has principal symbol a_{jk}, $a_{01} = a_{10}$, and

$$\sum_{j,k=0}^{1} a_{jk}(x, \xi')\xi_1^{j+k} \leqslant 0, \qquad x_1 \geqslant 0, \tag{6.3}$$

then

$$\operatorname{Re}\sum\left(A_{jk}(x, D')u_k, u_j\right)_X \leqslant C\sum \|u_j\|^2_{(0,-j)}, \qquad u_j \in C_0^\infty(\overline{\mathbf{R}}^n_+). \tag{6.4}$$

(We set

$$\|u\|_{(s,t)} = \left\|\left(1 + |D|^2\right)^{s/2}\left(1 + |D'|^2\right)^{t/2} u\right\|_{L^2}, \qquad u \in C_0^\infty(\mathbf{R}^n)$$

and use the same notation for the quotient norm defined for the restrictions to $\mathbf{R}^n_+$.) If $u_j = D_1^j u$, the right side is equivalent to $\|u\|^2_{(1,-1)}$. Estimate (6.4) follows from the sharp Gårding inequality for systems, since (6.3) implies

$$\sum a_{jk}(x, \xi')z_j\bar{z}_k \leqslant 0, \qquad (z_0, z_1) \in \mathbf{C}^2. \tag{6.3$'$}$$

However, it is not easy to construct q so that q is linear in ξ_1 and the sign of the right side of (6.2) is favorable everywhere, as one does when proving the interior propagation theorem. Indeed, if one makes a suitable choice of q, ignoring the linearity in ξ_1, one must then reduce modulo p to get a linear function of ξ_1, and that will spoil all information on $\{p, q\}$ outside the characteristic set. We must therefore examine how much (6.4) is weakened if (6.3) is only assumed in the characteristic set. To do so we need a lemma.

LEMMA 6.2. *Let* Y *be an open subset of* $\overline{\mathbf{R}}^n_+ = \{y \in \mathbf{R}^n;\ y_1 \geqslant 0\}$, *and let* $r \in C^\infty(Y)$. *Assume that* r *is real valued*, $dr \neq 0$ *when* $r = 0$, *and* $\partial r/\partial y_1 > 0$ *when* $r = y_1 = \partial r/\partial y_j = 0$ *for* $j \neq 1$. *Let*

$$A(t, y) = \sum_0^2 a_j(y)t^j$$

be a quadratic polynomial in t *with coefficients in* $C^\infty(Y)$ *such that*

$$A(t, y) = -\psi(t, y)^2 \quad \text{when } t^2 = r(y), \tag{6.5}$$

where $\psi \in C^\infty(\mathbf{R} \times Y)$. *Then one can find* $\psi_0, \psi_1, g \in C^\infty(Y)$ *with* $\psi_0(y) = \psi(0, y), \psi_1(y) = \partial\psi(0, y)/\partial t$ *when* $r(y) = 0$ *and*

$$(6.6)\quad A(t, y) + (\psi_0(y) + \psi_1(y)t)^2 \leqslant g(y)(t^2 - r(y)), \qquad t \in \mathbf{R}, y \in Y.$$

PROOF. If $0 < r(y) = s^2$ we choose $\psi_0(y), \psi_1(y)$ so that $\psi_0(y) + \psi_1(y)t = \psi(t, y)$ when $t^2 = r(y)$; that is, $t = \pm s$. Thus

$$\psi_0(y) = (\psi(s, y) + \psi(-s, y))/2, \qquad \psi_1(y) = (\psi(s, y) - \psi(-s, y))/2s.$$

The functions on the right are even smooth functions of s, so they can be written in the form $\Psi_j(s^2, y)$, where $\Psi_j \in C^\infty$. We extend the definition of ψ_j by $\psi_j(y) = \Psi_j(r(y), y)$. Now

$$B(t, y) = A(t, y) + (\psi_0(y) + \psi_1(y)t)^2 - (a_2(y) + \psi_1(y)^2)(t^2 - r(y))$$

is linear in t and vanishes when $t^2 = r(y)$. Hence, it follows from the implicit function theorem and the hypothesis on dr that the coefficients are $O(r^N)$ for any N on any compact subset of Y. Since $t^2 - r(y) \geqslant -r(y)$ and $t^2 - r(y) \geqslant 2|t|(-r(y))^{1/2}$ when $r(y) < 0$, we conclude that $B(t, y) \leqslant f(y)(t^2 - r(y))$ if $f(y) = 0$ when $r(y) > 0$ and if $f \in C^\infty$, but vanishes sufficiently slowly when $r(y) \to -0$.

LEMMA 6.3. *Let* $A_{jk} \in S^{1-j-k}(\mathbf{R}^n \times \mathbf{R}^{n-1})$, $j, k = 0, 1$, *have real homogeneous principal symbols* $a_{jk}(x, \xi')$ *vanishing for large* $|x|$, *and assume that*

$$\sum a_{jk}(x, \xi')\xi_1^{j+k} = -\psi(x, \xi)^2 \quad \text{when } \xi_1^2 = r(x, \xi'),$$

where $\psi \in C^\infty(\mathbf{R}^n \times (\mathbf{R}^n \setminus 0))$ *is homogeneous of degree* $\frac{1}{2}$. *Further assume that* $\partial r/\partial x_1 > 0$ *or that* $\partial r/\partial(x', \xi') \neq 0$ *in* $\bigcup \operatorname{supp} a_{jk}$. *Then one can choose* $\Psi_0(x, \xi') \in S^{1/2}(\mathbf{R}^n \times \mathbf{R}^{n-1})$ *and* $\Psi_1(x, \xi') \in S^{-1/2}(\mathbf{R}^n \times \mathbf{R}^{n-1})$ *with principal symbols equal to* $\psi(x, \xi)$ *and* $\partial\psi(x, \xi)/\partial\xi_1$ *when* $\xi_1 = r(x, \xi') = 0$, *such that for any* $u \in \overline{C}_0^\infty(\mathbf{R}^n_+)$ *and any* $\kappa \in \mathbf{R}$

$$(6.7)\quad \begin{aligned} &\operatorname{Re} \sum \big(A_{jk}(x, D')D_1^k u, D_1^j u\big)_X + \|\Psi_0(x, D') + \Psi_1(x, D')D_1 u\|_X^2 \\ &\leqslant C_\kappa\big(\|u\|_{(1,-1)}^2 + \|D_1 u(0, \cdot)\|_{(-\kappa-1)}\|u(0, \cdot)\|_{(\kappa)} + \|Pu\|_{(0,-1)}^2\big). \end{aligned}$$

The proof follows if one applies (6.4) to the sum of the terms on the left side and

$$-\operatorname{Re}\big(G(x, D')D_1 u, D_1 u\big) + \operatorname{Re}\big(G(x, D')R(x, D')u, u\big)$$

with the principal symbols g and ψ_j of G and Ψ_j given by Lemma 6.2, and then moves D_1 from the right to the left in the first term.

7. Sketch of proof of Theorem 5.1. Let $(0, \xi') \in G_d$ and choose an open conic neighborhood W of $(0, 0, \xi')$ in $X \times (\mathbf{R}^{n-1} \setminus 0)$ such that $\partial r/\partial x_1 > 0$ in W. Assume there is some point $\gamma_0 \notin \mathrm{WF}(u)$ on the bicharacteristic through $\gamma = (0, 0, 0, \eta')$. Because of the invariance of the wave front set under the bicharacteristic flow over X°, we may assume that the arc between γ_0 and γ lies over W. In what follows we assume that γ lies after γ_0 on the bicharacteristic; the proof is quite parallel if γ_0 lies after γ. We must then show that $(0, \eta') \notin \mathrm{WF}_b(u)$. Without restriction we may assume that u has compact support.

Choose a conic neighborhood Γ_0 of γ_0 in $T^*(X^\circ)\setminus 0$ such that $\Gamma_0 \cap \mathrm{WF}(u) = \varnothing$, and let W_0 be the set of all $(x, \xi') \in W$ such that $r(x, \xi') < 0$ or the backward bicharacteristics from $(x, \pm r(x, \xi')^{1/2}, \xi')$ go to Γ_0 (with at most one reflection or tangency at ∂X) while remaining over W. The results already mentioned on the propagation of singularities in the interior and in the hyperbolic set show that singularities occur in W_0 only in $G_d \cap W_0$ and in the forward H_p flowout from $G_d \cap W_0$. More precisely, if $\chi \in S^0$, cone supp $\chi \subset W_0$, and $r > 0$ in cone supp χ, then we have a factorization

$$P(x, D) = \big(D_1 - \Lambda_-(x, D')\big)\big(D_1 - \Lambda_+(x, D')\big) + \rho(x, D'), \qquad \rho \in S^{-\infty},$$

at cone supp χ with the principal symbol of $\Lambda_\pm$ equal to $\pm r^{1/2}$, and

$$\chi(x, D')\big(D_1 - \Lambda_+(x, D')\big)u \in C^\infty(X). \tag{7.1}$$

In fact, we can remove the factor $D_1 - \Lambda_-(x, D')$, since the bicharacteristics backwards from a point in cone supp χ go to Γ_0 with ξ_1 decreasing from a negative value; hence, r is increasing.

We shall prove that $\chi(x, D')u \in C^\infty$ for all $\chi \in S^0$ with cone supp $\chi \subset W_0$ by proving inductively the statement

$$(\mathrm{I})_s \qquad \begin{array}{l}\chi(x, D')u \in \overline{H}_{(1, s-1)} \text{ and } \chi(x, D')D_1 u_{|x_1=0} \in H_{(s-1)}(\mathbf{R}^{n-1}) \\ \text{for all } \chi \in S^0 \text{ with cone supp } \chi \subset W_0.\end{array}$$

By partial hypoellipticity this is true for s sufficiently large negative, so it suffices to show that $(\mathrm{I})_s \Rightarrow (\mathrm{I})_{s+1/2}$. The main step is the following

LEMMA 7.1. *Assume that* $q(x, \xi) = q_1(x, \xi')\xi_1 + q_0(x, \xi')$, *where* $q_j \in C^\infty(\mathbf{R}^n \times (\mathbf{R}^{n-1}\setminus 0))$ *have support in* W_0 *and are homogeneous of degree* $-j$, *and let* $q_1(0, x', \xi') = -t(x', \xi')^2$ *for some* $t \in C^\infty$. *Further assume that, for some constant* M,

$$\{p, q\} + qM|\xi'| = -\psi^2 + \rho\big(\xi_1 - r^{1/2}\big) \quad \text{and} \quad q = v^2 \qquad \text{when } p = 0, \tag{7.2}$$

where $\psi, v \in C^\infty(\mathbf{R}^n \times (\mathbf{R}^n \setminus 0))$ *have support over* W_0 *and are homogeneous of degree* $1/2$ *and* 0, *while* $\rho \in C^\infty(\mathbf{R}^n \times (\mathbf{R}^{n-1}\setminus 0))$ *is homogeneous of degree* 0, *and* $r > 0$ *in* supp ρ, *which is contained in* W_0. *If* u *satisfies* (1.1) *and* $(I)_s$, *and if* M *is larger than some number depending on* P *and* s, *it follows that* $T(x', D')D_1 u_{|x_1=0} \in H_{(s)}(\mathbf{R}^{n-1})$ *if* t *is the principal symbol of* T, *and* $\Psi_0(x, D')u + \Psi_1(x, D')D_1 \in \overline{H}_{(0, s)}$ *for some* $\Psi_j \in S^{1/2-j}$ *with cone supp* $\Psi_j \subset W_0$ *and principal symbols* $\partial^j\psi(0, x', 0, \xi')/\partial\xi_1^j$ *when* $x_1 = r(x, \xi') = 0$.

The lemma is proved by choosing $\chi \in S^0$ with cone supp $\chi \subset W_0$, so that $\chi = 1$ in a neighborhood of supp $q_0 \cup$ supp q_j, and applying (6.1) to

$$u_\varepsilon = \big(1 + \varepsilon^2|D'|^2\big)^{-1}\big(1 + |D'|^2\big)^{s/2}\chi(x, D')u.$$

An upper bound for the right side is obtained from Lemma 6.3.

To prove that $(\mathrm{I})_s \Rightarrow (\mathrm{I})_{s+1/2}$, we must also construct suitable functions to which Lemma 7.1 can be applied. Let $(y', \eta') \in G_d$, $(0, y', \eta') \in W_0$, and set

$$\varphi(x, \xi) = \varphi_1(\xi')\xi_1 + \varphi_0(x, \xi'),$$

$$\varphi_1(\xi') = 1/|\xi'|, \quad \varphi_0 = x_1^2 + |x' - y'|^2 + \left|\xi'/|\eta'\xi| - \eta'/|\eta'|\right|^2.$$

Then φ is an increasing function of ξ_1, and

$$H_p\varphi = \xi_1(2\partial\varphi_0/\partial x_1 - \{r, \varphi_1\}) - \{r, \varphi_0\} + \varphi_1 \partial r/\partial x_1 \geqslant c|\xi'|$$

in a conic neighborhood of $(0, y', 0, \eta')$ in the characteristic set, for the positive term $\varphi_1 \partial r/\partial x_1$ dominates completely there. Set $\chi_0(t) = \exp(-1/t)$, $t > 0$, and $\chi_0(t) = 0$ for $t \leqslant 0$. Then $\chi_0(1 - \varphi/\delta)$ is a decreasing function of ξ_1 and

$$\{p, \chi_0(1 - \varphi/\delta)\} = -\chi_0'(1 - \varphi/\delta) H_p\varphi\delta^{-1}.$$

Let $\chi_1 \in C^\infty(\mathbf{R})$ be a decreasing function equal to 1 in $(-\infty, 2)$ and 0 in $(3, \infty)$, and set

$$f = \chi_1(\varphi_0/\delta)\chi_0(1 - \varphi/\delta).$$

for small positive δ. In $\operatorname{supp} f \cap \operatorname{supp} d\chi_1(\varphi_0/\delta)$ we have $\varphi_0 \geqslant 2\delta$ and $\varphi_0 + \varphi_1\xi_1 \leqslant \delta$; hence, $\xi_1 \leqslant -\delta|\xi'|$ and $r \geqslant \delta^2|\xi'|^2$ if $p = 0$. Since $\varphi_0 \leqslant 3\delta$ in $\operatorname{supp} f$, it is clear that (x, ξ') is close to the ray through $(0, y', \eta')$ in $\operatorname{supp} f$. Hence, it is easy to see that (7.2) is satisfied by $q = f$. However, to obtain a polynomial we must take q equal to the remainder when f is divided by p using the Malgrange preparation theorem. This does not affect the validity of (7.2) on the characteristic set. Finally, one finds that $-q_1$ is the square of a smooth function by using the monotonicity of f with respect to ξ_1 and the properties of χ_0 as in Melrose–Sjöstrand [7]. Thus we obtain sufficiently many functions q to use in Lemma 7.1 to conclude that $(\mathrm{I})_{s+1/2}$ holds near $G_d \cap W_0$. The induction is completed by the well-known results on propagation of singularities in the interior.

8. Solutions with given singularities. There are many known constructions of distributions $u \in \mathscr{D}'(X^\circ)$ with $Pu \in C^\infty$ and $\mathrm{WF}(u)$ generated by a given bicharacteristic γ. One just has to choose a strictly positive Lagrangian Λ with real part γ such that the restriction of p to Λ vanishes of infinite order at γ. Then a recursive solution of transport equations gives the desired u as a Fourier integral distribution corresponding to Λ. If γ is a broken bicharacteristic, there is no difficulty in carrying this construction through *locally* so that $u \in \bar{\mathscr{D}}'(X^\circ)$, $Pu \in C^\infty(X)$, $u_{|\partial X} \in C^\infty$, and $\mathrm{WF}_b(u) = \gamma$. Well-known functional analytic arguments extend the result to generalized bicharacteristics that are limits of broken bicharacteristics. Through every point outside $\{(y', \eta') \in G\varphi(y', \eta'), H_{r_0} = 0\}$, there is such a generalized bicharacteristic (see [7]), which, consequently, can carry singularities. It is not known, though, if every generalized bicharacteristic is a limit of broken bicharacteristics in this situation. However, it is clear that no limit of broken bicharacteristics passes through a point in G_d such that $r_0 = 0$ in a neighborhood, for there are no hyperbolic points nearby then. This occurs for the Tricomi equation in the hyperbolic half-space, for example. Now P is microlocally hyperbolic of principal type in the larger set

$$\left\{(y', \eta'); r(x, \xi') \geqslant cx_1|\xi'|^2 \text{ in a neighborhood of } (0, y', 0) \text{ when } x_1 \geqslant 0\right\}.$$

For an operator with principal symbol (3.2) such that r satisfies this condition globally, we can solve the Cauchy problem just as in the strictly hyperbolic case apart from a slightly stronger regularity condition on the Cauchy data. If we solve the Cauchy problem with data $u = 0$, $D_1 u = \psi$ when $x_1 = 0$ such that $\mathrm{WF}(\psi)$ is generated by (y', η'), we also obtain in this case a solution which is singular precisely on the bicharacteristic through $(0, y', 0, \eta')$.

References

1. K. G. Andersson and R. B. Melrose, *The propagation of singularities along gliding rays*, Invent. Math. **41** (1977), 197–232.

2. G. I. Eskin, *Parametrix and propagation of singularities for the interior mixed hyperbolic problem*, J. Analyse Math. **32** (1977), 17–62.

3. L. Hörmander, *The analysis of linear partial differential operators*. III, Springer-Verlag, 1985, Chap. 24.

4. V. Ja. Ivrii, *Wave fronts for solutions of boundary value problems for a class of symmetric hyperbolic systems*, Sibirsk. Mat. Z. **21** (1980), 62–71; English transl. in Siberian Math. J. **21** (1980), 527–534.

5. R. B. Melrose, *Microlocal parametrices for diffractive boundary problems*, Duke Math. J. **42** (1975), 605–635.

6. ______, *Transformation of boundary problems*, Acta Math. **147** (1981), 149–236.

7. R. B. Melrose and J. Sjöstrand, *Singularities of boundary value problems*. I, II, Comm. Pure Appl. Math. **31** (1978), 593–617; ibid. **35** (1982), 129–168.

8. M. Taylor, *Grazing rays and reflection of singularities of solutions to wave equations*, Comm. Pure Appl. Math. **29** (1976), 1–38.

University of Lund, Sweden

Proceedings of Symposia in Pure Mathematics
Volume **43** (1985)

Application of the Microlocal Theory of Sheaves to the Study of $\mathcal{O}_X$

MASAKI KASHIWARA AND PIERRE SCHAPIRA

Abstract. We explain some constructions and results of [**5**] with emphasis on applications to the study of the microlocalization of the sheaf $\mathcal{O}_X$ of holomorphic functions on a complex manifold X along a real submanifold M of X.

I. Microsupport. Let X be a real manifold of class C^α, with $1 \leqslant \alpha \leqslant \infty$ or $\alpha = \omega$ (i.e., X real analytic). We denote the cotangent bundle to X by π: $T^*X \to X$ and the canonical 1-form on T^*X by ω_X. If Y is a submanifold of X, we denote the conormal bundle to Y by T^*_YX. In particular, T^*_XX denotes the zero section of T^*X, which one identifies with X. We denote by $D^+(X)$ the derived category of the category of complexes, bounded from below, of sheaves of abelian groups on X. Thus an object $F^\cdot$ of $D^+(X)$ is represented by a complex of sheaves

$$F^\cdot \cong \cdots \to F^i \to F^{i+1} \to \cdots,$$

with $F^i = 0$ for $i \ll 0$. Moreover, two complexes that are quasi-isomorphic are identified in $D^+(X)$, and any object $F^\cdot$ may be represented by a complex of flabby sheaves.

EXAMPLE 1.1. Let X be a complex manifold, $\mathcal{O}_X$ the sheaf of holomorphic functions on X, $\Omega_X^{(p)}$ the sheaf of holomorphic p-forms on X, $\overline{X}$ the complex conjugate of X, $X^{\mathbf{R}}$ the underlying real analytic manifold to X, $B_{X^{\mathbf{R}}}$ the sheaf of hyperfunctions on $X^{\mathbf{R}}$, and $B_{X^{\mathbf{R}}}^{(0,p)} = \Omega_{\overline{X}}^{(p)} \otimes_{\mathcal{O}_{\overline{X}}} B_{X^{\mathbf{R}}}$. In $D^+(X)$ the complexes

$$\cdots \to 0 \to \mathcal{O} \to 0 \to \cdots,$$
$$\cdots \to 0 \to B_{X^{\mathbf{R}}}^{(0,0)} \underset{\bar\partial}{\to} \cdots \to B_{X^{\mathbf{R}}}^{(0,n)} \to 0 \to \cdots$$

are identified. ($\mathcal{O}_X$ and $B_{X^{\mathbf{R}}}^{(0,0)}$ are of degree 0, $\bar\partial$ is the antiholomorphic differential on $X^{\mathbf{R}}$, and $n = \dim_{\mathbf{C}} X$.)

1980 *Mathematics Subject Classification*. Primary 32F99; 58G99.

EXAMPLE 1.2. Let Z be a locally closed subset of the real manifold X. One recalls that the sheaf on X satisfies

$$(\underline{\mathbf{Q}}_Z)_x = \begin{cases} \mathbf{Q} & \text{if } x \in Z, \\ 0 & \text{if } x \notin Z. \end{cases}$$

Assume Z is closed in X. We have an exact sequence

$$0 \to \underline{\mathbf{Q}}_{(X-Z)} \to \underline{\mathbf{Q}}_X \to \underline{\mathbf{Q}}_Z \to 0.$$

Thus we have an isomorphism in $D^+(X)$:

$$0 \to \underline{\mathbf{Q}}_{(X-Z)} \to \underline{\mathbf{Q}}_X \to 0 \cong \underline{\mathbf{Q}}_Z[-1]$$

(one identifies a sheaf F with the complex $\cdots \to 0 \to F \to 0 \to \cdots$, where F is in degree 0 and $F[d]$ is the shifted complex $F[d]^i = F^{i+d}$).

Now we return to the situation where X is a real manifold of class C^α.

DEFINITION 1.3. Let $F^\cdot \in \mathrm{Ob}(D^+(X))$. The microsupport of $F^\cdot$, denoted $\mathrm{SS}(F^\cdot)$, is the subset of T^*X defined by

> $p \notin \mathrm{SS}(F^\cdot) \Leftrightarrow$ there exists an open neighborhood U of p in T^*X such that, for any $x_1 \in X$, any real function ϕ of class C^α defined in a neighborhood of x_1 with $\phi(x_1) = 0$, $d\phi(x_1) \in U$, we have $(\mathbf{R}\Gamma_{\{x;\phi(x)\geqslant 0\}}(F^\cdot))_{x_1} = 0$.

Recall that if Z is a locally closed subset of X (here $Z = \{x; \phi(x) \geqslant 0\}$), the complex $\mathbf{R}\Gamma_Z(F^\cdot)$ is calculated by representing $F^\cdot$ by a complex of flabby sheaves and applying the functor $\Gamma_Z(\cdot)$, where $\Gamma_Z(F)$ is the subsheaf of F of sections with support in Z.

Roughly speaking, when F is a sheaf, $p \notin \mathrm{SS}(F)$ means that F has no section, and no "cohomology" supported by "half-spaces", whose conormal lies in a neighborhood of p.

Similarly, if $u: F^\cdot \to G^\cdot$ is a morphism in $D^+(X)$, one defines $\mathrm{SS}(u)$ by saying that $p \notin \mathrm{SS}(u)$ if, for U, x_1, ϕ, as in Definition 1.3, we have isomorphisms

$$\left(\mathbf{R}\Gamma_{\{x;\phi(x)\geqslant 0\}}(F^\cdot)\right)_{x_1} \xrightarrow{\sim} \left(\mathbf{R}\Gamma_{\{x;\phi(x)\geqslant 0\}}(G^\cdot)\right)_{x_1}.$$

It follows immediately by the definition that

—$\mathrm{SS}(F^\cdot)$ is a closed cone in T^*X;

—$\mathrm{SS}(F^\cdot) \cap T^*_XX = \mathrm{supp}(F^\cdot)$, where $\mathrm{supp}(F^\cdot) = \overline{\bigcup_j \mathrm{supp}\, H^j(F^\cdot)}$ is the support of the complex $F^\cdot$;

—if $0 \to F_1 \to F_2 \to F_3 \to 0$ is an exact sequence of sheaves (or more generally if we have a distinguished triangle $F_1 \to F_2 \to F_3 \to F_1[+1]$ in $D^+(X)$), then

$$\mathrm{SS}(F_i) \subset \mathrm{SS}(F_j) \cup \mathrm{SS}(F_k) \quad \text{if } \{i, j, k\} = \{1, 2, 3\}.$$

EXAMPLE 1.4. (i) Let Z be a closed submanifold of X, then $\mathrm{SS}(\underline{\mathbf{Q}}_Z) = T_Z^* X$.

(ii) Let ϕ be a real C^1-function, $\overline{Z}^+ = \{x; \phi(x) \geqslant 0\}$, $Z^+ = \{x; \phi(x) > 0\}$, and assume $d\phi \neq 0$ on $\{x; \phi(x) = 0\}$. Then

$$\mathrm{SS}(\underline{\mathbf{Q}}_{\overline{Z}^+}) = \left(T_X^*X \cap \overline{Z}^+\right) \cup \{(x; \lambda\, d\phi(x)); \phi(x) = 0, \lambda \geqslant 0\},$$

$$\mathrm{SS}(\underline{\mathbf{Q}}_{Z^+}) = \left(T_X^*X \cap \overline{Z}^+\right) \cup \{(x; \lambda\, d\phi(x)); \phi(x) = 0, \lambda \leqslant 0\}.$$

EXAMPLE 1.5. Let X be a complex manifold, $\mathscr{M}$ a coherent module over the sheaf of rings $\mathscr{D}_X$ of holomorphic differential operators (of finite order) (cf. [**9**]). Then

$$\mathrm{char}(\mathscr{M}) = \mathrm{SS}\left(\mathbf{R}\mathscr{H}om_{\mathscr{D}_X}(\mathscr{M}, \mathscr{O}_X)\right),$$

where $\mathrm{char}(\mathscr{M})$ denotes the characteristic variety of $\mathscr{M}$. Recall that locally on X, $\mathscr{M}$ admits a finite free presentation

$$0 \leftarrow \mathscr{M} \leftarrow \mathscr{D}_X^{N_0} \underset{P_0}{\leftarrow} \cdots \underset{P_{p-1}}{\leftarrow} \mathscr{D}_X^{N_p} \leftarrow 0$$

(the P_j's are matrices of differential operators), and then $\mathbf{R}\mathscr{H}om_{\mathscr{D}_X}(\mathscr{M}, \mathscr{O}_X)$ is represented by the complex

$$0 \to \mathscr{O}_X^{N_0} \underset{P_0}{\to} \cdots \underset{P_{p-1}}{\to} \mathscr{O}_X^{N_p} \to 0.$$

II. Contact transformations. In this section we assume $\alpha \geqslant 2$. Let Ω be a subset of T^*X and set

$$N(\Omega) = \{F^\cdot \in \mathrm{Ob}(D^+(X)); \mathrm{SS}(F^\cdot) \cap \Omega = \varnothing\}.$$

We introduce $D^+(X; \Omega)$, the localization of $D^+(X)$ by $N(\Omega)$ (cf. [**3**]). Then $\mathrm{Ob}(D^+(X;\Omega)) = \mathrm{Ob}(D^+(X))$, but a morphism $u\colon F^\cdot \to G^\cdot$ in $D^+(X)$ becomes an isomorphism in $D^+(X; \Omega)$ if $\mathrm{SS}(u) \cap \Omega = \varnothing$.

EXAMPLE 2.1. Let $X = \mathbf{R}$ and let x be a coordinate on X, (x, ξ) on T^*X. Let $\Omega = \{(x, \xi); \xi > 0\}$. Then

$$\underline{\mathbf{Q}}_{\{0\}} \cong \underline{\mathbf{Q}}_{\{x; x \geqslant 0\}} \cong \underline{\mathbf{Q}}_{\{x; x<0\}}[1] \quad \text{in } D^+(X; \Omega).$$

EXAMPLE 2.2. Let $f\colon X \to Y$ be a smooth map, and let V be the involutive manifold $X \times_Y T^*Y$ of T^*X, and $p \in V$. Let $F^\cdot \in \mathrm{Ob}(D^+(X))$ and assume $\mathrm{SS}(F^\cdot) \subset V$ in a neighborhood of p. Then there exists $G^\cdot \in \mathrm{Ob}(D^+(Y))$ such that $F^\cdot \cong f^{-1}G^\cdot$ in $D^+(X; p)$.

Now let X and Y be two manifolds of the same dimension, Ω_X and Ω_Y two conic open subsets of T^*X and T^*Y, respectively, and $\phi\colon \Omega_X \xrightarrow{\sim} \Omega_Y$ a (homogeneous) contact transformation. Let $\Lambda_\phi \subset \Omega_X^a \times \Omega_Y$ be the associated Lagrangian manifold; Ω_X^a is the image of Ω_X by the antipodal map of T^*X, and

$$\Lambda_\phi = \{(x, y; \xi, \eta) \in T^*(X \times Y); (y, \eta) = \phi(x, -\xi)\}.$$

Let q_1 and q_2 be the projection from $X \times Y$ to X and Y, respectively. To $K^\cdot \in \mathrm{Ob}(D^+(X \times Y))$ we associate the functor $\psi_{K^\cdot}$ from $D^+(X)$ to $D^+(Y)$:

$$\psi_{K^\cdot}(F^\cdot) = \mathbf{R}q_{2!}\left(K^\cdot \otimes^{\mathbf{L}} q_1^{-1}F^\cdot\right).$$

(Recall that $q_{2!}$ means the direct image by q_2 with proper supports, and $\mathbf{L}$ and $\mathbf{R}$ mean the left and right derived functors.)

THEOREM 2.3. *Let $p_X \in \Omega_X$, $p_Y = \phi(p_X) \in \Omega_Y$. There exists*
$$K^{\cdot} \in \mathrm{Ob}(D^+(X \times Y))$$
such that $\psi_{K^{\cdot}}$ induces an equivalence of categories
$$\psi_{K^{\cdot}} : D^+(X; p_X) \xrightarrow{\sim} D^+(Y; p_Y).$$
If $\Lambda_\phi = T^*_Z(X \times Y)$, where Z is a submanifold of $X \times Y$, the sheaf $K^{\cdot}$ satisfies
$$K^{\cdot} \cong \underline{\mathbf{Z}}_Z[d] \quad \text{in } D^+(X \times Y;\, p^a_X \times p_Y)$$
for some shift d.

As a corollary of Theorem 2.3, one can prove (using Example 2.2)

THEOREM 2.4. *Let $F^{\cdot} \in \mathrm{Ob}(D^+(X))$. Then $\mathrm{SS}(F^{\cdot})$ is an involutive subset of T^*X; i.e., if U is an open subset of T^*X and ϕ is a C^1-function on U vanishing on $\mathrm{SS}(F)$, then $\mathrm{SS}(F)$ is a union of integral curves of the Hamiltonian flow H_ϕ.*

We remark that this theorem gives a new proof of the involutivity of the characteristic variety of a coherent $\mathscr{D}_X$-module of **[9]** (cf. Example 1.5), and in fact the argument still holds for systems of microdifferential equations.

Now assume X and Y are complex manifolds of complex dimension n, and ϕ is a complex contact transformation. Assume for simplicity that $\Lambda_\phi = T^*_Z(X \times Y)$ for Z a complex submanifold.

THEOREM 2.5. *Assume $K^{\cdot} \cong \underline{\mathbf{Z}}_Z[n\text{-}\mathrm{codim}_{\mathbf{C}} Z]$ in $D^+(X \times Y;\, p^a_X \times p_Y)$. Then one can find (nonunique) isomorphisms $\psi_K(\mathcal{O}_X) \cong \mathcal{O}_Y$ in $D^+(Y, p_Y)$.*

We can describe this isomorphism as follows. Let $d = \mathrm{codim}_{\mathbf{C}} Z$.
$$\psi_{K^{\cdot}}(\mathcal{O}_X) \cong \mathbf{R}q_{2!}\big(q_1^{-1}\mathcal{O}_X|_Z\big)[n-d].$$
Let us choose a relative differential form v on Z above Y, $v \in \Omega_Z^{(2n-d)} \otimes_{\mathcal{O}_Y} \Omega_Y^{(n)\otimes -1}$. We have natural morphisms
$$q_1^{-1}\mathcal{O}_X|_Z \to \mathcal{O}_Z \quad (\text{restriction}),$$
$$\mathcal{O}_Z \to \Omega_Z^{(2n-d)} \otimes_{\mathcal{O}_Y} \Omega_Y^{(n)\otimes -1} \quad (\text{multiplication by } v),$$
$$\mathbf{R}q_{2!}\big(\Omega_Z^{(2n-d)} \otimes_{\mathcal{O}_Y} \Omega_Y^{(n)\otimes -1}\big)[n\text{-}\mathrm{codim}_{\mathbf{C}} Z] \to \mathcal{O}_Y \quad (\text{integration}).$$
This defines the morphism $\psi_{K^{\cdot}}(\mathcal{O}_X) \to \mathcal{O}_Y$. If Z is a hypersurface defined by the equation $\phi = 0$ (with $d\phi \neq 0$ on Z), we may take $v = w/d\phi$, where w is a volume element in $\Omega^{(2n)}_{X \times Y}$.

III. Microlocalization. We still assume X is of class C^α, $\alpha \geqslant 2$. Let M be a submanifold, $F^{\cdot} \in \mathrm{Ob}(D^+(X))$. Recall that Sato's microlocalization of $F^{\cdot}$ along M, denoted by $\mu_M(F^{\cdot})$, is an object of $D^+(T^*_M X)$ whose stalk at $(x^0, \xi^0) \in T^*_M X$ is given by
$$(\mu_M(F^{\cdot}))_{(x^0,\xi^0)} = \varinjlim_{U,G} \mathbf{R}\Gamma_{((M\cap U)+G)\cap U}(U, F),$$

where U runs over the set of open neighborhoods of x^0 in X, and G runs over the set of closed convex cones such that $\langle \gamma, \xi^0 \rangle > 0$ for $\gamma \in G \setminus \{0\}$ (we assume X is affine). We remark that $\mu_M(F^\cdot)|_{T_X^*X} \cong \mathbf{R}\Gamma_M(F^\cdot)$.

EXAMPLE 3.1. Assume $M = \{x \in X; \phi(x) = 0\}$ with $d\phi \neq 0$ on M. Then if $x^0 \in M$,

$$(\mu_M(F^\cdot))_{(x^0, d\phi(x^0))} = (\mathbf{R}\Gamma_{\{x; \phi(x) \geq 0\}}(F^\cdot))_{x^0}.$$

EXAMPLE 3.2. Let $X = \mathbf{C}$, $M = \{0\}$, $F^\cdot = \mathcal{O}_X$. Then

$$(\mathcal{H}^j(\mu_{\{0\}}(\mathcal{O}_X)))_{(0,\xi)} = 0, \qquad j \neq 1,$$

$$(\mathcal{H}^1(\mu_{\{0\}}(\mathcal{O}_X)))_{(0,\xi)} = \varinjlim_{U,G} \mathcal{O}(U \setminus G)/\mathcal{O}(U),$$

where U runs over the set of neighborhoods of 0, and G runs over the set of closed convex proper cones of R^2, such that $\langle \xi, v \rangle > 0$ for $v \in G \setminus \{0\}$.

In order to state our next result, let us recall the classical notion of "Maslov index" associated to three Lagrangian planes. Let (E, σ) be a (real) symplectic vector space, $\lambda_1, \lambda_2, \lambda_3$ three Langrangian planes (plane = linear subspace). The index $\tau(\lambda_1, \lambda_2, \lambda_3)$ is the signature of the quadratic form q on $\lambda_1 \oplus \lambda_2 \oplus \lambda_3$ defined by

$$q(x_1, x_2, x_3) = \sigma(x_1, x_2) + \sigma(x_2, x_3) + \sigma(x_3, x_1).$$

THEOREM 3.3. *Let X and Y be two manifolds of the same dimension, $\Omega_X \subset T^*X$ and $\Omega_Y \subset T^*Y$ two conic open subsets, and ϕ: $\Omega_X \to \Omega_Y$ a contact transformation. Let M and N be two submanifolds of X and Y, respectively, and assume that ϕ induces an isomorphism*

$$T_M^*X \cap \Omega_X \underset{\phi}{\tilde{\to}} T_N^*Y \cap \Omega_Y.$$

Let $p_X \in \Omega_X$, $p_Y = \phi(p_X) \in \Omega_Y$, and let $\psi_{K^\cdot}$ be the equivalence of categories given by Theorem 2.3. Then we have an isomorphism

$$\mu_M(F^\cdot)_{p_X} \cong \mu_N(\psi_{K^\cdot}(F^\cdot))_{p_Y}[d]$$

for some shift d.

Moreover, assume $\Lambda_\phi = T_Z^*(X \times Y)$, for Z a submanifold of $X \times Y$, and $K^\cdot = \underline{\mathbf{Z}}_Z$ in $D^+(X \times Y; p_X^a \times p_Y)$. Then

$$d = \tfrac{1}{2}(\dim M - \dim N - \dim X + \dim Z + \tau),$$

where $\tau = \tau(\lambda_1, \lambda_2, \lambda_3)$ and

$$\lambda_1 = T_{p_Y}\pi^{-1}\pi(p_Y), \quad \lambda_2 = d\phi(p_X) \cdot T_{p_X}\pi^{-1}\pi(p_X), \quad \lambda_3 = T_{p_Y}T_N^*Y.$$

When X and Y are complex manifolds and ϕ is a complex contact transformation, we may combine Theorems 2.5 and 3.3. First we introduce another index.

Let X be a complex manifold, M a real submanifold. Let $p \in T_M^*X$. We set $s(M, p) = \frac{1}{2}\tau(\lambda_1, \lambda_2, \lambda_3)$, where $\lambda_1 = T_pT_M^*X$, $\lambda_2 = i\lambda_1$, $\lambda_3 = T_p\pi^{-1}\pi(p)$, and τ is the index associated to the real symplectic structure of T^*X, that is, to $2\operatorname{Re} d\omega_X$.

COROLLARY 3.4. *Let X and Y be two complex manifolds and ϕ a complex contact transformation from $\Omega_X \subset T^*X$ to $\Omega_Y \subset T^*Y$. Let M and N be two real submanifolds of X and Y, respectively, and assume ϕ induces an isomorphism*

$$T_M^*X \cap \Omega_X \underset{\phi}{\xrightarrow{\sim}} T_N^*Y \cap \Omega_Y.$$

Then locally on $\Omega_X \cap T_M^*X$, we may find an isomorphism

$$\phi_*(\mu_M(\mathcal{O}_X)) \cong \mu_N(\mathcal{O}_Y)[d],$$

where

$$d = \tfrac{1}{2}\left[\dim M - \dim N + s(M, p_X) - s(N, \phi(p_X))\right], \qquad p_X \in \Omega_X \cap T_M^*X.$$

REMARK. d is a locally constant function of $p_X \in T_M^*X$.

IV. Applications. In this section we illustrate Corollary 3.4. Thus X is a complex manifold and M is a real submanifold of class C^2. Let $p \in T_M^*X$. We set

$$E(p) = T_pT^*X, \qquad \sigma(p) = d\omega(p),$$
$$\lambda_M(p) = T_pT_M^*X, \qquad \lambda_0(p) = T_p\pi^{-1}\pi(p),$$
$$\dim_{\mathbf{C}} X = n, \qquad \operatorname{codim}_{\mathbf{R}} M = d \quad (= \dim(\lambda_M(p) \cap \lambda_0(p))),$$
$$\dim_{\mathbf{C}}(\lambda_M(p) \cap i\lambda_M(p) \cap \lambda_0(p)) = \delta(p).$$

We have already defined the integer $s(M, p)$ as $\frac{1}{2}\tau(\lambda_M(p), i\lambda_M(p), \lambda_0(p))$, where τ is the index associated to the symplectic form $\operatorname{Re}\sigma(p)$ on $E(p)$. Now we define $s^+(M, p)$ and $s^-(M, p)$ by

$$s^+(M, p) - s^-(M, p) = s(M, p),$$
$$s^+(M, p) + s^-(M, p) = n - d + 2\delta(p) - \dim_{\mathbf{C}}(\lambda_M(p) \cap i\lambda_M(p)).$$

We remark that $\delta(p) = \operatorname{codim}_{\mathbf{C}}(T_{\pi(p)}M + iT_{\pi(p)}M)$. This number is, of course, equal to zero if M is a real hypersurface. More generally, $\delta(p) = 0$ is equivalent to saying that the submanifold M is noncharacteristic in $X^{\mathbf{R}}$ for the Cauchy–Riemann system $\bar{\partial}$.

EXAMPLE 4.1. Assume M is a real hypersurface and $\phi(x) = 0$ is an equation of M, with $d\phi \neq 0$ on M. Let

$$T_x^{\mathbf{C}}M = \{v \in T_xX;\ \langle v, \partial\phi(x)\rangle = 0\},$$

where $\partial\phi$ is the differential of ϕ with respect to the holomorphic variables. Let L_ϕ be the Levi form of ϕ on $T_x^{\mathbf{C}}M$. Recall that if $(x_1, \ldots, x_n)$ is a system of holomorphic coordinates on X, $(\bar{x}_1, \ldots, \bar{x}_n)$ the complex conjugate coordinates, then L_ϕ is represented by the matrix

$$\left(\frac{\partial^2\phi}{\partial x_i\,\partial\bar{x}_j}\right) \quad (1 \leqslant i, j \leqslant n) \qquad \text{on } T_x^{\mathbf{C}}M.$$

Then it can be proved that $s^+(M, d\phi(x))$ (resp. $s^-(M, d\phi(x))$) is equal to the number of positive (resp. negative) eigenvalues of L_ϕ (cf. [**10**]).

In order to formulate our next result, let us denote by $\nu(p)$ the complex line of T_pT^*X generated by the Euler vector field, i.e., by $H(\omega_X)$, where ω_X is the (complex) canonical 1-form on T^*X, and H is the symplectic isomorphism $T^*T^*X \cong TT^*X$.

PROPOSITION 4.2. *Let M be a real submanifold of class C^2 of X, $p \in T^*_MX$. Assume* $\dim_{\mathbf{R}}(T_pT^*_MX \cap \nu(p)) = 1$. *Then*

$$\mathscr{H}^j(\mu_M(\mathscr{O}_X))_p = 0 \quad \text{for } j \notin [d + s^-(M,p) - \delta(p), n - s^+(M,p) + \delta(p)].$$

SKETCH OF THE PROOF. We may interchange (T^*X, T^*_MX, p) with (T^*Y, T^*_NY, q) by a complex contact transformation, where now N is a real hypersurface of the complex manifold Y. Assume we know that

$$\mathscr{H}^j(\mu_N(\mathscr{O}_Y))_q = 0 \quad \text{for } j \notin [1 + s^-(N,q) - \alpha, n - s^+(N,q) + \beta].$$

Applying Corollary 3.4, we obtain that

$$\mathscr{H}^j(\mu_M(\mathscr{O}_X))_p = 0$$

for $j \notin [d + s^-(M,p) - \delta(p) - \alpha, n - s^+(M,p) + \delta(p) + \beta]$.

We can choose N such that $s^-(N,q) = 0$. Then we can take $\alpha = 0$, since we know by the principle of holomorphic extension that $\mathscr{H}^0(\mu_N(\mathscr{O}_Y))_q = 0$. Similarly, we can choose N such that $s^+(N,q) = 0$, and then take $\beta = 0$, since we know by a theorem of Malgrange [7] that $H^j_Z(Y, \mathscr{O}) = 0$ for $j > n$ for any locally closed subset Z of Y.

EXAMPLES 4.3. Let us denote the bundle $T^*X \setminus T^*_XX$ by $\dot{T}^*X$ and the projection $\dot{T}^*X \to X$ by $\dot{\pi}$.

Let M be a real submanifold of class C^2, of codimension d, and assume $\delta = 0$, that is, $TM + iTM = TX$. Let $\underline{\omega}_M$ be the orientation sheaf on M. We have a triangle (cf. [9])

$$\mathscr{O}_X|_M \to \mathbf{R}\Gamma_M(\mathscr{O}_X) \otimes \underline{\omega}_M[d] \to \mathbf{R}\dot{\pi}_*\mu_M(\mathscr{O}_X) \otimes \underline{\omega}_M[d] \to \mathscr{O}_X|_M[1].$$

Applying Proposition 4.2 we obtain

$$\mathscr{H}^j(\mu_M(\mathscr{O}_X)) = 0 \quad \text{for } j < d$$

(and, in particular, $\mathscr{H}^j_M(\mathscr{O}_X) = 0$ for $j < d$), and we have an exact sequence

$$0 \to \mathscr{O}_X|_M \to \mathscr{H}^d_M(\mathscr{O}_X) \otimes \underline{\omega}_M \to \dot{\pi}_*\mathscr{H}^d(\mu_M(\mathscr{O}_X)) \otimes \underline{\omega}_M \to 0.$$

As an application of Proposition 4.2, we get

(i) Assume $s^-(M,p) > 0\ \forall p \in \dot{T}^*_MX$. Then

$$\mathscr{H}^d_M(\mathscr{O}_X) \otimes \underline{\omega}_M \cong \mathscr{O}_X|_M.$$

(ii) Assume there exists a closed convex proper cone Γ in T^*_MX such that $s^-(M,p) > 0\ \forall p \notin \Gamma$. Then any section u of $\mathscr{H}^d_M(\mathscr{O}_X)$ is represented by (i.e., is the "boundary value of ") a function f holomorphic in a wedge along M whose polar is contained in Γ (f is unique modulo $\mathscr{O}_X|_M$).

(iii) Assume now that on an open subset U of $\dot{T}^*_MX$, we have $\lambda_M(p) \cap i\lambda_M(p) = 0$. Let γ denote the natural map $\dot{T}^*_MX \to S^*_MX = \dot{T}^*_MX/\mathbf{R}^+$. Then the sheaf $\gamma_*(\mathscr{H}^{d+s^-(M,p)}(\mu_M(\mathscr{O}_X)))$ is flabby on $\gamma(U)$ (here $p \in U$) since it is isomorphic to the sheaf of boundary values of holomorphic functions on the boundary of a strictly pseudoconvex open set (Corollary 3.4). This allows us to obtain results of the type "edge of the wedge theorem", but we leave the exact formulation to the reader (cf. e.g. [**2**]).

REMARK 4.4. Assume M is real analytic and $\delta = 0$. Then $\mathscr{H}^d_M(\mathscr{O}_X)$ is isomorphic to the sheaf of hyperfunction solutions of the induced Cauchy–Riemann system on M [**4**]. More generally, $\mu_M(\mathscr{O}_X)$ is isomorphic to the complex of solutions of the induced Cauchy–Riemann system, with values in the sheaf of Sato's microfunctions on $\sqrt{-1}\,T^*M$ [**4**]. This follows easily from the Cauchy–Kowalewski theorem.

PROPOSITION 4.5. *Let M be a real submanifold of class C^2, $p_0 \in T^*_M X$. Assume*

(i) $\dim_{\mathbf{R}}(T_{p_0}T^*_M X \cap \nu(p_0)) = 1$.

(ii) *$s^-(M, p) - \delta(p)$ is constant in a neighborhood of p_0.*

Set $j_0 = \operatorname{codim} M + s^-(M, p) - \delta(p)$. Then $\mathscr{H}^j(\mu_M(\mathscr{O}_X))_{p_0} = 0$ for $j \neq j_0$, and for $j = j_0$ this space is infinite dimensional.

SKETCH OF THE PROOF. By a complex contact transformation we interchange $T^*_M X$ (in a neighborhood of p_0) with the boundary of a (in general, nonstrict) pseudoconvex open set. Then the result is classical.

REMARK. Let us mention that some results stated in §4 are already well known, even if they are formulated differently. Since there is an extensive literature on this subject, perhaps starting with H. Lewy [**6**], we only quote some of the most recent papers on this subject, such as Nacinovich [**8**] or Baouendi–Chang–Treves [**1**].

REFERENCES

1. M. S. Baouendi, C. H. Chang and F. Treves, *Microlocal hypo-analyticity and extension of C. R. functions*, J. Differential Geom. **18** (1983), 331–391.

2. G. Bengel and P. Schapira, *Décomposition microlocale analytique des distributions*, Ann. Inst. Fourier (Grenoble) **29** (1979), 101–124.

3. R. Hartshorne, *Residues and duality*, Lecture Notes in Math., vol. 20, Springer-Verlag, 1966.

4. M. Kashiwara and T. Kawaï, *On the boundary value problem for ellliptic systems of linear differential equations*. I, II, Proc. Japan Acad. **48** (1972), 712–715; ibid **49** (1973), 164–168.

5. M. Kashiwara and P. Schapira, *Microlocal theory of sheaves*, Asterisque (1985) and C. R. Acad. Sci. Paris **295** (1982), 487–490; Proc. Japan Acad. **59** (1983), 349–351, 352–354.

6. H. Lewy, *On the local character of the solution of an atypical differential equation in three variables and a relatd problem for regular functions of two complex variables*, Ann. of Math. (2) **64** (1956), 514–522.

7. B. Malgrange, *Faisceaux sur des variétés analytiques réelles*, Bull. Soc. Math. France **85** (1957), 231–237.

8. M. Nacinovich, *Poincaré's lemma for tangential Cauchy–Riemann complexes*, preprint 41, Univ. di Pisa, 1983.

9. M. Sato, M. Kashiwara and T. Kawaï, *Hyperfunctions and pseudo-differential equations*, Lecture Notes in Math., vol. 287, Springer-Verlag, 1973, pp. 265–529.

10. P. Schapira, *Condition de positivité dans une variété symplectique complexe. Application à l'étude des microfonctions*, Ann. École Norm. Sup. **14** (1981), 121–139.

KYOTO UNIVERSITY

UNIVERSITÉ PARIS-NORD

Proceedings of Symposia in Pure Mathematics
Volume **43** (1985)

Recent Progress on Boundary-Value Problems on Lipschitz Domains

CARLOS E. KENIG[1]

Introduction. In this note we will describe and sketch the proofs of some recent developments on boundary-value problems on Lipschitz domains.

In 1977 B. E. J. Dahlberg was able to show the solvability of the Dirichlet problem for Laplace's equation on a Lipschitz domain D with $L^2(\partial D, d\sigma)$ data and optimal estimates. In fact, he proved that, given a Lipschitz domain D, there exists $\varepsilon = \varepsilon(D)$ such that this can be done for data in $L^p(\partial D, d\sigma)$, $2 - \varepsilon \leqslant p \leqslant \infty$ (see [**6**, **7**] and [**8**]). Also, simple examples show that, given $p < 2$, there exists a Lipschitz domain D where this fails in $L^p(\partial D, d\sigma)$. Dahlberg's method consisted of a careful analysis of the harmonic measure. His techniques relied on positivity, Harnack's inequality, and the maximum principle, and, thus, they were not applicable to the Neumann problem, to systems of equations, or to higher-order equations. In 1978 E. Fabes, M. Jodeit, Jr., and N. Riviere [**15**] were able to utilize A. P. Calderón's theorem [**1**] on the boundedness of the Cauchy integral on C^1 curves to extend the classical method of layer potentials to C^1 domains. They were thus able to resolve the Dirichlet and Neumann problems for Laplace's equation, with $L^p(\partial D, d\sigma)$ data and optimal estimates for C^1 domains. They relied on Fredholm theory, exploiting the compactness of the layer potentials in the C^1 case. In 1979 D. Jerison and C. Kenig [**20**, **21**] were able to give a simplified proof of Dahlberg's results, using an integral identity that goes back to Rellich [**33**]. However, the method still relied on positivity. Shortly afterwards, Jerison and Kenig [**22**] were also able to treat the Neumann problem on Lipschitz domains, with $L^2(\partial D, d\sigma)$ data and optimal estimates. To do so they combined the Rellich-type formulas with Dahlberg's results on the Dirichlet problem. This still relied on positivity and dealt only with the L^2 case, leaving the corresponding L^p theory open.

1980 *Mathematics Subject Classification*. Primary 35J25, 35J40, 35J55.
[1] Supported in part by the NSF, and the Alfred P. Sloan Foundation.

In 1981 R. Coifman, A. McIntosh, and Y. Meyer [**3**] established the boundedness of the Cauchy integral on any Lipschitz curve, opening the door to the applicability of the method of layer potentials to Lipschitz domains. This method is very flexible, does not rely on positivity, and does not in principle differentiate between a single equation or a system of equations. The difficulty then becomes the solvability of the integral equations, since, unlike the C^1 case, Fredholm theory is not applicable, because, on a Lipschitz domain, operators like the double-layer potential are not compact.

For the case of the Laplace equation, with $L^2(\partial D, d\sigma)$ data, this difficulty was overcome by G. C. Verchota [**36**] in 1982 in his doctoral dissertation. He made the key observation that the Rellich identities mentioned before are the appropriate substitutes for compactness in the case of Lipschitz domains. Thus, Verchota was able to recover the L^2 results of Dahlberg [**7**] and Jerison and Kenig [**22**] for Laplace's equation on a Lipschitz domain by using the method of layer potentials.

This paper is divided into two sections. The first, which consists of two parts, deals with Laplace's equation on Lipschitz domains: The first part explains the L^2 results of Verchota. The second part deals with a sketch of recent joint work of Dahlberg and Kenig (1984) [**9**]. We were able to show that, given a Lipschitz domain $D \subset \mathbb{R}^n$, there exists $\varepsilon = \varepsilon(D)$ such that one can solve the Neumann problem for Laplace's equation with data in $L^p(\partial D, d\sigma)$, $1 < p \leqslant 2 + \varepsilon$. Easy examples show that this range of p's is optimal. Moreover, we showed that the solution can be obtained by the method of layer potentials, and that Dahlberg's solution of the L^p Dirichlet problem can also be obtained by the method of layer potentials. We also obtained endpoint estimates for the Hardy space $H^1(\partial D, d\sigma)$, which generalize the results for $n = 2$ in [**25**] and [**26**] and for C^1 domains in [**16**]. The key idea in this work is that one can estimate the regularity of the so-called Neumann function for D by using the De Giorgi–Nash regularity theory for elliptic equations with bounded measurable coefficients. This, combined with the use of the so-called 'atoms', yields the desired results.

The second section, which consists of three parts, deals with higher-order problems. In Parts 1 and 2 we treat L^2 boundary-value problems for systems of equations. Part 1 deals with the systems of elastostatics; Part 2, with the Stokes system of hydrostatics. The results in Part 1 are joint work of Dahlberg, Kenig, and Verchota (see [**12**]); the results in Part 2 are joint work of E. Fabes, C. Kenig, and G. Verchota (see [**17**]). The results obtained had not been previously available for general Lipschitz domains, although a lot of work has been devoted to the case of piecewise linear domains (see [**27**, **28**] and their bibliographies). For the case of C^1 domains, our results for the systems of elastostatics had been previously obtained by A. Gutierrez [**19**], using compactness and Fredholm theory. This is, of course, not available for the case of Lipschitz domains. We are able to use once more the method of layer potentials. Invertibility is shown again by means of Rellich-type formulas. This works very well in the Dirichlet problem for the Stokes system (see Part 2), but serious difficulties occur for systems of elastostatics (see Part 1). These difficulties are overcome by proving a Korn-type inequality at the boundary. The proof of this inequality proceeds in three steps. One first establishes it for the case of small Lipschitz constant. One then proves an analogous inequality for nontangential maximal functions on any Lipschitz domain, by using the ideas of G. David [**13**] on increasing the Lipschitz constant. Finally, one can remove the nontangential maximal function, using the results on the Dirichlet problem for the Stokes system, which are established in Part 2. See Parts 1 and 2 for the details. Some partial results in this direction were previously announced in [**26**]. The third part of §2 deals with the Dirichlet problem for the biharmonic equation Δ^2 (a fourth-order elliptic equation) on an arbitrary Lipschitz domain in $\mathbb{R}^n$. This sketches joint work of Dahlberg, Kenig, and Verchota [**11**]. The case of C^1 domains in the plane was previously treated by J. Cohen and J. Gosselin [**2**], using layer potentials and compactness. We are able to reduce the problem, for an arbitrary Lipschitz domain in $\mathbb{R}^n$, to a bilinear estimate for harmonic functions. This is a Lipschitz-domain version of the paraproduct of J. M. Bony. See Part 3 of §2 for further details.

Complete proofs of the results explained in §1, Part 2 and §2 will appear in future publications.

ACKNOWLEDGEMENTS. As mentioned before, the results in §1, Part 2 are joint work with B. Dahlberg; the results in §2, Part 1, are joint work with B. Dahlberg and G. Verchota; the results in §2, Part 2 are joint work with E. Fabes and G. Verchota; and the results in §2, Part 3 are joint work with B. Dahlberg and G. Verchota. It is a great pleasure to express my gratitude to B. Dahlberg, E. Fabes, and G. Verchota for their contributions. I would also like to thank A. McIntosh for pointing out the applicability of the continuity method in §1, Part 1, and for pointing out to us the work of Nečas [**30**]. I also would like to thank G. David for making his unpublished result (Lemma 2.1.10 in §2, Part 1) available to us.

Part of this research was carried out while I was visiting the Center for Mathematical Analysis at the Australian National University, Princeton University, and the University of Paris, Orsay. I would like to thank these institutions for their kind hospitality.

1. Laplace's equation.

Part 1. *L^2 theory on a Lipschitz domain for Laplace's equation by the method of layer potentials.* A bounded Lipschitz domain $D \subset \mathbb{R}^n$ is one which is locally given by the domain above the graph of a Lipschitz function. Such domains satisfy both the interior and exterior cone conditions. For such a domain D the nontangential region of opening α at a point $Q \in \partial D$ is

$$\Gamma_\alpha(Q) = \{ X \in D: |X - Q| < (1 + \alpha)\mathrm{dist}(X, \partial D) \}.$$

All results in this paper are valid when suitably interpreted for all bounded Lipschitz domains in $\mathbb{R}^n$, $n \geqslant 2$, with the nontangential approach regions defined above. For simplicity in this exposition, we restrict ourselves to the case $n \geqslant 3$ (and sometimes even to the case $n = 3$) and to domains $D \subset \mathbb{R}^n$, $D = \{(x, y): y > \varphi(x)\}$, where $\varphi: \mathbb{R}^{n-1} \to \mathbb{R}$ is a Lipschitz function with Lipschitz constant M; i.e., $|\varphi(x) - \varphi(x')| \leqslant M|x - x'|$, $D^- = \{(x, y): y < \varphi(x)\}$. For fixed $M' < M$,

$$\Gamma_e(x) = \{(x, y): (y - \varphi(x)) < -M'|z - x|\} \subset D^-,$$

and

$$\Gamma_i(x) = \{(z, y): (y - \varphi(x)) > M'|z - x|\} \subset D.$$

Points in D will usually be denoted by X; points on ∂D, by $Q = (x, \varphi(x))$ or simply by x. N_x or N_Q will denote the unit normal to $\partial D = \Lambda$ at $Q = (x, \varphi(x))$. If u is a function defined on $\mathbb{R}^n \setminus \Lambda$, and $Q \in \partial D$, $u^{\pm}(Q)$ will denote $\lim_{X \to Q;\, X \in \Gamma_i(Q)} u(X)$ or $\lim_{X \to Q;\, X \in \Gamma_e(Q)} u(X)$, respectively. If u is a function defined on D, $N(u)(Q) = \sup_{X \in \Gamma_i(Q)} |u(X)|$.

We wish to solve the problems

$$\text{(D)} \quad \begin{cases} \Delta u = 0 & \text{in } D, \\ u|_{\partial D} = f \in L^2(\partial D, d\sigma), \end{cases} \qquad \text{(N)} \quad \begin{cases} \Delta u = 0 & \text{in } D, \\ \partial u / \partial N|_{\partial D} = f \in L^2(\partial D, d\sigma). \end{cases}$$

The results here are

THEOREM 1.1.1. *There exists a unique u solving (D), such that $N(u) \in L^2(\partial D, d\sigma)$, where the boundary values are taken nontangentially a.e. Moreover, u has the form*

$$u(X) = \frac{1}{\omega_n} \int_{\partial D} \frac{\langle X - Q, N_Q \rangle}{|Q - X|^n} g(Q)\, d\sigma(Q)$$

for some $g \in L^2(\partial D, d\sigma)$.

THEOREM 1.1.2. *There exists a unique u tending to 0 at ∞, such that $N(\nabla u) \in L^2(\partial D, d\sigma)$, solving (N) in the sense that $N_Q \cdot \nabla u(X) \to f(Q)$ as $X \to Q$ nontangentially a.e. Moreover, u has the form*

$$u(X) = \frac{-1}{\omega_n(n-2)} \int_{\partial D} \frac{1}{|X - Q|^{n-2}} g(Q)\, d\sigma(Q)$$

for some $g \in L^2(\partial D, d\sigma)$.

In order to prove the above theorems, we introduce

$$\mathscr{K} g(X) = \frac{1}{\omega_n} \int_{\partial D} \frac{\langle X - Q, N_Q \rangle}{|X - Q|^n} g(Q)\, d\sigma(Q)$$

and

$$Sg(X) = \frac{-1}{\omega_n(n-2)} \int_{\partial D} \frac{1}{|X - Q|^{n-2}} g(Q)\, d\sigma(Q).$$

If $Q = (x, \varphi(x))$, $X = (z, y)$, then

$$\mathscr{K}g(z, y) = \frac{1}{\omega_n}\int_{\mathbb{R}^{n-1}} \frac{y - \varphi(x) - (z - x)\cdot\nabla\varphi(x)}{\left[|x - z|^2 + [\varphi(x) - \varphi(z)]^2\right]^{n/2}} g(x)\,dx,$$

$$Sg(z, y) = \frac{-1}{\omega_n(n-2)}\int_{\mathbb{R}^{n-1}} \frac{\sqrt{1 + |\nabla\varphi(x)|^2}}{\left[|x - z|^2 + [\varphi(x) - y]^2\right]^{(n-2)/2}} g(x)\,dx.$$

THEOREM 1.1.3. (a) *If $g \in L^p(\partial D, d\sigma)$, $1 < p < \infty$, then $N(\nabla Sg)$, $N(\mathscr{K}g)$ also belong to $L^p(\partial D, d\sigma)$, and their norms are bounded by $C\|g\|_{L^p(\partial D, d\sigma)}$.*

(b) $$\lim_{\varepsilon\to 0}\frac{1}{\omega_n}\int_{|x-z|>\varepsilon} \frac{\varphi(z) - \varphi(x) - (z - x)\cdot\nabla\varphi(x)}{\left[|x - z|^2 + [\varphi(x) - \varphi(z)]^2\right]^{n/2}} g(x)\,dx = Kg(z)$$

exists a.e., and

$$\|Kg\|_{L^p(\partial D, d\sigma)} \leqslant C\|g\|_{L^p(\partial D, d\sigma)}, \qquad 1 < p < \infty;$$

$$\lim_{\varepsilon\to 0}\frac{-1}{\omega_n}\int_{|z-x|>\varepsilon} \frac{(z - x, \varphi(z) - \varphi(x))\sqrt{1 + |\nabla\varphi(x)|^2}}{\left[|2 - x|^2 + [\varphi(z) - \varphi(x)]^2\right]^{n/2}} g(x)\,dx$$

exists a.e. and in $L^p(\partial D, d\sigma)$, and its L^p norm is bounded by $C\|g\|_{L^p(\partial D, d\sigma)}$, $1 < p < \infty$.

(c) $$(\mathscr{K}g)^{\pm}(Q) = \pm\tfrac{1}{2}g(Q) + Kg(Q),$$

$$(\nabla Sg)^{\pm}(z) = \pm\tfrac{1}{2}g(z)N_z + \frac{1}{\omega_n}\lim_{\varepsilon\to 0}\int_{|z-x|>\varepsilon} \frac{(z - x, \varphi(z) - \varphi(x))\sqrt{1 + |\nabla\varphi(x)|^2}}{\left[|z - x|^2 + [\varphi(z) - \varphi(x)]^2\right]^{n/2}} g(x)\,dx.$$

COROLLARY 1.1.4. *$(N_z\nabla Sg)^{\pm}(z) = \pm\frac{1}{2}g(z) - K^*g(z)$, where K^* is the $L^2(\partial D, d\sigma)$-adjoint of K.*

The proof of Theorem 1.1.3 is an easy consequence of the deep results of Coifman–McIntosh–Meyer [**3**].

It is easy to see that (at least the existence part of) Theorems 1.1.1 and 1.1.2 will follow immediately if we can show that $\frac{1}{2}I + K$ and $\frac{1}{2}I + K^*$ are invertible on $L^2(\partial D, d\sigma)$. This is the result of G. Verchota [**36**].

THEOREM 1.1.5. *$\pm\frac{1}{2}I + K$, $\pm\frac{1}{2}I + K^*$ are invertible on $L^2(\partial D, d\sigma)$.*

In order to prove this theorem, it suffices to show that $\pm\frac{1}{2}I + K^*$ are invertible. In order to do so, we show that if $f \in L^2(\partial D, d\sigma)$, then

$$\left\|\left(\tfrac{1}{2}I + K^*\right)f\right\|_{L^2(\partial D, d\sigma)} \approx \left\|\left(\tfrac{1}{2}I - K^*\right)f\right\|_{L^2(\partial D, d\sigma)},$$

where the constants of equivalence depend only on the Lipschitz constant M. Let us take this for granted and show, for example, that $\frac{1}{2}I + K^*$ is invertible. To do this, note first that if $T = \frac{1}{2}I + K^*$, then $\|Tf\|_{L^2} \geqslant C\|f\|_{L^2}$, where C depends only on the Lipschitz constant M. For $0 \leqslant t \leqslant 1$ consider the operator $T_t = \frac{1}{2}I + K_t^*$, where K_t^* is the operator corresponding to the domain defined by $t\varphi$. Then $T_0 = \frac{1}{2}I$, $T_1 = T$, and $(\partial/\partial t)T_t\colon L^p(\mathbb{R}^{n-1}) \to L^p(\mathbb{R}^{n-1})$, $1 < p < \infty$, with bound independent of t, by the theorem of Coifman–McIntosh–Meyer. Moreover, for each t, $\|T_t f\|_{L^2} \geqslant C\|f\|_{L^2}$, C independent of t. The invertibility of T now follows from the continuity method.

LEMMA 1.1.6. *Suppose that $T_t\colon L^2(\mathbb{R}^{n-1}) \to L^2(\mathbb{R}^{n-1})$ satisfy*
(a) $\|T_t f\|_{L^2} \geqslant C_1\|f\|_{L^2}$,
(b) $\|T_t f - T_s f\|_{L^2} \leqslant C_2|t - s|\,\|f\|_{L^2}$, $0 \leqslant t, s \leqslant 1$,
(c) T_0: $L^2(\mathbb{R}^{n-1}) \to L^2(\mathbb{R}^{n-1})$ *is invertible.*
Then T_1 is invertible.

The proof of 1.1.6 is very simple. We are thus reduced to proving

$$\left\|\left(\tfrac{1}{2}I + K^*\right)f\right\|_{L^2(\partial D,\, d\sigma)} \approx \left\|\left(\tfrac{1}{2}I - K^*\right)f\right\|_{L^2(\partial D,\, d\sigma)}. \tag{1.1.7}$$

In order to prove (1.1.7) we use the following formula, which goes back to Rellich [**33**] (see also [**31**, **30**, **22**]).

LEMMA 1.1.8. *Assume that $u \in \operatorname{Lip}(\overline{D})$, $\Delta u = 0$ in D, and u and its derivatives are suitably small at ∞. Then if e_n is the unit vector in the direction of the y-axis,*

$$\int_{\partial D} \langle N_Q, e_n\rangle |\nabla u|^2\, d\sigma = 2\int_{\partial D} \frac{\partial u}{\partial y}\cdot\frac{\partial u}{\partial N}\, d\sigma.$$

PROOF. Observe that

$$\operatorname{div}\left(e_n|\nabla u|^2\right) = \frac{\partial}{\partial y}|\nabla u|^2 = 2\frac{\partial}{\partial y}\nabla u\cdot\nabla u,$$

and

$$\operatorname{div}\left(\frac{\partial u}{\partial y}\nabla u\right) = \frac{\partial}{\partial y}\nabla u\cdot\nabla u + \frac{\partial u}{\partial y}\cdot\operatorname{div}\nabla u = \frac{\partial}{\partial y}\nabla u\nabla u.$$

Stokes' theorem now gives the lemma.

We now deduce a few consequences of the Rellich identity. Recall that $N_x = (-\nabla\varphi(x), 1)/\sqrt{1 + |\nabla\varphi(x)|^2}$, so that $(1 + M^2)^{-1/2} \leqslant \langle N_x, e_n\rangle \leqslant 1$.

COROLLARY 1.1.9. *Let u be as in 1.1.8, and let $T_1(x), T_2(x), \ldots, T_{n-1}(x)$ be an orthogonal basis for the tangent plane to ∂D at $(x, \varphi(x))$. Let $|\nabla_t u(x)|^2 = \sum_{j=1}^{n-1}|\langle\nabla u(x), T_j(x)\rangle|^2$. Then*

$$\int_{\partial D}\left(\frac{\partial u}{\partial N}\right)^2 d\sigma \leqslant C\int_{\partial D}|\nabla_t u|^2\, d\sigma.$$

PROOF. Let $\alpha = e_n - \langle N_x, e_n\rangle N_x$, so that α is a linear combination of $T_1(x)$, $T_2(x), \ldots, T_{n-1}(x)$. Then

$$\partial u/\partial y = \langle N_x, e_n\rangle(\partial u/\partial N) + \langle\alpha, \nabla u\rangle.$$

Also,

$$|\nabla u|^2 = (\partial u/\partial N)^2 + |\nabla_t u|^2,$$

so

$$\int_{\partial D} \langle N_x, e_n \rangle \left(\frac{\partial u}{\partial N} \right)^2 d\sigma + \int_{\partial D} \langle N_x, e_n \rangle |\nabla_t u|^2 \, d\sigma$$
$$= 2\int_{\partial D} \langle N_x, e_n \rangle \left(\frac{\partial u}{\partial N} \right)^2 + 2\int_{\partial D} \langle \alpha, \nabla u \rangle \left(\frac{\partial u}{\partial N} \right) d\sigma.$$

Hence,

$$\int_{\partial D} \langle N_x, e_n \rangle \left(\frac{\partial u}{\partial N} \right)^2 d\sigma = \int_{\partial D} \langle N_x, e_n \rangle |\nabla_t u|^2 \, d\sigma - 2\int_{\partial D} \langle \alpha, \nabla u \rangle \frac{\partial u}{\partial N} d\sigma.$$

So

$$\int_{\partial D} \left(\frac{\partial u}{\partial N} \right)^2 d\sigma \leqslant C\int_{\partial D} |\nabla_t u|^2 \, d\sigma + C\left(\int_{\partial D} |\nabla_t u|^2 \, d\sigma \right)^{1/2} \left(\int_{\partial D} \left(\frac{\partial u}{\partial N} \right)^2 d\sigma \right)^{1/2},$$

and the corollary follows.

COROLLARY 1.1.10. *Let u be as in* 1.1.8. *Then*

$$\int_{\partial D} |\nabla_t u|^2 \, d\sigma \leqslant C\int_{\partial D} \left(\frac{\partial u}{\partial N} \right)^2 d\sigma.$$

PROOF.

$$\int_{\partial D} |\nabla u|^2 \, d\sigma \leqslant 2\left(\int_{\partial D} |\nabla u|^2 \, d\sigma \right)^{1/2} \left(\int_{\partial D} \left| \frac{\partial u}{\partial N} \right|^2 d\sigma \right)^{1/2},$$

and the corollary follows.

COROLLARY 1.1.11. *Let u be as in* 1.1.8. *Then*

$$\int_{\partial D} |\nabla_t u|^2 \, d\sigma \approx \int_{\partial D} \left| \frac{\partial u}{\partial N} \right|^2 d\sigma.$$

In order to prove (1.1.7) let $u = Sg$. Because of 1.1.3(c), $\nabla_t u$ is continuous across the boundary, and by 1.1.4,

$$(\partial u/\partial N)^{\pm} = (\pm \tfrac{1}{2} I - K^*) g.$$

We now apply 1.1.11 in D and D^- to obtain (1.1.7). This finishes the proofs of 1.1.1 and 1.1.2.

We now turn our attention to L^2 regularity in the Dirichlet problem.

DEFINITION 1.1.12. $f \in L_1^p(\Lambda)$, $1 < p < \infty$, if $f(x, \varphi(x))$ has a distributional gradient in $L^p(\mathbb{R}^{n-1})$. It is easy to check that if F is any extension to $\mathbb{R}^n$ of f, then $\nabla_x F(x, \varphi(x))$ is well defined and belongs to $L^p(\Lambda)$. We call this $\nabla_t f$. The norm in $L_1^p(\Lambda)$ will be $\|\nabla_t f\|_{L^p(\Lambda)}$.

THEOREM 1.1.13. *The single-layer potential S maps $L^2(\Lambda)$ into $L_1^2(\Lambda)$ boundedly and has a bounded inverse.*

PROOF. The boundedness follows from 1.1.3(a). L^2-Neumann theory and 1.1.11 give

$$\|\nabla_t S(F)\|_{L^2(\Lambda)} \geqslant C\left\|\frac{\partial S(f)}{\partial N}\right\|_{L^2(\Lambda)} \geqslant C\|f\|_{L^2(\Lambda)}.$$

The argument used in the proof of 1.1.5 now proves 1.1.13.

THEOREM 1.1.14. Given $f \in L_1^2(\Lambda)$, there exists a harmonic function u, with $\|N(\nabla u)\|_{L^2(\Lambda)} \leqslant C\|\nabla_t f\|_{L^2(\Lambda)}$, such that $\nabla_t u = \nabla_t f$ (a.e.) nontangentially on Λ. u is unique (modulo constants), and we can chose $u = S(g)$, where $g \in L^2(\Lambda)$.

The existence part of 1.1.14 follows directly from 1.1.13.

Part 2. *The L^p theory for Laplace's equation on a Lipschitz domain*. The main results in this section are the following:

THEOREM 1.2.1. *There exists $\varepsilon = \varepsilon(M) > 0$ such that, given $f \in L^p(\partial D, d\sigma)$, $2 - \varepsilon \leqslant p < \infty$, there exists a unique u harmonic in D, with $N(u) \in L^p(\partial D, d\sigma)$, such that u converges nontangentially a.e. to f. Moreover, the solution u has the form*

$$u(x) = \frac{1}{\omega_n}\int_{\partial D} \frac{\langle X - Q, N_Q\rangle}{|X - Q|^n} g(Q)\, d\sigma(Q)$$

for some $g \in L^p(\partial D, d\sigma)$.

THEOREM 1.2.2. *There exists $\varepsilon = \varepsilon(M) > 0$ such that, given $f \in L^p(\partial D, d\sigma)$, $1 < p \leqslant 2 + \varepsilon$, there exists a unique u harmonic in D, tending to 0 at ∞, with $N(\nabla u) \in L^p(\partial D, d\sigma)$, such that $N_Q \cdot \nabla u(X)$ converges nontangentially a.e. to $f(Q)$. Moreover, u has the form*

$$u(X) = \frac{-1}{\omega_n(n-2)}\int_{\partial D} \frac{1}{|X - Q|^{n-2}} g(Q)\, d\sigma(Q)$$

for some $g \in L^p(\partial D, d\sigma)$.

THEOREM 1.2.3. *There exists $\varepsilon = \varepsilon(M) > 0$ such that, given $f \in L_1^p(\Lambda)$, $1 < p \leqslant 2 + \varepsilon$, there exists a harmonic function u, with*

$$\|N(\nabla u)\|_{L^p(\Lambda)} \leqslant C\|\nabla_t f\|_{L^p(\Lambda)}$$

and $\nabla_t u = \nabla_t f$ (a.e.) nontangentially on Λ. u is unique (modulo constants). Moreover, u has the form

$$u(x) = \frac{-1}{\omega_n(n-2)}\int_{\partial D} \frac{1}{|X - Q|^{n-2}} g(Q)\, d\sigma(Q)$$

for some $g \in L^p(\partial D, d\sigma)$.

The case $p = 2$ of the above theorems was discussed in Part 1. The first part of 1.2.1 (i.e., without the representation formula) is due to Dahlberg (1977) **[7]**. Theorem 1.2.3 was first proved by Verchota (1982) **[36]**. The representation formula in 1.2.1, Theorem 1.2.2, and the proof that we present of 1.2.3 are due to Dahlberg and Kenig (1984) [**9**]. Just as in §1, 1.2.1–1.2.3 follow from

THEOREM 1.2.4. *There exists* $\varepsilon = \varepsilon(M) > 0$ *such that* $\pm \frac{1}{2}I - K^*$ *is invertible in* $L^p(\partial D, d\sigma)$, $1 < p \leqslant 2 + \varepsilon$, $\pm \frac{1}{2}I - K$ *is invertible in* $L^p(\partial D, d\sigma)$, $2 - \varepsilon \leqslant p < \infty$, *and* $S\colon L^p(\partial D, d\sigma) \to L_1^p(\partial D, d\sigma)$, $1 < p \leqslant 2 + \varepsilon$, *is invertible.*

In order to prove Theorem 1.2.4, just as in Part 1, it is enough to show that if $u = Sf$, f nice, then, for $1 < p \leqslant 2 + \varepsilon$,

$$\|\nabla_t u\|_{L^p(\partial D, d\sigma)} \approx \|\partial u/\partial N\|_{L^p(\partial D, d\sigma)}.$$

This will be done by proving the following two theorems:

THEOREM 1.2.5. *Let* $\Delta u = 0$ *in* D. *Then*

$$\|N(\nabla u)\|_{L^p(\partial D, d\sigma)} \leqslant C\|\partial u/\partial N\|_{L^p(\partial D, d\sigma)}, \qquad 1 < p \leqslant 2 + \varepsilon.$$

THEOREM 1.2.6. *Let* $\Delta u = 0$ *in* D. *Then*

$$\|N(\nabla u)\|_{L^p(\partial D, d\sigma)} \leqslant C\|\nabla_t u\|_{L^p(\partial D, d\sigma)}, \qquad 1 < p \leqslant 2 + \varepsilon.$$

We first turn our attention to the case $1 < p < 2$ of Theorem 1.2.5. In order to do so we introduce some definitions. A surface ball B in Λ is a set of the form $(x, \varphi(x))$, where x belongs to a ball in $\mathbb{R}^{n-1}$.

DEFINITION 1.2.7. An atom a on Λ is a function supported in a surface ball B, with $\|a\|_{L^\infty} \leqslant 1/\sigma(B)$ and $\int_\Lambda a\, d\sigma = 0$. Notice that atoms are, in particular, L^2 functions.

The following interpolation theorem is of importance to us.

THEOREM 1.2.8. *Let* T *be a linear operator such that* $\|Tf\|_{L^2(\Lambda)} \leqslant C\|f\|_{L^2(\Lambda)}$ *and for all atoms* a, $\|Ta\|_{L^1(\Lambda)} \leqslant C$. *Then, for* $1 < p < 2$, $\|Tf\|_{L^p(\Lambda)} \leqslant C\|f\|_{L^p(\Lambda)}$.

For a proof see [**5**]. Thus in order to establish the case $1 < p < 2$ of 1.2.5, it suffices to show that if $a = \partial u/\partial N$ is an atom, then $\|N(\nabla u)\|_{L^1(\Lambda)} \leqslant C$. By dilation and translation invariance we can assume that $\varphi(0) = 0$ and supp $a \subset B_1 = \{(x, \varphi(x))\colon |x| < 1\}$. Let B^* be a large ball centered at $(0,0)$ in $\mathbb{R}^n$ that contains $(x, \varphi(x))$, $|x| < 2$. The diameter of B^* depends only on M. Since $\|a\|_{L^p(\Lambda)} \leqslant 1/\sigma(B_1)^{1/2} = C$, by L^2-Neumann theory,

$$\int_{\partial D \cap B^*} N(\nabla u) \leqslant C \int_{\partial D \cap B^*} N(\nabla u)^2\, d\sigma \leqslant C.$$

Thus we need only estimate $\int_{B^* \cap \partial D} (\nabla u)\, d\sigma$. We do so by appealing to the regularity theory for divergence-form elliptic equations. Consider the bi-Lipschitzian mapping $\Phi\colon D \to D^-$ given by $\Phi(x, y) = (x, \varphi(x) - [y - \varphi(x)])$. Define u^* on D^- by $u^* = u \circ \Phi^{-1}$. A simple calculation shows that, in D^-, u^* verifies (in the weak sense) the equation $\operatorname{div}(A(x, y)\nabla u^*) = 0$, where

$$A(x, y) = \frac{1}{J\varphi(X)} \cdot (\Phi')^t(X) \cdot (\Phi')(X),$$

where $X = \Phi^{-1}(x, y)$. It is easy to see that $A \in L^\infty(D_-)$ and $\langle A(x, y)\xi, \xi\rangle \geqslant C|\xi|^2$. Notice also that $\operatorname{supp} \partial u/\partial N \subset B_1 \subset B^* \cap \partial D$. Now define

$$B(x, y) = \begin{cases} I & \text{for } (x, y) \in D, \\ A(x, y) & \text{for } (x, y) \in D^-, \end{cases}$$

and

$$\tilde{u}(x, y) = \begin{cases} u(x, y) & \text{for } (x, y) \in D, \\ u^*(x, y) & \text{for } (x, y) \in D^-. \end{cases}$$

Because $\partial u/\partial N = 0$ in $\partial D \setminus B^*$, it is easy to see that $\tilde{u}$ is a (weak) solution in $\mathbb{R}^n \setminus B^*$ of the divergence-form elliptic equation with bounded measurable coefficients, $L\tilde{u} = \operatorname{div} B(x, y)\nabla\tilde{u} = 0$. In order to estimate u (and hence ∇u) at ∞, we use the following theorem of J. Serrin and H. Weinberger [**34**].

THEOREM 1.2.9. *Let $\tilde{u}$ solve $L\tilde{u} = 0$ in $\mathbb{R}^n \setminus B^*$ and suppose $\|\tilde{u}\|_{L^\infty(\mathbb{R}^n \setminus B^*)} < \infty$. Let $g(X)$ solve $Lg = 0$ in $|X| > 1$, with $g(X) \approx |X|^{2-n}$. Then $\tilde{u}(X) = \tilde{u}_\infty + \alpha g(X) + v(X)$, where $Lv = 0$ in $\mathbb{R}^n \setminus B^*$ and $|v(X)| \leqslant C\|\tilde{u}\|_{L^\infty(\mathbb{R}^n \setminus B^*)} \cdot |X|^{2-n-\nu}$, where $\nu > 0$, $C > 0$ depend only on the ellipticity constants of L. Moreover, $\alpha = c\int B(X)\nabla\tilde{u}(X) \cdot \nabla\Psi(X)$, where $\Psi \in C^\infty(\mathbb{R}^n)$, $\Psi = 0$ for X in $2B^*$, and $\Psi \equiv 1$ for large X.*

Let us assume for now that u is bounded, and let us show that if α is as in 1.2.9, then $\alpha = 0$. Pick a Ψ as in 1.2.9. In D, $B(X) = I$, so

$$\int_D B\nabla u\nabla\Psi = \int_D \nabla u \cdot \nabla\Psi = \lim_{\varepsilon\to 0}\int_{D_\delta^\varepsilon} \nabla u \cdot \nabla\Psi,$$

where

$$D_\delta^\varepsilon = \{(x, y) \colon |(x, y)| < \rho,\ y > \varphi(x) + \varepsilon\},$$

and ρ is large. The right side equals

$$\lim_{\varepsilon\to 0}\int_{\partial D_\rho^\varepsilon} \Psi \cdot \frac{\partial u}{\partial N} = \lim_{\varepsilon\to 0}\int_{\partial D_\rho^\varepsilon} [\Psi - 1]\frac{\partial u}{\partial N},$$

since, by the harmonicity of u, $\int_{\partial D_\rho^\varepsilon} \partial u/\partial N = 0$. Let

$$\partial D_{\rho,1}^\varepsilon = \{(x, y) \in \partial D_\rho^\varepsilon \colon y > \varphi(x) + \varepsilon\}$$

and $\partial D_{\rho,2}^\varepsilon = \partial D_\rho^\varepsilon \setminus \partial D_{\rho,1}^\varepsilon$. Then

$$\lim_{\varepsilon\to 0}\int_{\partial D_\rho^\varepsilon} [\Psi - 1]\frac{\partial u}{\partial N} = \lim_{\varepsilon\to 0}\int_{\partial D_{\rho,1}^\varepsilon} [\Psi - 1]\frac{\partial u}{\partial N} + \lim_{\varepsilon\to 0}\int_{\partial D_{\rho,2}^\varepsilon} [\Psi - 1]\frac{\partial u}{\partial N}$$

$$= \int_{\partial D} [\Psi - 1]a = \int_{\partial D} \Psi a - \int_{\partial D} a = \int_{\partial D} \Psi a = 0,$$

since $\Psi = 0$ on $\operatorname{supp} a$. Moreover, $\int_{D_-} B\nabla\tilde{u}\nabla\Psi = \int_D \nabla u \cdot \nabla\Psi_*$, where $\Psi_* = \Psi \circ \Phi$ by our construction of B. The last term is also 0 by the same argument, so

$\alpha = 0$. We now show that u (and hence $\tilde{u}$) is bounded. We assume for simplicity that $n \geqslant 4$. Since $\|a\|_{L^2(\Lambda)} \leqslant C$, we know that

$$u(X) = C_n \int_{\partial D} \frac{f(Q)}{|X - Q|^{n-2}} d\sigma(Q),$$

with $\|f\|_{L^2(\Lambda)} \leqslant C$. Now, for $X \in D_1 = \{(x, y): y > \varphi(x) + 1\}$,

$$\frac{1}{|X - Q|^{n-2}} \leqslant \frac{C}{1 + |Q|^{n-2}} \in L^2(\Lambda),$$

so $u \in L^\infty(D_1)$. Now let B be any ball in $\mathbb{R}^n$ so that $2B \subset \mathbb{R}^n \setminus B^*$, B is of unit size, and such that a fixed fraction of B is contained in D_1. Since $N(\nabla u) \in L^2(\Lambda)$, with norm less than C, $\int_{2B \cap D} |\nabla u|^2 \leqslant C$, and, moreover, on $B \cap D_1$, $|u(x)| \leqslant C$. Therefore, by the Poincaré inequality, $\int_{2B} \tilde{u}^2 \leqslant C$. But, since $\tilde{u}$ solves $L\tilde{u} = 0$,

$$\max_B |\tilde{u}| \leqslant C\left(\int_{2B} |\tilde{u}|^2\right)^{1/2} \leqslant C$$

[**29**]. Therefore, $\tilde{u} \in L^\infty(\mathbb{R}^n \setminus B^*)$, $\|\tilde{u}\|_{L^\infty(\mathbb{R}^n \setminus B^*)} \leqslant C$. Hence, since $\alpha = 0$, $\nabla u = \nabla v$, and $|v(x, y)| \leqslant C/(|x| + |y|)^{n-2+\nu}$, $\nu > 0$. For $R \geqslant R_0 = \operatorname{diam} B^*$, set $b(R) = \int_{A_R} N(\nabla u)^2$, where $A_R = \{(x, \varphi(x)): R < |x| < 2R\}$. For each fixed R let

$$N_1(\nabla u)(x) = \sup\{|\nabla u(z, y)|: (z, y) \in \Gamma_i(x), \operatorname{dist}((z, y), \partial D) \leqslant \delta R\},$$

$$N_2(\nabla u)(x) = \sup\{|\nabla u(z, y)|: (z, y) \in \Gamma_i(x), \operatorname{dist}((z, y), \partial D) \geqslant \delta R\}.$$

In the set where the sup in N_2 is taken, u is harmonic, and the distance of any point X to the boundary is comparable to $|X|$. Thus, using our bound on v, we see that $N_2(\nabla u)(x) \leqslant C/|X|^{n-1+\nu} \approx C/R^{n-1+\nu}$, so $\int_{A_R} N_2(\nabla u)^2 \leqslant CR^{1-n-2\nu}$. Now let

$$\Omega_\tau = \{(x, y): \varphi(x) < y < \varphi(x) + CR, \tau R < |X| < \tau^{-1}R\}, \qquad \tau \in \left(\tfrac{1}{4}, \tfrac{1}{2}\right).$$

By L^2-Neumann theory in Ω_τ, $\int_{A_R} N_1(\nabla u)^2 \, d\sigma \leqslant C\int_{\partial\Omega_\tau} |\nabla u|^2 \, d\sigma$. Integrating in τ from $\frac{1}{4}$ to $\frac{1}{2}$ gives

$$\int_{A_R} N_1(\nabla u)^2 \, d\sigma \leqslant \frac{C}{R}\int_{\Omega_{1/4} \setminus \Omega_{1/2}} |\nabla u|^2 \, dX \leqslant \frac{C}{R^3}\int_{C_1 R < |X| < C_2 R} u^2,$$

since $L\tilde{u} = 0$ (see [**29**] for example). The right side is bounded by $(C/R^3)(1/R^{2(n-2)-2\nu})$. Then

$$\int_{A_R} N(\nabla u) \leqslant C\left(\int_{A_R} N(\nabla u)^2\right)^{1/2} R^{(n-1)/2} \leqslant CR^{-\nu}.$$

Choosing $R = 2^j$ and adding in j, we obtain the desired estimate.

We now turn to the case $1 < p < 2$ of 1.2.6. We need a further definition.

DEFINITION 1.2.10. A function a is an H_1^1 atom if $A = \nabla_t a$ satisfies (a) $\operatorname{supp} A \subset B$, a surface ball, (b) $\|A\|_{L^\infty} \leqslant 1/\sigma(B)$, (c) $\int A \, d\sigma = 0$.

We use the following interpolation result:

THEOREM 1.2.11. *Let T be a linear operator such that $\|Tf\|_{L^2(\Lambda)} \leqslant C\|f\|_{L_1^2(\Lambda)}$ and $\|Ta\|_{L^1(\Lambda)} < C$ for all H_1^1 atoms a. Then for $1 < p < 2$,*

$$\|Tf\|_{L^p(\Lambda)} \leqslant C\|f\|_{L_1^p(\Lambda)}.$$

Hence, all we need to show is that if $\Delta u = 0$, $\nabla_t u = \nabla_t a$, and a is a unit size H_1^1 atom, $N(\nabla u) \in L^1(\Lambda)$. But note that if we let

$$\tilde{u}(x, y) = \begin{cases} u(x, y), & (x, y) \in D, \\ -u^*(x, y), & (x, y) \in D_-, \end{cases}$$

then $\tilde{u}$ is a weak solution of $L\tilde{u} = 0$ in $\mathbb{R}^n \setminus B^*$ since $u|_{\partial D \setminus B^*} = 0$. Then $\tilde{u} = \tilde{u}_\infty + \alpha g + v$, but $\alpha = 0$ since $\tilde{u} - \tilde{u}_\infty$ must change sign at ∞. The argument is then identical to the one given before.

Before we pass to the case $2 < p < 2 + \varepsilon$, we would like to point out that, using the techniques described above, one can develop the Stein–Weiss Hardy-space theory on an arbitrary Lipschitz domain in $\mathbb{R}^n$. This generalizes the results for $n = 2$ obtained in [**24**] and [**25**] and the results for C^1 domains in [**16**].

Some of the results one can obtain are the following: Let

$$H^1_{\text{at}}(\partial D) = \left\{\sum \lambda_i a_i \colon \sum |\lambda_i| < +\infty, a_i \text{ is an atom}\right\},$$

$$H^1_{1,\text{at}}(\partial D) = \left\{\sum \lambda_i a_i \colon \sum |\lambda_i| < +\infty, a_i \text{ is an } H_1^1 \text{ atom}\right\}.$$

THEOREM 1.2.12. (a) *Given* $f \in H^1_{\text{at}}(\partial D)$, *there exists a unique harmonic function* u, *which tends to* 0 *at* ∞, *such that* $N(\nabla u) \in L^1(\partial D)$ *and* $N_Q \nabla u(X) \to f(Q)$ *nontangentially a.e. Moreover*, $u(X) = S(g)(X)$, $g \in H^1_{\text{at}}$. *Also*, $u|_{\partial D} \in H^1_{1,\text{at}}(\partial D)$. (b) *Given* $f \in H^1_{1,\text{at}}$, *there exists a unique (modulo constants) harmonic function* u *such that* $N(\nabla u) \in L^1(\partial D)$ *and* $\nabla_t u|_{\partial D} = \nabla_t f$ *a.e. Moreover*, $u = S(g)$, $g \in H^1_{\text{at}}$, *and* $\partial u/\partial N \in H^1_{\text{at}}(\partial D)$. (c) *If* u *is harmonic, and* $N(\nabla u) \in L^1(\partial D)$, *then* $\partial u/\partial N \in H^1_{\text{at}}(\partial D)$, $u|_{\partial D} \in H^1_{1,\text{at}}(\partial D)$. (d) $f \in H^1_{\text{at}}(\partial D)$ *if and only if* $N(\nabla Sf) \in L^1(\partial D)$ *if and only if* $(\frac{1}{2}I - K^*)f \in H^1_{\text{at}}(\partial D)$.

We turn now to L^p theory, $2 < p < 2 + \varepsilon$. In this case the results are obtained as automatic real-variable consequences of the fact that the L^2 results hold for all Lipschitz domains. We now show that $\|N(\nabla u)\|_{L^p(\Lambda)} \leqslant C\|\partial u/\partial N\|_{L^2(\Lambda)}$ for $2 < p < 2 + \varepsilon$.

The geometry will be clearer if we do it in $\mathbb{R}^n_+$ and transfer it to D by the bi-Lipschitzian mapping

$$\Phi\colon \mathbb{R}^n_+ \to D, \qquad \Phi(x, y) = (x, y + \varphi(x)).$$

We systematically ignore the distinction between sets in $\mathbb{R}^n_+$ and their images under Φ.

Let

$$\gamma = \{(x, y) \in \mathbb{R}^n_+ \colon |x| < y\}, \qquad \gamma^* = \{(x, y) \in \mathbb{R}^n_+ \colon \alpha|x| < y\},$$

where α is a small constant to be chosen. Let

$$m(x) = \sup_{(z, y) \in x + \gamma} |\nabla u(z, y)|, \qquad m^*(x) = \sup_{(z, y) \in x + \gamma^*} |\nabla u(z, y)|.$$

Our aim is to show that there is a small $\varepsilon_0 > 0$ such that

$$\int m^{2+\varepsilon}\, dx \leqslant C \int |f|^{2+\varepsilon}\, dx$$

for all $0 < \varepsilon \leqslant \varepsilon_0$, where $f = \partial u/\partial N$. Let $h = M(f^2)^{1/2}$, where M denotes the Hardy–Littlewood maximal operator. Let

$$E_\lambda = \{x \in \mathbb{R}^{n-1}: m^*(x) > \lambda\}.$$

We claim that

$$\int_{\{m^*>\lambda;\, h\leqslant\lambda\}} m^2 \leqslant C\lambda^2|E_\lambda| + C\alpha\int_{\{m^*>\lambda\}} m^2.$$

Let us assume the claim and prove the desired estimate. First note that

$$\int_{E_\lambda} m^2 \leqslant \int_{\{m^*>\lambda;\, h\leqslant\lambda\}} m^2 + \int_{\{h>\lambda\}} m^2 \leqslant C\lambda^2|E_\lambda| + C\alpha\int_{\{m^*>\lambda\}} m^2 + \int_{\{h>\lambda\}} m^2,$$

by the claim. Now choose and fix α so that $C\cdot\alpha < \frac{1}{2}$. Then

$$\int_{E_\lambda} m^2 \leqslant C\lambda^2|E_\lambda| + C\int_{\{h\geqslant\lambda\}} m^2.$$

For $\varepsilon > 0$

$$\int m^{2+\varepsilon} = \varepsilon\int_0^\infty \lambda^{\varepsilon-1}\int_{\{m>\lambda\}} m^2\,d\lambda \leqslant \varepsilon\int_0^\infty \lambda^{\varepsilon-1}\int_{E_\lambda} m^2\,d\lambda$$

$$\leqslant C\varepsilon\int_0^\infty \lambda^{1+\varepsilon}|\{m^*>\lambda\}|\,d\lambda + C\varepsilon\int_0^\infty \lambda^{\varepsilon-1}\left(\int_{h>\lambda} m^2\right)d\lambda.$$

By a well-known inequality (see [**18**] for example), $|E_\lambda| \leqslant C_\alpha|\{m>\lambda\}|$. Thus

$$\int m^{2+\varepsilon} \leqslant C\varepsilon\int_0^\infty \lambda^{1+\varepsilon}|\{m>\lambda\}|\,d\lambda + C\varepsilon\int_0^\infty \lambda^{\varepsilon-1}\left(\int_{h>\lambda} m^2\right)d\lambda$$

$$\leqslant C\varepsilon\int m^{2+\varepsilon} + c\int m^2h^\varepsilon.$$

If we now choose ε_0 so that $C\varepsilon_0 < 1/2$, for $\varepsilon < \varepsilon_0$, $\int m^{2+\varepsilon} \leqslant C\int m^2h^\varepsilon$. If we now use Hölder's inequality with exponents $(2+\varepsilon)/2$ and $(2+\varepsilon)/\varepsilon$, we see that

$$\int m^{2+\varepsilon} \leqslant C\left(\int m^{2+\varepsilon}\right)^{2/(2+\varepsilon)}\left(\int M(f^2)^{(2+\varepsilon)/2}\right)^{\varepsilon/(2+\varepsilon)},$$

and the desired inequality follows from the Hardy–Littlewood maximal theorem.

It remains to establish the claim. Let $\{Q_k\}$ be a Whitney decomposition of the set $E_\lambda = \{m^* > \lambda\}$ such that $3Q_k \subset E_\lambda$ and $\{3Q_k\}$ has bounded overlap. Fix k; we can assume there exists $x \in Q_k$ such that $h(x) \leqslant \lambda$, and, hence $\int_{2Q_k} f^2 \leqslant C\lambda^2|Q_k|$. For $1 \leqslant \tau \leqslant 2$, let $Q_{k,\tau} = \tau Q_k$ and

$$\tilde{Q}_{k,\tau} = \{(x, y): x \in \tau Q_k, 0 < y < \tau\,\text{length}(Q_k)\}.$$

$\tilde{Q}_{k,\tau}$ (and $\Phi(\tilde{Q}_{k,\tau})$) is a Lipschitz domain uniformly in k, τ. Also, by construction of Q_k, there exists x_k with $\text{dist}(x_k, Q_k) \approx \text{length}(Q_k)$ such that $m^*(x_k) \leqslant \lambda$. Let

$$A_{k,\tau} = \partial\tilde{Q}_{k,\tau} \cap x_k + \gamma^*, \qquad B_{k,\tau} = \partial\tilde{Q}_{k,\tau} \cap \mathbb{R}^n_+ \setminus A_{k,\tau},$$

so that

$$\partial \tilde{Q}_{k,\tau} = Q_{k,\tau} \cup A_{k,\tau} \cup B_{k,\tau}.$$

Note that the height of $B_{k,\tau}$ is dominated by $C\alpha\,\text{length}(Q_k)$, and that $|\nabla u| \leqslant \lambda$ on $A_{k,\tau}$. Let m_1 be the maximal function of ∇u corresponding to the domain $\tilde{Q}_{k,\tau}$ (i.e., where the cones are truncated at height $\approx l(Q_k)$). Then for $x \in Q_k$, $m(x) \leqslant m_1(x) + \lambda$. Also,

$$\begin{aligned}\int_{Q_k} m_1^2 &\leqslant \int_{\partial\tilde{Q}_{k,\tau}} m_1^2 \leqslant \quad (\text{using } L^2\text{-theory on } \tilde{Q}_{k,\tau}) \\ &\leqslant C\int_{B_{k,\tau}} |\nabla u|^2\,d\sigma + c\int_{A_{k,\tau}} |\nabla u|^2\,d\sigma + c\int_{2Q_k} f^2 \\ &\leqslant C\int_{B_{k,\tau}} |\nabla u|^2\,d\sigma + C\lambda^2|Q_k|.\end{aligned}$$

Integrating in τ between 1 and 2 we see that

$$\int_{Q_k} m_1^2 \leqslant \frac{C}{l(Q_k)}\int_0^{\alpha l(Q_k)}\int_{2Q_k} |\nabla u|^2 + C\lambda^2|Q_k| \leqslant C\alpha\int_{2Q_k} m^2 + C\lambda^2|Q_k|.$$

Thus

$$\int_{Q_k} m^2 \leqslant C\alpha\int_{2Q_k} m^2 + C\lambda^2|Q_k|.$$

Adding in k we see that

$$\int_{\{m^*>\lambda,\,h\leqslant\lambda\}} m^2 \leqslant C\lambda^2|E_\lambda| + C\alpha\int_{\{m^*>\lambda\}} m^2,$$

which is the claim. Note also that the same argument gives the estimate $\|N(\nabla u)\|_p \leqslant C\|\nabla_t u\|_p$, $2 < p < 2 + \varepsilon$, and the L^p theory is thus completed.

2. Higher-order boundary-value problems.

Part 1. *Systems of elastostatics*. We sketch the extension of the L^2 results for the Laplace equation to the systems of linear elastostatics on Lipschitz domains. These results are joint work of Dahlberg, Kenig, and Verchota and will be discussed in detail in a forthcoming paper [**12**]. Here we describe some of the main ideas in that work. For simplicity, we restrict our attention to domains D above the graph of a Lipschitz function $\varphi\colon \mathbb{R}^2 \to \mathbb{R}$.

Let $\lambda, \mu \geqslant 0$ be constants (Lamé moduli). We seek to solve the following boundary-value problems, where $\mathbf{u} = (u^1, u^2, u^3)$

$$\begin{aligned}&\mu\Delta\mathbf{u} + (\lambda+\mu)\nabla\,\operatorname{div}\mathbf{u} = 0 \quad \text{in } D,\\ &\mathbf{u}|_{\partial D} = \mathbf{f} \in L^2(\partial D, d\sigma);\end{aligned} \tag{2.1.1}$$

$$\begin{aligned}&\mu\Delta\mathbf{u} + (\lambda+\mu)\nabla\,\operatorname{div}\mathbf{u} = 0 \quad \text{in } D,\\ &\lambda(\operatorname{div}\mathbf{u})N + \mu\left\{\nabla\mathbf{u} + (\nabla\mathbf{u})^t\right\}N\Big|_{\partial D} = \mathbf{f} \in L^2(\partial D, d\sigma).\end{aligned} \tag{2.1.2}$$

(2.1.1) corresponds to knowing the displacement vector $\mathbf{u}$ on the boundary of D, and (2.1.2) corresponds to knowing the surface stresses on the boundary of D. We seek to solve (2.1.1) and (2.1.2) by the method of layer potentials. In order to do so we introduce the Kelvin matrix of fundamental solutions (see **[27]** for example),

$$\Gamma(X) = \big(\Gamma_{ij}(X)\big),$$

where

$$\Gamma_{ij}(X) = \frac{A}{4\pi}\frac{\delta_{ij}}{|X|} + \frac{C}{4\pi}\frac{X_i X_j}{|X|^3},$$

and

$$A = \frac{1}{2}\left[\frac{1}{\mu} + \frac{1}{2\mu+\lambda}\right], \qquad C = \frac{1}{2}\left[\frac{1}{\mu} - \frac{1}{2\mu+\lambda}\right].$$

We also introduce the stress operator T, where

$$T\mathbf{u} = \lambda(\operatorname{div}\mathbf{u})N + \mu\{\nabla\mathbf{u} + \nabla\mathbf{u}^t\}N.$$

The double-layer potential of a density $\mathbf{g}(Q)$ is then given by

$$\mathbf{u}(X) = \mathscr{K}\mathbf{g}(X) = \int_{\partial D}\{T(Q)\Gamma(X-Q)\}^t\mathbf{g}(Q)\,d\sigma(Q),$$

where the operator T is applied to each column of the matrix Γ.

The single-layer potential of a density $\mathbf{g}(Q)$ is

$$\mathbf{u}(X) = S\mathbf{g}(X) = \int_{\partial D}\Gamma(X-Q)\cdot\mathbf{g}(Q)\,d\sigma(Q).$$

Our main results here parallel those of §1, Part 1. They are

THEOREM 2.1.3. (a) *There exists a unique solution of problem* (2.1.1) *in* D *with* $N(\mathbf{u}) \in L^2(\partial D, d\sigma)$. *Moreover, the solution* u *has the form* $\mathbf{u}(x) = \mathscr{K}\mathbf{g}(x)$, $\mathbf{g} \in L^2(\partial D, d\sigma)$.

(b) *There exists a unique solution of* (2.1.2) *in* D *that is* 0 *at infinity, with* $N(\nabla\mathbf{u}) \in L^2(\partial D, d\sigma)$. *Moreover, the solution* $\mathbf{u}$ *has the form* $\mathbf{u}(X) = S\mathbf{g}(X)$, $\mathbf{g} \in L^2(\partial D, d\sigma)$.

(c) If the data $\mathbf{f}$ in (2.1.1) belongs to $L_1^2(\partial D, d\sigma)$, we can solve (2.1.1) with $N(\nabla\mathbf{u}) \in L^2(\partial D, d\sigma)$.

The proof of Theorem 2.1.3 starts out following the pattern we used to prove 1.1.1, 1.1.2, and 1.1.14. We first show, as in Theorem 1.1.3, that the following lemma holds:

LEMMA 2.1.4. *Let* $\mathscr{K}\mathbf{g}$, $S\mathbf{g}$ *be defined as above, so that they both solve* $\mu\Delta\mathbf{u} + (\lambda+\mu)\nabla\operatorname{div}\mathbf{u} = 0$ *in* $\mathbb{R}^3\setminus\partial D$. *Then*

(a)
$$\|N(\mathscr{K}\mathbf{g})\|_{L^p(\partial D, d\sigma)} \leqslant C\|\mathbf{g}\|_{L^p(\partial D, d\sigma)},$$
$$\|N(\nabla S\mathbf{g})\|_{L^p(\partial D, d\sigma)} \leqslant C\|\mathbf{g}\|_{L^p(\partial D, d\sigma)} \quad \text{for } 1 < p < \infty.$$

(b) $(\mathscr{K}\mathbf{g})^{\pm}(P) = \pm\mathbf{g}(P) + K\mathbf{g}(P)$,

$$\left(\frac{\partial}{\partial X_i}(S\mathbf{g})_j\right)^{\pm}(P) = \pm\left\{\frac{A+C}{2}n_i(P)g_j(P) - n_i(P)\cdot n_j(P)\langle N_P, g(P)\rangle\right\} + \left(\text{p.v.}\int_{\partial D}\frac{\partial}{\partial P_i}\Gamma(P-Q)\mathbf{g}(Q)\,d\sigma(Q)\right)_j,$$

where

$$K\mathbf{g}(P) = \text{p.v.}\int_{\partial D}\{T(Q)\Gamma(P-Q)\}^t\mathbf{g}(Q)\,d\sigma(Q),$$

and A, C are the constants in the definition of the fundamental solution.

Thus, just as in §1, Part 1 reduces to proving the invertibility of $\pm\frac{1}{2}I + K$, $\pm\frac{1}{2}I + K^*$ on $L^2(\partial D, d\sigma)$ and the invertibility of S from $L^2(\partial D, d\sigma)$ onto $L_1^2(\partial D, d\sigma)$. As before, using the jump relations, it suffices to show that if $\mathbf{u}(X) = S\mathbf{g}(X)$, then

$$\|T\mathbf{u}\|_{L^2(\partial D, d\sigma)} \approx \|\nabla_t\mathbf{u}\|_{L^2(\partial D, d\sigma)}.$$

Before explaining the difficulties in doing so, it is very useful to explain the stress operator T (and thus the boundary-value problem (2.1.2)), from the point of view of the theory of constant-coefficient, second-order, elliptic systems. We go back to working on $\mathbb{R}^n$ and use the summation convention.

Let a_{ij}^{rs}, $1 \leqslant r, s \leqslant m$, $1 \leqslant i, j \leqslant n$, be constants satisfying the ellipticity condition

$$a_{ij}^{rs}\xi_i\xi_j\eta^r\eta^s \geqslant C|\xi|^2|\eta|^2$$

and the symmetry condition $a_{ij}^{rs} = a_{ji}^{sr}$. Consider vector-valued functions $\mathbf{u} = (u^1, \ldots, u^m)$ on $\mathbb{R}^n$ satisfying the divergence-form system

$$\frac{\partial}{\partial X_i}a_{ij}^{rs}\frac{\partial}{\partial X_j}u^s = 0 \quad \text{in } D.$$

From variational considerations the most natural boundary conditions are Dirichlet conditions ($\mathbf{u}|_{\partial D} = \mathbf{f}$) or Neumann-type conditions, $\partial\mathbf{u}/\partial\nu = n_i a_{ij}^{rs}(\partial u^s/\partial X_j) = f_r$. The interpretation of problem (2.1.2) in this context is that we can find constants a_{ij}^{rs}, $1 \leqslant i, j \leqslant 3$, $1 \leqslant r, s \leqslant 3$, satisfying the ellipticity and symmetry conditions such that $\mu\Delta\mathbf{u} + (\lambda+\mu)\nabla\operatorname{div}\mathbf{u} = 0$ in D if and only if $(\partial/\partial X_i)a_{ij}^{rs}(\partial u^s/\partial X_j) = 0$ in D, and with $T\mathbf{u} = \partial\mathbf{u}/\partial\nu$. In order to obtain the equivalence between the tangential derivatives and the stress operator, we need an identity of Rellich type. Such identities are available for general, constant-coefficient systems (see [**32, 30**]).

LEMMA 2.1.5 (THE RELLICH, PAYNE–WEINBERGER, NEČAS IDENTITIES). *Suppose that $(\partial/\partial X_i)a_{ij}^{rs}(\partial/\partial X_j)u^s = 0$ in D, $a_{ij}^{rs} = a_{ji}^{sr}$, $\mathbf{h}$ is a constant vector in $\mathbb{R}^n$, and $\mathbf{u}$ and its derivatives are suitably small at ∞. Then*

$$\int_{\partial D} h_l n_l a_{ij}^{rs}\frac{\partial u^r}{\partial X_i}\frac{\partial u^s}{\partial X_j}\,d\sigma = 2\int_{\partial D} h_i\frac{\partial u^r}{\partial X_i}n_l a_{lj}^{rs}\frac{\partial u^s}{\partial X_j}\,d\sigma.$$

PROOF. Apply the divergence theorem to the formula

$$\frac{\partial}{\partial X_l}\left[\left(h_l a_{ij}^{rs} - h_i a_{lj}^{rs} - h_j a_{il}^{rs}\right)\frac{\partial u^r}{\partial X_i}\cdot\frac{\partial u^s}{\partial X_j}\right] = 0.$$

REMARK 1. Note that if we are dealing with the case $m = 1$, $a_{ij} = I$, and we choose $\mathbf{h} = e_n$, we recover the identity we previously used for Laplace's equation.

REMARK 2. Note that if we had the stronger ellipticity assumption that $a_{ij}^{rs}\xi_i^r\xi_j^s \geqslant C\Sigma_{l,t}|\xi_l^t|^2$, we would have, if $\partial D = \{(x, \varphi(x)) \colon \varphi\colon \mathbb{R}^{n-1} \to \mathbb{R}, \|\nabla\varphi\|_\infty \leqslant M\}$, that $\|\nabla_t u\|_{L^2(\partial D, d\sigma)} \approx \|\partial u/\partial\nu\|_{L^2(\partial D, d\sigma)}$. In fact, if we take $\mathbf{h} = e_n$, then

$$\sum_r \int_{\partial D} |\nabla u^r|^2\, d\sigma \leqslant C\int_{\partial D} h_l n_l a_{ij}^{rs}\frac{\partial u^r}{\partial X_i}\frac{\partial u^s}{\partial X_j}\, d\sigma = 2C\int_{\partial D} h_i\frac{\partial u^r}{\partial X_i}\cdot n_l a_{lj}^{rs}\frac{\partial u^s}{\partial X_j} d\sigma$$

$$\leqslant 2C\left(\sum_r \int_{\partial D} |\nabla u^r|^2\, d\sigma\right)^{1/2}\left(\int_{\partial D}\left|\frac{\partial u}{\partial\nu}\right|^2 d\sigma\right)^{1/2}.$$

Thus,

$$\sum_r \int_{\partial D} |\nabla u^r|^2\, d\sigma \leqslant C\int_{\partial D}\left|\frac{\partial u}{\partial\nu}\right|^2 d\sigma.$$

For the opposite inequality observe that, for each r, s, j fixed, the vector $h_i n_l a_{lj}^{rs} - h_l n_l a_{ij}^{rs}$ is perpendicular to N. Because of Lemma 2.1.5,

$$\int_{\partial D} h_l n_l a_{ij}^{rs}\frac{\partial u^r}{\partial X_i}\cdot\frac{\partial u^s}{\partial X_j}\, d\sigma = 2\int_{\partial D}\left(h_l n_l a_{ij}^{rs} - h_i n_l a_{lj}^{rs}\right)\frac{\partial u^l}{\partial X_i}\cdot\frac{\partial u^s}{\partial X_j}\, d\sigma.$$

Hence

$$\int_{\partial D} |\nabla u|^2\, d\sigma \leqslant C\left(\int_{\partial D} |\nabla_t u|^2\, d\sigma\right)^{1/2}\left(\int_{\partial D} |\nabla u|^2\, d\sigma\right)^{1/2},$$

so

$$\int_{\partial D}\left|\frac{\partial u}{\partial\nu}\right|^2 d\sigma \leqslant C\int_{\partial D} |\nabla u|^2\, d\sigma \leqslant C\int_{\partial D} |\nabla_t u|^2.$$

REMARK 3. In the case in which we are interested, i.e., the case of systems of elastostatics

$$a_{ij}^{rs}\frac{\partial u^s}{\partial X_i}\cdot\frac{\partial u^r}{\partial X_j} = \lambda(\operatorname{div}\mathbf{u})^2 + \frac{\mu}{2}\sum_{i,j}\left(\frac{\partial u^j}{\partial X_i} + \frac{\partial u^i}{\partial X_j}\right)^2,$$

which clearly does not satisfy

$$a_{ij}^{rs}\xi_i^r\xi_j^s \geqslant C\sum_{l,t}|\xi_l^t|^2,$$

since the quadratic form involves only the symmetric part of the matrix (ξ_i^r). In this case, of course,

$$\frac{\partial\mathbf{u}}{\partial\nu} = T\mathbf{u} = \lambda(\operatorname{div}\mathbf{u})N + \mu\{\nabla\mathbf{u} + \nabla\mathbf{u}^t\}N.$$

REMARK 4. The inequality

$$\|\nabla \mathbf{u}\|_{L^2(\partial D, d\sigma)} \leqslant C\|\nabla_t \mathbf{u}\|_{L^2(\partial D, d\sigma)}$$

holds in the general case, directly from Lemma 2.1.5, by a more complicated algebraic argument. In fact, as in Remark 2,

$$\int_{\partial D} h_l n_l a_{ij}^{rs} \frac{\partial u^r}{\partial X_i} \frac{\partial u^s}{\partial X_j}\, d\sigma = 2\int_{\partial D} \left(h_l n_l a_{ij}^{rs} - h_i n_l a_{lj}^{rs}\right) \frac{\partial u^l}{\partial X_i} \cdot \frac{\partial u^s}{\partial X_j}\, d\sigma,$$

and, for fixed r, s, j, $(h_l n_l a_{ij}^{rs} - h_i n_l a_{lj}^{rs})$ is a tangential vector.

Thus,

$$\int_{\partial D} h_l n_l a_{ij}^{rs} \frac{\partial u^r}{\partial X_i} \frac{\partial u^s}{\partial X_j}\, d\sigma \leqslant C\left(\int_{\partial D} |\nabla_t \mathbf{u}|^2\, d\sigma\right)^{1/2} \left(\int_{\partial D} |\nabla \mathbf{u}|^2\, d\sigma\right)^{1/2}.$$

Consider now the matrix $d_{rs} = (a_{ij}^{rs} n_i n_j)^{-1}$. This is a strictly positive matrix, since $a_{ij}^{rs}\xi_i\xi_j\eta^r\eta^s \geqslant C|\xi|^2|\eta|^2$. Moreover,

$$\begin{aligned} d_{rs}\left(\frac{\partial u}{\partial \nu}\right)_r \left(\frac{\partial u}{\partial \nu}\right)_s - a_{ij}^{rs} \frac{\partial u^r}{\partial X_i} \frac{\partial u^s}{\partial X_j} &= d_{rs} n_i a_{ij}^{rt} \frac{\partial u^t}{\partial X_j} \cdot n_l a_{lk}^{sm} \frac{\partial u^m}{\partial X_k} - a_{ij}^{rs} \frac{\partial u^r}{\partial X_i} \frac{\partial u^s}{\partial X_j} \\ &= d_{rs} n_k a_{kl}^{rt} \frac{\partial u^t}{\partial X_l} \cdot n_m a_{mv}^{s\tau} \frac{\partial u^\tau}{\partial X_v} - a_{vl}^{t\tau} \frac{\partial u^t}{\partial X_v} \frac{\partial u^\tau}{\partial X_l} \\ &= d_{rs} n_k a_{kv}^{rt} \frac{\partial u^t}{\partial X_v} \cdot n_m a_{ml}^{s\tau} \frac{\partial u^\tau}{\partial X_l} - a_{vl}^{t\tau} \frac{\partial u^t}{\partial X_v} \frac{\partial u^\tau}{\partial X_l} \\ &= \left\{ d_{rs} n_k a_{kv}^{rt} n_m a_{ml}^{s\tau} - a_{vl}^{t\tau} \right\} \frac{\partial u^t}{\partial X_v} \frac{\partial u^\tau}{\partial X_l}. \end{aligned}$$

Now, note that for t, τ, l fixed, $\{d_{rs} n_k a_{kv}^{rt} n_m a_{ml}^{s\tau} - a_{vl}^{t\tau}\}$ is perpendicular to N by our definition of d_{rs} and the symmetry of a_{ij}^{rs}:

$$\begin{aligned} d_{rs} n_k a_{kv}^{rt} n_m a_{ml}^{st} n_v - a_{vl}^{t\tau} n_v &= a_{kv}^{rt} n_k n_v d_{rs} a_{ml}^{st} n_m - a_{ml}^{t\tau} n_m \\ &= a_{vk}^{tr} n_v n_k d_{rs} a_{ml}^{st} n_m - a_{ml}^{t\tau} n_m = \delta_{ts} a_{ml}^{st} n_m - a_{ml}^{t\tau} n_m \\ &= a_{ml}^{t\tau} n_m - a_{ml}^{t\tau} n_m = 0. \end{aligned}$$

Therefore,

$$\int_{\partial D} h_l n_l d_{rs} \left(\frac{\partial \mathbf{u}}{\partial \nu}\right)_r \left(\frac{\partial \mathbf{u}}{\partial \nu}\right)_s \leqslant C\left(\int_{\partial D} |\nabla_t \mathbf{u}|^2\, d\sigma\right)^{1/2} \left(\int_{\partial D} |\nabla \mathbf{u}|^2\, d\sigma\right)^{1/2}.$$

Now,

$$\begin{aligned} \left(\frac{\partial \mathbf{u}}{\partial \nu}\right)_r - a_{kj}^{rs} n_k n_j \frac{\partial u^s}{\partial N} &= n_i a_{ij}^{rs} \frac{\partial u^s}{\partial X_j} - a_{kj}^{rs} n_k n_j n_i \frac{\partial u^s}{\partial X_i} \\ &= n_i a_{ij}^{rs} \frac{\partial u^s}{\partial X_j} - a_{ki}^{rs} n_k n_j n_i \frac{\partial u^s}{\partial X_j} = \left\{ n_i a_{ij}^{rs} - a_{ki}^{rs} n_k n_i n_j \right\} \frac{\partial u^s}{\partial X_j} \\ &= \left\{ n_i a_{ij}^{rs} - a_{ik}^{rs} n_k n_i n_j \right\} \frac{\partial u^s}{\partial X_j}. \end{aligned}$$

But, for i, r, s fixed, $a_{ij}^{rs} - a_{ik}^{rs} n_k n_j$ is perpendicular to N, so

$$\int_{\partial D} h_l n_l d_{rs} \left\{ a_{kj}^{rt} n_k n_j \frac{\partial u^t}{\partial N} \right\} \left\{ a_{il}^{s\tau} n_i n_l \frac{\partial u^\tau}{\partial N} \right\} d\sigma$$

$$\leqslant C \left\{ \left(\int_{\partial D} |\nabla_t \mathbf{u}|^2 \, d\sigma \right)^{1/2} \left(\int_{\partial D} |\nabla \mathbf{u}|^2 \, d\sigma \right)^{1/2} + \int_{\partial D} |\nabla_t \mathbf{u}|^2 \, d\sigma \right\}.$$

We now choose $\mathbf{h} = e_n$, so that $h_l n_l \geqslant C$, and recall that (d_{rs}) and $(a_{kj}^{rt} n_k n_j)$ are strictly positive-definite matrices. We then see that

$$\int_{\partial D} \left| \frac{\partial \mathbf{u}}{\partial N} \right|^2 d\sigma \leqslant C \left\{ \left(\int_{\partial D} |\nabla_t \mathbf{u}|^2 \, d\sigma \right)^{1/2} \left(\int_{\partial D} |\nabla \mathbf{u}|^2 \, d\sigma \right)^{1/2} + \int_{\partial D} |\nabla_t \mathbf{u}|^2 \, d\sigma \right\}.$$

Now, since $|\nabla \mathbf{u}|^2 = |\nabla_t \mathbf{u}|^2 + |\partial \mathbf{u}/\partial N|^2$, the remark follows.

REMARK 5. In order to show that $\int_{\partial D} |\nabla_t \mathbf{u}|^2 \, d\sigma \leqslant C \int_{\partial D} |T\mathbf{u}|^2 \, d\sigma$, it suffices to show that

$$\int_{\partial D} |\nabla \mathbf{u}|^2 \, d\sigma \leqslant C \int_{\partial D} \left| \lambda (\operatorname{div} \mathbf{u}) I + \mu \{ \nabla \mathbf{u} + \nabla \mathbf{u}^t \} \right|^2 d\sigma.$$

In fact, if this inequality holds, we would clearly have

$$\int_{\partial D} |\nabla \mathbf{u}|^2 \, d\sigma \leqslant C \int_{\partial D} |\nabla \mathbf{u} + \nabla \mathbf{u}^t|^2 \, d\sigma$$

(Korn-type inequality at the boundary). The Rellich–Payne–Weinberger–Nečas identity is, in this case (with $\mathbf{h} = e_n$),

$$\int_{\partial D} n_n \left\{ \frac{\mu}{2} |\nabla \mathbf{u} + \nabla \mathbf{u}^t|^2 + \lambda (\operatorname{div} \mathbf{u})^2 \right\} d\sigma$$

$$= 2 \int_{\partial D} \frac{\partial \mathbf{u}}{\partial y} \cdot \left\{ \lambda (\operatorname{div} \mathbf{u}) N + \mu \{ \nabla \mathbf{u} + \nabla \mathbf{u}^t \} N \right\} d\sigma.$$

But then

$$\int_{\partial D} |\nabla \mathbf{u}|^2 \, d\sigma \leqslant C \left(\int_{\partial D} |\nabla \mathbf{u}|^2 \, d\sigma \right)^{1/2}$$

$$\times \left(\int_{\partial D} \left| \lambda (\operatorname{div} \mathbf{u}) N + \mu \{ \nabla \mathbf{u} + \nabla \mathbf{u}^t \} N \right|^2 d\sigma \right)^{1/2}.$$

The rest of Part 1 is devoted to sketching the proof of the above inequality.

THEOREM 2.1.6. *Let* $\mathbf{u}$ *solve* $\mu \Delta \mathbf{u} + (\lambda + \mu) \nabla \operatorname{div} \mathbf{u} = 0$ *in* D, $\mathbf{u} = S(\mathbf{g})$, *where* $\mathbf{g}$ *is nice. Then there exists a constant* C *depending only on the Lipschitz constant of* φ *so that*

$$\int_{\partial D} |\nabla \mathbf{u}|^2 \, d\sigma \leqslant C \int_{\partial D} \left| \lambda (\operatorname{div} \mathbf{u}) I + \mu \{ \nabla \mathbf{u} + \nabla \mathbf{u}^t \} \right|^2 d\sigma.$$

The proof proceeds in two steps. They are

LEMMA 2.1.7. *Let* $\mathbf{u}$ *be as in Theorem* 2.1.6. *Then*

$$\int_{\partial D} N(\nabla \mathbf{u})^2 \, d\sigma \leqslant C \int_{\partial D} N\left(\lambda (\operatorname{div} \mathbf{u}) I + \mu \{ \nabla \mathbf{u} + \nabla \mathbf{u}^t \} \right)^2 d\sigma.$$

LEMMA 2.1.8. *Let* **u** *be as in Theorem* 2.1.6. *Then*

$$\int_{\partial D} N\big(\lambda(\operatorname{div}\mathbf{u})I + \mu\{\nabla\mathbf{u} + \nabla\mathbf{u}^t\}\big)^2\,d\sigma$$

$$\leqslant C\int_{\partial D} \big|\lambda(\operatorname{div}\mathbf{u})I + \mu\{\nabla\mathbf{u} + \nabla\mathbf{u}^t\}\big|^2\,d\sigma.$$

Lemma 2.1.7 is proved by first doing so in the case when the Lipschitz constant is small, and then passing to the general case by using the ideas of David [**13**]. Lemma 2.1.8 is proved by observing that if $\mathbf{v}$ is any row of the matrix $\lambda(\operatorname{div}\mathbf{u})I + \mu\{\nabla\mathbf{u} + \nabla\mathbf{u}^t\}$, then $\mathbf{v}$ is a solution of the Stokes system

$$\text{(S)}\qquad \begin{cases} \Delta\mathbf{v} = \nabla p & \text{in } D,\\ \operatorname{div}\mathbf{v} = 0 & \text{in } D,\\ \mathbf{v}|_{\partial D} = \mathbf{f} \in L^2(\partial D, d\sigma). \end{cases}$$

This is checked directly by using the system of equations $\mu\Delta\mathbf{u} + (\lambda + \mu)\nabla\operatorname{div}\mathbf{u} = 0$. One then invokes the following theorem of Fabes, Kenig, and Verchota, whose proof is presented in the next section.

THEOREM 2.1.9. *Given* $\mathbf{f} \in L^2(\partial D, d\sigma)$, *there exists a unique solution* $(\mathbf{v}, p)$ *to* (S) *with* p *tending to* 0 *at* ∞ *and* $N(\mathbf{v}) \in L^2(\partial D, d\sigma)$. *Moreover*,

$$\|N(\mathbf{v})\|_{L^2(\partial D, d\sigma)} \leqslant C\|\mathbf{f}\|_{L^2(\partial D, d\sigma)}.$$

We now turn to a sketch of the proof of Lemma 2.1.7. We will need the following, unpublished, real-variable lemma of David [**14**].

LEMMA 2.1.10. *Let* $F\colon \mathbb{R} \times \mathbb{R}^n \to \mathbb{R}$ *be a function of two variables*, $t \in \mathbb{R}$, $x = (x_1, \ldots, x_n) \in \mathbb{R}^n$. *Assume that for each* x, *the function* $t \to F(t, x)$ *is Lipschitz, with Lipschitz constant less than or equal to* M, *and for each* i, $1 \leqslant i \leqslant n$, *the function* $x_i \to F(t, x)$ *is Lipschitz, with Lipschitz constant less than or equal to* M_i, *for any choice of the other variables. Given an interval* $I \times J = I \times J_1 \times \cdots \times J_n$, *where the* J_i*'s and* I *are* 1-*dimensional compact intervals, there exists a function* $G(t, x)\colon \mathbb{R} \times \mathbb{R}^n \to \mathbb{R}$ *with the following properties*:

(a) $G(t, x) \geqslant F(t, x)$ *on* $I \times J$.

(b) *If* $E = \{(t, x) \in I \times J\colon F(t, x) = G(t, x)\}$ *then* $|E| \geqslant \frac{3}{8}|I|\,|J|$.

(c) *For each* i *the function* $G(t, x_1, x_2, \ldots, x_{i-1}, \ldots, x_{i+1}, \ldots, x_n)$ *is Lipschitz, with Lipschitz constant less than or equal to* M_i, *and one of the following statements is true*: *For each* x *either* $-M \leqslant (\partial G/\partial t)(t, x) \leqslant 4M/5$ *or* $-4M/5 \leqslant (\partial G/\partial t)(t, x) \leqslant M$.

The proof of this lemma is the same as in the 1-dimensional case, treating x as a parameter (see [**13**]).

Before we proceed with the proof of Lemma 2.1.7, we would like to point out that in the analogue of Lemma 2.1.7 for bounded domains, a normalization is necessary, since if $\mathbf{u}(X)$ solves the systems of elastostatics, so does $\mathbf{u}(X) + \mathbf{a} + BX$, where $\mathbf{a}$ is a constant vector and B is any antisymmetric 3×3 matrix. The right side of the inequality in the lemma, of course, remains unchanged, while the left side increases if B 'increases'. The most convenient normalization is that, for some fixed point X^* in the domain, $\nabla \mathbf{u}(X^*) - \nabla \mathbf{u}(X^*)^t = 0$. This also gives uniqueness modulo constants to problem (2.1.2) in bounded domains.

We now need to introduce some definitions. Let $D_0 \subset \mathbb{R}^n_+$ be a fixed C^∞ domain with

$$\{(x,0) : |\|x\|| = \max|x_i| \leqslant 1\} \subset \partial D_0,$$

$$\{(x, y) : 0 < y < 1, |\|x\|| \leqslant 1\} \subset D_0 \subset \{(x, y) : 0 < y < 2, |\|x\|| < 2\}.$$

If $\varphi\colon \mathbb{R}^{n-1} \to \mathbb{R}$ is Lipschitz, with $|\|\nabla\varphi\|| \leqslant M$, we construct the mapping T_φ: $\mathbb{R}^n_+ \to \mathbb{R}^n$ by $T_\varphi(x, y) = (x, cy + \eta_y * \varphi(x))$, where $\eta \in C_0^\infty(\mathbb{R}^{n-1})$ is radial, $\int \eta = 1$, and $c = c(M)$ is chosen so that $T_\varphi(\mathbb{R}^n_+) \subset \{(x, y) : y > \varphi(x)\}$, and so that T_φ is a bi-Lipschitzian mapping. Also, it is clear that T_φ is smooth for (x, y) with $y > 0$, and $T_\varphi(x,0) = (x, \varphi(x))$. We denote the point $T_\varphi(0,1)$ by A_φ. Lemma 2.1.7 is an easy consequence of

LEMMA 2.1.11. *Given $M > 0$ and φ with $|\|\nabla\varphi\|| \leqslant M$, there exists a constant $C = C(M)$ such that, for all functions $\mathbf{u}$ in D_φ that are Lipschitz in $\overline{D}_\varphi$ and satisfy $\mu\Delta\mathbf{u} + (\lambda + \mu)\nabla \operatorname{div}\mathbf{u} = 0$ in D_φ and $\nabla\mathbf{u}(A_\varphi) = \nabla\mathbf{u}(A_\varphi)^t$, we have*

$$\|N_\varphi(\nabla\mathbf{u})\|_{L^2(\partial D, d\sigma)} \leqslant C\|N_\varphi(\lambda(\operatorname{div}\mathbf{u})I + \mu\{\nabla\mathbf{u} + \nabla\mathbf{u}^t\})\|_{L^2(\partial D, d\sigma)}.$$

Here N_φ is the nontangential maximal operator corresponding to the domain D_φ.

This lemma will be proved by a series of propositions. Before we proceed we need to introduce one more definition. We say that proposition (M, ε) holds if, whenever φ is such that $|\|\nabla\varphi\|| \leqslant M$ and there exists a constant vector $\mathbf{a}$ with $|\|\mathbf{a}\|| \leqslant M$ so that $|\|\nabla\varphi - \mathbf{a}\|| \leqslant \varepsilon$, then for all Lipschitz functions $\mathbf{u}$ on $\overline{D}_\varphi$ with $\mu\Delta\mathbf{u} + (\lambda + \mu)\nabla \operatorname{div}\mathbf{u} = 0$ in D_φ and $\nabla\mathbf{u}(A_\varphi) = \nabla\mathbf{u}^t(A_\varphi)$, we have

$$\|N_\varphi(\nabla\mathbf{u})\|_{L^2(\partial D_\varphi, d\sigma)} \leqslant C\|N_\varphi(\lambda(\operatorname{div}\mathbf{u})I + \mu\{\nabla\mathbf{u} + \nabla\mathbf{u}^t\})\|_{L^2(\partial D_\varphi, d\sigma)},$$

where $C = C(M, \varepsilon)$.

Note that is proposition (M, ε) holds, then the corresponding estimates automatically hold for all translates, rotates, or dilates of the domains D_φ when φ satisfies the conditions in proposition (M, ε). In the rest of this section, a coordinate chart will be a translate, rotate, or dilate of a domain D_φ. The bottom B_φ of ∂D_φ will be $T_\varphi(\partial D_0 \cap (x,0) : x \in \mathbb{R}^{n-1})$.

PROPOSITION 2.1.12. *Given $M > 0$, there exists $\varepsilon = \varepsilon(M)$ so that proposition (M, ε) holds.*

We will not give the proof but will just make a few remarks about it. First, in this case the stronger estimate

$$\|N_\varphi(\nabla\mathbf{u})\|_{L^2(\partial D_\varphi, d\sigma)} \leqslant C\|\lambda(\operatorname{div}\mathbf{u})N + \mu\{\nabla\mathbf{u} + \nabla\mathbf{u}^t\}N\|_{L^2(\partial D_\varphi, d\sigma)}$$

holds, because in this case the domain D_φ is a small perturbation of the smooth domain $D_{\mathbf{ax}}$. For the smooth domain $D_{\mathbf{ax}}$, we can solve problem (2.1.2) by the method of layer potentials (see [**27**], for example). If ε is small, a perturbation analysis based on the theorem of Coifman–McIntosh–Meyer [**13**] shows that this is still the case. This easily gives the estimate claimed above.

PROPOSITION 2.1.13. *For all* M, $\varepsilon > 0$, $\alpha \in (0, 0.1)$, *if proposition* (M, ε) *holds, then proposition* $(1 - \alpha M, 1.1\varepsilon)$ *holds.*

We postpone the proof and show first how Propositions 2.1.12 and 2.1.13 yield lemma 2.1.11.

PROOF OF LEMMA 2.1.11. We show that proposition (M, ε) holds for any M, ε. Fix M, ε and choose N so large that if $\varepsilon(10M)$ is as in Proposition 2.1.12, then $(1.1)^N\varepsilon(10M) \geqslant \varepsilon$. Now pick $\alpha_j > 0$ so that $\prod_{j=1}^N (1 - \alpha_j) = 1/10$. Then, since proposition $(10M, \varepsilon(10M))$ holds by Proposition 2.1.12, applying Proposition 2.1.13 N times we see that proposition (M, ε) holds.

We now sketch the proof of Proposition 2.1.13. We first note that it suffices to show that

$$\|N_\varphi(\nabla \mathbf{u})\|_{L^2(\partial D, d\sigma)} \leqslant C\|\tilde{N}_\varphi(\lambda(\operatorname{div}\mathbf{u})I + \mu\{\nabla\mathbf{u} + \nabla\mathbf{u}^t\})\|_{L^2(\partial D, d\sigma)},$$

where $\tilde{N}_\varphi$ is the nontangential maximal operator with a wider opening of the nontangential region. This follows because of classical arguments relating nontangential maximal functions to different openings (see [**18**], for example). Now pick φ with $|\,\|\nabla\varphi\|\,| \leqslant (1 - \alpha)M$ and such that there exists $\mathbf{a}$ with

$$|\,\|\nabla\varphi - \mathbf{a}\|\,| \leqslant 1.1\varepsilon, \qquad |\,\|\mathbf{a}\|\,| \leqslant (1 - \alpha)M.$$

We choose $\tilde{N}_\varphi$ as follows: Since $\partial D_\varphi \setminus B_\varphi$ is smooth, it is easy to see that we can find a finite number of coordinate charts (i.e., rotates, translates, and dilates of D_Ψ) that are entirely contained in D_φ such that their bottoms B_Ψ are contained in ∂D_φ, $T_\Psi((x, 0)\colon |\,\|x\|\,| < 1/2)$ cover ∂D_φ, and the Ψ's involved satisfy $|\,\|\nabla\Psi\|\,| \leqslant (1 - \alpha/2)M$, and there exist $\mathbf{a}_\Psi$ such that

$$|\,\|\mathbf{a}_\Psi\|\,| \leqslant (1 - \alpha/2)M \quad \text{and} \quad |\,\|\nabla\Psi - \mathbf{a}_\Psi\|\,| \leqslant 1.11\varepsilon.$$

The nontangential region defining $\tilde{N}_\varphi$ on $T_\Psi((x, 0)\colon |\,\|x\|\,| < 1/2)$ is defined as follows: let $F \subset \{(x, 0)\colon |\,\|x\|\,| < 1/2\}$ be a closed set. Consider the cone on $\mathbb{R}^n_+$, $\gamma = \{(x, y) \in \mathbb{R}^n_+\colon b|x| < y\}$, where b is a small constant. Now consider the domain D_F on $\mathbb{R}^n_+$ given by $D_F = \bigcup_{x \in F}((x, 0) + \gamma)$. Then D_F is the domain above the graph of a Lipschitz function θ for which $|\,\|\nabla\theta\|\,| \leqslant cb$ for some absolute constant c (independent of F). It is also easy to see that we can now take b so small, depending only on M and ε, such that $T_\Psi(D_F)$ is the domain above the graph of a Lipschitz function $\tilde{\Psi}$, with $\tilde{\Psi} \geqslant \Psi$, that satisfies

$$|\,\|\nabla\tilde{\Psi}\|\,| \leqslant (1 - \alpha/10)M, \qquad |\,\|\nabla\tilde{\Psi} - \mathbf{a}_\Psi\|\,| \leqslant 1.111\varepsilon.$$

The nontangential region defining $\tilde{N}_\varphi$ for $Q \in T_\Psi((x, 0)\colon |\,\|x\|\,| < 1/2)$ is then the image under T_Ψ of $(x, 0) + \gamma$, with b chosen as above, suitably truncated, where $Q = T_\Psi((x, 0))$. To simplify notation let

$$m = N_\varphi(\nabla\mathbf{u}), \qquad \overline{m} = \tilde{N}_\varphi(\lambda(\operatorname{div}\mathbf{u})I + \mu\{\nabla\mathbf{u} + \nabla\mathbf{u}^t\}).$$

For $t > 0$ consider the open set $E_t = \{m > t\}$. We now produce a Whitney-type decomposition of E_t into a family of disjoint sets $\{U_j\}$ with the property that each U_j is contained in $T_\Psi((x,0)\colon |\,\|x\|\,| < 1/2)$ for a coordinate chart D_Ψ, and each U_j contains $T_\Psi(I_j)$, where I_j is a cube in $|\,\|x\|\,| < 1/2$, and is contained in $T_\Psi(\bar{I}_j)$, where $\bar{I}_j$ is a fixed multiple of I_j. Finally, we can also assume that there exists a constant η_0 such that, if $\operatorname{diam}(U_j) \leqslant \eta_0$, there exists a point Q_j in ∂D_φ, with $\operatorname{dist}(Q_j, U_j) \approx \operatorname{diam} U_j$, such that $m(Q_j) \leqslant t$. Now let $\beta > 1$ be given. We claim that there exists $\delta > 0$ so small that if $E_j = U_j \cap \{m > \beta t, \bar{m} \leqslant \delta t\}$, then

$$\sigma(E_j) \leqslant (1 - \eta_M)\sigma(U_j),$$

where $\eta_M > 0$. Assume the claim for now. Then

$$\begin{aligned}\int_{\partial D_\varphi} m^2\, d\sigma &= 2\int_0^\infty t\sigma(E_t)\, dt = 2\beta^2\int_0^\infty t\sigma(E_{\beta t})\, dt \\ &= \sum_j 2\beta^2\int_0^\infty t\sigma(U_j \cap E_{\beta t})\, dt \\ &\leqslant \sum_j 2\beta^2\int_0^\infty t\sigma(E_j)\, dt + 2\beta^2\int_0^\infty t\sigma(\bar{m} > \delta t)\, dt \\ &\leqslant \sum_j 2\beta^2(1 - \eta_M)\int_0^\infty t\sigma(U_j)\, dt + 2\frac{\beta^2}{\delta^2}\int_0^\infty t\sigma\{\bar{m} > t\}\, dt \\ &= \beta^2(1 - \eta_M)\int_{\partial D_\varphi} m^2\, d\sigma + \frac{\beta^2}{\delta^2}\int_{\partial D_\varphi} \bar{m}^2\, d\sigma.\end{aligned}$$

Thus if we choose $\beta > 1$, but so that $\beta^2 \cdot (1 - \eta_M) < 1$, the desired result follows. It remains to establish the claim. We argue by contradiction. Suppose the claim is false; then $\sigma(E_j) > (1 - \eta_M)\sigma(U_j)$. Let $\tilde{E}_j = T_\Psi^{-1}(E_j)$. If η_M is chosen sufficiently small, we can guarantee that $|\tilde{E}_j \cap I_j| \geqslant .99|I_j|$. Now let $F_j = \tilde{E}_j \cap I_j$ and construct the Lipschitz function $\tilde{\Psi}$ corresponding to it, as in the definition of $\tilde{N}_\varphi$. Thus $\tilde{\Psi} \geqslant \Psi$, $|\,\|\nabla\tilde{\Psi}\|\,| \leqslant (1 - \alpha/10)M$, and $|\,\|\nabla\tilde{\Psi} - \mathbf{a}_\Psi\|\,| \leqslant 1.111\varepsilon$. We now apply Lemma 2.1.10 to $\tilde{\Psi}$, one variable at a time, to find a Lipschitz function f, with $f \geqslant \tilde{\Psi}$ on I_j, such that if $\bar{F}_j = \{x \in I_j\colon f = \tilde{\Psi}\}$, then $|\bar{F}_j \cap F_j| \geqslant c\sigma(U_j)$, with $|\,\|\nabla f\|\,| \leqslant (1 - \alpha/10)M$, and such that there exists $\mathbf{a}_f$, with $|\,\|\mathbf{a}_f\|\,| \leqslant (1 - \alpha/10)M$ so that

$$|\,\|\nabla f - \mathbf{a}_f\|\,| \leqslant \tfrac{4}{5}(1.111\varepsilon) < \varepsilon.$$

We can also arrange the truncation of our nontangential regions in such a way that on the appropriate rotate, translate, and dilate of D_f (which, of course, is contained in the corresponding coordinate chart associated to D_Ψ, which is contained in D_φ),

$$\left|\lambda(\operatorname{div}\mathbf{u})I + \mu\{\nabla\mathbf{u} + \nabla\mathbf{u}^t\}\right| \leqslant \delta t.$$

To lighten the exposition we still denote the translate, rotate, and dilate of D_f by D_f. Note that proposition (M, ε) applies to it. We divide the sets U_j into two types. Type I sets are those with $\operatorname{diam} U_j \geqslant \eta_0$; type II sets are those for which $\operatorname{diam} U_j \leqslant \eta_0$. We first deal with type I. In this case D_f has diameter of order 1. Because of the solvability of problem (2.1.2) for balls, and our normalization, we see that on a ball $B \subset D_\varphi$, $\operatorname{diam} B \approx 1$, $A_\varphi \in B$, we have

$$\int_B |\nabla \mathbf{u}|^2 \leqslant C \int_B \left|\lambda \operatorname{div} \mathbf{u} I + \mu\{\nabla \mathbf{u} + \nabla \mathbf{u}^t\}\right|^2.$$

Joining A_f to A_φ by a finite number of balls and using interior regularity results for the system $\mu \Delta \mathbf{u} + (\lambda + \mu)\nabla \operatorname{div} \mathbf{u} = 0$, we see that $|\nabla \mathbf{u}(A_f)| \leqslant C\delta t$ for some absolute constant C. Then

$$\begin{aligned} C\sigma(U_j)\beta^2 t^2 &\leqslant \int_{T_\Psi(\bar{F}_j \cap F_j)} m^2\, d\sigma \leqslant C \int_{\partial D_f} N_f^2(\nabla u)\, d\sigma \\ &\leqslant C\sigma(U_j)\delta^2 t^2 + C\int_{\partial D_f} N_f^2\left(\nabla \mathbf{u} - \frac{[\nabla \mathbf{u}(A_f) - \nabla \mathbf{u}^t(A_f)]}{2}\right) d\sigma \\ &\leqslant C\sigma(U_j)\delta^2 t^2 + C\int_{\partial D_f} N_f^2\left(\lambda(\operatorname{div} \mathbf{u})I + \mu\{\nabla \mathbf{u} + \nabla \mathbf{u}^t\}\right)^2 d\sigma, \end{aligned}$$

by (M, ε). The last quantity is also bounded by $C\sigma(U_j)\delta^2 t^2$, which is a contradiction for small δ. Now assume that U_j is on type II. Note that, in this case, there exists $Q_j \in \partial D_\varphi$ with $\operatorname{dist}(Q_j, U_j) \approx \operatorname{diam} U_j$, and $|\nabla \mathbf{u}(x)| \leqslant t$ for all x in the nontangential region associated to Q_j. Therefore, it is easy to see, using the arguments we used to bound $|\nabla \mathbf{u}(A_f)|$ in case I, that for all X in a neighborhood of A_f and also on the top part of D_f, we have $|\nabla \mathbf{u}(X)| \leqslant t + C\delta t$. Since, for $Q \in T_\Psi(\bar{F}_j \cap F_j)$, $m(Q) \geqslant \beta t$ and $\beta > 1$, if δ is small enough, we see that we must have $N_f(\nabla \mathbf{u})(Q) \geqslant m(Q)$. Hence,

$$N_f\left(\nabla \mathbf{u} - \left[\frac{\nabla \mathbf{u}(A_f) - \nabla \mathbf{u}^t(A_f)}{2}\right]\right)(Q) \geqslant (\beta - 1 - C\delta)t \geqslant \frac{\beta - 1}{2} t$$

if δ is small and $Q \in T_\Psi(\bar{F}_j \cap F_j)$. Thus, applying (M, ε) to D_f, we see that

$$\begin{aligned} C(\beta - 1)^2 t^2 \sigma(U_j) &\leqslant \int_{T_\Psi(\bar{F}_j \cap F_j)} N_f\left(\nabla \mathbf{u} - \left[\frac{\nabla \mathbf{u}(A_f) - \nabla \mathbf{u}^t(A_f)}{2}\right]\right)^2 d\sigma \\ &\leqslant \int_{\partial D_f} N_f\left(\nabla \mathbf{u} - \left[\frac{\nabla \mathbf{u}(A_f) - \nabla \mathbf{u}^t(A_f)}{2}\right]\right)^2 d\sigma \\ &\leqslant C\sigma(U_j)\delta^2 t^2, \end{aligned}$$

a contradiction if δ is small. This finishes the proofs of Proposition 2.1.13 and, hence, Lemma 2.1.11.

Part 2. *The Stokes system of linear hydrostatics.* In this part I will sketch the proof of the L^2 results for the Stokes system of hydrostatics. These results are joint work of E. Fabes, C. Kenig, and G. Verchota [**17**]. We keep the notation introduced in Part 1.

We seek a vector-valued function $\mathbf{u} = (u^1, u^2, u^3)$ and a scalar-valued function p satisfying

$$(2.2.1) \qquad \begin{cases} \Delta \mathbf{u} = \nabla p & \text{in } D, \\ \operatorname{div} \mathbf{u} = 0 & \text{in } D, \\ \mathbf{u}|_{\partial D} = \mathbf{f} \in L^2(\partial D, d\sigma) & \text{in the nontangential sense.} \end{cases}$$

THEOREM 2.2.2 (ALSO THEOREM 2.1.9). *Given* $\mathbf{f} \in L^2(\partial D, d\sigma)$, *there exists a unique solution* $(\mathbf{u}, p)$ *to* (2.2.1), *with p tending to* 0 *at* ∞, *and* $N(\mathbf{u}) \in L^2(\partial D, d\sigma)$. *Moreover*, $\mathbf{u}(X) = \mathscr{K}\mathbf{g}(X)$, *with* $\mathbf{g} \in L^2(\partial D, d\sigma)$. ($\mathscr{K}$ *will be defined below.*)

In order to sketch the proof of 2.2.2, we introduce the matrix $\Gamma(X) = (\Gamma_{ij}(X))$ of fundamental solutions (see the book of Ladyzhenskaya [**28**]), where

$$\Gamma_{ij}(X) = \frac{1}{8\pi}\frac{\delta_{ij}}{|X|} + \frac{1}{8\pi}\frac{X_i X_j}{|X|^3},$$

and its corresponding pressure vector

$$q(X) = \left(q^i(X)\right), \quad \text{where } q^i(X) = X_i/4\pi|X|^3.$$

Our solution of (2.2.2) will be given in the form of a double-layer potential

$$\mathbf{u}(X) = \mathscr{K}\mathbf{g}(X) = -\int_{\partial D} \{H'(Q)\Gamma(X-Q)\}\mathbf{g}(Q)\, d\sigma(Q),$$

where

$$(H'(Q)\Gamma(X-Q))_{il} = \delta_{ij}q^l(X-Q)n_j(Q) + (\partial\Gamma_{il}/\partial Q_j)(X-Q)n_j(Q).$$

We also use the single-layer potential

$$\mathbf{u}(X) = S\mathbf{g}(X) = \int_{\partial D} \Gamma(X-Q)\mathbf{g}(Q)\, d\sigma(Q).$$

In the same way as one establishes 2.1.4, we have

LEMMA 2.2.3. *Let* $\mathscr{K}\mathbf{g}$, $S\mathbf{g}$ *be defined as above, with* $\mathbf{g} \in L^2(\partial D, d\sigma)$. *Then they both solve* $\Delta\mathbf{u} = \nabla p$ *in D and* D_-, $\operatorname{div}\mathbf{u} = 0$ *in D and* D_-. *Also,*

(a) $\|N(\mathscr{K}\mathbf{g})\|_{L^2(\partial D, d\sigma)} \leqslant C\|\mathbf{g}\|_{L^2(\partial D, d\sigma)}$,

(b) $(\mathscr{K}\mathbf{g})^{\pm}(P) = \pm\frac{1}{2}\mathbf{g}(P) - \text{p.v.} \int_{\partial D} \{H'(Q)\Gamma(P-Q)\}\mathbf{g}(Q)\, d\sigma(Q)$,

(d) $\|N(\nabla S\mathbf{g})\|_{L^2(\partial D, d\sigma)} \leqslant C\|\mathbf{g}\|_{L^2(\partial D, d\sigma)}$,

(d) $$\left(\frac{\partial}{\partial X_i}(S\mathbf{g})_j\right)^{\pm}(P) = \pm\left\{\frac{n_i(P)g_j(P)}{2} - \frac{n_i(P)n_j(P)}{2}\langle N_P, \mathbf{g}(P)\rangle\right\} + \text{p.v.} \int_{\partial D} \frac{\partial}{\partial p_i}\Gamma(P, Q)\mathbf{g}(Q)\, d\sigma(Q),$$

(e) $(HS\mathbf{g})^{\pm}(P) = \pm\frac{1}{2}\mathbf{g}(P) + \text{p.v.} \int_{\partial D} \{H(P)\Gamma(P-Q)\}\mathbf{g}(Q)\, d\sigma(Q)$, *where*

$$(H(X)\Gamma(X-Q))_{il} = n_j(x)\frac{\partial \Gamma_{il}}{\partial X_j}(X-Q) - \delta_{ij}q^l(X-Q)n_j(X).$$

For the proof in the case of smooth domains, see [**28**].

The proof of Theorem 2.2.2 (at least the existence part of it) reduces to the invertibility in $L^2(\partial D, d\sigma)$ of the operator $\frac{1}{2}I + K$, where

$$K\mathbf{g}(P) = -\text{p.v.}\int_{\partial D}\{H'(Q)\Gamma(P-Q)\}\mathbf{g}(Q)\,d\sigma(Q).$$

As in previous cases, it is enough to show

(2.2.4) $$\|(\tfrac{1}{2}I - K^*)\mathbf{g}\|_{L^2(\partial D,\, d\sigma)} \approx \|(\tfrac{1}{2}I + K^*)\mathbf{g}\|_{L^2(\partial D,\, d\sigma)}.$$

This is shown by using the following two integral identities:

LEMMA 2.2.5. *Let* $\mathbf{h}$ *be a constant vector in* $\mathbb{R}^n$, *and suppose that* $\Delta \mathbf{u} = \nabla p$, $\operatorname{div}\mathbf{u} = 0$, D, *and* $\mathbf{u}$, p *and their derivatives are suitably small at* ∞. *Then*

$$\int_{\partial D} h_l n_l \frac{\partial u^s}{\partial X_j}\cdot\frac{\partial u^s}{\partial X_j}\,d\sigma = 2\int_{\partial D}\frac{\partial u^s}{\partial N}\cdot h_l\frac{\partial u^s}{\partial X_l}\,d\sigma - 2\int_{\partial D} p n_s h_l \frac{\partial u^s}{\partial X_l}\,d\sigma.$$

LEMMA 2.2.6. *Let* $\mathbf{h}$, p *and* $\mathbf{u}$ *be as in* 2.2.5. *Then*

$$\int_{\partial D} h_l n_l p^2\,d\sigma = 2\int_{\partial D} h_r \frac{\partial u^r}{\partial N}p\,d\sigma - 2\int_{\partial D} h_r\frac{\partial u^r}{\partial X_i}\frac{\partial u^i}{\partial N}\,d\sigma + 2\int_{\partial D} h_r n_s \frac{\partial u^s}{\partial X_j}\frac{\partial u^r}{\partial X_i}\,d\sigma.$$

The proofs of 2.2.5 and 2.2.6 are simple applications of the properties of $\mathbf{u}$, p and the divergence theorem.

Choosing $\mathbf{h} = e_3$, we see that, from 2.2.6, we obtain

COROLLARY 2.2.7. *Let* $\mathbf{u}$, p *be as in* 2.2.5. *Then* $\int_{\partial D} p^2\,d\sigma \leqslant C\int_{\partial D}|\nabla \mathbf{u}|^2\,d\sigma$, *where* C *depends only on* M.

A consequence of Corollary 2.2.7 and Lemma 2.2.5 is that if $\partial\mathbf{u}/\partial\nu = \partial\mathbf{u}/\partial N - p\cdot N$, then we have

COROLLARY 2.2.8. *Let* $\mathbf{u}$, p *be as in* 2.2.5. *Then*

$$\int_{\partial D}\left|\frac{\partial \mathbf{u}}{\partial \nu}\right|^2 d\sigma \approx \int_{\partial D}|\nabla_t \mathbf{u}|^2\,d\sigma + \sum_j\int_{\partial D}\left|n_s\frac{\partial u^s}{\partial X_j}\right|^2 d\sigma,$$

where the constants of equivalence depend only on M.

PROOF. Lemma 2.2.5 clearly implies, by Schwartz's inequality, that

$$\int_{\partial D}|\nabla \mathbf{u}|^2\,d\sigma \leqslant C\int_{\partial D}\left|\frac{\partial \mathbf{u}}{\partial \nu}\right|^2 d\sigma.$$

Moreover, arguing as in the second part of Remark 2 after 2.1.5, we see that 2.2.5 shows that

$$\int_{\partial D} |\nabla \mathbf{u}|^2 \, d\sigma \leqslant C \int_{\partial D} |\nabla_t \mathbf{u}|^2 \, d\sigma + \left| \int_{\partial D} p n_s h_l \frac{\partial u^s}{\partial X_l} \, d\sigma \right|.$$

By Corollary 2.2.7 the right side is bounded by

$$C\left(\int_{\partial D} |\nabla \mathbf{u}|^2 \, d\sigma \right)^{1/2} \left(\sum_j \int_{\partial D} \left| n_s \frac{\partial u^s}{\partial X_j} \right|^2 d\sigma \right)^{1/2} + c\int_{\partial D} |\nabla_t u|^2 \, d\sigma.$$

Corollary 2.2.8 now follows, using 2.2.7 once more.

To prove 2.2.4 let $\mathbf{u} = S(\mathbf{g})$. By 2.2.3(d), $\nabla_t \mathbf{u}$ and $n_s(\partial u^s/\partial X_j)$ are continuous across ∂D. Using this fact, 2.2.3(e), and Corollary 2.2.8, (2.2.4) follows.

In closing this part we would like to point out another boundary-value problem for the Stokes system that is of physical significance: the so-called slip boundary condition

$$(2.2.9) \qquad \begin{cases} \Delta \mathbf{u} = \nabla p & \text{in } D, \\ \operatorname{div} \mathbf{u} = 0 & \text{in } D, \\ \left((\nabla \mathbf{u} + \nabla \mathbf{u}^t) N - p \cdot N \right)\big|_{\partial D} = \mathbf{f} \in L^2(\partial D, d\sigma). \end{cases}$$

This problem is very similar to (2.1.2). Using the techniques introduced in Part 1, together with the observation that if $\Delta \mathbf{u} = \nabla p$, $\operatorname{div} \mathbf{u} = 0$ in D, the same is true for each row $\mathbf{v}$ of the matrix $[\nabla \mathbf{u} + \nabla \mathbf{u}^t - pI]$, we have obtained

THEOREM 2.2.10. *Given* $\mathbf{f} \in L^2(\partial D, d\sigma)$ *there exists a unique solution* $(\mathbf{u}, p)$ *to* (2.2.9) *tending to* 0 *at* ∞, *with* $N(\nabla \mathbf{u}) \in L^2(\partial D, d\sigma)$. *Moreover*, $\mathbf{u}(X) = S(\mathbf{g})(X)$, *with* $\mathbf{g} \in L^2(\partial D, d\sigma)$.

Part 3. *The Dirichlet problem for the biharmonic equation on Lipschitz domains.* This part deals with the Dirichlet problem for Δ^2 on an arbitrary Lipschitz domain in $\mathbb{R}^n$. The results are joint work of B. Dahlberg, C. Kenig, and G. Verchota [**11**]. We continue using the notation previously introduced.

We seek a function u defined in D such that

$$(2.3.1) \qquad \begin{cases} \Delta^2 u = 0 & \text{in } D, \\ u|_{\partial D} = f \in L_1^2(\partial D, d\sigma), \\ \partial u/\partial N|_{\partial D} = g \in L^2(\partial D, d\sigma), \end{cases}$$

where the boundary values are taken nontangentially a.e.

THEOREM 2.3.2. *There exists a unique u solving* (2.3.1), *with* $N(\nabla u) \in L^2(\partial D, d\sigma)$ *and*

$$\|N(\nabla u)\|_{L^2(\partial D, d\sigma)} \leqslant C\left\{ \|g\|_{L^2(\partial D, d\sigma)} + \|f\|_{L_1^2(\partial D, d\sigma)} \right\},$$

where C depends only on M.

We only discuss existence. By 1.1.14 we may assume $f = 0$ on ∂D. Let $G(X, Y)$ be the Green function for Δ on D. Then, since $u|_{\partial D} = 0$, we have $u(X) = \int_D G(X, Y)\Delta u(y)\,dy$. Notice that $w(y) = \Delta u(y)$ is harmonic in D. We claim that $w(Y) = (\partial/\partial y)v(Y)$, where v is a harmonic function in D with $L^2(\partial D, d\sigma)$ Dirichlet data, and that the operator T: $v|_{\partial D} \to \partial u/\partial N|_{\partial D}$ is an invertible map from $L^2(\partial D, d\sigma)$ onto $L^2(\partial D, d\sigma)$. This would establish 2.3.2. In fact, by using the Green's potential representation, Fubini's theorem, and the fact that $(\partial/\partial N)G(-, Y)$ is the density of harmonic measure at $Y \in D$,

$$\int_{\partial D} vTv\,d\sigma = \int_D v(Y)\frac{\partial}{\partial y}v(Y)\,dY = \frac{1}{2}\int_{\mathbb{R}^{n-1}} v(x, \varphi(x))^2\,dx \geqslant C\int_{\partial D} v^2\,d\sigma.$$

This shows that if T: $L^2(\partial D, d\sigma) \to L^2(\partial D, d\sigma)$ is bounded, it will have a bounded inverse. To establish the boundedness of T, note that if h is harmonic in D, then the argument given above shows that

$$\int_{\partial D} hTv\,d\sigma = \int_D \frac{\partial v}{\partial y}(Y)h(Y)\,dY.$$

All we need, therefore, is the following bilinear estimate.

THEOREM 2.3.3. *If v, h are harmonic in D and tend to 0 at ∞, then*

$$\left|\int_D \frac{\partial v}{\partial y}(Y)\cdot h(Y)\,dY\right| \leqslant C\|v\|_{L^2(\partial D, d\sigma)}\cdot\|h\|_{L^2(\partial D, d\sigma)}.$$

PROOF. This theorem is a generalization to Lipschitz domains of the fact that the paraproduct of two L^2 functions is in L^1 (see [**4**]).

To establish the inequality we can assume, due to the invertibility of the double-layer potential (the representation formula in Theorem 1.1.1), that

$$h(Y) = \frac{1}{\omega_n}\int_{\partial D} \frac{\langle Y - Q, N_Q\rangle}{|Y - Q|^{n-2}} g(Q)\,d\sigma(Q),$$

with

$$\|g\|_{L^2(\partial D, d\sigma)} \leqslant C\|h\|_{L^2(\partial D, d\sigma)}.$$

Thus, since

$$\frac{\langle Y - Q, N_Q\rangle}{|Y - Q|^{n-2}} = C_n\frac{\partial}{\partial N_Q}\left(\frac{1}{|Y - Q|^{n-2}}\right),$$

it suffices to show that

$$\left\|\frac{\partial}{\partial N_Q}\int_D \frac{1}{|Y - Q|^{n-2}}\frac{\partial v}{\partial y}(Y)\,dY\right\|_{L^2(\partial D, d\sigma)} \leqslant C\|v\|_{L^2(\partial D, d\sigma)}.$$

In order to do so we obtain a representation formula for

$$\frac{\partial}{\partial N_Q}\int_D \frac{1}{|Y - Q|^{n-2}}\frac{\partial v}{\partial y}(Y)\,dY.$$

Fix $Q \in \partial D$ and let B satisfy $\Delta_Y B(Y-Q) = 1/|Y-Q|^{n-2}$; i.e., B is the fundamental solution for Δ^2 (for example, if $n \geqslant 5$, $B(Y) = C_n|Y|^{4-n}$). We recall the definition of the Riesz transforms $v_j = R_j v$, $j = 1, \ldots, n-1$. They are harmonic functions that together with v satisfy the generalized Cauchy–Riemann equations (see **[35]**); i.e.,

$$\frac{\partial v}{\partial X_j} = \frac{\partial}{\partial y} R_j v \quad \text{and} \quad \frac{\partial v}{\partial y} = -\sum_{j=1}^{n-1} \frac{\partial}{\partial x_j} R_j v.$$

If $Y = (x, y)$, then

$$\frac{1}{|Y-Q|^{n-2}} \frac{\partial}{\partial y} v(Y) = \Delta_Y B \frac{\partial v}{\partial y}.$$

Using the summation convention, we have

$$\begin{aligned}\Delta_Y B \frac{\partial v}{\partial y} &= \left(\frac{\partial^2}{\partial x_j^2} B + \frac{\partial^2}{\partial y^2} B \right) \frac{\partial}{\partial y} v \\ &= \frac{\partial^2}{\partial x_j^2} B \frac{\partial}{\partial y} v - \frac{\partial^2 B}{\partial x_j \partial y} \frac{\partial v}{\partial y} + \frac{\partial^2}{\partial x_j \partial y} B \frac{\partial R_j v}{\partial y} - \frac{\partial^2 B}{\partial y^2} \cdot \frac{\partial}{\partial x_j} R_j v.\end{aligned}$$

Now let $e_1, e_2, \ldots, e_{n-1}, e_n$ be the standard basis of $\mathbb{R}^n$, with e_n pointing in the direction of the y-axis. Then we can rewrite the right side as

$$\left\langle \left(-\frac{\partial^2}{\partial x_1 \partial y} B, \frac{-\partial^2 B}{\partial x_2 \partial y}, \ldots, \frac{-\partial^2 B}{\partial x_{n-1} \partial y}, \sum_{j=1}^{n-1} \frac{\partial^2 B}{\partial x_j^2} \right), \nabla v \right\rangle$$
$$+ \sum_{j=1}^{n-1} \left\langle \frac{\partial^2 B}{\partial x_j \partial y} e_n, \nabla R_j v \right\rangle - \sum_{j=1}^{n-1} \left\langle \frac{\partial^2 B}{\partial y^2} e_j, \nabla R_j v \right\rangle.$$

Let

$$\boldsymbol{\alpha} = \left(-\frac{\partial^2 B}{\partial x_1 \partial y}, \frac{-\partial^2 B}{\partial x_2 \partial y}, \ldots, \frac{-\partial^2 B}{\partial x_{n-1} \partial y}, \sum_{j=1}^{n-1} \frac{\partial^2 B}{\partial x_j^2} \right), \qquad \boldsymbol{\beta}_j = \frac{\partial^2 B}{\partial x_j \partial y} e_n - \frac{\partial^2 B}{\partial y^2} e_j.$$

Note that $\operatorname{div} \boldsymbol{\beta}_j = 0$, $\operatorname{div} \boldsymbol{\alpha} = 0$, and

$$\begin{aligned}\int_D \frac{1}{|Y-Q|^{n-2}} \frac{\partial}{\partial y} v(Y)\, dY &= \int_D \langle \boldsymbol{\alpha}, \nabla v \rangle + \sum_{j=1}^{n-1} \int_D \langle \boldsymbol{\beta}_j, \nabla R_j v \rangle \\ &= \int_{\partial D} v(P) \cdot \langle \boldsymbol{\alpha}(P), N_P \rangle \, d\sigma(P) + \sum_{j=1}^{n-1} \int_{\partial D} R_j v(P) \cdot \langle \boldsymbol{\beta}_j(P), N_P \rangle \, d\sigma(p)\end{aligned}$$

by the divergence theorem. This can be rewritten as

$$\int_{\partial D} \left[-n_j(P) \frac{\partial}{\partial P_j} \frac{\partial}{\partial P_n} B(P-Q) + n_n(P) \frac{\partial}{\partial P_j} \frac{\partial}{\partial P_j} B(P-Q) \right] v(P)\, d\sigma(P)$$
$$+ \sum_{j=1}^{n-1} \int_{\partial D} \left[n_n(P) \frac{\partial}{\partial P_j} \frac{\partial}{\partial P_n} B(P-Q) - n_j(P) \frac{\partial^2}{\partial P_n^2} B(P-Q) \right] R_j v(P)\, d\sigma(P).$$

Hence,

$$\frac{\partial}{\partial N_Q}\int_D \frac{1}{|Y-Q|^{n-2}}\frac{\partial}{\partial y}v(Y)\,dY$$

$$= \int_{\partial D}\left[-n_j(P)\frac{\partial}{\partial P_j}\frac{\partial}{\partial P_n}\langle \nabla B(P-Q), N_Q\rangle + n_n(P)\frac{\partial^2}{\partial P_n^2}\langle \nabla B(P-Q), N_Q\rangle\right]v(P)\,d\sigma(P)$$

$$+\sum_{j=1}^{n-1}\int_{\partial D}\left[n_n(P)\frac{\partial}{\partial P_j}\frac{\partial}{\partial P_n}\langle \nabla B(P-Q), N_Q\rangle - n_j(P)\frac{\partial^2}{\partial P_n^2}\langle \nabla B(P-Q), N_Q\rangle\right]R_jv(P)\,d\sigma(P).$$

But, by the Coifman–McIntosh–Meyer theorem [**3**],

$$\frac{\partial}{\partial P_j}\frac{\partial}{\partial P_i}\frac{\partial}{\partial Q_k}B(P-Q)$$

is the kernel of a bounded operator in $L^2(\partial D, d\sigma)$. Thus,

$$\left\|\frac{\partial}{\partial N_Q}\int_D \frac{1}{|Y-Q|^{n-2}}\frac{\partial v}{\partial y}(Y)\,dY\right\| \leqslant C\left\{\|v\|_{L^2(\partial D,d\sigma)} + \sum_{j=1}^{n-1}\|R_jv\|_{L^2(\partial D,d\sigma)}\right\}.$$

Finally, we invoke the result of Dahlberg [**8**] that

$$\|R_jv\|_{L^2(\partial D,d\sigma)} \leqslant C\|v\|_{L^2(\partial D,d\sigma)}.$$

This concludes the proof of Theorem 2.3.3.

As a final comment we would like to point out that in this exposition we have emphasized nontangential maximal function estimates, but that optimal Sobolev space estimates also hold. For example, the solution **u** of (2.1.1) is in the Sobolev space $H^{1/2}(D)$, that of (2.1.2) in the Sobolev space $H^{3/2}(D)$, and the same is true for **u** in 2.1.3(c). The solution of (2.2.1) is in $H^{1/2}(D)$, while that of (2.2.9) is in $H^{3/2}(D)$. Finally, the solution u of (2.3.1) is in $H^{3/2}(D)$. All of these results can be proved in a unified fashion, using a variant of the proof of Lemma 2.1.11. The details will appear in a forthcoming paper of Dahlberg and Kenig [**10**].

REFERENCES

1. A. P. Calderon, *Cauchy integrals on Lipschitz curves and related operators*, Proc. Nat. Acad. Sci. U.S.A. **74** (1977), 1324–1327.

2. J. Cohen and J. Gosselin, *The Dirichlet problem for the biharmonic equation in a bounded C^1 domain in the plane*, Indiana Univ. Math. J. **32** (1983), 635–685.

3. R. R. Coifman, A. McIntosh and Y. Meyer, *L'integrale de Cauchy definit un operateur borne sur L^2 pour les courbes lipschitziennes*, Ann. of Math. (2) **116** (1982), 361–387.

4. R. R. Coifman and Y. Meyer, *Au-delà des opérateurs pseudo-différentials*, Astérisque **57** (1978).

5. R. R. Coifman and G. Weiss, *Extensions of Hardy spaces and their use in analysis*, Bull. Amer. Math. Soc. **83** (1977), 569–645.

6. B. E. J. Dahlberg, *On estimates of harmonic measure*, Arch. Rational Mech. Anal. **65** (1977), 272–288.

7. ______, *On the Poisson integral for Lipschitz and C^1 domains*, Studia Math. **66** (1979), 13–24.

8. ______, *Weighted norm inequalities for the Lusin area integral and the non-tangential maximal functions for functions harmonic in a Lipschitz domain*, Studia Math. **67** (1980), 297–314.

9. B. E. J. Dahlberg and C. E. Kenig, *Hardy spaces and the L^p Neumann problem for Laplace's equation in a Lipschitz domain* (in preparation).

10. ______, *Area integral estimates for higher order boundary value problems on Lipschitz domains* (to appear).

11. B. E. J. Dahlberg, C. E. Kenig and G. C. Verchota, *The Dirichlet problem for the biharmonic equation in a Lipschitz domain* (in preparation).

12. ______, *Boundary value problems for the systems of elastostatics on a Lipschitz domain* (in preparation).

13. G. David, *Operateurs integraux singuliers sur certaines courbes du plan complex*, Ann. Sci. École Norm. Sup **17** (1984), 157–189.

14. ______, Personal communication, 1983.

15. E. Fabes, M. Jodeit, Jr. and N. Riviere, *Potential techniques for boundary value problems on C^1 domains*, Acta. Math. **141** (1978), 165–186.

16. E. Fabes and C. E. Kenig, *On the Hardy space H^1 of a C^1 domain*, Ark. Mat. **19** (1981), 1–22.

17. E. Fabes, C. E. Kenig and G. C. Verchota, *The Stokes system on a Lipschitz domain* (in preparation).

18. C. Fefferman and E. Stein, *H^p spaces of several variables*, Acta. Math. **129** (1972), 137–193.

19. A. Gutierrez, *Boundary value problems for linear elastostatics on C^1 domains*, Contributions to Nonlinear Partial Differential Equations, Res. Notes in Math., vol. 89, Pitman, 1983, pp. 90–97.

20. D. S. Jerison and C. E. Kenig, *An identity with applications to harmonic measure*, Bull. Amer. Math. Soc. (N.S.) **2** (1980), 447–451.

21. ______, *The Dirichlet problem in nonsmooth domains*, Ann. of Math. (2) **113** (1981), 367–382.

22. ______, *The Neumann problem on Lipschitz domains*, Bull. Amer. Math. Soc. **4** (1981), 203–207.

23. D. S. Jerison and C. E. Kenig, *Boundary value problems on Lipschitz domains*, Studies in Partial Differential Equations (W. Littmann, ed.), MAA Studies in Mathematics, Vol. 23, Math. Assoc. Amer., Washington, D. C., 1982, 1–68.

24. C. E. Kenig, *Weighted H^p spaces on Lipschitz domains*, Amer. J. Math. **102** (1980), 129–163.

25. ______, *Weighted Hardy spaces on Lipschitz domains*, Proc. Sympos. Pure Math., vol. 35, part 1, Amer. Math. Soc., Providence, R. I., 1979, pp. 263–274.

26. ______, *Boundary value problems of linear elastostatics and hydrostatics on Lipschitz domains*, Seminaire Goulaouic–Meyer–Schwartz, 1983–1984, Exposé no. XXI, École Polytechnique, Palaiseau, France.

27. V. D. Kupradze, *Three dimensional problems of the mathematical theory of elasticity and thermoelasticity*, North-Holland, New York, 1979.

28. C. A. Ladyzhenskaya, *The mathematical theory of viscous incompressible flow*, Gordon and Breach, New York, 1963.

29. J. Moser, *On Harnack's theorem for elliptic differential equations*, Comm. Pure Appl. Math., **14** (1961), 577–591.

30. J. Nečas, *Les methodes directes en theorie des equations elliptiques*, Academia, Prague, 1967.

31. L. Payne and H. Weinberger, *New bounds in harmonic and biharmonic problems*, J. Math. Phys. **33** (1954), 291–307.

32. ______, *New bounds for solutions of second order elliptic partial differential equations*, Pacific J. Math. **8** (1958), 551–573.

33. F. Rellich, *Darstelling der eigenwerte von $\Delta u = \lambda u$ durch ein randintegral*, Math. Z. **46** (1940), 635–646.

34. J. Serrin and H. Weinberger, *Isolated singularities of solutions of linear elliptic equations*, Amer. J. Math. **88** (1966), 258–272.

35. E. Stein and G. Weiss, *On the theory of harmonic functions of several variables*. I, Acta Math. **103** (1960), 25–62.

36. G. C. Verchota, *Layer potentials and boundary value problems for Laplace's equation in Lipschitz domains*, Thesis, Univ. of Minnesota, 1982, J. Funct. Anal. **59** (1984), 572–611.

UNIVERSITY OF MINNESOTA

Proceedings of Symposia in Pure Mathematics
Volume **43** (1985)

Estimates for $\bar{\partial}_b$ on Pseudoconvex CR Manifolds

J. J. KOHN[1]

1. The 3-dimensional case. Let M be a compact CR manifold with $\dim_{\mathbb{R}} M = 3$. The CR structure is given by a subbundle $T^{1,0}(M)$ of the complexified tangent bundle $\mathbb{C}T(M)$ such that

(a) $\dim_{\mathbb{C}} T_x^{1,0}(M) = 1$, where $T_x^{1,0}(M)$ is the fiber of $\mathbb{C}T(M)$ over $x \in M$;

(b) $T^{1,0}(M) \cap \overline{T^{1,0}(M)} = \{0\}$.

We set $T^{0,1}(M) = \overline{T^{1,0}(M)}$, and denote by $B^{0,1}(M)$ the dual bundle of $T^{0,1}(M)$, and by $T_U^{0,1}$, $T_U^{0,1}$, and $B_U^{0,1}$ the spaces of C^∞ sections over an open set $U \subset M$ of $T^{0,1}(M)$, $T^{0,1}(M)$, and $B^{0,1}(M)$, respectively. We define $\bar{\partial}_b$: $C^\infty(U) \to B_U^{0,1}$ by

$$\langle \bar{\partial}_b u, \bar{L} \rangle = \bar{L}u, \tag{1.1}$$

where $L \in T_U^{0,1}$, $u \in C^\infty(U)$, and $\langle \ , \ \rangle$ is the pairing between $T_U^{0,1}$ and $B_U^{0,1}$. If L does not vanish at any point of U, we let ω be the dual to L; i.e., $\langle \omega, L \rangle = 1$, and we have

$$\bar{\partial}_b u = \bar{L}(u)\bar{\omega}. \tag{1.1$'$}$$

The subbundle of 1-forms which annihilate $T^{1,0}(M) \oplus T^{0,1}(M)$ has fibers of complex dimension one and is closed under conjugation. Thus, in a neighborhood U of $x_0 \in M$ we can choose a real 1-form γ that does not vanish and annihilates $T_U^{1,0} \oplus T_U^{0,1}$. The quadratic form on $T_{x_0}^{1,0}(M)$ defined by

$$\mathscr{L}_{x_0}(L, L') = \sqrt{-1}\left\langle (d\gamma)_{x_0}, L \wedge \bar{L}' \right\rangle \tag{1.2}$$

is called the *Levi form*. M is called *pseudoconvex* (*strongly pseudoconvex*) if γ can be chosen so that the Levi form is nonnegative (positive definite). (In the 3-dimensional case the Levi form is a scalar.)

On M we define a Riemannian metric which induces a hermitian metric on $T_p^{1,0} \oplus T_p^{0,1}$. Let dV denote the volume element by this metric, and let $L_2(M)$ and

---

1980 *Mathematics Subject Classification*. Primary 35N15, 58G30.

[1] The research for this paper was done while the author was partially supported by NSF grant DMS 821307.

$L_2^{0,1}(M)$ denote the Hilbert spaces of square-integrable functions and sections of $B^{0,1}(M)$, respectively. We also denote by $\bar\partial_b$ the L_2-closure of $\bar\partial_b$, by $\bar\partial_b^*$ the L_2-adjoint of $\bar\partial_b$, by $\mathscr{H}_b$ the null space of $\bar\partial_b$ (i.e., the square-integrable CR functions), and by S: $L_2(M) \to \mathscr{H}_b$ the orthogonal projection operator. Further (as in [**K1**]), on functions we define the Laplacian by $\Box_b = \bar\partial_b^*\bar\partial_b$.

1.3. THEOREM. *Suppose M is a compact CR manifold of dimension 3 with the following properties*:

(a) *the range of $\bar\partial_b$ is closed*;

(b) *M is pseudoconvex*;

(c) *the Levi form is positive at $x_0 \in M$.*

If U is a neighborhood of x_0, on whose closure the Levi form is positive, if $u \perp \mathscr{H}_b$, and if $\bar\partial_b u|_U \in C^\infty(U)$, then $u|_U \in C^\infty(U)$. More precisely, if $\zeta, \zeta' \in C_0^\infty(U)$ such that $\zeta' = 1$ on a neighborhood of the support of ζ, then for each $s \in \mathbb{R}$, $s \geqslant 0$, there exists $C_s > 0$ such that if $\zeta'\bar\partial_b u \in H_s^{0,1}$ then $\zeta u \in H_{s+1/2}$ and

$$\|\zeta u\|_{s+1/2} \leqslant C_s\left(\|\zeta'\bar\partial_b u\|_s + \|\bar\partial_b u\|\right). \tag{1.4}$$

Here, $H_s^{0,1}$ and H_s denote the Sobolev spaces based on sections of $B^{0,1}(M)$ and $C^\infty(M)$, respectively, and $\| \quad \|_s$ denotes the Sobolev norm.

In [**BM**] Boutet de Monvel proves that if M is a compact strongly pseudoconvex CR manifold and if $\dim_{\mathbb{R}} M > 3$, then, for some N, M can be embedded in $\mathbb{C}^N$. In [**B**] D. Burns shows that if the operator $\bar\partial_b u \mapsto u$, with $u \perp \mathscr{H}_b$, is continuous in the C^∞-topology and M is strongly pseudoconvex with $\dim M = 3$, then Boutet de Monvel's construction works, and M can also be embedded. By an embedding of M we mean that there exists a diffeomorphism such that the CR structure induced on the image is the same as the given CR structure.

COROLLARY. *If M is a strongly pseudoconvex compact CR manifold such that* $\dim M = 3$ *and the range of $\bar\partial_b$ is closed in $L_2(M)$, then M is embeddable in $\mathbb{C}^N$.*

The hypothesis of closed range is necessary, according to a result of H. Rossi [**R**]. He proves that if $S^3 = \{z \in \mathbb{C}^2 \mid |z| = 1\}$ and, for each small t, we let

$$L_t = \bar z_2\frac{\partial}{\partial z_1} - \bar z_1\frac{\partial}{\partial z_2} + t\left(z_2\frac{\partial}{\partial \bar z_1} - z_1\frac{\partial}{\partial \bar z_2}\right), \tag{1.5}$$

then the CR manifold M_t, which is S^3 with the CR structure given by (1.5), is not embeddable whenever $t \neq 0$. Since M_t is strongly pseudoconvex, we conclude the following.

COROLLARY. *If $t \neq 0$, then the operator $\bar\partial_b$ on $L_2(M_t)$ does not have a closed range.*

Recently, M.-C. Shaw (see [**S**]) showed that if $\dim_{\mathbb{R}} M > 3$ and M is the boundary of a pseudoconvex domain $\Omega \subset \mathbb{C}^n$, then the range of $\bar\partial_b$ is closed in L_2.

Before proceeding with the proof of Theorem 1.3, we define the notion of a regularizing family of zero-order pseudodifferential operators, which will be useful in our proof. Let $\mathscr{P}$ be the set of all zero-order pseudodifferential operators on M such that, if $P \in \mathscr{P}$ and p is the symbol of P, then

$$\overline{\{x \in U \mid \text{there exists } \xi \text{ with } p(x, \xi) \neq 0\}} \subset U.$$

Let $\mathscr{A} \subset \mathscr{P}$. A *regularizing family in* $\mathscr{A}$ is a set $\{P_\delta\}$, with $P_\delta \in \mathscr{A}$ for each $\delta \in [0, 1]$, such that, if p_δ denotes the symbol of P_δ, we have the following:

(a) There exists $f \in C\,(0, 1\,]$, $f > 0$, satisfying $\lim_{\delta \to 0} f(\delta) = \infty$ such that, for $\delta > 0$,

$$\operatorname{supp} p_\delta \subset \{(x, \xi) \mid |\xi| \leqslant f(\delta)\}.$$

(b) P_δ is of order zero uniformly in δ; that is,

$$\left|D_x^\alpha D_\xi^\beta p_\delta(x, \xi)\right| \leqslant C_{\alpha\beta}(1 + |\xi|)^{-|\beta|},$$

where $C_{\alpha\beta}$ is independent of δ.

If $\{P_\delta\}$ and $\{P'_\delta\}$ are regularizing families, we say that $\{P'_\delta\}$ *dominates* $\{P_\delta\}$ if $p'_\delta(x, \xi) = 1$ on a neighborhood of $\operatorname{supp} p_\delta$.

PROOF OF THEOREM 1.3. To prove the theorem it suffices to prove that (1.4) holds on some small neighborhood U of x^0. Let $L \in T_U^{1,0}$, with $|L| = 1$; choose $T \in \mathbf{C}T_U$ so that L, $\bar{L}$, T is an orthonormal basis of $\mathbf{C}T_x$ for all $x \in U$ and T is normalized so that $\bar{T} = -T$. It then follows that, if $\lambda = \mathscr{L}(L, \bar{L})$,

$$[L, \bar{L}] = \lambda T + aL + b\bar{L}.$$

On U we choose coordinates x_1, x_2, x_3, with origin at x^0, so that

$$\partial/\partial x_3 = \sqrt{-1}\, cT,$$

where $c \in C^\infty(U)$, $c > 0$, and

$$L|_{x^0} = \frac{1}{2}\left(\frac{\partial}{\partial x_1} - \sqrt{-1}\,\frac{\partial}{\partial x_2}\right).$$

Correspondingly, we denote the dual coordinates by ξ_1, ξ_2, ξ_3. Then $\sigma(L)$, $\sigma(\bar{L})$, and $\sigma(T)$, the symbols of L, $\bar{L}$, and T, are given by

$$\sigma(L) = \frac{1}{2}\left(\sqrt{-1}\,\xi_1 + \xi_2\right) + \sum_1^3 g_j \xi_j, \tag{1.6}$$

$$\sigma(\bar{L}) = \frac{1}{2}\left(\sqrt{-1}\,\xi_1 - \xi_2\right) + \sum_1^3 \bar{g}_j \xi_j,$$

and

$$\sigma(T) = (1/c)\xi_3,$$

where $c, g_j \in C^\infty(U)$ and $g_j(0) = 0$.

We define

$$\mathscr{P}^0 = \left\{P \in \mathscr{P} \mid p(x, \xi) = 0 \text{ when } \xi_3^2 \geqslant 2\left(\xi_1^2 + \xi_2^2\right) \text{ and } |\xi| > 1\right\},$$

$$\mathscr{P}^+ = \left\{P \in \mathscr{P} \mid p(x, \xi) = 0 \text{ when } \xi_3 \leqslant \left(\xi_1^2 + \xi_2^2\right)^{1/2} \text{ and } |\xi| > 1\right\},$$

and

$$\mathscr{P}^{-} = \left\{ P \in \mathscr{P} \mid p(x,\xi) = 0 \text{ when } \xi_3 \geqslant -\left(\xi_1^2 + \xi_2^2\right)^{1/2} \text{ and } |\xi| > 1 \right\}.$$

We prove that if $\{P_\delta\}$ and $\{P'_\delta\}$ are regularizing families in $\mathscr{P}^0$, $\mathscr{P}^+$, or $\mathscr{P}^-$ and $\{P'_\delta\}$ dominates $\{P_\delta\}$, then for each $s \geqslant 0$ there is a constant $C_s > 0$, independent of δ, such that, for $\delta > 0$,

$$\|P_\delta u\|_{s+1/2} \leqslant C_s\left(\left\|P'_\delta(\bar\partial_b u)\right\|_s + \|\bar\partial_b u\|\right) \tag{1.7}$$

for all $u \perp \mathscr{H}_b$ with $u \in \operatorname{dom} \bar\partial_b$. Here $P'_\delta(\bar\partial_b u) = (P'_\delta \bar L(u))\bar\omega$ so that $\|P'_\delta \bar\partial_b u\| = \|P'_\delta \bar L u\|$. To prove that (1.4) follows from (1.7), first observe that $\mathscr{P} = \mathscr{P}^0 + \mathscr{P}^+ + \mathscr{P}^-$. Let $\chi \in C_0^\infty(\mathbb{R}^3)$, with

$$\chi(\xi) = \begin{cases} 1 & \text{if } |\xi| \leqslant 1, \\ 0 & \text{if } |\xi| > 2. \end{cases}$$

Then we let $\{R_\delta\}$ be the family whose symbols are $r_\delta(x,\xi) = \chi(\delta\xi)\zeta(x)$. We can write $R_\delta = P_\delta^0 + P_\delta^+ + P_\delta^-$ so that $\{P_\delta^0\}$, $\{P_\delta^+\}$, and $\{P_\delta^-\}$ are regularizing families in $\mathscr{P}^0$, $\mathscr{P}^+$, and $\mathscr{P}^-$ respectively, with

$$\operatorname{supp} p_\delta^0 \cup \operatorname{supp} p_\delta^+ \cup \operatorname{supp} p_\delta^- \subset \left\{ (x,\xi) \mid x \in \operatorname{supp}\zeta \text{ and } |\xi| < 2/\delta \right\}.$$

Let $\{P_\delta^{0'}\}$, $\{P_\delta^{+'}\}$, and $\{P_\delta^{-'}\}$ be regularizing families that dominate such that $\chi(\delta\xi/2)\xi'(x) = 1$ in a neighborhood of $\operatorname{supp} p_\delta^{0'} \cup \operatorname{supp} p_\delta^{+'} \cup \operatorname{supp} p_\delta^{-'}$. Then from (1.7) we get, with a different C_s, for $\delta > 0$,

$$\|R_\delta u\|_{s+1/2} \leqslant C_s\left(\left\|R'_\delta \bar\partial_b u\right\|_s + \|\bar\partial_b u\|\right),$$

where R'_δ has symbol $r'_\delta(x,\xi) = \chi(\delta\xi/2)\zeta'(x)$. Inequality (1.4) then follows by letting $\delta \to 0$ and observing that $\|R_\delta u\|_{s+1/2}$ is bounded independently of δ, which implies that $\zeta u \in H^{s+1/2}$.

We now prove (1.7). We begin by noting that, on U,

$$\bar\partial_b u = (\bar L u)\bar\omega \quad \text{and} \quad \bar\partial_b^*(v\omega) = -Lv + gv,$$

where $g \in C^\infty(U)$. Integration by parts gives

$$\|\bar\partial_b f\|^2 = \|\bar L f\|^2 = \|Lf\|^2 - \left(f, [L,\bar L]f\right) + O\left(\|Lf\|\,\|f\| + \|f\|^2\right), \tag{1.8}$$

where $f \in C_0^\infty(U)$.

To prove (1.7) for $\{P_\delta\} \subset \mathscr{P}^0$ we observe that

$$\begin{aligned}
\|\bar\partial_b f\|^2 &\geqslant \frac{1}{2}\left(\|Lf\|^2 + \|\bar L f\|^2\right) + O\left(\|f\|_1\|f\| + \|f\|^2\right) \\
&\geqslant \frac{1}{2}\sum_1^2 \left\|\frac{\partial f}{\partial x_j}\right\|^2 + O\left(\varepsilon\|f\|_1^2 + \|f\|_1\|f\| + \|f\|^2\right),
\end{aligned}$$

where ε is the maximum of the oscillations of the g_j on U and, hence, can be made small by choosing U small. Since $P_\delta \in \mathscr{P}^0$, we have

$$\begin{aligned}
\sum_1^2 \left\|\frac{\partial}{\partial x_j}(\Lambda^s P_\delta)\right\|^2 &= \text{const}\left|\left(\Lambda^2 P_\delta u, \left(\xi_1^2 + \xi_1^2\right)\widehat{\Lambda^s P_\delta u}\right)\right| \\
&\geqslant \text{const}\|P_\delta u\|_{s+1}^2 + O\left(\|u\|^2\right),
\end{aligned}$$

where Λ^s is the pseudodifferential operator with symbol $\theta(x)(1+|\xi|^2)^{s/2}$, where $\theta \in C_0^\infty(U)$, and $\theta = 1$ in a neighborhood of $\overline{\bigcup_{\xi,\delta} \operatorname{supp} p_\delta(\cdot,\xi)}$. Then substituting $\Lambda^s P_\delta$ for f we obtain, after choosing U sufficiently small,

$$\|P_\delta u\|_{s+1}^2 \leqslant C_s\left(\|\bar\partial_b(\Lambda^s P_\delta u)\|^2 + \|u\|^2\right). \tag{1.9}$$

Since

$$\|\bar\partial_b(\Lambda^s P_\delta u)\| \leqslant \|P_\delta\, \bar\partial_b u\|_s + \|\Lambda^{-s}[\bar L, \Lambda^s P_\delta]\, u\|_s \tag{1.10}$$

and $\{\Lambda^{-s}[\bar L, \Lambda^s P_\delta]\}$ is a regularizing family in $\mathscr{P}^0$, we use (1.9) with $\{P_\delta\}$ replaced by the above family and s replaced by $s-1$. Repeating this procedure $[s]+1$ times, we obtain, with a different C_s,

$$\|P_\delta u\|_{s+1} \leqslant C_s\left(\|P_\delta\, \bar\partial_b u\|_s + \|P'_\delta\, \bar\partial_b u\|_{s-1} + \|u\|\right).$$

Then (1.7) for $P_\delta, P'_\delta \in \mathscr{P}^0$ follows, since $\|w\|_{s+1/2} \leqslant \|w\|_{s+1}$ and $\|u\| \leqslant c\|\bar\partial_b u\|$ for $u \perp \mathscr{H}_b$, because the range of $\bar\partial_b$ is closed.

Next, we prove (1.7) for $\{P_\delta\}$, $\{P'_\delta\}$ in $\mathscr{P}^-$. From (1.8) and the formula $[L, \bar L]$ we have

$$\|\bar\partial_b f\|^2 = \|\bar L f\|^2 = \|Lf\|^2 - (f, \lambda T f) + O(\|Lf\|\,\|f\| + \|f\|^2), \tag{1.11}$$

where $f \in C_0^\infty(U)$. Since $P_\delta \in \mathscr{P}^-$, the symbol of λT (which is $\lambda(x)c(x)\xi_3$, with $\lambda > 0$ and $c > 0$) is nonpositive on $\operatorname{supp} p_\delta \cap \{(x,\xi) \mid |\xi| \geqslant 1\}$; hence, by Gårding's inequality,

$$(\Lambda^s P_\delta u, \lambda T \Lambda^s P_\delta u) \leqslant \text{const}\|P_\delta u\|_s^2. \tag{1.12}$$

From (1.11) we then obtain

$$\|L\Lambda^s P_\delta u\|^2 + \|\bar L \Lambda^s P_\delta u\|^2 \leqslant c\left(\|\bar\partial_b \Lambda^s P_\delta u\|^2 + \|P_\delta u\|_s^2\right).$$

Furthermore, since $\lambda > 0$, we have

$$T = \lambda^{-1}(L\bar L - \bar L L - aL - b\bar L);$$

hence,

$$|(f, Tf)| \leqslant \text{const}\left(\|Lf\|^2 + \|\bar L f\|^2 + \|f\|^2\right),$$

which implies that

$$\|f\|_{1/2}^2 \leqslant \text{const}\left(\|Lf\|^2 + \|\bar L f\|^2 + \|f\|^2\right). \tag{1.13}$$

Setting $f = \Lambda^s P_\delta u$ and combining this with (1.13), we obtain

$$\|P_\delta u\|_{s+1/2}^2 \leqslant C\left(\|\bar\partial_b \Lambda^s P_\delta u\|^2 + \|P_\delta u\|_s^2\right).$$

Now, proceeding as before, we apply the above to the regularizing family $\{\Lambda^{-s}[\bar\partial_b, \Lambda^s P_\delta]\} \subset \mathscr{P}^-$, with s replaced by $s - 1/2$. Continuing this process $[2s]+1$ times, we obtain

$$\|P_\delta u\|_{s+1/2} \leqslant C_s\left(\|P_\delta\, \bar\partial_b u\|_s + \|P'_\delta\, \bar\partial_b u\|_{s-1/2} + \|P'_\delta u\|\right). \tag{1.14}$$

The above inequality holds for all u in the domain of $\bar\partial_b$; if, in addition, $u \perp \mathscr{H}_b$, then we obtain (1.7) for P_δ, $P'_\delta \in \mathscr{P}^-$, since $\|u\| \leqslant \text{const}\|\bar\partial_b u\|$ and $\|P'_\delta u\| \leqslant \text{const}\|u\|$.

To complete the proof of Theorem 1.3, we must show that (1.8) holds for $\{P_\delta\}$, $\{P'_\delta\}$ in $\mathscr{P}^+$. To do this we recall that $\bar\partial_b^*$, the L_2-adjoint of $\bar\partial_b$, is expressed on $B_U^{0,1}$ by $\bar\partial_b^*(v\bar\omega) = -Lu + gv$. So setting $f = v$ in (1.8) and (1.11), we obtain

$$(1.15)\qquad \begin{aligned} \|\bar\partial_b^*(v\bar\omega)\|^2 &= \|Lv\|^2 + O(\|v\|^2) \\ &= \|\bar Lv\|^2 + (v, \lambda Tv) + O(\|Lv\|\,\|v\| + \|v\|^2). \end{aligned}$$

In this expression the sign of $(v, \lambda T\lambda)$ is the opposite of the corresponding term in (1.11). Since $P_\delta \in \mathscr{P}^+$, the symbol of λT is nonnegative on the support of p_δ for $|\xi| > 1$, and, hence, by Gårding's inequality, we obtain

$$(1.16)\qquad \|P_\delta v\|_{s+1/2}^2 \leqslant C\left(\|\bar\partial_b^*((\Lambda^s P_\delta v)\bar\omega)\|^2 + \|P_\delta v\|_s^2\right).$$

Again applying this inequality successively to commutators, we obtain

$$(1.17)\qquad \|P_\delta v\|_{s+1/2}^2 \leqslant C_s\left(\|P_\delta \bar\partial_b^*(v\bar\omega)\|_s^2 + \|P'_\delta \bar\partial_b^*(v\bar\omega)\|_{s-1/2} + \|P'_\delta v\|^2\right),$$

which holds for all v for which $v\bar\omega$ is in $\operatorname{dom}\bar\partial_b^*$.

The fact that the range of $\bar\partial_b$ is closed implies that the range of $\bar\partial_b^*$ is closed and, hence, equal to the orthogonal complement of $\mathscr{H}_b$. Therefore, if $u \perp \mathscr{H}_b$, there exists $v\bar\omega \in \operatorname{dom}\bar\partial_b^*$ so that $u = \bar\partial_b^*(v\bar\omega)$. We may assume that $v\bar\omega$ is orthogonal to the null space of $\bar\partial_b^*$. We also have

$$\|v\| \leqslant \text{const}\|\bar\partial_b^*(v\bar\omega)\|$$

for all $v\bar\omega$ orthogonal to the null space of $\bar\partial_b^*$ that belong to the domain of $\bar\partial_b^*$. Then for all $u \perp \mathscr{H}_b$, $u \in \operatorname{dom}\bar\partial_b$, we have

$$\begin{aligned} \|P_\delta u\|_{s+1/2}^2 &= \left(P_\delta u, \Lambda^{2s+1}P_\delta \bar\partial_b^*(v\bar\omega)\right) = \left(P_\delta \bar\partial_b u, (\Lambda^{2s+1}P_\delta v)\bar\omega\right) \\ &\quad + O\left(\|P_\delta u\|_{s+1/2}\|P''_\delta v\|_{s+1/2} + \|P''_\delta u\|_s\|P_\delta v\|_s\right), \end{aligned}$$

where $\{P''_\delta\}$ is a regularizing family that dominates $\{P_\delta\}$ and is dominated by $\{P'_\delta\}$. Hence, we obtain

$$\begin{aligned} \|P_\delta u\|_{s+1/2}^2 &\leqslant \text{l.c.}\left(\|P_\delta \bar\partial_b u\|_s^2 + \|P''_\delta v\|_{s+1/2}^2 + \|P''_\delta u\|_s^2\right) \\ &\quad + \text{s.c.}\left(\|P_\delta v\|_{s+1}^2 + \|P_\delta u\|_{s+1/2}^2 + \|P_\delta v\|_s^2\right), \end{aligned}$$

where s.c. denotes a constant that can be chosen as small as we wish, provided l.c. is chosen sufficiently large. Then, applying (1.17), we have

$$\begin{aligned} \|P_\delta u\|_{s+1/2}^2 \leqslant \text{l.c.}\Big\{&\|P_\delta \bar\partial_b u\|_s^2 + C_s\Big(\|P''_\delta \bar\partial_b^*(v\bar\omega)\|_s^2 \\ &+ \|P'_\delta \bar\partial_b^*(v\bar\omega)\|_{s-1/2}^2 + \|P'_\delta v\|_s^2\Big) + \|P'_\delta u\|_s^2\Big\} \\ &+ \text{s.c.}\left(\|P_\delta \bar\partial_b^*(v\bar\omega)\|_{s+1/2}^2 + \|P'_\delta \bar\partial_b^*(v\bar\omega)\|_s^2 + \|P'_\delta v\|^2\right). \end{aligned}$$

Since $u = \bar{\partial}_b^*(v\bar{\omega})$,

$$\|P_\delta u\|_{s+1/2}^2 \leqslant \tilde{C}_s\left(\|P_\delta' \bar{\partial}_b u\|_s^2 + \|P_\delta' u\|_s^2 + \|P_\delta' v\|^2\right).$$

Applying this inequality repeatedly to the term $\|P_\delta' u\|_s^2$, we obtain

$$\|P_\delta u\|_{s+1/2}^2 \leqslant \tilde{\tilde{C}}_s\left(\|P_\delta' \bar{\partial}_b u\|_s^2 + \|P_\delta' u\|^2 + \|P_\delta' v\|^2\right).$$

Then (1.8) follows, since

$$\|P_\delta' u\| \leqslant \text{const}\|u\| \leqslant \text{const}\|\bar{\partial}_b u\|$$

and

$$\|P_\delta' v\| \leqslant \text{const}\|v\| \leqslant \text{const}\|\bar{\partial}_b^*(v\bar{\omega})\| = \text{const}\|u\| \leqslant \text{const}\|\bar{\partial}_b u\|.$$

This concludes the proof of Theorem 1.3.

We now consider the case of M pseudoconvex but not strongly pseudoconvex. As in [**K3**], we define finite type as follows.

DEFINITION. If M is a 3-dimensional pseudoconvex manifold and $x_0 \in M$, we denote by $\mathbb{C}T_{x_0}$ set of germs of complex vector fields at x_0. For each integer $k \geqslant 0$ we define $\mathscr{L}^k(x_0) \subset \underline{\mathbb{C}T_{x_0}}$ inductively:

$$\mathscr{L}^0(x_0) = \left\{ fL + g\bar{L} \mid L \in \underline{T_{x_0}^{1,0}} \text{ and } f, g \in C^\infty \right\},$$

$$\mathscr{L}^{k+1}(x_0) = \mathscr{L}^k(x_0) + \left[\mathscr{L}^k(x_0), \mathscr{L}^0(x_0)\right].$$

We say that x_0 is of *finite type* if, for some k, $\mathscr{L}^k(x_0) = \underline{\mathbb{C}T_{x_0}}$. The least such k is called the *type* of x_0.

Now we generalize Theorem 1.3 to points of finite type as follows.

1.18. THEOREM. *Suppose M is a compact CR manifold of dimension 3 with the following properties*:

(a) *the range of $\bar{\partial}_b$ is closed*;

(b) *M is pseudoconvex*;

(c) *$x_0 \in M$ is of type k.*

Let U be a neighborhood of x_0 such that each point of $\bar{U}$ is of type at most k. If $u \perp \mathscr{H}_b$, $u \in \operatorname{dom}\bar{\partial}_b$, and $\bar{\partial}_b u|_U \in C^\infty(U)$, then $u|_U \in C^\infty(U)$. More precisely, if $\zeta, \zeta' \in C_0^\infty(U)$, with $\zeta' = 1$ on a neighborhood of $\operatorname{supp}(\zeta)$, then for each $s \in \mathbb{R}$, $s \geqslant 0$, there exists C_s such that, if $\zeta'\bar{\partial}_b u \in H_s^{0,1}$, then $\zeta u \in H_{s+1/k}$ and

$$\|\zeta u\|_{s+1/k} \leqslant C_s\left(\|\zeta'\bar{\partial}_b u\|_s + \|\bar{\partial}_b u\|\right). \tag{1.19}$$

PROOF. This theorem is established by the same argument as Theorem 1.3 with (1.13) replaced by

$$\|f\|_{1/k}^2 \leqslant \text{const}\left(\|Lf\|^2 + \|\bar{L}f\|^2 + \|f\|^2\right) \tag{1.20}$$

for all $f \in C_0^\infty(U)$. If we write $X_1 = \mathrm{Re}(L)$ and $X_2 = \mathrm{Im}(L)$, then we see that (1.20) is equivalent to

$$\|f\|_{1/k}^2 \leqslant \mathrm{const}\left(\sum_1^2 \|X_j f\| + \|f\|^2\right), \tag{1.21}$$

where the X_j satisfy the Hörmander condition (see [**H, K4**] and [**RS**]).

2. The case $\dim M > 3$. Let M be a compact CR manifold of dimension $2n - 1$. The CR structure on M is given by subbundle $T^{1,0}(M) \subset \mathbb{C}T(M)$ with the properties

(a) $\dim_{\mathbb{C}} T_x^{1,0}(M) = n - 1$;
(b) $T^{1,0}(M) \cap \overline{T^{1,0}(M)} = \{0\}$;
(c) if $L, L' \in T_U^{1,0}$, then $[L, L'] \in T_U^{1,0}$.

Note that the integrability condition (c) is automatically satisfied when $n = 1$ (i.e., $\dim M = 3$).

Let $L_1, \ldots, L_{n-1}, \overline{L}_1, \ldots, \overline{L}_{n-1}, T$ be an orthonormal basis of $\mathbb{C}T_U$ with $L_1, \ldots, L_{n-1} \in T_U^{1,0}$. Then the Levi form is represented by the hermitian matrix (c_{ij}) given by

$$\left[L_i, \overline{L}_j\right] = c_{ij} T \mod\left(L_1, \ldots, L_{n-1}, \overline{L}_1, \ldots, \overline{L}_{n-1}\right).$$

We denote by $\omega_1, \ldots, \omega_{n-1}$ the basis of $B_U^{1,0}$ dual to $L_1, \ldots, L_{n-1}$. In terms of these bases we have

$$\bar\partial_b u = \sum \left(\overline{L}_i u\right) \bar\omega_i \quad \text{and} \quad \bar\partial_b^* \varphi = -\sum_i \left(L_i \varphi_i + a_i \varphi_i\right),$$

where $\varphi = \sum \varphi_i \bar\omega_i$ and $a_i \in C^\infty(U)$.

If $\{U_\nu\}$ is a finite covering of M by coordinate neighborhoods on which local bases of $T_{U_\nu}^{1,0}$ are defined, and $\{\zeta_\nu\}$ is a corresponding partition of unity, then we define

$$\|u\|_s^2 = \sum_\nu \|\zeta_\nu u\|_s^2 \quad \text{for } u \in C^\infty(M)$$

and

$$\|\varphi\|_s^2 = \sum_{i,\nu} \|\zeta_\nu \varphi_i^{(\nu)}\|_s^2 \quad \text{for } \varphi \in B_M^{0,1}.$$

It is shown in [**KN**] that if there exist $\varepsilon > 0$ and $C > 0$ such that

$$\|\varphi\|_\varepsilon^2 \leqslant C\left(\|\bar\partial_b \varphi\|^2 + \|\bar\partial_b^* \varphi\|^2 + \|\varphi\|^2\right) \tag{2.1}$$

for all $\varphi \in B_M^{0,1}$, then the ranges of $\bar\partial_b$—both are functions and (0, 1)-forms— are closed. Furthermore, if $\zeta, \zeta' \in C^\infty(M)$ and $\zeta' = 1$ on a neighborhood of supp(ζ), then for all $u \perp \mathscr{H}_b$, with $u \in \operatorname{dom} \bar\partial_b$ and $\zeta' \bar\partial_b u \in H_s^{0,1}$, we have $\zeta u \in H_{s+\varepsilon}$ and

$$\|\zeta u\|_{s+\varepsilon} \leqslant C_s\left(\|\zeta' \bar\partial_b u\|_s + \|\bar\partial_b u\|\right). \tag{2.2}$$

In [**K**] it is shown that (2.1) holds with $\varepsilon = 1/2$ if M is strongly pseudoconvex of dimension bigger than 3.

To study the estimates of (2.1) and (2.2) when M is pseudoconvex but not strongly pseudoconvex, we use the type of ideals introduced in [**K5**] in order to study the $\bar\partial$-Neumann problem on weakly pseudoconvex domains (see also [**K6**] for an application of these methods in more general situations). D. Catlin [**C**] has established necessary conditions for subelliptic estimates for the $\bar\partial$-Neumann problem on weakly pseudoconvex domains. Recently, he has proved that these conditions are also sufficient. This work makes essential use of the results of D'Angelo in [**D'A**]. The analogous problem for CR manifolds is to prove (2.1). R. Diaz [**D**] has generalized Catlin's necessary proof to the case of embedded CR manifolds.

Here we show how the methods developed in §1 and in [**K5**] can be used to establish (2.1) and (2.2). For $x^0 \in M$ we define inductively two sequences of ideals of germs of C^∞ functions at x^0 as follows:

$$I_1 = \sqrt[\mathbb{R}]{(\det(c_{ij}))}, \tag{2.3}$$

$$I_{k+1} = \sqrt[\mathbb{R}]{\left(I_k, \left\{\operatorname{coeff}\left(\partial_b f_1 \wedge \cdots \wedge \partial_b f_p \wedge \left(\sum c_{ij}\omega_i \wedge \bar\omega_j\right)^{n-p-1}\right) \middle| f_1,\ldots,f_p \in I_k\right\}\right)},$$

and

$$\begin{aligned} J_1 &= \sqrt[\mathbb{R}]{\left(\det\left(\delta_{ij}\sum c_{qq} - c_{ij}\right)\right)}, \\ J_{k+1} &= \sqrt[\mathbb{R}]{\left(J_k, \left\{\det A_{ij}^{(k)}\right\}\right)}, \end{aligned} \tag{2.4}$$

where the $(A_{ij}^{(k)})$ are $(n-1)\times(n-1)$ matrices whose ith row is either $\delta_{ij}\sum L_q(f)\overline{L_q(f)} - L_i(f)\overline{L_j(f)}$, with $f \in I_k$, or $\delta_{ij}\sum c_{qq} - c_{ij}$.

2.5. THEOREM. *Suppose that M is a compact pseudoconvex CR manifold of dimension $2n-1$ with $n > 2$ and that there is an integer k so that $1 \in I_k(x^0)$ for all $x^0 \in M$. Then there exist an $\varepsilon > 0$ and $C > 0$ so that* (2.1) *holds for all $\varphi \in B_M^{0,1}$.*

2.6. THEOREM. *Let M be a pseudoconvex CR manifold and assume that*

(a) *$\bar\partial_b$ has a closed range,*

(b) *for some $x^0 \in M$ there exists k with $1 \in I_k(x_0)$.*

Then there exist a neighborhood U of x^0 and an $\varepsilon > 0$ such that if $\zeta, \zeta' \in C_0^\infty(U)$, with $\zeta' = 1$ in a neighborhood of $\operatorname{supp}\zeta$, and $u \perp \mathscr{H}_b$, with $\zeta'\bar\partial_b u \in H_s$, then $\zeta \in H_{s+\varepsilon}$ and (2.2) *holds.*

OUTLINE OF PROOFS OF THEOREMS 2.5 AND 2.6. Integration by parts (as in [**K5**]) gives, for φ with support in U,

$$\sum\left(c_{ij}T\varphi_i, \varphi_j\right) + \sum\left\|\bar L_i\varphi_j\right\|^2 \leqslant CQ_b(\varphi, \varphi) \tag{2.7}$$

and

$$(2.8) \qquad -\sum\left(\left(\delta_{ij}\sum_q c_{qq}-c_{ij}\right)T\varphi_i,\varphi_j\right)+\sum\|L_i\varphi_j\|^2\leqslant CQ_b(\varphi,\varphi),$$

where

$$Q_b(\varphi,\varphi)=\|\bar\partial_b\varphi\|^2+\|\bar\partial_b^*\varphi\|^2+\|\varphi\|^2.$$

Using the notation of the first section, we obtain, for $P\in\mathscr{P}^0$,

$$(2.9) \qquad \|P\varphi\|_1^2\leqslant CQ_b(P\varphi,P\varphi)\leqslant C'Q_b(\varphi,\varphi).$$

If $P\in\mathscr{P}^+$ we have, by (2.7),

$$(2.10) \qquad \left|\sum(c_{ij}TP\varphi_i,P\varphi_j)\right|+\sum\|\bar L_iP\varphi_j\|^2\leqslant CQ_b(\varphi,\varphi).$$

The arguments of **[K5]** prove that (2.10) and $1\in I_k$ imply

$$(2.11) \qquad \|P\varphi\|_\varepsilon^2\leqslant CQ_b(\varphi,\varphi).$$

For $P\in\mathscr{P}^-$ we have

$$(2.12) \qquad \left|\sum\left(\left(\delta_{ij}\sum_q c_{qq}-c_{ij}\right)TP\varphi_i,P\varphi_j\right)\right|+\sum\|L_iP\varphi_j\|^2\leqslant CQ_b(\varphi,\varphi).$$

It is easy to see that, for $n>2$, $I_k\subset J_k$, so that $1\in I_k$ implies $1\in J_k$. Again, $1\in J_k$, together with (2.12), imply (2.11). Hence, Theorem 2.5 follows by combining the above with a suitable partition of unity.

To prove Theorem 2.6 we first observe that $1\in I_k(x^0)$ implies that the Lie algebra generated by $L_1,L_2,\ldots,L_{n-1},\bar L_1,\ldots,\bar L_{n-1}$ contains all the germs of vector fields at x^0. We also have

$$(2.13) \qquad \begin{aligned}\|\bar\partial_bu\|^2&=\sum_j\|\bar L_ju\|^2\\&=\sum_j\|L_ju\|^2-\left(\sum c_{ij},Tu,u\right)+O\left(\sum\|L_ju\|\,\|u\|+\|u\|^2\right).\end{aligned}$$

As in §1 we then obtain, for $\{P_\delta\}\subset\mathscr{P}^0$,

$$(2.14) \qquad \|P_\delta u\|_{s+1}\leqslant C_s\left(\|P'_\delta\bar\partial_bu\|_s+\|P'_\delta u\|\right).$$

For $\{P_\delta\}\subset\mathscr{P}^-$ we have, as in §1,

$$(2.15) \qquad \|P_\delta u\|_{s+\varepsilon}\leqslant C_s\sum\left(\|L_j\Lambda^sP_\delta u\|+\|\bar L_j\Lambda^sP_\delta u\|\right)\leqslant C'_s\left(\|P'_\delta\bar\partial_bu\|_s+\|P'_\delta u\|\right).$$

Finally, for $\{P_\delta\}\subset\mathscr{P}^+$, we have $u=\bar\partial_b^*\varphi$, since $u\perp\mathscr{H}_b$ and the range of $\bar\partial_b$ is closed. We may choose φ orthogonal to the null space of $\bar\partial_b^*$; hence, φ is in the range of $\bar\partial_b$, so that $\bar\partial_b\varphi=0$ and $Q_b(\varphi,\varphi)=\|\bar\partial_b^*\varphi\|^2+\|\varphi\|^2$. Thus, substituting $P_\delta\varphi$ in (2.7), we obtain

$$\left|\sum(c_{ij}TP_\delta\varphi_i,P_\delta\varphi_j)\right|+\sum\|\bar L_iP_\delta\varphi\|^2\leqslant C\left(\|P'_\delta\bar\partial_b^*\varphi\|^2+\|P'_\delta\varphi\|^2\right).$$

Since $1 \in I_k$, we again obtain

$$\|P_\delta \varphi\|_\varepsilon^2 \leqslant C\left(\|P'_\delta \bar\partial_b^* \varphi\|^2 + \|P'_\delta \varphi\|^2\right)$$

and

$$\|P_\delta \varphi\|_{s+\varepsilon} \leqslant C_s\left(\|P'_\delta \bar\partial_b^* \varphi\|_s + \|P'_\delta \varphi\|\right).$$

Now as in §1 we estimate $\|P_\delta u\|_{s+\varepsilon}$ by setting

$$\|P_\delta u\|_{s+\varepsilon}^2 = \left(P_\delta u, \Lambda^{2s+2\varepsilon} P_\delta \bar\partial_b^* \varphi\right).$$

We then have, arguing as the previous section,

$$\|P_\delta u\|_{s+\varepsilon} \leqslant C_s\left(\|P'_\delta \bar\partial_b u\|_s + \|\bar\partial_b u\|\right).$$

Piecing these estimates together, we conclude the proof of Theorem 2.6.

References

[BM] L. Boutet De Monvel: *Integration des equations de Cauchy–Riemann induites formelles*, Seminaire Goulaouic–Lions–Schwartz, Exposé IX, 1974–1975.

[B] D. Burns: *Global behavior of some tengential Cauchy–Riemann equations*, Partial Differential Equations and Geometry (Proc. Conf., Park City, Utah, 1977), Dekker, New York, 1979, pp. 51–56.

[C] D. W. Catlin: *Necessary conditions for the subellipticity of the $\bar\partial$-Neumann problem*, Ann. of Math. (2) **117** (1982), 147–171.

[D'A] J. P. D'Angelo: *Real hypersurfaces, orders of contact, and applications*, Ann. of Math. (2) **115** (1982), 615–637.

[D] R. L. Diaz: *Necessary conditions for the subellipticity of the boundary laplacian for the tangential Cauchy–Riemann complex on pseudo-convex boundaries*, Ph.D. thesis, Princeton, 1983.

[H] L. Hörmander: *Hypoelliptic second-order differential equations*, Acta Math. **119** (1967), 147–171.

[K1] J. J. Kohn: *Boundaries of complex manifolds*, Proc. Conf. Complex Analysis (Minneapolis, 1964), Springer-Verlag, 1965, pp. 81–94.

[K2] ______: *Microlocalization of CR structures*, Several Complex Variables (Proc. 1981 Hangzhou Conf.), Birkhäuser, Boston, 1984, pp. 29–36.

[K3] ______: *Boundary behavior of $\bar\partial$ on weakly pseudo-convex manifolds of dimension two*, J. Differential Geom. **6** (1972), 523–542.

[K4] ______: *Pseudo-differential operators and non-elliptic problems*, Pseudo-differential Operators, C.I.M.E. Stresa, Italy, 1968, pp. 157–165.

[K5] ______: *Subellipticity of the $\bar\partial$-Neumann problem on pseudo-convex domains: sufficient conditions*, Acta Math. **142** (1979), 79–122.

[K6] ______: *Subelliptic estimates*, Proc. Conf. Linear Partial Differential Equations (Saint Jean des Monts, 1981).

[KN] J. J. Kohn and L. Nirenberg: *Non coersive boundary value problems*, Comm. Pure Appl. Math. **18** (1965), 443–492.

[R] H. Rossi: *Attaching analytic spaces to a space along a psuedo-convex boundary*, Proc. Conf. Complex Analysis (Minneapolis, 1965), Springer-Verlag, 1965, pp. 242–256.

[RS] L. P. Rothschild and E. M. Stein: *Hypoelliptic differential operators and nilpotent groups*, Acta Math. **137** (1976), 247–320.

[S] M.-C. Shaw: *L^2 estimates and existence theorems for the tangential Cauchy–Riemann complex*, (1984 preprint).

Princeton University

Proceedings of Symposia in Pure Mathematics
Volume **43** (1985)

Real Analysis and Operator Theory

YVES MEYER

Introduction. In the 1960s A. P. Calderón was defining and initiating a scientific program that renewed harmonic analysis and yielded a new operator theory adapted to the needs of partial differential equations.

The first problem raised by Calderón was to define new operations on functions of real or complex variables. More precisely, Calderón wanted real-variable substitutes for usual algorithms on holomorphic functions (like the ordinary product or the product twisted by differentiation, introduced by Calderón) that would preserve the striking estimates of the complex case.

The second problem is related to the first. Those new algorithms should be given by singular integral operators in the broad sense (proposed by Calderón).

The results described here are due to G. David, J. L. Journé, and P. G. Lemarié. Applications to partial differential equations have been obtained by J. M. Bony (Orsay) and the Minneapolis team around E. Fabes, D. Jerison, and C. Kenig.

1. Classical facts about Hardy spaces. Let us begin with the definition of holomorphic Hardy spaces. Let Ω be the (open) upper half-plane, and let $H^p(\Omega)$ be the vector space of all holomorphic functions $f\colon \Omega \to C$ for which $\sup_{y>0}(\int_{-\infty}^{+\infty} |f(x+iy)|^p\, dx)^{1/p} < +\infty$. The Banach algebra of bounded holomorphic functions on Ω is denoted by $H^\infty(\Omega)$ and can be viewed as an operator algebra on $H^p(\Omega)$ by pointwise multiplication.

When f belongs to $H^p(\Omega)$, its trace on the boundary $R = \partial\Omega$ is defined by $\lim_{\varepsilon\to 0} f(x+i\varepsilon)$. This limit exists for the L^p norm when $1 \leqslant p < +\infty$ and should be taken in the distributional sense when $0 < p < 1$.

The function $f \in H^p(\Omega)$ is uniquely defined by its trace (denoted $f(x)$), and for that reason the spaces $H^p(\Omega)$ can be identified with spaces of functions (or distributions), denoted by $H^p(\mathbb{R})$, $0 < p \leqslant +\infty$, on the real line.

Before raising the problem of extending this action of H^∞ on H^p $(0 < p \leqslant 1)$ to $L^\infty(\mathbb{R}^n)$ acting on the generalized Hardy spaces $\mathscr{H}^p(\mathbb{R}^n)$ as defined by Stein

---

1980 *Mathematics Subject Classification*. Primary 42B20, 47G05, 47-02; Secondary 47D30, 22E30.

0082-0717/85 $1.00 + $.25 per page

and Weiss, we would like to describe a second operation created by Calderón. When $f \in H^p(\Omega)$ and $g \in H^q(\Omega)$, Calderón defines the holomorphic function h: $\Omega \to C$ by $h(+i\infty) = 0$ and $h'(z) = f'(z)g(z)$ (where f' is the derivative at f).

In 1965 Calderón proved that f, $g \in H^2$ imply $h \in H^1$, as if h were the ordinary product of f and g. In 1978 C. Pommerenke extended Calderón theorem to the case where $f \in$ BMOA and $g \in H^2(\Omega)$; then h belongs to $H^2(\Omega)$, and this is surprising because the ordinary product between a function in BMOA and a function in H^2 cannot yield a function in H^2. Roughly speaking, f belongs to BMOA when f satisfies a growth condition like $f(z) = O(\log|z|)$ as $|z| \to +\infty$ and $\operatorname{Im} z \geqslant \varepsilon > 0$ and when $f(x+iy)$ belongs to BMO (the space introduced by John and Nirenberg) uniformly in $y > 0$.

The Stein and Weiss $\mathscr{H}^p$-spaces have an atomic description, which will now be recalled for $0 < p \leqslant 1$. When $1 < p < +\infty$, $\mathscr{H}^p = L^p$. A p-atom is a function $a(x)$ of the real variable $x \in \mathbb{R}^n$ with the following three properties: $a(x)$ is supported by a ball B (whose finite volume is denoted by $|B|$); $\|a\|_\infty \leqslant |B|^{-1/p}$; and $\int_B x^\alpha a(x)\,dx = 0$ for all $\alpha \in N^n$ with $|\alpha| \leqslant 1/p - 1$. A distribution S belongs to $\mathscr{H}^p(\mathbb{R}^n)$ if and only if S has an atomic decomposition: $S = \sum_0^\infty \lambda_k a_k(x)$, where $a_k(x)$ are p-atoms and $\sum_0^\infty |\lambda_k|^p < +\infty$. The latter condition ensures the convergence of the series in the distributional sense.

The linear properties of $\mathscr{H}^p$-spaces are well understood by now and show that these spaces are the correct substitute for the holomorphic Hardy spaces $H^p(\mathbb{R}^n)$. In contrast, the algebraic properties of the holomorphic Hardy spaces cannot be generalized without drastic modifications to the $\mathscr{H}^p$-setting.

For instance, what should be the action of $L^\infty(\mathbb{R}^n)$ on $\mathscr{H}^p(\mathbb{R}^n)$ that would generalize the obvious action (by pointwise multiplication) of $H^\infty(\Omega)$ on $H^p(\Omega)$. It is fairly obvious that the pointwise multiplication does not work in general: when $0 < p < 1$, the nasty distributions $S \in \mathscr{H}^p(\mathbb{R}^n)$ cannot be multiplied by the nonsmooth functions $b(x) \in L^\infty(\mathbb{R}^n)$. The problem still exists when $p = 1$, since $b(x) \in L^\infty$ multiplied by $f(x) \in \mathscr{H}^1(\mathbb{R}^n)$ does not yield an $\mathscr{H}^1$-function unless b is a constant. Following Calderón's hint, we see that the construction of these new algebraic operations is intimately related to the existence of some new singular integral operators.

We have to confess that the new algebraic operations between functions that will be introduced are more similar to $h' = f'g$ than to $h = fg$ if we return to the holomorphic case. It means that commutativity and associativity will be lost and that the Banach space of the operators $T \in \mathscr{L}(\mathscr{H}^p, \mathscr{H}^p)$, $0 < p \leqslant 1$, obtained by these actions of $L^\infty(\mathbb{R}^n)$ will be isomorphic to $L^\infty(\mathbb{R}^n)$ as a Banach space but not as a Banach algebra.

2. Generalized singular integral operators. In the following theory we start with an operator that is weakly defined. It means that T: $\mathscr{D}(\mathbb{R}^n) \to \mathscr{D}'(\mathbb{R}^n)$ is linear and continuous (where $\mathscr{D} = C_0^\infty$ is the usual space of test functions, and $\mathscr{D}'$ is the space of all distributions). We are looking for extra assumptions on the distributional kernel $K(x, y)$ of T that would ensure the continuity of T on given functional spaces like $L^2(\mathbb{R}^n)$, $L^p(\mathbb{R}^n)$, $\mathcal{H}^p(\mathbb{R}^n)$, Λ^s, or B^s (the definitions of the homogeneous Hölder space or of the homogeneous Sobolev space will be given later).

This program is obviously too broad to receive something other than a tautological answer. Therefore, the class of operators under consideration will be restricted by some rules given by Calderón and Zygmund.

We assume that, for any $\varphi \in \mathcal{D}(\mathbb{R}^n)$, the distribution $T(\varphi)$ will be given, outside the support of φ, by an integral representation

$$(2.1) \qquad T(\varphi)(x) = \int K(x, y)\varphi(y)\,dy,$$

where the function $K(x, y)$ is, with a slight abuse of notation, the restriction of the distribution kernel to $x \neq y$, $x \in \mathbb{R}^n$, $y \in \mathbb{R}^n$ and satisfies the following conditions:

(2.2) *there exists a constant C such that* $|K(x, y)| \leqslant C|x - y|^{-n}$;

(2.3) *there exists an exponent γ, $0 < \gamma \leqslant 1$, and a constant C such that*

$$|K(x', y) - K(x, y)| \leqslant C|x' - x|^{\gamma}|x - y|^{-n-\gamma}$$

whenever $|x' - x| \leqslant \frac{1}{2}|x - y|$, $x \neq y$, $x, y \in \mathbb{R}^n$;

(2.4) *condition* (2.3) *is also satisfied by the transposed kernel* $L(x, y) = K(y, x)$.

These conditions give a precise description of the function $T(\varphi)$ outside the support of φ, but do not give any information about the distribution $T(\varphi)$ on a neighborhood of this support. It means that the knowledge of $K(x, y)$ is not sufficient for the construction of the operator T.

The extra information will be given by the so-called distributional kernel of T, also denoted by $K(x, y)$, which is the unique distribution in $\mathcal{D}'(\mathbb{R}^n \times \mathbb{R}^n)$ such that

$$(2.5) \qquad \langle K, \varphi_1 \otimes \varphi_2 \rangle = \langle T\varphi_1, \varphi_2 \rangle$$

for any $\varphi_1, \varphi_2 \in \mathcal{D}(\mathbb{R}^n)$.

This notation will be kept throughout the paper; $\langle \cdot\,, \cdot \rangle$ denotes the duality between testing functions and distributions.

The connection between (2.1) and (2.5) is obvious. The open subset of $\mathbb{R}^n \times \mathbb{R}^n$ defined by $y \neq x$ will be denoted by U, and the function $K\colon U \to C$ satisfying (2.2)–(2.4) is the restriction to U of the distributional kernel of T.

The lower bound of the C's appearing on the right side of (2.2)–(2.4) will be denoted by $\|K\|_\gamma$, and our goal will be a priori inequality of the following kind:

$$(2.6) \qquad \|T\|_{L^2, L^2} \leqslant C\|K\|_\gamma + \text{“something”},$$

where "something" should be defined. With this goal in mind, conditions (2.2)–(2.4) can be better understood. They have a remarkable invariance under the action of the group $\mathcal{G}$ of transformations of $\mathbb{R}^n$ of the type $g(x) = \delta\rho(x) + x_0$, where $\delta > 0$ corresponds to a dilation, $\rho \in SO(n; \mathbb{R})$, and $x_0 \in \mathbb{R}^n$ is a translation. The notation $|g| = \delta^n$ will be of later use. The action of $\mathcal{G}$ on $\mathcal{D}(\mathbb{R}^n)$ is defined by $R_g\varphi = \varphi \circ g^{-1}$ ($g \in \mathcal{G}$) and will be extended canonically to $\mathcal{D}'(\mathbb{R}^n)$.

The spaces B that will be used are functional homogeneous Banach spaces. This means that $\mathcal{D}(\mathbb{R}^n) \subset B \subset \mathcal{D}'(\mathbb{R}^n)$ and both injections are continuous. The homogeneity means that an exponent $t \in \mathbb{R}$ exists such that, for any $g \in \mathcal{G}$ and any $\varphi \in B$, $\|R_g(\varphi)\|_B = |g|^t \|\varphi\|_B$.

Those properties imply the following fact. If $T: B \to B$ is linear and continuous, with norm $\|T\|$, then $R_g T R_g^{-1}: B \to B$ is continuous and $\|R_g T R_g^{-1}\| = \|T\|$.

If the distributional kernel of T is denoted by $K(x, y)$, the corresponding kernel of $R_g T R_g^{-1}$ is $K_g(x, y) = |g| K(gx, gy)$.

We return to our main goal and observe that $\|R_g T R_g^{-1}\|_{L^2, L^2} = \|T\|_{L^2, L^2}$ and $\|K_g\|_\gamma = \|K\|_\gamma$. This means that the norm $\|\cdot\|_\gamma$ has the correct homogeneity under the group action of $\mathcal{G}$.

If $|x - y|^{-n}$ were to be modified into $\omega(x, y)$ on the right side of (2.2)–(2.4), this invariance would be lost (unless $\omega(x, y) = c|x - y|^{-n}$ for some positive constant c). This invariance under the $\mathcal{G}$-action has a geometrical meaning. It appears when the operator T that is being studied arises from a true geometrical problem (complex analysis, potential theory for Lipschitz domains, etc.).

The missing "something" on the right side of (2.6) will be partially unveiled now. For $m \in \mathbb{N}$ and $\varphi \in \mathcal{S}(\mathbb{R}^n)$ we define

$$q_m(\varphi) = \sup_{x \in \mathbb{R}^n} \sum_{|\alpha| \leqslant m} (1 + |x|)^m |\partial^\alpha \varphi(x)|. \tag{2.7}$$

If, for instance, $m > n/2$, then $\|\varphi\|_2 \leqslant C q_m(\varphi)$. Therefore, any continuous operator $T: L^2(\mathbb{R}^n) \to L^2(\mathbb{R}^n)$ is weakly continuous, which means that

$$|\langle T(u), v \rangle| \leqslant C q_m(\varphi) q_m(\varphi) \tag{2.8}$$

for some $m = m(T)$ and all $u, v \in \mathcal{D}$.

Since we would like our hypothesis to be invariant under the $\mathcal{G}$-action, we are led to the following definition.

DEFINITION 1. *A continuous linear operator $T: \mathcal{D} \to \mathcal{D}'$ has the weak boundedness property* (*WBP*) *if there exists an integer $m \in \mathbb{N}$ and a constant C such that, for every $g \in \mathcal{G}$, for any $u \in \mathcal{D}(\mathbb{R}^n)$, and for any $v \in \mathcal{D}(\mathbb{R}^n)$, we have*

$$\left|\left\langle R_g T R_g^{-1}(u), v \right\rangle\right| \leqslant C q_m(\varphi) q_m(\varphi). \tag{2.9}$$

This condition becomes weaker as m increases.

DEFINITION 2. *A continuous linear operator $T: \mathcal{D} \to \mathcal{D}'$ is called a singular integral operator when the following conditions are satisfied*: *T has the WBP, and its kernel fulfills* (2.2)–(2.4).

Let B be a homogeneous functional Banach space whose norm $\|\cdot\|_B$ satisfies $\|u\|_B \leqslant C q_m(\varphi)$ for some integer m, some constant C, and all $u \in \mathcal{D}$. Let us assume that the dual norm $\|\cdot\|_{B^*}$ satisfies a similar inequality. Then any continuous linear operator $T: B \to B$ has the WBP.

Definition 1 can be completed by denoting by $\|T\|_{(m)}$ the lower bound of all possible constants appearing on the right side of (2.9). Our fundamental problem can be restated precisely as

$$\|T\|_{L^2,L^2} \leqslant C\|K\|_\gamma + \|T\|_{(m)} + \text{“something”}. \tag{2.10}$$

The missing term in (2.10) will now be defined.

3. The definition of $T(1)$ for a singular integral operator T. We would like to emphasize that, indeed, (2.4) is not needed in order to define $T(1)$. Defining $T(1)$ amounts to giving a meaning to the divergent integral $\int_{\mathbf{R}^n} K(x, y)\, dy$, and the problem arises at infinity. A summation process should be used. Let $\varphi \in \mathscr{D}(R)$ be a testing function such that $\varphi(0) = 1$. For $\varepsilon > 0$, $\varphi^{(\varepsilon)}(x) = \varphi(\varepsilon x)$, and $S_\varepsilon(x) = \int_{\mathbb{R}^n} K(x, y)\varphi(\varepsilon y)\, dy = T(\varphi^{(\varepsilon)})$ is a distribution in $\mathscr{D}'(\mathbb{R}^n)$. With these definitions we have

PROPOSITION 1. *There exists a family $c(\varepsilon)$, $\varepsilon > 0$, of complex constants such that the difference $S_\varepsilon - c(\varepsilon)$ converges in the distributional sense to a distribution $T(1)$. This distribution is only defined modulo constant functions and is independent of φ.*

The transposed operator T^*: $\mathscr{D}(\mathbb{R}^n) \to \mathscr{D}'(\mathbb{R}^n)$ is defined by its distributional kernel $L(x, y) = K(y, x)$. For a singular integral operator T, $T^*(1)$ is also defined by Proposition 1.

The preceding definitions can be sharpened if two new Banach spaces are introduced: the Besov spaces $\Lambda^0_{1,1}$ and $\Lambda^0_{\infty,\infty}$. These spaces are easily defined in terms of the Littlewood-Paley decomposition, whose definition will be recalled.

The Fourier transformation is denoted by $\mathscr{F}$, and $\theta \in \mathscr{D}(\mathbb{R}^n)$ is a radial function such that $\theta(\xi) = 1$ for $|\xi| \leqslant 1$ and $\theta(\xi) = 0$ for $|\xi| \geqslant 2$. Then for $k \in \mathbb{Z}$, S_k: $\mathscr{S}'(\mathbb{R}^n) \to \mathscr{S}'(\mathbb{R}^n)$ is defined by $\mathscr{F}(S_k(f)) = \theta_k \mathscr{F}(f)$, where $\theta_k(\xi) = \theta(2^{-k}\xi)$.

A functional Banach space B is adapted to the Littlewood–Paley decomposition when $\mathscr{S}(\mathbb{R}^n) \subset B \subset \mathscr{S}'(\mathbb{R}^n)$ (the two inclusions being continuous) and if

$$\lim_{k \to -\infty} \|S_k(f)\|_B = 0, \qquad \lim_{k \to +\infty} \|f - S_k(f)\|_B = 0 \quad \text{for every } f \in B.$$

Then every $f \in B$ can be written as a telescopic series $f = \sum_{-\infty}^{+\infty} \Delta_k(f)$, where $\Delta_k = S_{k+1} - S_k$.

For instance, $B = L^1(\mathbb{R}^n)$ is not adapted to the Littlewood–Paley decomposition, whereas all $L^p(\mathbb{R}^n)$, $1 < p < +\infty$, are adapted, and $L^\infty(\mathbb{R}^n)$ is not adapted. The Hardy space $\mathscr{H}^1(\mathbb{R}^n)$ is adapted to the Littlewood–Paley decomposition, and $\mathrm{BMO}(\mathbb{R}^n)$ is also adapted if we decide that the topology for BMO should be the $\sigma(\mathrm{BMO}, \mathscr{H}^1)$-topology.

The Besov spaces $\Lambda^s_{p,q}$, $s \in \mathbb{R}$, $1 \leqslant p \leqslant +\infty$, $1 \leqslant q \leqslant +\infty$, are equipped with the norm $(\|2^{ks}\Delta_k(f)\|_p)_{l^q(Z)}$ and are either spaces of functions (with some vanishing moments) or spaces of distributions (modulo polynomials of degree $\leqslant m$).

For instance, $\Lambda^0_{1,1}$ is the subspace of $L^1(\mathbb{R}^n)$ defined by $\sum_{-\infty}^{+\infty} \|\Delta_k(f)\|_1 < +\infty$. It is easily checked that $\Lambda^0_{1,1}$ is a dense subspace of the Stein–Weiss space

$\mathscr{H}^1(\mathbb{R}^n)$. The dual space $\Lambda^0_{\infty,\infty}$ of $\Lambda^0_{1,1}$ can also be defined using the Littlewood–Paley decomposition. A tempered distribution $S \in \mathscr{S}'(\mathbb{R}^n)$ (only defined modulo constant functions) belongs to $\Lambda^0_{\infty,\infty}$ if and only if there exists a constant C such that $\|\Delta_k(S)\|_\infty \leqslant C$.

In dimension one, S belongs too $\Lambda^0_{\infty,\infty}$ if and only if S is the distributional derivative of a function in the Zygmund class. The holomorphic version of $\Lambda^0_{\infty,\infty}$ is the Bloch space consisting of all holomorphic functions $f\colon \Omega \to \mathbb{C}$ such that $\sup_{y>0} y|f'(x+iy)| \leqslant C < +\infty$.

We can sharpen Proposition 1.

PROPOSITION 2. *Let $T\colon \mathscr{D}(\mathbb{R}^n) \to \mathscr{D}'(\mathbb{R}^n)$ be a continuous linear operator enjoying the WBP. Assume that the distributional kernel $K(x, y)$ satisfies* (2.2) *and* (2.3). *Then the range of $T\colon \mathscr{D}(\mathbb{R}^n) \to \mathscr{D}'(\mathbb{R}^n)$ is contained in $\Lambda^0_{\infty,\infty}$, and $T(\varphi^{(\varepsilon)}) = S_\varepsilon \in \Lambda^0_{\infty,\infty}$ converges to $T(1) \in \Lambda^0_{\infty,\infty}$ in the $\sigma(\Lambda^0_{\infty,\infty}, \Lambda^0_{1,1})$-topology.*

If, moreover, $T\colon \mathscr{D} \to \mathscr{D}'$ commutes with translations, it forces $T(1) = 0$ and $T^*(1) = 0$.

4. The David–Journé theorem.

THEOREM 1. *Let $T\colon \mathscr{D}(\mathbb{R}^n) \to \mathscr{D}'(\mathbb{R}^n)$ be a continuous linear operator whose distributional kernel $K(x, y)$ satisfies* (2.2)–(2.4).

Then a necessary and sufficient condition for the L^2-continuity of T is the collection of the following properties:

(4.1) *T has the WBP*;

(4.2) $T(1) \in \mathrm{BMO}(\mathbb{R}^n)$;

(4.3) $T^*(1) \in \mathrm{BMO}(\mathbb{R}^n)$.

A function $b(x)$ in $L^1_{\mathrm{loc}}(\mathbb{R}^n)$ belongs to $\mathrm{BMO}(\mathbb{R}^n)$ if a constant C exists such that, for all cubes $Q \subset \mathbb{R}^n$ (with volume Q), one has

$$\frac{1}{|Q|}\int_Q \left|b(x) - \beta_Q\right| dx \leqslant C,$$

where β_Q is a constant depending on Q. In fact, $\mathrm{BMO}(\mathbb{R}^n)$ is a quotient space since constant functions have a zero norm. Then $\mathrm{BMO}(\mathbb{R}^n) \subset \Lambda^0_{\infty,\infty}(\mathbb{R}^n)$, and the spaces are (obviously) distinct. We see that (4.1) and the hypotheis of Theorem 1 imply that $T(1) \in \Lambda^0_{\infty,\infty}$ and $T^*(1) \in \Lambda^0_{\infty,\infty}$. The L^2-continuity of T demands that these conditions be sharpened into $T(1) \in \mathrm{BMO}$ and $T^*(1) \in \mathrm{BMO}$.

The fact that the L^2-continuity of T implies (4.1) has already been explained. The fact that $T(1)$ should belong to BMO is a corollary of a more general fact due to J. Peetre.

PROPOSITION 3. *Let $T\colon L^2(\mathbb{R}^n) \to L^2(\mathbb{R}^n)$ be continuous and linear and assume that the kernel $K(x, y)$ of T satisfies the celebrated Calderón–Zygmund condition*

$$\int_{|x-y|\geqslant 2|x'-x|} \left|K(x', y) - K(x, y)\right| dy \leqslant C < +\infty. \tag{4.4}$$

Then T can be extended to a bounded linear operator $T\colon L^\infty(\mathbb{R}^n) \to \mathrm{BMO}(\mathbb{R}^n)$.

This means that the strength of Theorem 1 comes from the fact that $1 \in L^\infty(\mathbb{R}^n)$ is so special as an L^∞-function that one hopes that $T(1)$ should be easily computed.

Before giving examples, a special case of Theorem 1 should be described. Let $U \subset \mathbb{R}^n \times \mathbb{R}^n$ be the complement of the diagonal, and let $K: U \to C$ be a function fulfilling (2.2)–(2.4) and $K(y, x) = -K(x, y)$ on U. Then the distribution p.v. K, defined by

$$\langle \text{p.v.} K, \varphi \rangle = \lim_{\varepsilon \downarrow 0} \iint_{|x-y| \geqslant \varepsilon} K(x, y) \varphi(x) \varphi(y)\, dx\, dy,$$

defines an operator $T: \mathcal{D}(\mathbb{R}^n) \to \mathcal{D}'(\mathbb{R}^n)$.

COROLLARY 1. *With the preceding notations a necessary and sufficient condition for T to be bounded on L^2 is* $T(1) \in \mathrm{BMO}(\mathbb{R}^n)$.

The WBP comes by easy calculation from the antisymmetry of the kernel. Since $T^*(1) = -T(1)$, the only remaining condition is $T(1) \in \mathrm{BMO}$.

We can return now to the new algebraic operations predicted (or found) by Calderón. The most famous and most studied is the following: Let $a_1(x), \ldots, a_k(x)$ be k functions of a real variable x in $L^\infty(\mathbb{R})$, let $A_1(x), \ldots, A_k(x)$ be the primitives of $a_1(x), \ldots, a_k(x)$, and let $f(x)$ be a function in $L^2(\mathbb{R})$. Then the H-product $H(a_1, \ldots, a_k; f) = g$ is defined by (and one of the problems has been to prove that this definition makes sense)

$$(4.5) \quad g(x) = \frac{1}{\pi} \text{p.v.} \int_{-\infty}^{\infty} \frac{A_1(x) - A_1(y)}{x-y} \cdots \frac{A_k(x) - A_k(y)}{x-y} \frac{f(y)}{x-y}\, dy.$$

It could be tempting to replace each ratio by the value of the derivative. But we then hesitate between choosing $a_j(x)$ or $a_j(y)$, and it gives the vague feeling that $g(x)$ has something to do with expressions like $\prod_{i \in I} a_i(x) H(\prod_{j \in J} a_j f)$, where H is the Hilbert transform and $I \cup J$ is a partition of $\{1, 2, \ldots, k\}$. Dreams of this type are unrealistic, but their failure motivated the fascination that (4.5) exerted on many people.

Theorem 1 can be applied to the operator $T_k(f) = H(a_1, \ldots, a_k; f): \mathcal{D}(\mathbb{R}) \to \mathcal{D}'(\mathbb{R})$. Since the kernel is antisymmetric, we are reduced to computing $T_k(1)$, and a simple integration by parts gives

$$T_k(1) = k^{-1}\big(T_{k-1}^{(1)}(a_1) + \cdots + T_{k-1}^{(k)}(a_k)\big),$$

where the above index corresponds to the missing-term index in the kernel of T_{k-1}. J. Peetre's theorem, together with an induction on k, gives $T_k(1) \in \mathrm{BMO}$. Therefore, T_k is bounded on L^2, since T_0 is the Hilbert transform. A careful examination of this proof gives an estimate of the type $\|H(a_1, \ldots, a_k); f\|_2 \leqslant C^k \|a_1\|_\infty \cdots \|a_k\|_\infty \|f\|_2$, where $C > 1$ is some unspecified constant. This C^k can be improved into $C(1+k)^4$ by a different approach [**13**].

Let us restate Theorem 1 in a quantitative version.

COROLLARY 2. *For each exponent* $\gamma \in\ ]0,1]$ *and each integer* $m \geqslant 1$, *there exists a constant* $C = C(m, n, \gamma)$ *such that, for any operator* $T\colon \mathscr{D} \to \mathscr{D}'$,

$$\|T\|_{L^2, L^2} \leqslant C\left(\|K\|_\gamma + \|T\|_{(m)} + \|T(1)\|_{\mathrm{BMO}} + \|T^*(1)\|_{\mathrm{BMO}}\right). \tag{4.6}$$

A second algebraic operation, intimately related to the proof of Theorem 1, will now be described. This operation is a linear action of $\mathrm{BMO}(\mathbb{R}^n)$ on $L^2(\mathbb{R}^n)$ that extends the one discovered by Pommerenke in 1978.

For harmonic functions in $\mathbb{R}^{n+1}_+$ vanishing at infinity, we have at our disposal two fundamental operators. The first is the trace operator on the boundary; the second is the extension operator that rebuilds the harmonic function from its trace. The second operator is given by the Poisson semigroup P_t ($t \geqslant 0$): if $f(x) \in L^2(\mathbb{R}^n)$ is the given trace, then $P_t f(x) = u(x,t)$ is the solution of the Dirichlet problem, and the first operator is given by the formula

$$u(x,0) = -\int_0^\infty \frac{\partial}{\partial t} u(x,t)\, dt = \int_0^\infty Q_t f\, \frac{dt}{t},$$

where $Q_t = -t\partial P_t/\partial t$. When $a \in \mathrm{BMO}(\mathbb{R}^n)$ and $f \in L^2(\mathbb{R}^n)$, we define

$$\pi(a,f) = 4\int_0^\infty Q_t\{(Q_t a)(P_t f)\}\, \frac{dt}{t} \tag{4.7}$$

(this is meaningful due to the theorem that follows). We could study this operation using Theorem 1, but this would be unfair because the existence of (4.7) is an ingredient in the proof of Theorem 1.

Let us sketch the proof of the fundamental inequality

$$\|\pi(a,f)\|_2 \leqslant C\|a\|_{\mathrm{BMO}}\|f\|_2. \tag{4.8}$$

The identity

$$\left(\int_0^\infty \|Q_t f\|_2^2 \frac{dt}{t}\right)^{1/2} = \frac{1}{2}\|g\|_2$$

is a straightforward application of Plancherel's formula. Then the trilinear form $\langle \pi(a,f), g\rangle = I(a,f,g)$ is studied when $\|g\|_2 \leqslant 1$. We obtain

$$I(a,f,g) = \int_{\mathbb{R}^{n+1}} (Q_t g)(Q_t a)(P_t f)\, \frac{dx\, dt}{t},$$

and the Cauchy–Schwartz inequality yields

$$|I(a,f,g)| \leqslant \frac{1}{2}\left(\int_{\mathbb{R}^{n+1}_+} |u(x,t)|^2\, d\mu(x,t)\right)^{1/2},$$

where $u(x,t) = P_t f$ is the harmonic extension of f and $d\mu(x,t) = |Q_t a|^2\, dx\, dt/t$. But Fefferman and Stein proved that this measure is a Carleson measure, which means that

$$\int_{\mathbb{R}^{n+1}_+} |u(x,t)|^2\, d\mu(x,t) \leqslant C\|f\|_2^2.$$

Let us rename T_a as the operator defined by $T_a(f) = \pi(a, f)$. Its distributional kernel satisfies

$$|K_a(x, y)| \leqslant C\|a\|_{\Lambda^0_{\infty,\infty}}|x - y|^{-n},$$

$$|(\partial/\partial x_j)K_a(x, y)| \leqslant C\|a\|_{\Lambda^0_{\infty,\infty}}|x - y|^{-n-1},$$

and

$$|(\partial/\partial y_j)K_a(x, y)| \leqslant C\|a\|_{\Lambda^0_{\infty,\infty}}|x - y|^{-n-1}$$

for $1 \leqslant j \leqslant n$. Finally, $T_a(1) = a$ and $T_a^*(1) = 0$.

If we return to the holomorphic one-dimensional case ($a \in$ BMOA, $f \in H^2$), we obtain that $g = \pi(a, f)$ satisfies Calderón's equation $g'(z) = a'(z)f(z)$ and $(g + i\infty) = 0$.

Let us return to the proof of Theorem 1. We peel off the given operator T in order to find its deep part. We know that both $T(1) = a$ and $T^*(1) = b$ belong to BMO, and therefore T is peeled by subtracting T_a (where $T_a(f) = \pi(a, f)$) and T_b^*. Then $R = T - T_a - T_b^*$ is a singular integral operator such that $R(1) = R^*(1) = 0$. These operators have remarkable boundedness properties that can be obtained by elementary computations. For instance, R is bounded on all homogeneous Sobolev spaces B^s for $-\gamma < s < \gamma$. In particular, R is bounded on L^2.

This section could not be closed without saying a few words about Bony's paraproduct. This operation uses the Littlewood–Paley decomposition and is defined as

$$\beta(a, f) = \sum_{-\infty}^{\infty} (\Delta_k a)(S_{k-3} f).$$

Let us rename B_a as the operator that takes f into $\beta(a, f)$. then B_a is a singular integral operator if and only if $a \in \Lambda^0_{\infty,\infty}$, and B_a is bounded on $L^2(\mathbb{R}^n)$ if and only if $a \in$ BMO. We have $B_a(1) = a$ and $B_a^*(1) = 0$ in such a way that B_a could be used instead of T_a to peel off T.

5. Continuity of singular integral operators on other homogeneous functional spaces. We recall that a singular integral operator $T\colon \mathcal{D} \to \mathcal{D}'$ is a linear and continous operator with the WBP whose kernel satisfies (2.2)–(2.4).

We would like to know under what conditions such an operator can be extended to a linear and continuous operator $T\colon B \to B$, where B is a classical functional Banach space.

In what follows B will be $L^p(\mathbb{R}^n)$ for $1 < p < +\infty$, $\mathcal{H}^1$, BMO, and, furthermore, the $\mathcal{H}^p$ spaces for $0 < p < 1$, the homogeneous Hölder spaces Λ^s ($s > 0$), and the homogeneous Sobolev spaces $\Lambda^s_{2,2}$ ($0 < s < 1$).

The first result is a variant of the David–Journé theorem.

THEOREM 2. *Let T be a singular integral operator. Then the L^p-continuity of T ($1 < p < +\infty$) is equivalent to the L^2-continuity. The continuity of T:* BMO$(\mathbb{R}^n) \to$ BMO$(\mathbb{R}^n)$ *requires L^2-continuity and $T(1) = 0$. The continuity of $T\colon \mathcal{H}^1(\mathbb{R}^n) \to \mathcal{H}^1(\mathbb{R}^n)$ is equivalent to L^2-continuity together with $T^*(1) = 0$.*

Theorem 2 means that the BMO continuity of a singular integral operator is a rather exceptional situation. For instance, Calderón's first commutator $T: L^2(\mathbb{R}) \to L^2(\mathbb{R})$, whose kernel is

$$\text{p.v.}\frac{A(x)-A(y)}{(x-y)^2},$$

where $A' \in L(\mathbb{R})$, is never bounded on BMO unless T is a constant multiple of the Hilbert transform.

On the other hand, the operator T whose kernel is

$$\text{p.v.}\frac{A(x)-A(y)-(x-y)A'(y)}{(x-y)^2}$$

is always bounded on BMO when $A' \in L^\infty$ (and it is even true when $A' \in \text{BMO}$).

This observation leads to the study of the vector space $\mathscr{E}_\gamma$ of all bounded operators $T: L^2(\mathbb{R}^n) \to L^2(\mathbb{R}^n)$ whose kernel satisfies (2.2) and (2.3) such that $T(1) = 0$. Then J. Peetre's theorem yields the BMO continuity of any $T \in \mathscr{E}_\gamma$. But what is definitively missing is a criterion, like the David–Journé theorem, giving the L^2-continuity of an operator $T: \mathscr{D} \to \mathscr{D}'$ with the WBP and whose kernel satisfies (2.2) and (2.3). We already know that $T^*(1)$ does not make sense for such operators, and $T^*(1) = 0$ would not suffice.

The examples of the Cauchy kernel and the double-layer potential operator show that those operators bounded on BMO with nonsmooth kernel in y are more natural from a geometrical viewpont.

Let $\Gamma \subset \mathbb{C}$ be the graph of a Lipschitz function $\varphi: \mathbb{R} \to \mathbb{R}: \|\varphi'\|_\infty \leqslant M < +\infty$, and let $z(x) = x + i\varphi(x)$ be a parametrization of Γ by $x \in \mathbb{R}$.

One way of writing the Cauchy kernel is

$$K(x, y) = \text{p.v.}\frac{1}{z(x)-z(y)} \in \mathscr{S}'(\mathbb{R}^2).$$

Another way is

$$\tilde{K}(x, y) = \text{p.v.}\frac{z'(y)}{z(x)-z(y)},$$

which is better adapted to the usual Cauchy formula.

In principle, the first operator T can be studied by Theorem 1. Indeed, this theorem applies only when $\|\varphi'\|_\infty \leqslant \eta$, and $\eta > 0$ is small enough. The extension to the general case can be obtained by the real-variable methods introduced by David [**1**].

Since $z'(y) = 1 + i\varphi'(y) \in L^\infty(\mathbb{R})$, the continuity of the true Cauchy kernel follows immediately, and this operator $\tilde{T}$ is bounded on BMO. Moreover $\tilde{T}$ is an isomorphism of BMO, and it follows that T is never bounded on BMO unless Γ is a straight line.

Another example illustrates the same philosophy. Let $D \subset \mathbb{R}^{n+1}$ be the open set defined by $t > \varphi(x)$, $x \in \mathbb{R}^n$, φ a Lipschitz function. Then the double-layer potential is the operator T: $\mathscr{D}(\mathbb{R}^n) \to \mathscr{D}'(\mathbb{R}^n)$ defined by the kernel

$$K(x, y) = \text{p.v.}\frac{1}{\omega_n}\frac{\varphi(x) - \varphi(y) - (x - y)\cdot\nabla\varphi(y)}{\left(|x - y|^2 + (\varphi(x) - \varphi(y))^2\right)^{(n+1)/2}}.$$

Then $T(1) = 0$ (mod constant functions), and T is bounded on $L^2(\mathbb{R}^n)$. Therefore, T is bounded on BMO, but the David–Journé theorem does not apply because the y-smoothness of the kernel is missing.

The last example was proposed by T. Kato. Let $A(x) = ((a_{j,k}(x)))_{1 \leqslant j,\, k \leqslant n}$ be a complex $n \times n$ matrix whose entries belong to $L^\infty(\mathbb{R}^n)$ and satisfy

$$\operatorname{Re}\sum_1 \sum_1 a_{j,k}(x)\xi_j\overline{\xi_k} \geqslant \eta|\xi|^2$$

for a given constant $\eta > 0$.

Then the accretive form

$$J(u, v) = \sum_1^n \sum_1^n a_{j,k}(x)\frac{\partial u}{\partial x_j}\frac{\partial \bar{v}}{\partial x_k}dx$$

defines the accretive operator $T(f) = -\operatorname{div}(A(x)\operatorname{Grad} f)$, whose domain $V \subset L^2(\mathbb{R}^n)$ is nonlinear in $A(x)$.

Kato conjectured that the domain $W \subset L^2(\mathbb{R}^n)$ of the accretive square root $S = \sqrt{T}$ would be the Sobolev space $H^1(\mathbb{R}^n)$. This conjecture is not yet fully established, but two distinct teams **[4]** could prove the following partial answer. There exists a positive constant $\varepsilon_n > 0$ such that, for $A(x) = I + B(x)$ and $\|B(x)\|_\infty < \varepsilon_n$,

$$S = \sqrt{T} = \sum_1^n R_j(B)\partial_j,$$

where $\partial_j = \partial/\partial x_j$: $H^1(\mathbb{R}^n) \to L^2(\mathbb{R}^n)$, and $R_j(B)$: $L^2(\mathbb{R}^n) \to L^2(\mathbb{R}^n)$ is bounded. Moreover, the kernel of $R_j(B)$ satisfies (2.2) and (2.3) but not (2.4). Therefore, the L^2-boundedness of $R_j(B)$ cannot be established through the David–Journé theorem. Nevertheless, this difficulty can be avoided by using a modification of $R_j(B)$, whose L^2-boundedness is equivalent to that of $R_j(B)$ and to which the David–Journé theorem applies.

These three examples advocate a theorem giving a necessary and sufficient condition for the L^2-boundedness of an operator T: $\mathscr{D}(\mathbb{R}^n) \to \mathscr{D}'(\mathbb{R}^n)$ with the WBP and a kernel satisfying (2.2) and (2.3).

6. The boundedness of singular integral operators on $\mathscr{H}^p$, $0 < p < 1$, Λ^s $(s > 0)$, and $B^s = \Lambda^s_{2,2}$ $(0 < s < 1)$. A new concept is needed for writing the criteria.

PROPOSITION 4. *Let $m \in \mathbb{N}$ be an integer and T: $\mathscr{D}(\mathbb{R}^n) \to \mathscr{D}'(\mathbb{R}^n)$ an operator with the WBP, whose distributional kernel satisfies*

$$(6.1) \qquad |\partial_x^\alpha K(x, y)| \leqslant C|x - y|^{-n-|\alpha|}$$

for all $\alpha \in \mathbb{N}^n$ *such that* $|\alpha| \leqslant m+1$. *Then for each* $\beta \in \mathbb{N}^n$, *such that* $|\beta| \leqslant m$, *and each* $\varphi \in \mathcal{D}(\mathbb{R}^n)$, *such that* $\varphi(0) = 1$, *there exists a family* $P_{\varepsilon,\beta}$ *of polynomials of degrees* $\leqslant |\beta|$ *such that*

$$\lim_{\varepsilon \downarrow 0}\left\{\int K(x,y)y^\beta\varphi(\varepsilon y)\,dy - P_{\varepsilon,P}(x)\right\} \tag{6.2}$$

exists in the distributional sense.

This limit is denoted by $T(x^\beta)$ *and is defined modulo polynomials of degrees* $\leqslant |\beta|$. *If* $T(x^\gamma) = 0$ *(mod polynomials of degrees* $\leqslant |\gamma|$*) whenever* $\gamma < \beta$, *then* $T(x^\beta) \in \Lambda^{|\beta|}_{\infty,\infty}$.

In this statement $\gamma < \beta$ means $\gamma_j \leqslant \beta_j$ for all $j = 1,\dots,n$, with equality for all j's excluded. The homogeneous Hölder space $\Lambda^m_{\infty,\infty}$ is also a quotient space modulo polynomials of degrees $< m$.

With this definition in mind, a criterion for the Λ^s continuity is

THEOREM 3. *If* $m < s < m+1$ *(*$m \in \mathbb{N}$ *and* $s > 0$*), and if the kernel* $K(x,y)$ *of* $T\colon \mathcal{D}(\mathbb{R}^n) \to \mathcal{D}'(\mathbb{R}^n)$ *satisfies* (6.1), *then a necessary and sufficient condition for the* Λ^s*-continuity of* T *is the WBP together with* $T(x^\alpha) = 0$ *for all* $\alpha \in \mathbb{N}^n$, *such that* $|\alpha| \leqslant m$.

The lack of smoothness in y of the kernel does not affect the Λ^s-continuity. The same property holds for the homogeneous Sobolev spaces.

THEOREM 4. *Let* $T\colon \mathcal{D}(\mathbb{R}^n) \to \mathcal{D}'(\mathbb{R}^n)$ *be an operator with the WBP whose kernel satisfies* (2.2) *and* (2.3). *If, moreover,* $T(1) = 0$, *then* $T\colon B^s \to B^s$ *is continuous when* $0 < s < \gamma$.

Finally, the continuity of a singular integral operator on the Stein–Weiss $\mathcal{H}^p$-space can be investigated when $0 < p < 1$. The integer $m \in \mathbb{N}$ is defined by $n/(n+m-1) \geqslant p > n/(n+m)$, and we assume that the kernel $K(x,y)$ of the singular integral operator T satisfies $|\partial_y^\alpha K(x,y)| \leqslant C|x-y|^{-n-m}$ whenever $|\alpha| \leqslant m$. Then the criterion for the $\mathcal{H}^p$-continuity of T is $T(1) \in \mathrm{BMO}(\mathbb{R}^n)$, together with $T^*(x^\alpha) = 0$ for $|\alpha| \leqslant m-1$.

The condition $T(1) = 0$ in Theorem 4 is not necessary. The correct condition has been found in joint work with Martin Meyer, using a notion due to Stegenga. If $0 < s < 1$, BMO_s denotes the subspace of $\mathrm{BMO}(\mathbb{R}^n)$ consisting of all $\beta \in \mathrm{BMO}$ for which there exists a constant C such that

$$\sum_{-\infty}^{\infty} 4^{ks}\|\Delta_k(\beta)S_{k-3}(f)\|_2^2 \leqslant C\sum_{-\infty}^{\infty} 4^{ks}\|\Delta_k(f)\|_2^2$$

for all test functions $f \in \mathcal{S}(\mathbb{R}^n)$.

Then Theorem 4 can be sharpened. If $T\colon \mathcal{D}(\mathbb{R}^n) \to \mathcal{D}'(\mathbb{R}^n)$ has a kernel $K(x,y)$ satisfying (2.2) and (2.3), and if $0 < s < \gamma$, then the following properties of T are equivalent:

(6.3) $T\colon B^s \to B^s$ is bounded;

(6.4) T has the WBP and $T(1) \in \mathrm{BMO}_s$.

An amusing example is the case studied by Stegenga, where T is pointwise multiplication by a given function $m(x)$. Then (2.2) and (2.3) are automatically satisfied since $K(x, y) = 0$ for $x \neq y$, and the WBP amounts to $m(x) \in L^\infty(\mathbb{R}^n)$. This explains why, in Stegenga, the pointwise multipliers of B^s are the functions $m(x) \in L^\infty(\mathbb{R}^n) \cap \mathrm{BMO}_s(\mathbb{R}^n)$.

7. Algebras of Calderón–Zygmund operators.

DEFINITION 3. *A Calderón–Zygmund operator T is a bounded operator T: $L^2(\mathbb{R}^n) \to L^2(\mathbb{R}^n)$ whose kernel satisfies* (2.2)–(2.4) *for some* $\gamma \in]0, 1]$.

For $\gamma \in]0, 1]$ we define the Banach space E_γ of Calderón–Zygmund operators for which (2.3) and (2.4) hold with exponent γ. This space is equipped with the norm $\|T\|_\gamma = \|K\|_\gamma + \|T\|_{L^2, L^2}$, but it is not an algebra. More precisely, we have

THEOREM 5. *Let S and T belong to E_γ. If $S(1) = 0$ and $T^*(1) = 0$, then ST belongs to E_η for any $\eta < \gamma$.*

Conversely, if $T \in E_\gamma$ is fixed and it is assumed that, for every $S \in E_\gamma$ such that $S(1) = 0$, ST is still a Calderón–Zygmund operator, then $T^*(1) = 0$.

Similarly, if $S \in E_\gamma$ is such that ST is a Calderón–Zygmund operator for all $T \in E_\gamma$ such that $T^*(1) = 0$, this forces $S(1) = 0$.

Let us denote by $\mathscr{A}_\gamma$ the collection of all $T \in E_\eta$, $\eta > \gamma$, such that $T(1) = T^*(1) = 0$. Then $\mathscr{A}_\gamma$ is a selfadjoint Fréchet subalgebra of $\mathscr{L}(L^2(\mathbb{R}^n), L^2(\mathbb{R}^n))$.

An even simpler subalgebra is provided by the collection of operators $\mathscr{A}_\infty$ that preserve "functional dyadic grids"–a notion which will now be defined.

The subspace $\mathscr{S}_0(\mathbb{R}^n) \subset \mathscr{S}(\mathbb{R}^n)$ is defined by the vanishing of all moments: $\int x^\alpha f(x)\,dx = 0$ for all $\alpha \in \mathbb{N}^n$. A subset $\mathscr{B} \subset \mathscr{S}_0(\mathbb{R}^n)$ is bounded when, for each integer m, a constant C_m exists such that $\sup_{\varphi \in \mathscr{B}} q_m(\varphi) \leqslant C_m$ (q_m is defined by (2.7)). The collection of all dyadic cubes

$$Q = \left\{x;\ p_j 2^{-k} \leqslant x_j < (p_j - 1)2^{-k}\right\}, \qquad p_j \in \mathbb{Z}, k \in \mathbb{Z}, 1 \leqslant j \leqslant n,$$

is denoted by $\mathscr{Q}$, and we arrive at the following definition.

DEFINITION 4. *A functional dyadic grid is a collection ψ_Q, $Q \in \mathscr{Q}$, of functions in $\mathscr{S}_0(\mathbb{R}^n)$ for which a bounded subset $\mathscr{B} \subset \mathscr{S}_0(\mathbb{R}^n)$ exists with the following property: for each dyadic cube as above, $\psi_Q(x) = \tilde{\Psi}_Q(2^k x - p)$, where $\tilde{\psi}_Q \in \mathscr{B}$ and $p = (p_1, \ldots, p_n)$.*

The Besov spaces can be built with pieces belonging to functional dyadic grids. Let us give two examples.

If ψ_Q, $Q \in \mathscr{Q}$, is any functional dyadic grid, then we have the one-sided Plancherel inequality

$$\left\| \sum_{\mathscr{Q}} c_Q 2^{kn/2} \psi_Q \right\|_2 \leqslant C(\mathscr{B}) \left(\sum_{\mathscr{Q}} |c_Q|^2 \right)^{1/2}, \tag{7.1}$$

which means that the pieces inside a functional dyadic grid are almost orthogonal. But, conversely, there exists a functional dyadic grid $\psi_Q^{(0)}$ such that any function $f \in L^2(\mathbb{R}^n)$ can be represented as a series $\Sigma_Q c_Q 2^{kn/2} \psi_Q^{(0)}$ such that $\|f\|_{L^2(\mathbb{R}^n)}$ and $(\Sigma_Q |c_Q|^2)^{1/2}$ be equivalent norms in $L^2(\mathbb{R}^n)$.

The same is true for $\Lambda^0_{1,1}$. There exists a functional dyadic grid $\psi_Q^{(0)}$ such that every function in $\Lambda^0_{1,1}$ can be written as

$$f = \sum_{\mathcal{Q}} \gamma_Q 2^{kn} \psi_Q^{(0)}, \quad \text{with } \sum_{\mathcal{Q}} |\gamma_Q| \leqslant C\|f\|_{\Lambda^0_{1,1}},$$

while for any functional dyadic grid ψ_Q, $Q \in \mathcal{Q}$, there exists a constant $C(\mathscr{B})$ such that

$$\left\| \sum_{\mathcal{Q}} c_Q 2^{kn} \psi_Q \right\|_{\Lambda^0_{1,1}} \leqslant C(\mathscr{B}) \sum_{\mathcal{Q}} |c_Q|. \tag{7.2}$$

This is the substitute for the atomic decomposition of functions in the Stein–Weiss $\mathscr{H}^1$-space.

DEFINITION 5. *We denote by $\mathscr{A}_\infty$ the collection of all operators T: $\mathscr{S}_0(\mathbb{R}^n) \to \mathscr{S}_0(\mathbb{R}^n)$ such that for any functional dyadic grid ψ_Q, $Q \in \mathcal{Q}$, $T(\psi_Q)$ is also a functional dyadic grid.*

From what we have seen, those operators preserve $L^2(\mathbb{R}^n)$. By the definition of $\mathscr{A}_\infty$ it is obvious that this collection is an algebra. But what is more interesting is the fact that those operators $T \in \mathscr{A}_\infty$ can be characterized by their kernels.

THEOREM 6. *A continuous linear operator T: $\mathscr{S}(\mathbb{R}^n) \to \mathscr{S}'(\mathbb{R}^n)$ belongs to $\mathscr{A}_\infty$ if and only if its kernel $K(x, y)$ satisfies $|\partial_x^\alpha \partial_y^\beta K(x, y)| \leqslant C_{\alpha,\beta}|x - y|^{-n-|\alpha|-|\beta|}$ if T has the WBP, and, for each $m \in \mathbb{N}$, $T(E_m) \subset E_m$ and $T^*(E_m) \subset E_m$, where E_m is the space of polynomials of degree $\leqslant m$.*

The conditions are symmetric in T and T^*, and therefore $\mathscr{A}_\infty$ is a selfadjoint subalgebra of $\mathscr{L}(L^2(\mathbb{R}^n), L^2(\mathbb{R}^n))$.

Inside $\mathscr{A}_\infty$ lies the collection $\mathscr{B}_\infty$ of Bony's paradifferential operators of order 0. We recall that $T \in \mathscr{B}_\infty$ when T is a pseudodifferential operator whose symbol $\sigma(x, \xi) \in S^0_{1,1}(\mathbb{R}^n \times \mathbb{R}^n)$, which is the forbidden Hörmander class, and

$$\sigma(x, \xi) = 0 \quad \text{for } |\xi| \leqslant 1, \tag{7.3}$$

(7.4) the partial Fourier transform (in x) of $\sigma(x, \xi)$ vanishes when $|\eta| \geqslant r|\xi|$ ($0 < r < 1$, r is fixed, and η is the dual variable of x).

The hypothesis $\sigma(x, \xi) \in S^0_{1,1}(\mathbb{R}^n \times \mathbb{R}^n)$ implies the estimates on the kernel $K(x, y)$ of T, (7.3) implies $T(x^\alpha) = 0$ and (7.4) implies $T^*(x^\alpha) = 0$ (but this would not be the case for $r > 1$).

8. Actions of $L^2(\mathbb{R}^n)$ on $\mathscr{H}^p(\mathbb{R}^n)$, $0 < p \leqslant 1$. We first let $L^\infty(\mathbb{R}^n)$ act on $\mathscr{H}^1(\mathbb{R}^n)$ by

$$\chi(a, f) = \pi(a, f) + \pi(f, a) = T_a(f), \tag{8.1}$$

where $\pi(a, f)$ is defined by (4.7).

Theorem 2 applies to T_a and yields $\chi(a, f) \in \mathscr{H}^1$ whenever $a \in L^\infty(\mathbb{R}^n)$ and $f \in \mathscr{H}^1(\mathbb{R}^n)$.

When $n = 1$, $a \in H^\infty(\mathbb{R})$, and $f \in H^1(\mathbb{R})$, then $\chi(af) = af$ and (8.1) is the Leibniz rule.

This action of $L^\infty(\mathbb{R}^n)$ on $\mathscr{H}^1(\mathbb{R}^n)$ is not compatible with the Banach algebra structure of $L^\infty(\mathbb{R}^n)$. It can be guessed that such a compatible action does not exist.

A technical modification is needed for extending this action to $\mathscr{H}^p(\mathbb{R}^n)$ when $0 < p < 1$. We denote by $\psi \in \mathscr{S}(\mathbb{R}^n)$ a radial function such that $\hat{\psi}$ is supported by $1 \leqslant |\xi| \leqslant 2$ and

$$\int_0^\infty t|\xi|e^{-t|\xi|}\hat{\psi}(t\xi)\,\frac{dt}{t} = 1.$$

We denote the convolution with $\psi_t(x) = t^{-n}\psi(x/t)$ by $\tilde{Q}_t$, and we modify (4.7) to get

$$\tilde{\pi}(a, f) = \int_0^\infty \tilde{Q}_t\{(Q_t a)(P_t f)\}\,\frac{dt}{t}.$$

This leads to defining $\tilde{\chi}(a, f) = \tilde{\pi}(a, f) + \tilde{\pi}(f, a)$. Then if $a \in L^\infty(\mathbb{R}^n)$ and $f \in \mathscr{H}^p(\mathbb{R}^n)$, $0 < p \leqslant 1$, we have $\tilde{\pi}(a, f) \in \mathscr{H}^p(\mathbb{R}^n)$, and this action extends the trivial action of $H^\infty(\mathbb{R})$ on $H^p(\mathbb{R})$.

9. The nilpotent Lie group setting. Let $\mathscr{N}$ be a stratified (simply connected) nilpotent Lie group, and let $N = V_1 \oplus \cdots \oplus V_m$ be its graded Lie algebra: $[V_i, V_j] \subset V_{i+j}$ (with $V_j = \{0\}$ for $j > m$) and $[V_1, V_j] = V_{j+1}$ for all j's.

The dilations δ_t: $N \to N$ $(t > 0)$ are defined by $\delta_t(X) = t^j X$ when $X \in V_j$ and are automorphisms of the Lie algebra structure of N. Via the exponential mapping, we are led to define $\delta_t(\exp X) = \exp(\delta_t, X)$, and $(\delta_t)_{t>0}$ is now a one-parameter group of automorphisms of $\mathscr{N}$.

The Schwartz class $\mathscr{S}(N)$ and the subspace $\mathscr{S}_0(N)$ are defined in terms of the underlying vector-space structure of N, and $\mathscr{S}_0(N)$ is lifted up to $\mathscr{S}_0(\mathscr{N})$ by the exponential mapping ($g \in \mathscr{S}_0(\mathscr{N})$ means $g(\exp X) = f(X)$, where $f \in \mathscr{S}_0(N)$).

For each bounded subset $\mathscr{B} \subset \mathscr{S}_0(\mathscr{N})$, $\mathscr{M}(\mathscr{B}; x, t)$ denotes the collection of functions f: $\mathscr{N} \to C$ of the type $f(y) = g(\delta_t^{-1}(x^{-1}y))$, where $g \in \mathscr{B}$. Such a function $f \in \mathscr{M}(\mathscr{B}; x, t)$ is called a $\mathscr{B}$-molecule centered at x with width $t > 0$. If $f \in \mathscr{M}(B; x, t)$, then $f(ay) \in \mathscr{M}(\mathscr{B}; ax, t)$ such that the definition is compatible with left translations.

DEFINITION 6. *$\mathscr{A}_\infty$ is the algebra of all operators T: $\mathscr{S}_0(\mathscr{N}) \to \mathscr{S}_0(\mathscr{N})$ with the property that, for each bounded subset $\mathscr{B} \subset \mathscr{S}_0(\mathscr{N})$, another bounded subset $\mathscr{B}' \subset \mathscr{S}_0(\mathscr{N})$ would exist such that, for all $x \in \mathscr{N}$ and all $t > 0$, T maps any $\mathscr{B}$-molecule centered at x with width t into a $\mathscr{B}'$-molecule centered at x with width t.*

These operators extend to bounded operators T: $L^2(\mathscr{N}) \to L^2(\mathscr{N})$. Their kernel satisfies

$$(9.1) \qquad \left|X_x^I X_y^J K(x, y)\right| \leqslant C(I, J)\left|x^{-1}y\right|^{-Q-|I|-|J|},$$

where the notations will now be explained.

We denote a basis of the vector space $V_1 \subset N$, by $X_1, \ldots, X_p$, $I = \{i_1, \ldots, i_k\}$ with $i_j \in \{1, \ldots, p\}$, and $X^I = X_{i_1} \cdots X_{i_k}$ with $|i| = k$.

Moreover, in (9.1), $|x|$ is the homogeneous norm satisfying $|\delta_t(x)| = t|x|$, and Q is the homogeneous dimension of $\mathcal{N}$ characterized by $d(\delta_t(x)) = t^Q\,dx$. For stating the converse, we need to define the WBP.

The action L_x of $\mathcal{N}$ on $\mathcal{S}(\mathcal{N})$ is defined by $(L_x f)(y) = f(x^{-1}y)$, while the action δ_t of $]0, +\infty]$ on $\mathcal{S}(\mathcal{N})$ is simply $(\delta_t f)(y) = f(\delta_t^{-1}y)$.

DEFINITION 7. *A linear and continuous operator T: $\mathcal{S}(\mathcal{N}) \to \mathcal{S}'(\mathcal{N})$ has the WBP if there exists an integer $m \in \mathbb{N}$ and a constant C such that*

$$(9.2) \qquad \left|\left\langle \delta_t L_x T L_x^{-1} \delta_t^{-1} u, v\right\rangle\right| \leqslant C q_m(u) q_m(v)$$

for all $x \in \mathcal{N}$, all $t > 0$, and any pair (u, v) of two functions of $\mathcal{S}(\mathcal{N})$.

The norms q_m are defined by (2.7) on the Lie algebra N and are then lifted on $\mathcal{N}$ by the exponential mapping. For instance, a continuous linear operator T: $\mathcal{S}(\mathcal{N}) \to \mathcal{S}'(\mathcal{N})$ commuting with translations and dilations has the WBP. With these definitions in mind, we have

THEOREM 7. *A continuous linear operator T: $\mathcal{S}(\mathcal{N}) \to \mathcal{S}'(\mathcal{N})$ belongs to $\mathcal{A}_\infty$ if and only if the following properties of T are satisfied*:

(9.3) *the kernel $K(x, y)$ of T satisfies* (9.1);

(9.4) *T has the WBP*;

(9.5) *for each integer m, let E_m be the space of polynomials on $\mathcal{N}$ of (homogeneous) degree $\leqslant m$; then $T(E_m) \subset E_m$.*

The first examples of such operators are due to Knapp and Stein. They studied the convolution case where T: $\mathcal{S}(\mathcal{N}) \to \mathcal{S}'(\mathcal{N})$ commutes with left translations. Then the kernel of T has the form $K(x, y) = k(y^{-1}x)$, where $k \in \mathcal{S}'(\mathcal{N})$.

Assuming $k \in C^\infty(\mathcal{N}\backslash\{1\})$, together with $t^Q k(\delta_t y) = k(y)$, $t > 0$, $Q =$ homogeneous dimension of $\mathcal{N}$, yields the size and smoothness conditions on $K(x, y)$. Any convolution operator preserves all spaces E_m, $m \in N$. Then the key condition is the WBP, which takes the form

$$(9.6) \qquad \int_{r \leqslant |y| \leqslant R} k(y)\,dy = 0 \quad \text{for } R > r > 0.$$

If instead of assuming $k(y)$ homogeneous, we release this condition to the size constraints

$$(9.7) \qquad |X^I k(y)| \leqslant C(I) y^{-Q-|I|},$$

then the WBP takes the form

$$(9.8) \qquad \left|\int_{r \leqslant |y| \leqslant R} k(y)\,dy\right| \leqslant C$$

(for a given constant C and all $R > r > 0$). An example is $k(y) = |y|^{-Q+i\gamma}$, where $\gamma \neq 0$.

References

1. Guy David, *Opérateurs intégraux singuliers sur certaines courbes du plan complexe*, Ann. Sci. École Norm. Sup. (4) **17** (1984), 157–189.

2. Guy David and Jean-Lin Journé, *A boundedness criterion for generalized Calderón–Zygmund operators*, Ann. of Math. (2) **120** (1984), 371–397.

3. Pierre-Gilles Lemarié, Thèse (Publications Mathématiques, Batîment 425, Faculté des Sciences, 91405–Orsay, France).

4. Yves Meyer, *Intégrales singulières, opérateurs mltilinéaires, analyse complexe et équations aux derivées partielles*, Actes Congrés Internat. Math. (Varsovie, 1983).

Centre de Mathématique, Ecole Polytechnique, France

Proceedings of Symposia in Pure Mathematics
Volume **43** (1985)

Integrability and Holomorphic Extendibility for Rigid CR Structures

LINDA PREISS ROTHSCHILD[1]

We consider here surfaces M in $\mathbb{C}^{n+l}$ defined locally by

$$M = \{(z, w) \in \mathbb{C}^{n+l}: z \in \mathbb{C}^n, w \in \mathbb{C}^l \text{ and } \operatorname{Im} w = \Phi(z, \bar{z})\}, \tag{1}$$

where Φ: $\mathbb{R}^n \times \mathbb{R}^n \to \mathbb{R}^l$ is smooth; such a surface will be called *rigid*. We give an invariant characterization of CR (Cauchy–Riemann) structures giving rise to rigid surfaces and give sufficient conditions for holomorphic extendibility of solutions of the tangential Cauchy–Riemann equations. The results announced here represent joint work with M. S. Baouendi and F. Treves; details will appear elsewhere.

Extendibility of CR functions has been studied extensively, beginning with the classical paper of Hans Lewy [**11**]. Later results were obtained by R. Nirenberg [**12**], Hill and Taiani [**9**], Boggess and Polking [**7**], and others, using the method of analytic disks. In [**3**]–[**5**], and in [**2**] with Chang, Baouendi and Treves obtained new results on extendibility by using microlocal methods—in particular, the Fourier–Bros–Iagolnitzer (FBI) integral (see [**13**]). The FBI integral is again a major technical tool in the present work.

Let $\Omega \subset \mathbb{R}^{2n+l}$ be open. A generic (abstract) CR structure on Ω is a subbundle $\mathscr{V} \subset \mathbb{C}T(\Omega)$ such that

$$\mathscr{V} \cap \bar{\mathscr{V}} = (0), \tag{2}$$

and

$$[\mathscr{V}, \mathscr{V}] \subset \mathscr{V}. \tag{3}$$

Let $\mathbb{L} = C^\infty(\Omega, \mathscr{V})$, and let $L_1, L_2, \ldots, L_n$ be a basis of $\mathbb{L}$. The inclusion (3) means that $[L, L'] \in \mathbb{L}$ for any $L, L' \in \mathbb{L}$. The structure $\mathscr{V}$ (or $\mathbb{L}$) is called *locally*

1980 *Mathematics Subject Classification*. Primary 32F25.
[1] Partially supported by NSF grant #DMS-8319819.

integrable at ω_0 if there exist C^∞ functions $\zeta_1, \ldots, \zeta_{n+l}$, $\zeta_i\colon U \to \mathbb{C}$, $\omega_0 \in U \subset \Omega$, with $\{d\zeta_i(\omega_0)\}$ linearly independent and

$$\mathbb{L}\zeta_i = 0, \qquad i = 1, 2, \ldots, n. \tag{4}$$

If the ζ_i exist, they define a local embedding of Ω onto a submanifold M of $\mathbb{C}^{n+l}$; M is then called a *CR manifold*.

THEOREM 1. *Let $\mathscr{V}$ be an abstract generic CR structure of dimension n. Suppose there exists an l-dimensional real subspace $\mathbb{T} \subset C^\infty(\Omega, \mathbb{C}T(M))$ such that*

$$[\mathbb{T}, \mathbb{T}] = 0, \tag{5}$$

and

$$[\mathbb{L}, \mathbb{T}] \subset \mathbb{L}. \tag{6}$$

Then $\mathscr{V}$ is integrable and there exist coordinates (z, w) on $\mathbb{C}^{n+l}$ such that

$$\zeta_j = z_j, \quad j = 1, 2, \ldots, n, \qquad \textit{and} \qquad \zeta_{n+j} = \operatorname{Re} w_j + i\Phi_j(z, \bar{z}), \tag{7}$$

where Φ_j is a real-valued smooth function on $\mathbb{R}^{2n}$ for $j = 1, \ldots, n$.

Conversely, if $\mathscr{V}$ is a generic integrable CR structure for which the ζ_i can be taken to be in the form (7), *then $\mathbb{T}$, satisfying* (4) *and* (5), *exists.*

A complex-valued function h on a CR manifold M is called *CR* if $\zeta^*(\mathbb{L})h = 0$ in a neighborhood of $\zeta(\omega_0)$, where $\zeta^*(L)$ is the push forward of $L \in \mathbb{L}$ from Ω to M. Now suppose M is rigid. By a *wedge* in $\mathbb{C}^{n+l}$ we mean an open set of the form

$$W = \{(z, w)\colon \operatorname{Im} w = \Phi(z, \bar{z}) + v, v \in V_\delta, z \in U\}, \tag{8}$$

where $V_\delta = \{v \in \Gamma\colon |v| < \delta\}$, where $\Gamma \subset \mathbb{R}^l - \{0\}$ is a strictly conic set and U is a neighborhood of 0 in $\mathbb{C}^n$. If H is holomorphic in the wedge W and h is a CR function on M, we write $h = bH$ if h is (locally) the boundary value of H.

The following theorem generalizes a result of Andreotti–Hill [**1**].

THEOREM 2. *Let M be rigid, and let h be a CR function on M. Then we may write $h = h_1 + h_2 + \cdots + h_k$ such that, for each h_i, there is a wedge W_i and a holomorphic function H_i defined in W_i such that $h_i = bH_i$.*

In order to guarantee extendibility for CR functions, it is necessary to add the following hypothesis. M is of *finite type* if there is a number m such that the set of all commutators of $\mathbb{L} \oplus \overline{\mathbb{L}}$ of length $\leqslant m$ spans the tangent space to Ω. The following generalizes Theorem I.1 of [**5**] for hypersurfaces in the rigid case.

THEOREM 3. *If M is a rigid CR manifold of finite type, then for any CR function h on M there is a wedge W of the form* (8) *in $\mathbb{C}^{n+l}$ and a holomorphic function H on W such that $h = bH$ near the origin.*

We now recall the definition of the hypoanalytic wave-front set [**2**] for CR functions on M. If h is a CR function, let $\tilde{h} = h \circ \zeta$, $\tilde{h}\colon \Omega \to \mathbb{C}$, with $\mathbb{L}\tilde{h} = 0$ near ω_0. We let $\tilde{h}_0(x, s) = \tilde{h}(x, 0, s)$ and $\zeta_0(x, s) = \xi(x, 0, s)$. For $u(x, s)$ compactly supported let

$$F^K(u; v, \eta) = \int e^{-i\eta \cdot \zeta_0(x,s) - K\langle\eta\rangle[v - \zeta_0(x,s)]^2} u(x, s)\, d\zeta_0(x, s),$$

where $v, \eta \in \mathbf{C}^{n+l}$, with $|\mathrm{Im}\,\eta| < |\mathrm{Re}\,\eta|$, $\langle\eta\rangle = (\eta_1^2 + \cdots + \eta_{n+l}^2)^{1/2}$, and $[v]^2 = v_1^2 + \cdots + v_{n+l}^2$. Then h is *hypoanalytic* at $(0;\ 0, \eta_0)$, $\eta_0 \in \mathbf{R}^l$, if there exists $\varphi \in C_0^\infty(\mathbf{R}^{n+l})$, $R > 0$, $K > 0$, and an open cone $\mathscr{C}$ in $C^{n+l} \setminus \{0\}$ containing η_0 such that

$$|F^K(\varphi h_0, v, \eta)| \leqslant Ce^{-|\eta|/R} \tag{9}$$

for all $\eta \in \mathscr{C}$ and all $v \in V^{\mathbf{C}}$, a neighborhood of 0 in $\mathbf{C}^m$. Then or any $\eta \in \mathbf{R}^l$,

$$(0;\ 0, \eta) \notin \mathrm{WF}_{\mathrm{ha}}\, h \quad \text{iff} \quad h \text{ is hypoanalytic at } (0;\ 0, \eta).$$

The connection of $\mathrm{WF}_{\mathrm{ha}}$ with holomorphic extendibility is the following. Let $\Gamma \subset \mathbf{R}^l - \{0\}$ be a strictly convex cone with Γ^0 its polar; i.e., $\Gamma^0 = \{v \in \mathbf{R}^l\colon v \cdot \xi > 0 \text{ for all } \xi \in \Gamma\}$. Then if h is a CR function, $\mathrm{WF}_{\mathrm{ha}} \subset \mathrm{h}\{(0;\ 0, \eta)\colon \eta \in \Gamma\}$ if and only if $h = bH$, where H is holomorphic in a wedge W of the form (8) with Γ replaced by Γ^0. Our final result gives a sufficient condition for implying that $(0;\ 0, \xi) \notin \mathrm{WF}_{\mathrm{ha}}\, h$.

In [**5**] Baouendi and Treves introduced the notion of sector property in order to give a sufficient condition for extendibility for CR functions on a hypersurface. A real-valued polynomial $p_m(c, \bar{c})$, $c \in \mathbf{C}$, homogeneous of degree m satisfies the *sector property* if there exists $q_m(c, \bar{c})$, harmonic and also of degree m, such that

$$p_m(c, \bar{c}) + q_m(c, \bar{c}) < 0 \text{ on a sector of angle } > \pi/m \text{ in the plane.} \tag{10}$$

THEOREM 4. *Let M be a rigid CR manifold defined by* (1) *and* $\gamma\colon \mathbf{C} \to \mathbf{C}^n$ *a* (*possibly singular*) *holomorphic curve. For $\xi \in \mathbf{R}^l - \{0\}$ let $\Phi(\gamma(c)) \cdot \xi = p_m(c, \bar{c}) + O(m + 1)$, where $O(m+1)$ is of degree $\geqslant m + 1$. Then if $p_m(c, \bar{c})$ satisfies the sector property*, $(0;\ (0, \xi)) \notin \mathrm{WF}_{\mathrm{ha}}$ *h for any CR function h on M.*

REMARK. E. Bedford [**6**] has recently defined a weaker property, called the *amalgamated sector property*, which implies extendibility on some classes of hypersurfaces. It is possible that Theorem 4 might extend to this case also.

REFERENCES

1. A. Andreotti and C. D. Hill, *E. E. Levi convexity and the Hans Lewy problem*. I, II, Ann. Scuola Norm. Sup. Pisa. Sci. Fis. Mat. **26** (1972), 325–363, 747–806.

2. M. S. Baouendi, C. H. Chang and F. Treves, *Microlocal hypo-analyticity and extension of CR functions*, J. Differential Geom. **18** (1983), 331–391.

3. M. S. Baouendi and F. Treves, *A property of the functions and distributions annihilated by a locally integrable system of complex vector fields*, Ann. of Math. (2) **113** (1981), 387–421.

4. ______, *A microlocal version of Bochner's tube theorem*, Indiana Univ. Math. J. **31** (1982), 885–895.

5. ______, *About the holomorphic extension of CR functions on real hypersurfaces in complex space*, Duke Math. J. **51** (1984), 77–107.

6. E. Bedford, *Local and global envelopes of holomorphy in* $\mathbf{C}^2$ (preprint).

7. A. Boggess and J. Polking, *Holomorphic extension of CR functions* (preprint).

8. G. Henkin, *Solution des equations de Cauchy–Riemann tangentielles sur des varietés de Cauchy–Riemann q-concaves*, C. R. Acad Sc. Paris **292** (1981), 27–31.

9. C. D. Hill and G. Taiani, *Families of analytic discs in* $\mathbb{C}^n$ *with boundaries on a prescribed CR submanifold*, Ann. Scuola Norm. Sup. Pisa **4-5** (1978), 327–380.

10. L. R. Hunt and R. O. Wells, *Extensions of CR functions*, Amer. J. Math. **98** (1976), 805–820.

11. H. Lewy, *On the local character of the solution of an atypical differential equation in three variables and a related problem for regular functions of two complex variables*, Ann. of Math. (2) **64** (1956), 514–522.

12. R. Nirenberg, *On the H. Lewy extension phenomenon*, Trans. Amer. Math. Soc. **168** (1972), 337–356.

13. J. Sjöstrand, *Singularities analytiques microlocales*, Astérisque **95** (1982), 1–166.

UNIVERSITY OF CALIFORNIA, SAN DIEGO

Proceedings of Symposia in Pure Mathematics
Volume **43** (1985)

Multiple Wells and Tunneling

JOHANNES SJÖSTRAND

We present some of the results obtained jointly with B. Helffer [**2**]–[**5**]. Let M be a compact Riemannian manifold of dimension n (or possibly $\mathbb{R}^n$), and let $V \in C^\infty(M; \mathbb{R})$. (If $M = \mathbb{R}^n$ we make the additional assumption that $\underline{\lim}_{|x|\to\infty} V(x) > 0$. Then all our results are essentially valid in this case also.) We are interested in the eigenvalues close to 0 of the semiclassical Schrödinger operator $P = -h^2\Delta + V(x)$ and, particularly, in the exponentially small splitting effects that may appear, for instance, in symmetric situations. Among the mathematically rigorous works in the field, we have been particularly stimulated by those of E. M. Harrell [**1**] and B. Simon [**8**, **9**], who treat mainly the case of double wells, and Jona-Lasinio, Martinelli, and Scoppola [**7**]. As noted by Simon [**9**], the Agmon metric $\max(V, 0)dx^2$ plays a very important role for these problems. Here dx^2 denotes the given Riemannian metric.

Suppose $\{x \in M; V(x) \leqslant 0\} = U_1 \cup \cdots \cup N_N$, where U_j are compact, disjoint, and of diameter 0 with respect to the (degenerate) distance d associated with the Agmon metric. In our approach to the problem, we make systematic use of certain reference problems, obtained by decreasing M into smaller manifolds with boundary, that contain precisely one of the wells. (We can also obtain a reference problem for each well by increasing the potential near the other wells until they are destroyed.) For $\eta > 0$ very small but independent of h, we temporarily define $M_j = M \setminus \bigcup_{k\neq j} B(U_k, \eta)$, and we let P_{M_j} denote the Dirichlet realization of $P = -h^2\Delta + V$ as a selfadjoint operator in $L^2(M_j)$. Here $B(U_k, \eta)$ denotes the open ball of center U_k and radius η with respect to the Agmon distance d. (If ∂M_j is not C^∞, we make a small modification of M_j, so that we obtain this property.) Let $I(h)$ be a compact interval tending to $\{0\}$ as $h \to 0$. We assume that P and P_{M_j} have no eigenvalues in $(I(h) + [-a(h), a(h)]) \setminus I(h)$, where $a(h) > 0$ and

1980 *Mathematics Subject Classification*. Primary 81C12, 85B40, 35Y10, 85P05, 35P20.

$\log a(h) = o(1/h)$. Here h varies in some set $J \subset]0,1]$, with $0 \in \bar{J}$. It turns out that this assumption is essentially independent of the choice of $\eta > 0$ (sufficiently small), because the eigenvalues of P_{M_j} near $I(h)$ only change by an exponentially small quantity when η is changed. If $I(h) \cap \mathrm{Sp}(P_{M_k}) = \varnothing$, we call U_k a *nonresonant* well; otherwise U_k is a *resonant* well. Let $U_1, \dots, U_{\bar{N}}$, $\bar{N} \leqslant N$, be the resonant wells. We now modify the definition of the M_j, so that, $1 \leqslant j \leqslant \bar{N}$,

$$M_j = M \setminus \bigcup_{j \neq k \leqslant \bar{N}} B(U_k, \eta). \tag{1}$$

Assume for simplicity that P_{M_j} has precisely one eigenvalue μ_j in $I(h)$ for each j. If φ_j is the corresponding normalized eigenfunction, then

$$\varphi_j = O\big(\exp\big(-\big(d(x, U_j)/h\big)\big)\big),$$

and similarly for all the derivatives of φ_j. Here, O means that there is some function $\varepsilon(\eta)$ tending to 0 as η tends to 0 such that

$$\varphi_j(x,h) = O\big(\exp\big(\big(\varepsilon(\eta) - d(x, U_j)\big)/h\big)\big)$$

uniformly for $x \in M_j$, $h \in J$. This Agmon-type estimate is due to B. Simon [**9**] in the present context. (More precise forms of this kind of estimate, with the factor $\exp(\varepsilon(\eta)/h)$ replaced by some power of h, play an important role in ou works:for instance, when justifying various BKW expansions.)

Let $F \subset L^2(M)$ be the sum of the eigenspaces of P corresponding to $\mathrm{Sp}(P) \cap I(h)$, and let π_F be the corresponding orthogonal projection. Let $\theta_k \in C_0^\infty(B(U_k, 2\eta))$ be equal to 1 near $\overline{B(U_k, \eta)}$ for $1 \leqslant k \leqslant \bar{N}$, and put $\psi_j = \chi_j \varphi_j$, where $\chi_j = 1 - \sum_{k \neq j} \theta_k$, $1 \leqslant j \leqslant \bar{N}$.

THEOREM 1 [**3**]. *The vectors $v_j = \pi_F \psi_j$ form a basis in F, and if $\mathbf{e} = \mathbf{v}((v_j|v_k))^{-1/2}$ is the orthonormalization of this basis, then the corresponding matrix of $P|_F$ is of the form*

$$\mathrm{diag}(\mu_j) + (\hat{w}_{j,k}) + O(\mathscr{D}'^2 + \mathscr{D}'^3), \tag{2}$$

where

$$\hat{w}_{j,k} = (w_{j,k} + w_{k,j})/2, \quad w_{j,j} = 0,$$

$$w_{j,k} = h^2 \int \chi_j(\varphi_k \nabla \varphi_j - \varphi_j \nabla \varphi_k) \nabla \chi_k \, dx, \qquad j \neq k,$$

and $\mathscr{D}'$ is the matrix $((1 - \delta_{j,k}) \exp(-d(U_j, U_k)/h))$.

The meaning of (2) is this: the first two terms give the matrix of $P|_F$ up to an error that at (j,k) is $O(1)$ times the element at (j,k) of $\mathscr{D}'^2 + \mathscr{D}'^3$. In the one-dimensional case both φ_j and $w_{j,k}$ admit, in principle, asymptotic expansions. The same is true in the multidimensional case if we assume that the U_j are points where $V'' > 0$, and if we make suitable nondegeneracy assumptions on the minimal geodesics (with respect to the Agmon metric) between the resonant wells. (See [**2, 3**] for more details.) Then if $I(h) = [C_1 h, C_2 h]$,

$$\tilde{w}_{j,k} = a_{j,k}(h) \exp(-d(U_j, U_k)/h), \tag{3}$$

(4) $$a_{j,k}(h) \sim h^{m_{j,k}}(b_0 + b_1 h + \cdots).$$

If the minimal geodescs pass through certain of the nonresonant wells, then (3), with suitable assumptions, is still valid, but the asymptotic expansion (4) is more complicated, involving terms of the form $h^k(\log h)^m$ for $k \in \overline{\mathbb{R}}_+$, $m \in \mathbb{N}$. (See [**4**] for more details.)

Witten's paper [**10**] contains many beautiful ideas about how to study the topology of a compact oriented Riemannian manifold with semiclassical methods; in [**5**] we managed to give mathematical answers to some of his statements and open questions. His basic idea is to replace the ordinary de Rahm complex d by a conjugated one $d_f = e^{-f/h}\tilde{d}e^{f/h}$, where $\tilde{d} = hd$ and f is a smooth, real Morse function on M. The associated Laplacian on q-forms is then

(5) $$P^{(q)} = d_f^* d_f + d_f d_f^* = -h^2\Delta^{(q)} + \|df\|^2 + h\left(\mathscr{L}_{\nabla f} + \mathscr{L}^*_{\nabla f}\right).$$

Here $\Delta^{(q)}$ is the Hodge Laplacian on q-forms, and $\mathscr{L}_{\nabla f}$ is the Lie derivative along the gradient field ∇f, so the last term in (5) is a matrix. The operator $P^{(q)}$ then looks very much like a scalar Schrödinger operator with potential $\|df\|^2$, and the methods of [**2**]–[**4**] apply.

Let $U_j, j = 1, \ldots, N$, be the critical points of f, and let $C^{(q)} = \{j;\ U_j$ is of index $q\}$. Here the index of U_j is defined to be the number of negative eigenvalues of $f''(U_j)$. If we take $I(h) = [0, \varepsilon_0 h]$ for ε_0 small enough, it turns out that the resonant wells for $P^{(q)}$ are precisely those of class $C^{(q)}$. If we define M_k as before, then for $k \in C^{(q)}$, $P^{(q)}_{M_k}$ has precisely one eigenvalue $\mu_k^{(q)}$ in $I(h)$ and, moreover, $\mu_k^{(q)} = O(h^2)$. Let $\varphi_k^{(q)} = O(\exp(-d(x, U_k^{(q)})/h)) \in L^2(M_j)$ be the corresponding normalized eigenform. Observing that $P^{(q+1)}d_f\varphi_k^{(q)} = \mu_k^{(q)}d_f\varphi_k^{(q)}$ and the resonant wells for $P^{(q+1)}$ are the ones of class $C^{(q+1)}$, we show (using suitable L^2-estimates) that $d_f\varphi_k^{(q)}$ and all its derivatives are exponentially small near $U_k^{(q)}$. The same result holds for $d_f^*\varphi_k^{(q)}$, so $P^{(q)}\varphi_k^{(q)}$ is exponentially small near $U_k^{(q)}$. Since $\varphi_k^{(q)}$ has most of its L^2 norm concentrated near that point, we conclude that $\mu_k^{(q)}$ is actually exponentially small as $h \to 0$.

It then follows that $P^{(q)}$ has precisely $\#(C^{(q)})$ exponentially small eigenvalues, and if $F^{(q)}$ is the corresponding sum of eigenspaces, then the Witten complex d_f splits into one exact part, $\cdots \to F^{(q)\perp} \to F^{(q+1)\perp} \to \cdots$, and one interesting part, $\cdots \to F^{(q)} \to F^{(q+1)} \to \cdots$, where the cohomology sits. This statement implies the Morse inequalities. A proof for these was outlined by Witten [**10**], and a rigorous proof along these lines was given by B. Simon [**8**]. (See also G. Henniart [**6**]. Unpublished proofs have also been given by L. Boutet de Monvel and R. Melrose.)

The proofs of the Morse inequalities are more elementary than the general machinery outlined above, and they do not require any study of the interaction between potential wells. As pointed out by Witten, the situation is different if one really wants to compute the cohomology. The main result of [**5**] is that this is possible under generic assumptions. As in the scalar case, we construct an orthonormal basis $e_k^{(q)}$, $k \in C^{(q)}$, in $F^{(q)}$, and we then have a general "abstract" result, similar to Theorem 1, that describes the matrix $N^{(q)} = (n_{j,k}^{(q)})$ of d_f: $F^{(q)} \to F^{(q+1)}$. A more explicit result can be obtained if we introduce two generic assumptions.

(H1) If $f(U_j) - f(U_k) = d(U_j, U_k)$, then $\operatorname{ind}(U_j) \geqslant \operatorname{ind}(U_k) + 1$.

Here d is the Agmon distance associated to $\|df\|^2\, dx^2$, and we notice that (a) $|f(x) - f(y)| \leqslant d(x, y)$ for all $x, y \in M$, and (b) if $|f(x) - f(y)| = d(x, y)$, then the ∇f integral curves from x to y (possibly "generalized" in the sense that they are allowed to pass through a finite number of the critical points of f, spending an infinite time to get through each such point) coincide with the (generalized) minimal Agmon geodesics from x to y.

If $U_k^{(q)}$ is a critical point of index q, then the vector field ∇f has two naturally defined stable incoming (outgoing) manifolds V_k^- (V_k^+) of dimensions q and $n - q$, respectively, that intersect orthogonally at $U_k^{(q)}$. Our second assumption is

(H2) If $f(U_j) - f(U_k) = d(U_j, U_k)$ and $\operatorname{ind}(U_j) = \operatorname{ind}(U_k) + 1$, then there are only finitely many minimal Agmon geodesics from U_k to U_j, and V_k^+ and V_j^- intersect transversally along each one of these.

Following Witten, we fix an orientation on each of the V_k^- and define a certain inersection number $\operatorname{ind}(\gamma) = \pm 1$ for each of the minimal geodesics γ as in (H2). We also put $\beta_{j,k}^{(q)} = \Sigma_\gamma \operatorname{ind}(\gamma)$ for j, k as in (H2), with $\operatorname{ind}(U_k) = q$. If $j \in C^{(q+1)}$, $k \in C^{(q)}$, and $f(U_j) - f(U_k) < d(U_j, U_k)$, we put $\beta_{j,k}^{(q)} = 0$.

THEOREM 2 **[5]**. *There are classical elliptic symbols $a_k^{(q)}(h)$ of order* 0 *and a number $\varepsilon_0 > 0$ such that, for all $q \in \{0, \ldots, n-1\}$, $k \in C^{(q)}, j \in C^{(q+1)}$,*

$$n_{j,k}^{(q)} = \left(\frac{h}{\pi}\right)^{1/2} \frac{a_k^{(q)}(h) \exp\big(f\big(U_k^{(q)}\big)/h\big)}{a_j^{(q+1)}(h) \exp\big(f\big(U_j^{(q+1)}\big)/h\big)} \beta_{j,k}^{(q)} + \big(\exp\big(-\big(\varepsilon_0 + f(U_j) - f(U_k)\big)/h\big)\big). \tag{6}$$

Witten gives a formula of this type, with $a_k^{(q)} = a_j^{(q+1)} = 1$, but omits $(h/\pi)^{1/2}$, any remainder estimate, and assumptions (H1), (H2).

A corresponding weaker version of (6), where the remainder term is only one power of h smaller than the leading order of magnitude, can be proved by combining BKW-computations and weighted L^2 estimates with the general result similar to Theorem 1 that we mentioned above. This weaker result already shows that the matrices $\beta^{(\beta)}$ form a complex whose Betti numbers are larger than or equal to the Betti numbers of M. The full result (6) is proved by suitably deforming the metric and f into a simplified case, together with the observation that $N^{(q)}$ changes in a very pleasant way under such deformations. Using the full strength of (6), we can also estimate the Betti numbers of M from below and obtain an analyst's proof of the following result, which appears to be known to topologists.

THEOREM 3. *The Betti numbers of M and the complex β are equal in each degree.*

REFERENCES

1. E. M. Harrell, *Double wells*, Comm. Math. Phys. **75** (1980), 239–261.

2. B. Helffer and J. Sjöstrand, *Multiple wells in the semiclassical limit*. I, Comm. Partial Differential Equations **9** (1984), 337–408.

3. ______, *Puits multiples en limite semiclassique*. II, *Intersection moleculaire, symétries, perturbation*, Ann. Inst. H. Poincaré (to appear).

4. ______, *Multiple wells in the semiclassical limit*. III, *Intersection through non-resonant wells*, Math. Nachr. (to appear).

5. ______, *Puits multiples en limite semiclassique*. IV, *Étude du complexe de Witten*, Comm. Partial Differential Equations. (to appear).

6. G. Henniart, *Les inégalités de Morse* (d'apres E. Witten), Sém. Bourbaki 1983–1984, no. 617.

7. G. Jona-Lasinio, F. Martinelli and E. Scoppola, *New approach in the semiclassical limit of quantum mechanics*. I: *Multiple tunnelings in one dimension*, Comm. Math. Phys. **80** (1981), 223.

8. B. Simon, *Semi-classical analysis of low eigenvalues*. I, Ann. Inst. Poincaré **38** (1983), 295–307.

9. ______, *Semi-classical analysis of low eigenvalues*. II, *Tunneling* (preprint); see also Bull. Amer. Math. Soc. (N.S.) **8** (1983), 323–326.

10. E. Witten, *Supersymmetry and Morse theory*, J. Differential Geom. **17** (1982), 661–692.

UNIVERSITÉ PARIS-SUD, FRANCE

Proceedings of Symposia in Pure Mathematics
Volume **43** (1985)

The Real Analytic and Gevrey Regularity of the Heat Kernel for $\Box_b$

NANCY K. STANTON[1] AND DAVID S. TARTAKOFF[2]

Introduction. In [ST] we gave an explicit construction of the fundamental solution of the heat equation for the $\bar\partial_b$-Laplacian on a compact strictly pseudoconvex CR manifold equipped with a Levi metric. Beals, Greiner, and Stanton [**BGS**, Appendix] extended the method of [ST] to arbitrary Hermitian metrics and to compact CR manifolds satisfying $Y(q)$. In this note we study the regularity of the fundamental solution in the case of real-analytic, nondegenerate, CR manifolds.

Let M be a compact, oriented, nondegenerate, analytic, CR manifold of dimension $2n + 1$ equipped with an analytic Hermitian metric. Thus, there is an analytic, orthogonal splitting of the complexified tangent bundle of M,

$$\mathbf{C}TM = T^{1,0} \oplus T^{0,1} \oplus E,$$

where $T^{1,0} = \overline{T^{0,1}}$ has dimension n, and E has dimension 1. The subbundle $T^{1,0}$ satisfies the integrability condition $[T^{1,0}, T^{1,0}] \subset T^{1,0}$. Let θ be a real, nonvanishing one-form which annihilates $T^{1,0}$. Nondegeneracy means that the Levi form L_θ, defined by

$$L_\theta(X, Y) = -\mathrm{i}d\theta(X, \bar{Y}), \qquad X, Y \in T^{1,0},$$

is nondegenerate. Condition $Y(q)$ holds if L_θ has $\max(q + 1, n + 1 - q)$ eigenvalues of the same sign or $\min(q + 1, n + 1 - q)$ pairs of eigenvalues of opposite signs at every point. Let $\Lambda^{p,0}$ denote the p-forms on M that annihilate $T^{0,1} \oplus E$, and let $\Lambda^{p,q} = \Lambda^{p,0} \wedge \Lambda^{0,q}$, where $\Lambda^{0,q} = \overline{\Lambda^{q,0}}$. The tangential Cauchy–Riemann operator is $\bar\partial_b = \pi_{p,q+1} \circ d\colon C^\infty(\Lambda^{p,q}) \to C^\infty(\Lambda^{p,q+1})$, where $\pi_{p,q+1}$ is orthogonal projection onto $\Lambda^{p,q+1}$. The $\bar\partial_b$-Laplacian is

$$\Box_b = \bar\partial_b\bar\partial_b^* + \bar\partial_b^*\bar\partial_b\colon C^\infty(\Lambda^{p,q}) \to C^\infty(\Lambda^{p,q}).$$

1980 *Mathematics Subject Classification*. Primary 32F20, 35N15.

[1] The first author was partially supported by NSF grant DMS 8200442-01 and the Alfred P. Sloan Foundation.

[2] The second author was partially supported by NSF grant MCS 8301650.

A (p, q)-form $F(x, t)$ on $M \times \mathbf{R}^+$ solves the heat equation for $\square_b$ if

$$(\partial/\partial t + \square_b)F = 0.$$

The initial-value problem for the heat equation is the following. Given $f \in C^0(\Lambda^{p,q})$, find a solution F of the heat equation for $\square_b$ that satisfies $\lim_{t \to 0} F(x, t) = f(x)$. If $Y(q)$ is satisfied, then there is a unique *fundamental solution* $p(x, y, t) \in C^\infty(M \times M \times \mathbf{R}^+; \Lambda^{p,q} \otimes \Lambda^{q,p})$ of this problem [**S**]. The solution F is given by

$$F(x, t) = \int_M p(x, y, t) \wedge *f(y).$$

The fundamental solution or *heat kernel* p satisfies

$$(\partial/\partial t + \square_b^x)p(x, y, t) = 0. \tag{1.1}$$

In [**ST**] and [**BGS**], p is shown to be C^∞ on $M \times M \times \mathbf{R}^+ \cup ((M \times M \setminus \Delta) \times \overline{\mathbf{R}^+})$, where Δ is the diagonal of $M \times M$, and at $t = 0$, p vanishes to infinite order off Δ. Beals, Greiner, and Stanton [**BGS**] proved the existence of an asymptotic expansion of $p(x, x, t)$ as $t \to 0$ in integer powers of t, and Stanton and Tartakoff [**ST**] gave another proof of this result in the special case of strictly pseudoconvex CR manifolds with Levi metrics. These results about the heat kernel hold without the assumptions of analyticity and nondegeneracy.

In order to state our theorem we introduce some notation. We cover M by open sets $U_1, \ldots, U_S$, with $T^{1,0}(U_j)$ trivial, $j = 1, \ldots, S$. We fix an orthonormal basis $\{\omega_j^i\}$ of $\Lambda^{1,0}(U_j)$ and a partition of unity $\{\phi_j\}$ subordinate to $\{U_j\}$. We also fix real analytic vector fields $W_1, \ldots, W_L$ that span $T(M)$ at every point. If $w \in C^\infty(M \times M \times \mathbf{R}^+; \Lambda^{p,q} \otimes \Lambda^{r,s})$, we write

$$w(x, y, t) = \sum \phi_j(x)\phi_k(y) w_{IJ,I'J'}^{j,k}(x, y, t)\omega_j^I(x) \wedge \bar{\omega}_j^J(x) \otimes \omega_k^{I'}(y) \wedge \bar{\omega}_k^{J'}(y).$$

We apply W_i to w by applying it to the coefficients $w_{IJ,I'J'}^{jk}$. We let W_j^x denote W_j acting in the variable x and let

$$D_x^\alpha = W_1^{x^{\alpha_1}} \cdots W_L^{x^{\alpha_L}}, \qquad \alpha = (\alpha_1, \ldots, \alpha_L).$$

Our main result is

THEOREM 1.1. *Let M be a compact real analytic nondegenerate CR manifold equipped with a real-analytic Hermitian metric. If $Y(q)$ holds, then the fundamental solution $p(x, y, t)$ of the heat equation for $\square_b$ on (p, q)-forms satisfies*

(i) *$p(x, y, t)$ is real analytic on $M \times M \times \mathbf{R}^+$;*

(ii) *for each sufficiently small $\varepsilon > 0$, there is a constant C_ε such that for any multi-indices α, β and integer k,*

$$\left|D_x^\alpha D_y^\beta D_t^k p(x, y, t)\right| \leqslant C_\varepsilon^{|\alpha+\beta|+k+1}(|\alpha + \beta| + 2k)! \tag{1.2}$$

uniformly on $(M \times M)_\varepsilon \times (0,1)$, where $(M \times M)_\varepsilon = \{(x, y) \in M \times M$: $d(x, y) > \varepsilon\}$, and d is the Riemannian distance.

We prove the theorem using the L^2-method developed by Tartakoff to prove the local analytic hypoellipticity of $\Box_b$ [**Ta 1, 2, 3**]. Our results and proofs also hold for the heat kernels for the more general class of differential operators studied in [**Ta 2**]. Unfortunately, we do not see a way to apply the results of Trèves [**Tr 1**] and Tartakoff [**Ta 1**] on the local analytic hypoellipticity of $\Box_b$ directly to the heat kernel. We also do not see a way to use the explicit construction of the kernel, which involves the use of cutoff functions and partitions of unity. The theory of holomorphic semigroups would yield analyticity in t for $t > 0$, but not the uniform estimates we want as $t \to 0$.

The regularity in Theorem 1.1 is clearly optimal, since it is as good as the behavior of the Euclidean heat kernel.

In §2 we prove the key estimates—an a priori estimate and L^2-estimates of high t-derivatives and high x-derivatives of $p(x, y, t)$. We use these in §3 to prove L^2-estimates for all derivatives of p, with appropriate growth. The theorem then follows by the Sobolev Lemma.

2. The key estimates.

(a) *The a priori estimate*. We assume throughout that M satisfies the hypotheses of Theorem 1.1. In all our estimates we allow the value of constants to change from line to line, as long as the new constant depends only on the allowable parameters. All our L^2-estimates rely on the following *a priori* inequality.

LEMMA 2.1. *Let X^x and X'^x be smooth sections of $T^{1,0} \oplus T^{0,1}$ in the variable x, and let T^x be a smooth section of E in the variable x. Let $T_0 > 0$. There is a constant C such that if $v \in C^\infty(M \times M \times \mathbf{R}; \Lambda^{p,q} \otimes \Lambda^{q,p})$ and* $\operatorname{supp} v \subset M \times M \times [0, T_0]$, *then*

$$\|X^x X'^x v\|^2_{L^2_{x,y,t}} + \|T^x v\|^2_{L^2_{x,y,t}} + \|\partial v/\partial t\|^2_{L^2_{x,y,t}} \leqslant C\left(\|(\Box^x_b + \partial/\partial t)v\|^2_{L^2_{x,y,t}} + \|v\|^2_{L^2_{x,y,t}}\right). \tag{2.1}$$

PROOF. By Kohn's a priori estimate [**FK**],

$$\|X^x v\|^2_{L^2_x} + \|v\|^2_{(1/2),x} \leqslant C\left(\operatorname{Re}(\Box^x_b v, v)_{L^2_x} + \|v\|^2_{L^2_x}\right), \tag{2.2}$$

uniformly in y and t. Since $\operatorname{supp} v \subset M \times M \times [0, T_0]$, integration in t gives

$$\|X^x v\|^2_{L^2_{x,t}} + \|v\|^2_{(1/2)_x, L^2_t} \leqslant C\left[\int_0^{T_0} \left|\operatorname{Re}\left(\left(\Box^x_b + \frac{\partial}{\partial t}\right)v, v\right)_{L^2_x}\right| dt + \|v\|^2_{L^2_{x,t}}\right]. \tag{2.3}$$

The argument of [**Ta 2**, Appendix] applies to show

$$\|X^x X'^x v\|^2_{L^2_{x,t}} + \|T^x v\|^2_{L^2_{x,t}} \leqslant C\left(\|(\Box^x_b + \partial/\partial t)v\|^2_{L^2_{x,t}} + \|v\|^2_{L^2_{x,t}}\right). \tag{2.4}$$

Now

$$\|\partial v/\partial t\|_{L^2_{x,t}} \leqslant \|(\Box_b + \partial/\partial t)v\|_{L^2_{x,t}} + \|\Box_b v\|_{L^2_{x,t}}. \tag{2.5}$$

Let $U \subset M$ be an open set such that $T^{1,0}$ is trivial over U. Let $\{\omega^1, \ldots, \omega^n, \bar\omega^1, \ldots, \bar\omega^n, \omega\}$ be an orthonormal basis of $\mathbf{C}T^*U$, with $\omega^i \in \Lambda^{1,0}$ and ω

a real annihilator of $T^{1,0}$. Let $\{Z_1,\dots,Z_n, \overline{Z}_1,\dots,\overline{Z}_n, T\}$ be the dual basis. If $\phi = \Sigma\phi_{I\bar{J}}\omega^I \wedge \bar{\omega}^J \in C^\infty(\Lambda^{p,q})$ over U,

$$\begin{aligned}\Box_b\phi = &-\Big(\sum_{j\notin J} Z_j\overline{Z}_j + \sum_{j\in J} \overline{Z}_jZ_j\Big)\phi_{I\bar{J}}\ \omega^I \wedge \bar{\omega}^J \\ &-\sum_{j\neq k}\big[\overline{Z}_j, Z_k\big]\phi_{I\bar{J}}\ \omega^I \wedge \bar{\omega}^j \wedge (\bar{\omega}^k \lrcorner \bar{\omega}^J) + L\phi\end{aligned} \tag{2.6}$$

where L is a first-order differential operator in Z_j and $\overline{Z}_j$ only. Using (2.6) and (2.3) to estimate the last term of (2.5), we see that $\|\partial v/\partial t\|^2_{L^2_{x,t}}$ can be estimated by the right side of (2.4). The lemma now follows by integrating in y. ■

(b) *High t-derivatives away from $t = 0$.* Our argument for t-derivatives is the elliptic argument. We require appropriate cutoff functions in t. The existence of such functions is given by

LEMMA 2.2. *There is a constant K depending only on m such that if $\Omega_1 \Subset \Omega_2 \subset \mathbf{R}^m$, Ω_i open, and d is the distance from Ω_1 to Ω_2^c, then for any N there exists $\psi_N \in C_0^\infty(\Omega_2)$ with $\psi_N \equiv 1$ on a neighborhood of $\overline{\Omega}_1$ and*

$$|\psi_N^{(\alpha)}| \leqslant K^{|\alpha|+1}d^{-|\alpha|}N^{|\alpha|} \tag{2.7}$$

for all multi-indices α with $|\alpha| \leqslant 3N$.

This is proved in [**Tr 2**, Volume 1, Chapter 5, Lemma 1.1]. We fix N, take $m = 1$, $\Omega_1 = (t_0, T_0')$, $\Omega_2 = (0, T_0)$, $T_0' < T_0$, and let ϕ_N denote the function given by the lemma. We apply Lemma 2.1 to $v = D_t^k\phi_N(t)p(x, y, t)$ with $k \leqslant N$. By (2.1)

$$\begin{aligned}&\|X^\alpha X'^\alpha D_t^k\phi_N p\|^2_{L^2_{x,y,t}} + \|T^\alpha D_t^k\phi_N p\|^2_{L^2_{x,y,t}} + \|D_t^{k+1}\phi_N p\|^2_{L^2_{x,y,t}} \\ &\quad\leqslant C\Big[\|(\Box_b^x + \partial/\partial t)D_t^k\phi_N p\|^2_{L^2_{x,y,t}} + \|D_t^k\phi_N p\|^2_{L^2_{x,y,t}}\Big] \\ &\quad\leqslant C\Big[\|D_t^k\phi_N(\Box_b + \partial/\partial t)p\|^2_{L^2_{x,y,t}} + \|D_t^k\phi_N'(t)p\|^2_{L^2_{x,y,t}} + \|D_t^k\phi_N p\|^2_{L^2_{x,y,t}}\Big] \\ &\quad= CI_{k,1},\end{aligned} \tag{2.8}$$

where

$$I_{l,j} = \sup_{\substack{l'\leqslant l\\ j'\leqslant j}} \|D_t^{l'}\phi_N^{(j')}(t)p\|^2_{L^2_{x,y,t}}.$$

Here we have used (1.1). Now

$$I_{k+1,0} \leqslant CI_{k,1} \leqslant C^2I_{k-1,2} \leqslant \cdots,$$

so

$$I_{k,1} \leqslant C^k \sup_{k'\leqslant k+1} \|\phi_N^{(k')}p\|^2_{L^2_{x,y,t}} \leqslant C^{k+1}(N^{k+1})^2 \sup_{t\in \operatorname{supp}\phi_N} |p(x, y, t)|^2.$$

Thus there is a C independent of N such that, for $k \leqslant N$,

$$\left\|X^x X'^x D_t^k \phi_N p\right\|^2_{L^2_{x,y,t}} + \left\|T^x D_t^k \phi_N p\right\|^2_{L^2_{x,y,t}} + \left\|D_t^{k+1} \phi_N p\right\|^2_{L^2_{x,y,t}} \leqslant C^{k+1}\left(N^{k+1}\right)^2 \sup_{t \in \operatorname{supp} \phi_N} |p|^2. \tag{2.9}$$

PROPOSITION 2.3. *For any $t_0 > 0$, $T_0' < T_0$, there is a C such that, for every k,*

$$\left\|X^x X'^x D_t^k p\right\|_{L^2_{x,y,T \in [\tau_0, T_0']}} + \left\|T^x D_t^k p\right\|_{L^2_{x,y,t \in [t_0, T_0']}} + \left\|D_t^{k+1} p\right\|_{L^2_{x,y,t \in [t_0, T_0']}} \leqslant C^{k+1}(k+1)!. \tag{2.10}$$

PROOF. This follows immediately from (2.9) with $k = N$, using the facts that $\phi_N \equiv 1$ on $[t_0, T_0']$ and $N^{N+1} \leqslant C^{N+1}(N+1)!$. ■

(c) *High x-derivatives.* We fix $x_0 \in M$. Because the CR structure on M is nondegenerate and real-analytic, there is an underlying real-analytic contact structure. By Darboux's Theorem there are a coordinate neighborhood U of x_0 and real-analytic coordinates $x^0, \ldots, x^{2n}$ on U such that the vector fields

$$X_j = \partial/\partial x^j - x^{n+j}(\partial/\partial x^0), \quad X_{n+j} = \partial/\partial x^{n+j}, \qquad 1 \leqslant j \leqslant n, \tag{2.11}$$

are a basis of $T^{1,0} \oplus T^{0,1}$ over U. The vector field

$$T = \partial/\partial x^0 \tag{2.12}$$

is complementary. By (2.6), $\square_b$ is a second-order differential operator in the X's alone. We write X^q for $X^x_{i_1} \cdots X^x_{i_{q'}}$, $q' \leqslant q$, X'^q if $i_1, \ldots, i_{q'} \leqslant n$, and X''^q if $i_1, \ldots, i_{q'} > n$. Here X^x_i means X_i acting in the variable x.

We localize as follows. Let $V \Subset U$ be a neighborhood of x_0, and let ε, $T_0 > 0$. (Later we will pick V carefully.) Fix N. By Lemma 2.2 there are functions $\phi_{1,N} \in C_0^\infty((-\varepsilon, 2\varepsilon))$, $\phi_{1,N} \equiv 1$ on $[0, \varepsilon]$, $\phi_{2,N} \in C_0^\infty((-\varepsilon, T_0 + \varepsilon))$, $\phi_{2,N} \equiv 1$ on $[0, T_0]$, and $\phi_{3,N} \in C_0^\infty(U)$, $\phi_{3,N} \equiv 1$ on $\overline{V}$, satisfying (2.7). Let

$$\psi_N(x, y, t) = \phi_{2,N}(t)\left(1 - \phi_{1,N}\left(d^2(x, y)\right)\phi_{1,N}(t)\right)\phi_{3,N}(x).$$

Then

$$\operatorname{supp} \psi_N \subset \tilde{U} = \left(U \times M \times (-\varepsilon, T_0 + \varepsilon)\right) \cap \left\{(x, y, t) \colon d^2(x, y) \leqslant \varepsilon, 0 \leqslant t \leqslant \varepsilon\right\}^c$$

and $\psi_N \equiv 1$ on

$$\tilde{V} = V \times M \times [0, T_0] \cap \left\{(x, y, t) \colon d^2(x, y) > 2\varepsilon \text{ or } t > 2\varepsilon\right\}.$$

Thus, ψ_N localizes in a neighborhood of x_0 away from the singularity of $p(x, y, t)$. By (2.7), for ε sufficiently small, there is a K (depending on ε but not on N) such that the (x, t)-derivatives of ψ_N satisfy

$$|\psi_N^{(\alpha)}| \leqslant K^{|\alpha|+1} d^{-|\alpha|} N^{|\alpha|} \tag{2.13}$$

for $|\alpha| \leqslant 3N$, d the distance from $\{\psi_N \equiv 1, t \geqslant 0\}$ to $\{\psi_N \equiv 0, t \geqslant 0\}$. Here we have used the fact that $d^2(x, y)$ is analytic in a neighborhood of the diagonal.

Our goal in the remainder of this section is to estimate $\|X^a T^s p\|_{L^2(\psi_N \equiv 1)}$ with analytic growth (Proposition 2.8). To simplify the notation we drop the subscript

N and note that vector fields act in x only. Since $\square_b$ is not elliptic in the T direction, our main tool will be the estimate on $\|X^2 v\|_{L^2_{x,y,t}}$ in Lemma 2.1. To iterate this estimate we must take advantage of the fact that $(\partial/\partial t + \square_b^x)p = 0$, and thus we must commute a localization of T^s past $\square_b$. To obtain good enough commutation relations to allow us to handle the nonellipticity in T, we follow [**Ta 1, 2, 3**]. We take as our localization of T^s

$$(2.14)\qquad (T^s)_\psi = \sum_{|\alpha+\beta|\leqslant s} \frac{(-1)^{|\alpha|}}{\alpha!\beta!}\left(X'^{\alpha}X''^{\beta}\psi\right)X'^{\beta}X''^{\alpha}T^{s-|\alpha+\beta|}.$$

Then on V, $(T^s)_\psi = \psi T^s = T^s$. A straightforward calculation, using (2.11) and (2.12), yields the commutation relations

$$(2.15)\qquad \left[(T^s)_\psi, X'_j\right] \equiv 0 \quad\text{and}\quad \left[(T^s)_\psi, X''_j\right] \equiv -(T^{s-1})_{T\psi}X''_j$$

modulo C^s terms of the form $(X^{s+1}\psi)X^s/s!$. The commutators of the coefficients of $\square_b^x$ with $(T^s)_\psi$ are more complicated.

LEMMA 2.4. *With $(T^s)_\psi$ defined by* (2.14) *and $g, v \in C_0^\infty(U)$,*

$$\left[(T^s)_\psi, g\right]v = \sum_{0<i+2j+k\leqslant s} \frac{(s-i-j)!}{i!j!k!(s-i-j-k)!}C^{i+j} \text{ terms of the form}$$

$$g^{(i+j+k)}X''^{\mu'+\nu'+\nu''}X'^{\nu-\nu'-\nu''}\cdot\left(T^{s-i-2j-k}\right)_{\psi^{(i+j+|\nu''|)}}\cdot X''^{\nu'}v$$

for some $|\mu'| \leqslant i$, $|\nu'| + |\nu''| \leqslant j = |\nu|$.

This is proved in [**Ta 2**, Lemma 2.1].

To estimate $\|X^a T^s p\|_{L^2(\psi\equiv 1)}$ we localize T^s by $(T^s)_\psi$ and iterate. This leads us to expressions of the form $\|X^a T^b (T^s)_{\psi^{(r)}} X^e T^d p\|_{L^2_{x,y,t}}$. We introduce some convenient notation for handling such expressions. Let

$$(2.16)\qquad G_{A,\psi} = X^a T^b (T^s)_{\psi^{(r)}} X^e T^d,$$

where the vector fields operate in the variable x and

$$(2.17)\qquad \begin{aligned} &A = (s, r, a, b, d, e), \qquad s, b, d \in \mathbf{Z},\ a, e \in \mathbf{Z}^{2n},\ r \in \mathbf{Z}^{2n+2},\\ &\sum_{i>n} a_i \leqslant 2, \qquad s, r, b, d, a, e \geqslant 0.\end{aligned}$$

Then $G_{A,\psi}$ is a differential operator of order $|A|$ with at most two of the X's in X^a being X''s. Here

$$(2.18)\qquad |A| = s + |a| + b + d + |e|.$$

Since $T = [X_j, X_{n+j}]$, we sometimes want to assign double weight to T^b and T^d, so we let

$$(2.19)\qquad \|A\| = |A| + b + d.$$

We define norms of constant coefficient linear combinations of the $G_{A,\psi}$ by

$$(2.20)\qquad \left|\left|\left|\sum c_A G_{A,\psi}\right|\right|\right|_M = \sum |c_A| C_0^s M^{|A|+|r|+s} s!$$

with $C_0 \geqslant 2n-2$. These norms are a useful bookkeeping device. The key step in our iteration is

PROPOSITION 2.5. *Let $G_{A,\psi}$ be as in (2.16), (2.17), with $|a| \geqslant 2$. There is a constant C such that, for $M \geqslant |A| + |r|$,*

$$\|G_{A,\psi}p\|_{L^2_{x,y,t}} \leqslant C \sup_{\substack{|||c_{A'}G_{A',\psi}|||_M \leqslant c|||G_{A,\psi}|||_M \\ |A'|<|A|,\ \|A'\|\leqslant\|A\|, \\ |A'|+|r'|\leqslant|A|+|r| \text{ if } s'>0, \\ |A'|+|r'|\leqslant 2|A|+|r| \text{ if } s'=0 \\ \sum_{i>n} a_i' \leqslant 2 \\ |a'|+|e'|\geqslant 1 \text{ if } s'>0 \text{ unless} \\ |A'|<|A|-1,\ \|A'\|<\|A\|}} \|c_{A'}G_{A',\psi}p\|_{L^2_{x,y,t}}. \tag{2.21}$$

Note that the effect of (2.21) is to estimate $G_{A,\psi}p$ by a supremum of similar expressions with *smaller* $|A'|$ and no greater $\|A'\|$. In (2.35) we iterate this until $|A'|$ is very small or $|a'| + |e'| < 2$; if we start with $|a| \sim 2$, s large, we may iterate until $|a'| + |e'| \leqslant 2$, $s' = 0$, and (since $\|A'\| \leqslant \|A\|$) $b' + d' \sim |A|/2$, thereby effectively reducing the total degree by one-half.

PROOF. We can write

$$G_{A,\psi} = X^2X'^{a-2}T^b(T^s)_{\psi^{(r)}}X^eT^d \tag{2.22}$$
$$+ \text{ at most two terms of the form } X^{a-2}T^{b+1}(T^s)_{\psi^{(r)}}X^eT^d,$$

where $a-2$ denotes a multi-index $\tilde{a}$ with $|\tilde{a}| = |a| - 2$, $\sum_{i>n}\tilde{a}_i \leqslant 2$. Hence, it suffices to consider

$$G_{A,\psi} = X^2X'^{a-2}T^b(T^s)_{\psi^{(r)}}X^eT^d. \tag{2.23}$$

By (2.1)

$$\|G_{A,\psi}p\|_{L^2_{x,y,t}} \leqslant C\Big(\big\|(\partial/\partial t + \square_b^x)X'^{a-2}T^b(T^s)_{\psi^{(r)}}X^eT^dp\big\|_{L^2_{x,y,t}} \tag{2.24}$$
$$+ \big\|X'^{a-2}T^b(T^s)_{\psi^{(r)}}X^eT^dp\big\|_{L^2_{x,y,t}}\Big).$$

Since $(\partial/\partial t + \square_b^x)p = 0$, it suffices to show that

$$\big\|\big[X'^{a-2}T^b(T^s)_{\psi^{(r)}}X^eT^d, \square_b^x + \partial/\partial t\big]\,p\big\|_{L^2_{x,y,t}} \tag{2.25}$$

can be estimated by the right side of (2.21). Now

$$\big[X'^{a-2}T^b(T^s)_{\psi^{(r)}}X^eT^d, \partial/\partial t\big] = -G_{A',\psi}, \tag{2.26}$$

with $A' = (s, r+1, a-2, b, d, e)$. By (2.6), $\square_b$ is a sum of terms of the form gX^c, $|c| \leqslant 2$ (acting on coefficients of a form). We write

$$\big[X'^{a-2}T^b(T^s)_{\psi^{(r)}}X^eT^d, gX^c\big] = \sum_{j=1}^{6} H_j, \tag{2.27}$$

where

(2.28)

$$H_1 = X'^{a-2}T^b(T^s)_{\psi^{(r)}}\big[X^eT^d, g\big]X^c, \qquad H_4 = gX'^{a-2}T^b(T^s)_{\psi^{(r)}}\big[X^eT^d, X^c\big],$$
$$H_2 = X'^{a-2}T^b\big[(T^s)_{\psi^{(r)}}, g\big]X^{e+c}T^d, \qquad H_5 = gX'^{a-2}T^b\big[(T^s)_{\psi^{(r)}}, X^c\big]X^eT^d,$$
$$H_3 = \big[X'^{a-2}T^b, g\big](T^s)_{\psi^{(r)}}X^{e+c}T^d, \qquad H_6 = g\big[X'^{a-2}T^b, X^c\big](T^s)_{\psi^{(r)}}X^eT^d.$$

Observe that for $|c| = 2$, each term contains at least one X or is 0. By Lemma 2.4,

(2.29)

$$\begin{aligned} &H_1 + H_2 + H_3 \\ &= \sum_{\substack{0 < i+2j+k+|a'|+b'+d'+|e'|, \\ i+2j+k \leqslant s,\, |a'| \leqslant |a|-2, \\ b' \leqslant b,\, d' \leqslant d,\, |e'| \leqslant |e|}} \frac{(s-i-j)!(a-2)!b!d!e!}{(s-i-j-k)!(a-2-a')!(b-b')!(d-d')!(e-e')!} C^{i+j} \\ &\qquad \cdot \text{terms of the form } \frac{g^{(i+j+k+a'+b'+d'+e')}}{i!j!k!a'!b'!d'!e'!} x''^{\mu'+\nu'+\nu''} X'^{a-2-a'+\nu-\nu'-\nu''} \\ &\qquad \cdot T^{b-b'}\left(T^{s-i-2j-k}\right)_{\psi^{(r+i+j+|\nu''|)}} X^{e-e'+\nu'+|c|} T^{d-d'} \end{aligned}$$

for some $|\mu'| \leqslant i$, $|\nu'| + |\nu''| \leqslant j = |\nu|$. By (2.15), if $|c| = 1$ or 2,

(2.30)

$$\begin{aligned} H_5 &= (|c|-1) g X'^{a-2} X T^b \left(T^{s-1}\right)_{\psi^{(r+1)}} X^{e+|c|-1} T^d \\ &\quad + g X'^{a-2} T^b \left(T^{s-1}\right)_{\psi^{(r+1)}} X^{e+|c|} T^d \\ &\quad + (C^s/s!) \text{ terms } g(x) X'^{a-2} T^b \left\{ X^{|c|-1} \psi^{(s+r+1)} + \psi^{(s+r+1)} X^{|c|-1} \right\} X^{s+e} T^d. \end{aligned}$$

By (2.11) and (2.12), if $|c| = 1$ or 2,

$$\begin{aligned} H_4 + H_6 &= |c|\,|e| g(x) X'^{a-2} T^b (T^s)_{\psi^{(r)}} X^{e+c-2} T^{d+1} \\ &\quad + (|a|-2) g(x) X'^{a-3} X^{|c|-1} T^{b+1} (T^s)_{\psi^{(r)}} X^e T^d \\ &\quad + (|c|-1)(|a|-2) g(x) X^{|c|-1} X'^{a-3} T^{b+1} (T^s)_{\psi^{(r)}} X^e T^d. \end{aligned} \tag{2.31}$$

Taking L^2-norms and using the analyticity of g, so $|g^{(l)}| \leqslant D^{|l|} l!$ in $\operatorname{supp} \psi$ for some D independent of l, (2.29) gives

$$\begin{aligned} &\|(H_1 + H_2 + H_3) p\|_{L^2_{x,y,t}} \\ &\qquad \leqslant \sum_{\substack{i+j+k+|a'|+b'+d'+|e'| > 0 \\ |a'| \leqslant |a|-2,\, b' \leqslant b,\, d' \leqslant d,\, |e'| \leqslant |e| \\ i+2j+k \leqslant s}} (DC)^{i+j+k+|a'|+b'+d'+|e'|} \\ &\qquad\qquad \sup_{|\nu'| = |\nu''| \leqslant |\nu| = j} \| M^{|a'|+b'+d'+|e'|+k} X'^{a-2-a'+\nu-\nu'-\nu''} T^{b-b'} \\ &\qquad\qquad\qquad \cdot \left(T^{s-i-2j-k}\right)_{\psi^{(r+i+j+|\nu''|)}} X^{e-e'+\nu'+|c|} T^{d-d'} p \|_{L^2_{x,y,t}}. \end{aligned} \tag{2.32}$$

By dilation we may assume DC is so small that (2.32) implies

$$\|(H_1 + H_2 + H_3) p\|_{L^2_{x,y,t}} \leqslant \operatorname*{supremum}_{\substack{|||c_{A'} G_{A',\psi}|||_M \leqslant |||G_{A,\psi}|||_M \\ |A'| < |A|,\, ||A'|| \leqslant ||A|| \\ |A'| + |r'| \leqslant |A| + |r|}} \|c_{A'} G_{A',\psi} p\|_{L^2_{x,y,t}}. \tag{2.33}$$

Similarly, by (2.30) and (2.31) we can estimate $\|(H_4 + H_5 + H_6) p\|_{L^2_{x,y,t}}$ by the right side of (2.21), and, thus, using (2.26) and (2.27), we can estimate (2.25) by the right side of (2.21). ■

REMARK. The proof of Proposition 2.5 is essentially that of [**Ta 2**]. However, because we have to commute $\psi^{(\alpha)}$ and $\partial/\partial t$, and because $\Box_b p$ is not known *a priori* to be analytic, we cannot just quote [**Ta 2**].

Next we iterate (2.21) to prove

PROPOSITION 2.6. *Let $G_{A,\psi}$ be as in Proposition 2.5; i.e., $|a| \geqslant 2$. There is a C such that, for $M \geqslant |A| + r$,*

$$\|G_{A,\psi}p\|_{L^2_{x,y,t}} \leqslant C^M \sup_{\substack{|||c_{A'}G_{A',\psi}|||_M \leqslant |||G_{A,\psi}|||_M \\ |a'|+|e'|\leqslant 1,\, s'=0, \\ \|A'\|\leqslant\|A\|,\, |A'|+|r'|\leqslant 2|A|+|r|}} \|c_{A'}G_{A',\psi}p\|_{L^2_{x,y,t}}. \tag{2.34}$$

COROLLARY 2.7. *There is a constant C such that for $|a| + s \leqslant M$ and $\tilde{\psi} \in C_0^\infty(M \times M \times \mathbf{R})$ with $\tilde{\psi} \equiv 1$ in a neighborhood of* supp ψ,

$$\frac{\|X^aT^sp\|_{L^2(\psi\equiv 1)}}{M^{|a|+s}} \leqslant C^M \sup_{\substack{|a'|+2b'\leqslant|a|+s+1 \\ |a'|\leqslant 1,\, |a'|+b'+|r'|\leqslant 2(|a|+s+1)}} \frac{|\psi^{(r')}|}{M^{|r'|}} \frac{\|X^{a'}T^{b'}p\|_{L^2(\tilde{\psi}\equiv 1)}}{\tilde{M}^{|a'|+b'}}, \tag{2.35}$$

where $\tilde{M} = M/2$.

PROOF OF COROLLARY 2.7. If $|a| \geqslant 2$, we apply (2.34) with $A = (s, 0, a, 0, 0, 0)$, $M \geqslant |a| + s$. Since $M^s/s! \leqslant C^M$, this yields

$$\begin{aligned} \|X^aT^sp\|_{L^2(\psi=1)} &\leqslant \|G_{A,\psi}p\|_{L^2_{x,y,t}} \\ &\leqslant C^M \sup_{\substack{\|A'\|\leqslant|a|+s,\, |a'|+|e'|\leqslant 1, \\ |a'|+b'+|r'|\leqslant 2(|a|+s),\, s'=0}} M^{|a|+s-(|a'|+b'+|r'|+d'+|e'|)}\|G_{A',\psi}p\|_{L^2_{x,y,t}} \\ &\leqslant C^M \sup_{\substack{|a'|+2b'\leqslant|a|+s \\ |a'|\leqslant 1 \\ |r'|\leqslant 2(|a|+s)}} \frac{|\psi^{(r')}|}{M^{r'}} \frac{M^{a+s}}{M^{a'+b'}} \|X^{a'}T^{b'}p\|_{L^2(\operatorname{supp}\psi^{(r')})}. \end{aligned} \tag{2.36}$$

The last inequality comes from commuting the $\psi^{(r')}$ in $G_{A',\psi}$ to the left. Now (2.35) follows from (2.36). For $|a| \leqslant 1$ and $s \geqslant 1$ we write $X^aT^s = X^a[X_1, X_{n+1}]T^{s-1}$, and then apply (2.36) to the two resulting terms to obtain (2.35). For $s = 0$, $|a| = 1$, (2.35) holds trivially. ■

PROOF OF PROPOSITION 2.6. We want to iterate (2.21) until we have $s' = 0$. To do this we must have $|a'| \geqslant 2$, where $A' = (s', r', a', b', d', e')$ is as on the right side of (2.21). If $|a'| < 2$, but $|e'| \geqslant 1$, we may commute an X from $X^{e'}$ in $G_{A',\psi}$ to the left. By (2.15) the errors produced are of the type $c_{A''}G_{A'',\psi}$ as on the right of (2.21), with $|A''| < |A'|$, $\|A''\| < \|A'\|$ or $s'' = 0$. If $|a'| + |e'| < 2$ and $s' > 0$, we may assume $|e'| = 0$ by commuting $X^{e'}$ to the left if necessary. Integrating by parts and applying Cauchy–Schwarz if $|a'| = 1$, we see that for $|a'| \leqslant 1$, $|e'| = 0$,

$$\begin{aligned} &\|X^{a'}T^{b'}(T^{s'})_{\psi^{(r')}}T^{d'}p\|^2_{L^2_{x,y,t}} \\ &\leqslant \|T^{b'}(T^{s'})_{\psi^{(r')}}T^{d'}p\|^2_{L^2_{x,y,t}} + |a'|\,\|X^2T^{b'}(T^{s'})_{\psi^{(r')}}T^{d'}p\|^2_{L^2_{x,y,t}} \\ &= \|G_{A_1,\psi}p\|^2_{L^2_{x,y,t}} + |a'|\,\|G_{A_2,\psi}p\|^2_{L^2_{x,y,t}}. \end{aligned} \tag{2.37}$$

Recall that $A' = (s', r', a', b', d', 0)$ has $|A'| \leqslant |A| - 1$, $\|A'\| \leqslant \|A\|$.

For the first term on the right,

$$G_{A_1,\psi} = T^{b'}(T^{s'})_{\psi^{(r')}}T^{d'},$$

we have $|A_1| < |A| - 1$, $\|A_1\| < \|A\|$. We may assume $b' \geqslant 1$, for, by (2.14),

$$(2.38) \qquad (T^s)_\phi \equiv TT_\phi^{s-1} - T_{T\phi}^{s-1} \quad \text{modulo } (C^s/s!) \text{ terms } \phi^{(s)}X^s,$$

and the errors committed are of the type $c_{A''}G_{A'',\psi}$ encountered above. If we use (2.38), A_1 now satisfies $|A_1| < |A| - 1$, $\|A_1\| \leqslant \|A\|$. We create X's by writing $T = [X', X'']$. Now A_1 satisfies $|A_1| < |A|$, $\|A_1\| \leqslant \|A\|$. Thus

$$\left\|G_{A_1,\psi}p\right\|^2_{L^2_{x,y,t}} \leqslant 2\left\|G_{A_1',\psi}p\right\|^2_{L^2_{x,y,t}},$$

where $|A_1'| < |A|$, $\|A_1'\| \leqslant \|A\|$ and $G_{A_1',\psi}$ has two X's; hence, it may be subjected to (2.21) again.

The second term on the right in (2.37),

$$G_{A_2,\psi} = X^2T^{b'}(T^{s'})_{\psi^{(r')}}T^{d'},$$

has $A_2 = (s', r', 2, b', d', 0)$, with $|A_2| \leqslant |A|$, $\|A_2\| \leqslant \|A\| + 1$. While this term has two X's, the norms $|A_2|$ and $\|A_2\|$ are both too high. Nonetheless, we apply (the proof of) (2.21) to such a term, which contains exactly two X's; after this second application we see that the norms become acceptable and the essential features of this term are preserved. We note in passing that using (2.21) itself on $G_{A_2,\psi}$ will not suffice.

In applying the proof of (2.21) to $G_{A_2,\psi}$, which has $|a_2| = 2$, $e_2 = 0$, we note that in (2.28), when $|a| = 2$, $e = 0$, $H_1 \cdots H_6$ all contain two X's when $|c| = 2$ (or are zero), and no new T's are generated. Thus not only will the $|\ |$-norm drop for these terms, but also the $\|\ \|$-norm as well. Thus, application of that part of the proof of (2.21) with $|c| = 2$ (the principal part of $\square_b$) to $G_{A_2,\psi}$ yields only terms $G_{A',\psi}$, with $|a'| + |e'| \geqslant 2$, $|A'| < |A_2|$, $\|A'\| < \|A_2\|$, as desired. When $|c| < 2$ in (2.28), we must take some care. Again, for $|a| = 2$, $e = 0$, each term will result in a $G_{\tilde{A}_c,\psi}$ with $|c|$ X's, no new T's, and, hence, $|\tilde{A}_c| < |A_2| - 1$, $\|\tilde{A}_c\| < \|A_2\| - 1$ if $|c| = 1$ and $|\tilde{A}_c| < |A_2| - 2$, $\|\tilde{A}_c\| < \|A_2\| - 2$ for $c = 0$; but now these terms have fewer than two X's. In both cases we have $|\tilde{A}_c| < |A| - (2 - |c|)$, $\|\tilde{A}_c\| < \|A\| - (2 - |c|)$. As above, we may take $\tilde{A}_c = (\tilde{s}, \tilde{r}, \tilde{a} = c, \tilde{b}', \tilde{d}', \tilde{e} = 0)$ modulo the errors $c_{A''}G_{A'',\psi}$ encountered earlier. Once again we apply (2.38), as in the treatment of $G_{A_1,\psi}$ above, but note that both $|\tilde{A}_c|$ and $\|\tilde{A}_c\|$ are already at least one lower than the norms in that case. Thus (2.38) will yield two terms, $G_{B_1,\psi}$ and $G_{B_2,\psi}$, where $\|G_{B_1,\psi}p\|^2_{L^2_{x,y,t}} \leqslant 2\|G_{B_1',\psi}p\|^2_{L^2_{x,y,t}}$ as above, $G_{B_1',\psi}$ contains two X's, $|B_1'| < |A| - 1$ and $\|B_1'\| \leqslant \|A\| - 1$, and G_{B_2} also has two X's, as above, $|B_2| \leqslant |A| - 1$ and $\|B_2\| \leqslant \|A\|$. Thus, both $G_{B_1',\psi}$ and $G_{B_2,\psi}$ contain two X's and satisfy the restrictions on the right of (2.21); hence, they may be resubjected to (2.21).

Iteration of this process at most $|A|$ times will yield only expressions such as those on the right in (2.21) but with $s' = 0$, $|a'| + |e'| < 2$. ■

On the right side of (2.35), $|a'| + b' \leqslant (|a| + s + 2)/2$, which is approximately one-half the value of $|A|$ on the left. We take $\log_2 N$ nested open sets in $M \times M \times \overline{\mathbf{R}^+}$,

$$\tilde{V} = V_1 = V \times M \times [0, T_0] \cap \{d^2(x, y) > 2\varepsilon \text{ or } t > 2\varepsilon\} \Subset V_2 \Subset \cdots \Subset \tilde{U}$$
$$= V_{1+\log_2 N} = U \times M \times [0, T_0 + \varepsilon) \cap \{d^2(x, y) > \varepsilon \text{ or } t > \varepsilon\}$$

with separations $d_j = d/2^j$ between them, where d is the distance from V_1 to $(V_{1+\log_2 N})^c$. (We neglect $t < 0$ since p vanishes for $t < 0$.) We fix N and choose cutoff functions $\psi_{i,N}$, $\{\psi_{i,N} \equiv 1\} \supset V_i$, $\operatorname{supp} \psi_{i,N} \subset V_{i+1}$ satisfying (2.13), with d replaced by d_i. Let $M_1 = N$, $M_2 = N/2 + 1$, $M_3 = N/4 + 1, \ldots.$ Iteration of (2.35), starting with M_1, ψ_1 on the left and ψ_2 on the right, yields

$$\text{(2.39)} \qquad \sup_{|a|+s \leqslant N} \frac{\|X^a T^s p\|_{L^2\{\psi_1 = 1\}}}{N^{|a|+s}} \leqslant (KC)^{2\Sigma(N/2^j+1)} K^{\log_2 N} \prod (2^j)^{2(N/2^j+1)} C_p$$
$$\leqslant C'^N C_p,$$

where

$$C_p = \sup_{|a'|+b' \leqslant 1} \|X^{a'} T^{b'} p\|_{L^2(\operatorname{supp} \psi_{\log_2 N})}.$$

Since we can cover

$$A_\delta = (M \times M \times [0, T_0]) \setminus \{d^2(x, y) < \varepsilon, t < \varepsilon\}$$

by a finite number of sets of the type of V_1, we have proved

PROPOSITION 2.8. *There is a constant C such that, for all N,*

$$\sup_{|\alpha| \leqslant N} \|D_x^\alpha p(x, y, t)\|_{L^2(A_\delta)} \leqslant C^N N!.$$

3. Mixed derivatives and proof of the theorem. Theorem 1.1 will follow from the appropriate L^2-estimates. We review this briefly. To prove (i) it suffices to prove that, for any $0 < t_0 < T_0$, there is a constant C such that on $M \times M \times [t_0, T_0]$,

$$\text{(3.1)} \qquad \left|D_x^\alpha D_y^\beta D_t^k p(x, y, t)\right| \leqslant C^{|\alpha+\beta|+k+1} (|\alpha + \beta| + k)!.$$

By the Sobolev Lemma, estimates (3.1) and (1.2) follow from the same estimates on slightly larger sets with the sup-norm replaced by the L^2-norm. So far we have proved the L^2-norm estimates for high t-derivatives off $t = 0$ (Proposition 2.3) and high x-derivatives down to $t = 0$ off the diagonal (Proposition 2.8). This immediately gives us the analogous estimate for high y-derivatives, since $p(x, y, t) = \overline{p(y, x, t)}$. In Proposition 3.1 we derive the L^2-version of (3.1) from Proposition 2.3 and 2.8. Finally, we use Proposition 2.8 and the heat equation to prove the L^2-version of (1.2) in Proposition 3.2.

PROPOSITION 3.1. *Let $0 < t_1 < T_1$. There is a constant C such that for all α, β, and k,*

$$\text{(3.2)} \qquad \left\|D_x^\alpha D_y^\beta D_t^k p(x, y, t)\right\|_{L^2(M \times M \times [t_1, T_1])} \leqslant C^{|\alpha+\beta|+k+1} (|\alpha + \beta| + k)!.$$

PROOF. By Lemma 2.2 there is a K independent of N and, for each N, a function $\phi_N \in C_0^\infty(\mathbf{R}^+)$, with $\phi_N \equiv 1$ on $[t_1, T_1]$ and

$$|D_t^m \phi(t)| \leqslant K^{m+1} N^m, \qquad m \leqslant N. \tag{3.3}$$

Fix N and let $|\alpha + \beta| + k \leqslant N$. Integration by parts gives

$$\begin{aligned}
\left\|D_x^\alpha D_y^\beta D_t^k p(x, y, t)\right\|^2_{L^2(M\times M\times[t_1, T_1])} &\leqslant \left\|\phi_N(t) D_x^\alpha D_y^\beta D_t^k p(x, y, t)\right\|^2_{L^2(M\times M\times \mathbf{R}^+)} \\
&\leqslant \left|\left(D_t^k \phi_N^2(t) D_t^k p(x, y, t), \tilde{D}_x^{2\alpha} \tilde{D}_y^{2\beta} p(x, y, t)\right)_{L^2(M\times M\times\mathbf{R}^+)}\right| \qquad (3.4)\\
&\leqslant \left\|D_t^k \phi_N^2(t) D_t^k p(x, y, t)\right\|_{L^2(M\times M\times\mathbf{R}^+)} \left\|\tilde{D}_x^{2\alpha} \tilde{D}_y^{2\beta} p(x, y, t)\right\|_{L^2(\operatorname{supp}\phi_N)}.
\end{aligned}$$

Here we write $\tilde{D}^{2\alpha}$ for a product of $2|\alpha|$ W_j and W_k^*s, with the notation of Theorem 1.1. Also,

$$\begin{aligned}
\left\|D_x^{2\alpha} D_y^{2\beta} p\right\|^2_{L^2(\operatorname{supp}\phi_N)} &\leqslant \left|\left(\tilde{D}_x^{4\alpha} p, \tilde{D}_y^{4\beta} p\right)_{L^2(\operatorname{supp}\phi_N)}\right| \qquad (3.5)\\
&\leqslant \left\|\tilde{D}_x^{4\alpha} p\right\|_{L^2(\operatorname{supp}\phi_N)} \left\|\tilde{D}_y^{4\beta} p\right\|_{L^2(\operatorname{supp}\phi_N)} \\
&\leqslant C^{4|\alpha+\beta|}(4|\alpha|)!(4|\beta|)! \leqslant \left(C_1^{2|\alpha+\beta|}(2|\alpha+\beta|)!\right)^2
\end{aligned}$$

by Proposition 2.8. We can write $D_t^k \phi_N^2(t) D_t^k p$ as a sum of C^k terms of the form $\phi_N^{(k_1)} \phi_N^{(k_2)} D_t^{2k - k_1 - k_2} p$. Using this, estimate (2.10), and the bounds (3.3) on ϕ_N, we have, for $k \leqslant N$,

$$\begin{aligned}
\left\|D_t^k \phi_N^2(t) D_t^k p\right\|_{L^2} &\leqslant C^k \sup_{k_1 + k_2 \leqslant k} K^{k_1 + k_2 + 2} N^{k_1 + k_2} C^{2k - k_1 - k_2}(2k - k_1 - k_2)! \qquad (3.6)\\
&\leqslant C_2^k N^{2k},
\end{aligned}$$

with C_2 independent of k and N. Using (3.5) and (3.6) in (3.4) gives

$$\begin{aligned}
\left\|D_x^\alpha D_y^\beta D_t^k p(x, y, t)\right\|_{L^2(M\times M\times[t_1, T_1])} &\leqslant C_1^{|\alpha+\beta|} C_2^{k/2} N^k (2|\alpha + \beta|!)^{1/2} \qquad (3.7)\\
&\leqslant C^N N^N.
\end{aligned}$$

Taking $N = |\alpha + \beta| + k$, we obtain (3.2). ■

PROPOSITION 3.2. *For each sufficiently small* ε *there is a constant* C_ε *such that for any* α, β, *and* k,

(3.8)

$$\left\|D_x^\alpha D_y^\beta D_t^k p(x, y, t)\right\|_{L^2((M\times M\times[0,1])\cap\{d^2(x,y) > \varepsilon\})} \leqslant C_\varepsilon^{|\alpha+\beta|+k+1}(|\alpha + \beta| + 2k)!.$$

PROOF. Since $(\square_b^x + \partial/\partial t)p = 0$,

$$D_t^k p = \left(-\square_b^x\right)^k p. \tag{3.9}$$

Hence, it suffices to prove (3.8) with $k = 0$. We fix N and take cutoff functions $\phi_{1,N} \in C_0^\infty((-\varepsilon, 1 + \varepsilon))$, $\phi_{1,N} \equiv 1$ on $[0,1]$, and $\phi_{2N} \in C_0^\infty(-\varepsilon, \varepsilon)$, $\phi_{2,N} \equiv 1$ on $[-\varepsilon/2, \varepsilon/2]$, satisfying $|\phi_{j,N}^{(i)}| \leqslant K^i N^i$ for $i \leqslant N$, $j = 1, 2$, where K is independent of N. We let

$$\phi_N(x, y, t) = \phi_{1,N}(t)\left(1 - \phi_{2,N}\left(d^2(x, y)\right)\right).$$

As in § 2(c), for ε sufficiently small, we have

$$|D_x^\alpha D_y^\beta D_t^k \phi_N| \leqslant K^{|\alpha+\beta|+k+1} N^{|\alpha+\beta|+k} \tag{3.10}$$

for $|\alpha + \beta| + k \leqslant 3N$, K independent of N. We integrate by parts, as in the proof of Proposition 3.1. Now $\|\phi_N D_x^\alpha D_y^\beta p\|_{L^2}^2 = (\phi_N D_x^\alpha D_y^\beta p, \phi_N D_x^\alpha D_y^\beta p)_{L^2}$,

$$D_x^\alpha \phi_N^2 D_y^\beta = \sum_{\substack{\alpha' \leqslant \alpha \\ \beta' \leqslant \beta}} \binom{\alpha}{\alpha'}\binom{\beta}{\beta'} \text{ terms } D_y^{\beta-\beta'}(\phi_N^2)^{(\alpha'+\beta')} D_x^{\alpha-\alpha'},$$

and (3.10) implies

$$|(\phi_N^2)^{(r)}| \leqslant K'^r N^r.$$

Hence, we have, for $|\alpha + \beta| \leqslant N$,

$$\|\phi_N(x, y, t) D_x^\alpha D_y^\beta p\|_{L^2_{x,y,t}}^2 \leqslant (C^N N^N)^2 \tag{3.11}$$

in view of Proposition 2.8 (as in the proof of (3.6)). Now (3.8) follows from (3.11). ■

References

[BGS] R. Beals, P. C. Greiner and N. K. Stanton, *The heat equation on a CR manifold*, J. Differential Geom. (to appear).

[FK] G. B. Folland and J. J. Kohn, *The Neumann problem for the Cauchy–Riemann complex*, Ann. of Math. Stud., No. 75, Princeton Univ. Press, Princeton, N. J., 1972.

[S] N. K. Stanton, *The fundamental solution of the heat equation associated with the $\bar\partial$-Neumann problem*, J. Analyse Math. **34** (1978), 265–274.

[ST] N. K. Stanton and D. S. Tartakoff, *The heat equation for the $\bar\partial_b$-Laplacian*, Comm. Partial Differential Equations **9** (1984), 597–686.

[Ta1] D. S. Tartakoff, *Local analytic hypoellipticity for $\Box_b$ on nondegenerate Cauchy–Riemann manifolds*, Proc. Nat. Acad. Sci. U.S.A., **75** (1978), 3027–3028.

[Ta2] ______, *The local real analyticity of solutions to $\Box_b$ and the $\bar\partial$-Neumann problem*, Acta Math. **145** (1980), 177–204.

[Ta3] ______, *Elementary proofs of analytic hypoellipticity for the $\Box_b$ and $\bar\partial$-Neumann problem*, *Analytic solutions of partial differential equations*, Astérisque, **89-90** (1981), 85–116.

[Tr1] F. Trèves, *Analytic hypo-ellipticity of a class of pseudodifferential operators with double characteristics and applications to the $\bar\partial$-Neumann problem*, Comm. Partial Differential Equations **3** (1978), 475–642.

[Tr 2] ______, *Introduction to pseudifferential and Fourier integral operators*, Vols. 1, 2, Plenum, New York, 1980.

University of Notre Dame

University of Illinois at Chicago

Proceedings of Symposia in Pure Mathematics
Volume **43** (1985)

Fefferman–Phong Inequalities in Diffraction Theory[1]

MICHAEL E. TAYLOR

0. Introduction. This paper deals with some methods of obtaining regularity estimates and propagation of singularities results for certain classes of boundary problems

$$Lu = 0 \quad \text{on } \Omega, \qquad Bu = f \quad \text{on } \partial\Omega \tag{0.1}$$

in the case of grazing rays. We begin with a brief overview of the subject. Recall that grazing rays are null bicharacteristics of L that hit $\partial\Omega$ tangentially with exactly second-order contact and remain in $\overline{\Omega}$. The basic form of the grazing ray parametrix produced by [**19**] and [**27**] is

$$u = \int [gA(\zeta) + ihA'(\zeta)]\, A(\zeta_0)^{-1} e^{i\theta} \hat{F}(\xi)\, d\xi. \tag{0.2}$$

We assume for simplicity that L is a second-order, scalar, hyperbolic operator. Then the amplitudes $g \in S^0$, $h \in S^{-1/3}$ are given by certain transport equations, the phase functions (ζ, θ)—homogeneous in ξ of degrees 2/3 and 1, respectively satisfy certain eikonal equations and, by virtue of results of [**21**], one can arrange that

$$\zeta\big|_{\partial\Omega} = \zeta_0 = |\xi|^{-1/3}\xi_n, \qquad h\big|_{\partial\Omega} = 0. \tag{0.3}$$

$A(\zeta)$ is an Airy function. The quantity F in (0.2) is related to f via the boundary condition $Bu = f$ on $\partial\Omega$.

For the Dirichlet problem with boundary condition $u|_{\partial\Omega} = f$, we take (0.2) and evaluate on $\partial\Omega$, using (0.3) to get

$$u|_{\partial\Omega} = \int g_0 e^{i\theta_0} \hat{F}(\xi)\, d\xi = JF, \tag{0.4}$$

1980 *Mathematics Subject Classification*. Primary 35L20.
[1]Research partially supported by NSF grant MCS 820176A01.

where $\theta_0 = \theta|_{\partial\Omega}$, $g_0 = g|_{\partial\Omega}$. It can be arranged that J is an elliptic Fourier integral operator. Denoting a microlocal parametrix of J by J^{-1}, we see that it follows that setting $F = J^{-1}f$ and plugging into (0.2) yield a parametrix for the Dirichlet problem.

Of fundamental importance—equal to the Dirichlet problem—is the Neumann problem

$$\partial u/\partial\nu = g \quad \text{on } \partial\Omega, \tag{0.5}$$

where $\partial/\partial\nu$ is the normal derivative with respect to the Lorentz metric on $\overline{\Omega}$ given by the principal symbol of L. A brief calculation gives

$$\partial u/\partial\nu|_{\partial\Omega} = \int (\zeta_\nu g_0 + ih_\nu)e^{i\theta_0}A'(\zeta_0)A(\zeta_0)^{-1}\hat{F}(\xi)\,d\xi + \int g_\nu e^{i\theta_0}\hat{F}(\xi)\,d\xi, \tag{0.6}$$

if we use the fact that $\theta_\nu|_{\partial\Omega} = 0$, which follows from the eikonal equation together with (0.3). Note the Fourier multiplier

$$\Phi(\zeta_0) = A'(\zeta_0)/A(\zeta_0). \tag{0.7}$$

It is easy to show that

$$\Phi(\zeta_0) \in S^{1/3}_{1/3,0}, \tag{0.8}$$

and applying (0.4) and (0.6) we get, for "outgoing" solutions described by (0.2),

$$u|_{\partial\Omega} = f \Rightarrow \partial u/\partial\nu|_{\partial\Omega} = Nf, \tag{0.9}$$

where

$$N = J(A\Phi + B)J^{-1}, \tag{0.10}$$

with

$$A \in OPS^{2/3} \text{ elliptic}, \qquad B \in OPS^0. \tag{0.11}$$

Inverting N is equivalent to inverting $A\Phi + B = (A + B\Phi^{-1})\Phi$. Note that $\Phi(\zeta_0)^{-1} \in S^0_{1/3,0}$. Thus $A + B\Phi^{-1}$ is an elliptic element of $OPS^{2/3}_{1/3,0}$ and has a parametrix in $OPS^{-2/3}_{1/3,0}$. Thus a parametrix for N is given by

$$N^{-1} = J\Phi^{-1}(A + B\Phi^{-1})^{-1}J^{-1}, \tag{0.12}$$

where

$$\Phi^{-1}(A + B\Phi^{-1})^{-1} \in OPS^{-2/3}_{1/3,0}. \tag{0.13}$$

Consequently, we are able to construct a parametrix for the Neumann boundary problem (0.5) by letting $f = N^{-1}g$ and solving the Dirichlet problem.

Maxwell's equations in a region bounded by a perfect conductor are amenable to an analysis mixing the Dirichlet and Neumann problems; see [**26**].

An oblique derivative problem of the form

$$(\partial/\partial\nu + X)u|_{\partial\Omega} = g \tag{0.14}$$

gives rise to an operator of the form

$$P = N + iQ, \qquad Q \in OPS^1. \tag{0.15}$$

In view of (0.10) we have

$$P = J(A\Phi + i\tilde{Q})J^{-1}, \tag{0.16}$$

where

$$\tilde{Q} = J^{-1}QJ - iB \in OPS^1 \tag{0.17}$$

is given by Egorov's theorem. If Q is elliptic on the grazing set—i.e., if $\tilde{Q}$ is elliptic on $\xi_n = 0$—then, microlocally near $\xi_n = 0$, $A\Phi + i\tilde{Q}$ is an elliptic element of $OPS^1_{1/3,0}$. In the elliptic case this operator and its parametrix are constructable as *Airy operators*; generally an element of $\mathscr{A}^{+,m}$ is an operator of the form

$$Pu(x) = \int [a(x, y, \xi) + b(x, y, \xi)\Phi(\zeta)] e^{i(x-y)\cdot\xi} u(y)\, dy\, d\xi, \tag{0.18}$$

where

$$a \in S^m, \qquad b \in S^{m-1/3}; \tag{0.19}$$

from now on we drop the subscript from ζ_0, setting

$$\zeta = |\xi|^{-1/3}\xi_n. \tag{0.20}$$

If $P \in \mathscr{A}^{+,m}$ is elliptic, then it has a parametrix in $\mathscr{A}^{+,-m}$. See [**20**] for this and [**30, 23**] for a simplified discussion. These last two references also develop a theory of Airy operators for gliding ray problems, but we will not say more about that here. Cases where $\tilde{Q}$ may have the characteristic set intersecting $\xi_n = 0$ can give rise to some further problems, some of which are studied in this paper.

One result about the special function $\Phi(\zeta)$, in addition to its asymptotic behavior

$$\Phi(\zeta) \sim (\zeta)^{1/2} + c_1(\zeta)^{-1} + c_2(\zeta)^{-5/2} + \cdots, \qquad \zeta \to \pm\infty, \tag{0.21}$$

that has played an important role in several investigations, such as transmission problems [**29**], as well as other boundary problems [**14, 15, 22**], is the fact that

$$\zeta \in R \Rightarrow \operatorname{Im}\Phi(\zeta) > 0 \quad \text{and} \quad \operatorname{Re}\Phi(\zeta) > 0. \tag{0.22}$$

We refer to Appendix A of [**23**] for a complete treatment of Airy functions and Airy quotients; the result (0.22) was proved in [**22**] and strongly suggested by a graph in [**25**].

Operators in $OPS^m_{1/3,0}$ are much more singular than operators in $OPS^m_{1,0}$, and obtaining desired a priori estimates becomes correspondingly more difficult. We make substantial use of the fact that $\mathscr{A}^{+,m}$ belongs to a much smaller class of operators than $OPS^m_{1/3,0}$—in fact a certain Beals–Fefferman class, studied in §1. Fefferman–Phong inequalities are among the most powerful estimates for pseudodifferential operators available today, and we show how they can be exploited to give a systematic treatment of a priori estimates, leading to hypoellipticity and propagation of singularities results for classes of operators of which $\mathscr{A}^{+,m}$ is a special case. This is not the only route to all our results, and we mention that some of the results given here, as well as numerous other interesting results, have been derived by Eskin [**3, 4**] by different means.

We also mention that only the simplest Fefferman–Phong inequalities, derived from [**6**], are used here. One might expect more recent developments, which have produced incisive analyses of operators in OPS^m, also to have implications for more singular classes of operators.

1. Airy operators and Beals–Fefferman calculus. Airy operators, defined in the introduction, include operators with a symbol of the form

$$A(x,\xi) = q(x,\xi) + p(x,\xi)\Phi(\zeta) \in \mathscr{A}^{+,m}, \tag{1.1}$$

where

$$\Phi(\zeta) = A'(\zeta)/A(\zeta), \qquad \zeta = |\xi|^{-1/3}\xi_n, \tag{1.2}$$

and

$$q(x,\xi) \in S^m, \qquad p(x,\xi) \in S^{m-1/3}. \tag{1.3}$$

It is easy to verify that

$$\mathscr{A}^{+,m} \subset S^m_{1/3,0}, \tag{1.4}$$

where, as usual, $S^m_{\rho,\delta}$ denotes the symbol classes of Hörmander [**9**]. A more precise estimate on such symbols is given by

$$\left|D^{\alpha'}_{\xi'} D^{\alpha_n}_{\xi_n} D^\beta_x A(x,\xi)\right| \leqslant C_{\alpha\beta}\langle\xi\rangle^{m-|\alpha'|}\left(\langle\xi\rangle^{1/3} + |\xi_n|\right)^{-\alpha_n}, \tag{1.5}$$

where $\xi' = (\xi_1,\ldots,\xi_{n-1})$. Such a class was considered in [**18, 26, 28**], where it was denoted by $\mathscr{N}^m_{1/3}$. It is equivalent to a class defined by Beals–Fefferman weight vectors.

Here we confine our attention to symbol classes defined by a pair of weight functions; for the most part the action of D_{ξ_n} is not distinguished from that of any other D_{ξ_j} (an exception is the derivation of (4.20)). Thus, we use the fact that any $A(x,\xi) \in \mathscr{A}^{+,m}$ satisfies

$$\left|D^\alpha_\xi D^\beta_x A(x,\xi)\right| \leqslant C_{\alpha\beta}\langle\xi\rangle^m\left(\langle\xi\rangle^{1/3} + |\xi_n|\right)^{-|\alpha|} \tag{1.6}$$

to place $\mathscr{A}^{+,m}$ inside the Beals–Fefferman class

$$S^{m\lambda}_{\Psi,\psi}, \tag{1.7}$$

where

$$\Psi(x,\xi) = |\xi|^{1/3}\langle\zeta\rangle \approx \langle\xi\rangle^{1/3} + |\xi_n|, \tag{1.8}$$

$$\psi(x,\xi) = 1, \tag{1.9}$$

and

$$e^{m\lambda} = |\xi|^m. \tag{1.10}$$

We recall that, generally, if Ψ, ψ is a pair of Beals–Fefferman weight functions, satisfying say conditions (1.1)—(1.5) of [**1**], and if μ belongs to the class $\mathscr{O}(\Psi,\psi)$, specified by (3.1)—(3.2) of [**1**], then we say

$$p(x,\xi) \in S^\mu_{\Psi,\psi}, \tag{1.11}$$

provided

$$\left|D_\xi^\alpha D_x^\beta p(x, \xi)\right| \leqslant C_{\alpha\beta} e^{\mu(x,\xi)} \Psi(x, \xi)^{-|\alpha|} \psi(x, \xi)^{-|\beta|}. \tag{1.12}$$

Some particularly important symbol classes are denoted

$$S_{\Psi,\psi}^{(k,l)}, \tag{1.13}$$

where

$$e^{(k,l)} = \Psi^k \psi^l. \tag{1.14}$$

Note that one assumption on Beals–Fefferman weight functions is the "uncertainty principle" $\Psi\psi \geqslant 1$.

We have used the fact that a first-order ξ-derivative of an element $A(x, \xi)$ of $\mathscr{A}^{+,m}$ has order 2/3 lower; $\nabla_\xi A \in S_{1/3,0}^{m-2/3}$; more precisely, as is easy to check, for $A(x, \xi) \in \mathscr{A}^{+,m}$ we have

$$\left|D_\xi^\alpha D_x^\beta A(x, \xi)\right| \leqslant C_{\alpha\beta} |\xi|^{m-1/3} \langle \zeta \rangle^{1/2} \left(|\xi|^{1/3} \langle \zeta \rangle\right)^{-|\alpha|} \quad \text{for } |\alpha| \geqslant 1. \tag{1.15}$$

In particular, using the weight functions (1.8)–(1.9),

$$A \in \mathscr{A}^{+,m} \Rightarrow \nabla_\xi A \in S_{\Psi,\psi}^{m\lambda - \gamma - (1,0)}, \tag{1.16}$$

where

$$e^\gamma = \langle \zeta \rangle^{-1/2} |\xi|^{1/3}. \tag{1.17}$$

Note that this is an improvement over the usual implication

$$A \in S_{\Psi,\psi}^{\mu} \Rightarrow \nabla_\xi A \in S_{\Psi,\psi}^{\mu - (1,0)}. \tag{1.18}$$

Note that, since $\psi(x, \xi) = 1$ in (1.9), we have $e^{(1,0)} = e^{(1,1)}$ in this case.

We now prove a simple general result that implies (1.6) and will be useful in subsequent sections.

LEMMA 1.1. *With* $\zeta = |\xi|^{-1/3} \xi_n$, *let*

$$F(\xi) = q(\zeta), \tag{1.19}$$

where $q \in C^\infty(R)$ *satisfies the estimates*

$$\left|q^{(k)}(\zeta)\right| \leqslant C_k \langle \zeta \rangle^{m-k}. \tag{1.20}$$

Then

$$F(\xi) \in S_{\Psi,\psi}^{(2/3)m\lambda} \quad \text{if } m \geqslant 0, \qquad F(\xi) \in S_{\Psi,\psi}^{0} \quad \text{if } m \leqslant 0. \tag{1.21}$$

More precisely, if we set $\langle \zeta \rangle = e^{Z(\xi)} \in S_{\Psi,\psi}^{Z}$, *then*

$$F(\xi) \in S_{\Psi,\psi}^{mZ}. \tag{1.22}$$

If $q \in C^\infty(R)$ *satisfies the estimates*

$$\left|q^{(k)}(\zeta)\right| \leqslant C_k \langle \zeta \rangle^{-k} |q(\zeta)|, \tag{1.23}$$

then

$$F(\xi) = e^{f(\xi)} \in S_{\Psi,\psi}^{f}. \tag{1.24}$$

PROOF. We have

$$D_\xi^\alpha F(\xi) = \sum_{\sigma_1 + \cdots + \sigma_k = \alpha} C_\sigma \left(D_\xi^{\sigma_1}\zeta\right) \cdots \left(D_\xi^{\sigma_k}\zeta\right) q^{(k)}(\zeta).$$

It is easy to see that, with Ψ given by (1.8),

$$\langle\zeta\rangle^{-1}\left|D_\xi^{\sigma_j}\zeta\right| \leqslant C_{\sigma_j}\Psi^{-|\sigma_j|},$$

so each term in the sum above is dominated in absolute value by $C\Psi^{-|\alpha|}|q^{(k)}(\zeta)|\langle\zeta\rangle^k$. From here the conclusions are apparent.

We shall make use of both the Kohn–Nirenberg calculus and the Weyl calculus of pseudodifferential operators. These associate operators to symbols by the following rules:

$$p(x, D)u = (2\pi)^{-n}\int p(x, \xi)e^{i(x-y)\cdot\xi}u(y)\, dy\, d\xi, \tag{1.25}$$

$$p(X, D)u = (2\pi)^{-n}\int p\left(\tfrac{1}{2}(x + y), \xi\right)e^{i(x-y)\cdot\xi}u(y)\, dy\, d\xi. \tag{1.26}$$

We record some rules for the Weyl calculus (1.26); see **[12]** for details. If $p_j \in S^{\mu_j}_{\Psi,\psi}$, then $p_1(X, D)p_2(X, D) = (p_1 \circ p_2)(X, D)$, with

$$p_1 \circ p_2(x, \xi) \sim p_1 p_2 + \sum_{j \geqslant 1}\left(\frac{1}{j!}\right)\{p_1, p_2\}_j(x, \xi), \tag{1.27}$$

where

$$\begin{aligned}\{p_1, p_2\}_j(x, \xi)&\\ &= \left(\frac{-i}{2}\right)^j \sum_{k=1}^n \left(\frac{\partial^2}{\partial y_k \partial \xi_k} - \frac{\partial^2}{\partial x_k \partial \eta_k}\right)^j p_1(x, \xi)p_2(y, \eta)\Bigg|_{\substack{y=x\\ \eta=\xi}}.\end{aligned} \tag{1.28}$$

The meaning of (1.27) is that the difference between $p_1 \circ p_2$ and the sum over $j < N$ belongs to $S^{\mu_1+\mu_2-(N,N)}_{\Psi,\psi}$. In particular,

$$p_1 \circ p_2 = p_1 p_2 - \tfrac{1}{2}i\{p_1, p_2\} \mod S^{\mu_1+\mu_2-(2,2)}_{\Psi,\psi}, \tag{1.29}$$

where $\{p_1, p_2\}$ is the usual Poisson bracket. It follows that, for scalar operators,

$$\begin{aligned}p_1 \circ p_2 \circ p_3 = p_1 p_2 p_3 &- \tfrac{1}{2}ip_1\{p_2, p_3\} - \tfrac{1}{2}ip_2\{p_1, p_3\}\\ &-\tfrac{1}{2}ip_3\{p_1, p_2\} \mod S^{\mu_1+\mu_2+\mu_3-(2,2)}_{\Psi,\psi}.\end{aligned} \tag{1.30}$$

If $p_1 = p_3 = q$ then we get

$$q \circ p \circ q = q^2 p \mod S^{\mu-(2,2)}_{\Psi,\psi}; \qquad \mu = \mu_1 + \mu_2 + \mu_3; \tag{1.31}$$

if $p_1 = q = p_3^{-1}$, with $p_1 \in S^{\mu_1}_{\Psi,\psi}$ and $p_3 \in S^{-\mu_1}_{\Psi,\psi}$, we get

$$q \circ p \circ q^{-1} = p + i\{p, q\}q^{-1} \mod S^{\mu-(2,2)}_{\Psi,\psi} \quad \text{if } p \in S^\mu_{\Psi,\psi}. \tag{1.32}$$

These formulas will be used in the following sections.

We close this section with a brief discussion of the assumptions made on Beals–Fefferman weight functions in **[1]**, which are a bit more general than those made in **[2]**. The main assumptions, for some constants c, C, $\delta > 0$, are

$$\psi \leqslant C, \tag{1.33}$$

$$\Psi\psi \geqslant c, \tag{1.34}$$

$$|x-y| \leqslant c\psi(x,\xi), |\xi-\eta| \leqslant c\Psi(x,\xi) \Rightarrow \Psi(x,\xi) \sim \Psi(y,\eta) \text{ and } \psi(x,\xi) \sim \psi(y,\eta), \tag{1.35}$$

where $A \sim b$ means $c \leqslant A/B \leqslant C$; we also give the following hypotheses on

$$R(x,\xi) = \Psi(x,\xi)/\psi(x,\xi): \tag{1.36}$$

namely,

$$R(x,0) \leqslant C\langle x\rangle^{C}, \tag{1.37}$$

and

(1.38)

$$|x-y| \leqslant cR(x,\xi)^{\delta}R(y,\eta)^{-1/2}, |\xi-\eta| \leqslant cR(x,\xi)^{\delta+1/2} \Rightarrow R(x,\xi) \sim R(y,\eta).$$

In [**1**] it is shown that (1.35) and (1.38) hold if Ψ, ψ satisfy

$$\left|D_\xi^\alpha D_x^\beta \Psi\right| \leqslant C_{\alpha\beta}\Psi\Psi^{-|\alpha|}\psi^{-|\beta|}, \tag{1.39}$$

$$\left|D_\xi^\alpha D_x^\beta \psi\right| \leqslant C_{\alpha\beta}\psi\Psi^{-|\alpha|}\psi^{-|\beta|}, \tag{1.40}$$

and

$$\left|D_\xi^\alpha D_x^\beta R\right| \leqslant C_{\alpha\beta}R(R^{\delta+1/2}+\Psi)^{-|\alpha|}(R^{\delta-1/2}+\psi)^{-|\beta|}. \tag{1.41}$$

Also, any weight functions Ψ, ψ satisfying (1.33)–(1.38) can be altered in a minor way to satisfy (1.39)–(1.41). Such weight functions are called smooth weight functions.

Note that when $\Psi = \Psi(x,\xi)$, $\psi = 1$, we have $R = \Psi$, and (1.34)–(1.35) (plus (1.37)) are the only nontrivial conditions; (1.38) follows from (1.35) with $\delta = 1/2$. A special class of weight functions is

$$\Psi_a = \langle\xi\rangle^a + \langle\xi_n\rangle, \qquad \psi_a = 1, \tag{1.42}$$

where $a \in (0,1)$.

In the following sections we restrict our attention to weight functions satisfying the further requirement

$$\Psi(x,\xi) \leqslant C\langle\xi\rangle, \tag{1.43}$$

which implies that $S^0_{\Psi,\psi}$ contains the Kohn–Nirenberg–Hörmander class $S^0_{1,0}$:

$$S^0_{1,0} \subset S^0_{\Psi,\psi}. \tag{1.44}$$

2. Fefferman–Phong inequalities and sharp Gårding inequalities for Airy operators. The sharp Gårding inequality of Hörmander states that $p(x,D)$ is semibounded for any $p(x,\xi) \geqslant 0$ provided $p(x,\xi) \in S^1_{1,0}$. This was extended to symmetric matrix valued $p(x,\xi)$ by Lax and Nirenberg [**17**]. As shown by Kumano-go [**16**], such semiboundedness holds for all positive $p(x,\xi) \in S^{\rho-\delta}_{\rho,\delta}$, provided $0 \leqslant \delta < \rho \leqslant 1$. This in turn is a special case of the Beals–Fefferman result [**2**] that semiboundedness holds for all positive $p(x,\xi) \in S^{(1,1)}_{\Phi,\varphi}$ for any pair of Beals–Fefferman weight functions Φ,φ. In 1978 Fefferman and Phong [**6**] showed that $p(x, D)$ is semibounded for any positive (scalar) $p(x, \xi) \in S^2_{1,0}$. The key ingredient in the proof is the result stated as Lemma A below.

Before stating the lemma we introduce some notation from **[6]**. If $p(x, \xi)$ is supported on $|x| \leqslant 1$, $|\xi| \leqslant M$, we say

$$p(x, \xi) \in S^m(1 \times M) \tag{2.1}$$

provided

$$\left|D_x^\beta D_\xi^\alpha p(x, \xi)\right| \leqslant C_{\alpha\beta} M^{m-|\alpha|}. \tag{2.2}$$

Then the main technical result from **[6]** is

LEMMA A. *We have the semiboundedness estimate*

$$\operatorname{Re}(p(x, D)u, u) \geqslant -K\|u\|^2_{L^2}, \qquad u \in C_0^\infty(R^n), \tag{2.3}$$

for any scalar $p(x, \xi) \geqslant 0$ *belonging to* $S^2(1 \times M)$, *i.e., satisfying* (2.1)–(2.2) *with* $m = 2$. *The constant* K *in* (2.3) *is independent of* M *and depends only on the* $C_{\alpha\beta}$ *in* (2.2).

The fact that any positive scalar $p(x, \xi) \in S^2_{1,0}$ defines a semibounded operator $p(x, D)$ follows from Lemma A via Beals–Fefferman calculus. In fact, a more general statement of the Fefferman–Phong inequality is

PROPOSITION 2.1. *For any Beals–Fefferman pair of weight functions* Ψ, ψ, *we have the semiboundedness estimate* (2.3) *for any positive scalar* $p(x, \xi)$ *belonging to* $S^{(2,2)}_{\Psi,\psi}$.

PROOF. It is possible to cover $R^{2n} = R^n_x \times R^n_\xi$ by rectangles Q_j, centered at points (x^j, ξ^j), of side $\Psi(x^j, \xi^j)$ in the ξ direction and $\psi(x^j, \xi^j)$ in the x directions, and produce a partition of unity

$$1 = \sum \varphi_j^2(x, \xi), \qquad \varphi_j \in C_0^\infty(Q_j), \varphi_j(x, \xi) \geqslant 0, \tag{2.4}$$

such that the l^2-valued function $\varphi(x, \xi) = (\varphi_1(x, \xi), \varphi_2(x, \xi), \ldots)$ defines a Hilbert-space-valued symbol

$$\varphi \in S^0_{\Psi,\psi}(R^n, l^2). \tag{2.5}$$

Thus $\varphi(X, D)$ is an element of $OPS^0_{\Psi,\psi}(R^n, l^2)$, and

$$I = \varphi(X, D)^*\varphi(X, D) = \sum \varphi_j(X, D)^2 \mod OPS^{(-2,-2)}_{\Psi,\psi}. \tag{2.6}$$

Here we use the Weyl calculus described at the end of §1. Note that

$$\varphi(X, D)^* p(X, D)\varphi(X, D) = \sum_j \varphi_j(X, D)p(X, D)\varphi_j(X, D). \tag{2.7}$$

Now by the Weyl calculus we have

$$\varphi(X, D)^* p(X, D)\varphi(X, D) = a(X, D) \tag{2.8}$$

with

$$\begin{aligned} a(x, \xi) \sim p(x, \xi) &+ i\{\varphi^* p, \varphi\}_1 - i\{\varphi^*, p\}_1\varphi \\ &- \{\varphi^*, p\}_2\varphi - \{\varphi^* p, \varphi\}_2 - \{\{\varphi^*, p\}_1, \varphi\}_1 + \cdots, \end{aligned} \tag{2.9}$$

where

$$(2.10)\quad \{p, q\}_k(x, \xi) = (-2)^{-k} \sum_{l=1}^{n} \left(\frac{\partial^2}{\partial y_l \partial \xi_l} - \frac{\partial^2}{\partial x_l \partial \eta_l} \right)^k p(x, \xi) q(y, \eta) \Big|_{\substack{y=x \\ \eta=\xi}}.$$

Note that $\{\varphi^* p, \varphi\}_1 - \{\varphi^*, p\}_1 \varphi = 0$. It follows that, given any $p \in S^{\mu}_{\Psi,\psi}$, the remainder terms in (2.9) after $p(x, \xi)$ belong to $S^{\mu-(2,2)}_{\Psi,\psi}$. In particular,

$$p \in S^{(2,2)}_{\Psi,\psi} \Rightarrow p(X, D) - \sum_j \varphi_j(X, D) p(X, D) \varphi_j(X, D) \in OPS^0_{\Psi,\psi}.$$

Thus, to prove semiboundedness of $p(X, D)$, it suffices to prove semiboundedness

$$(2.11)\qquad (\chi_j p)(X, D) \geqslant -C, \qquad C \text{ independent of } j,$$

where χ_j is a cutoff function equal to 1 on $\operatorname{supp} \varphi_j$ and supported in Q_j. This desired semiboundedness follows from Lemma A. For $p(x, \xi) \in S^{(2,2)}_{\Psi,\psi}$ we have $p(X, D) - p(x, D) \in OPS^{(1,1)}_{\Psi,\psi}$, but if $p(x, \xi)$ is real, the selfadjoint part of this difference belongs to $OPS^0_{\Psi,\psi}$. Thus semiboundedness of $p(X, D)$ is equivalent to semiboundedness of $p(x, D)$, and Proposition 2.1 is proved.

For the special case of Hörmander's symbol classes $S^m_{\rho,\delta}$, we have

COROLLARY 2.2. *If $0 \leqslant \delta < \rho \leqslant 1$, then any positive scalar $p(x, \xi) \in S^{2(\rho-\delta)}_{\rho,\delta}$ defines a semibounded operator $p(x, D)$.*

We will specialize from general Beals–Fefferman weight functions Ψ, ψ to those of the form

$$(2.12)\qquad \Psi = \Psi(\xi), \qquad \psi = 1.$$

The weight functions (1.9) for Airy operators are of this form. We set

$$(2.13)\qquad \beta(\xi) = \langle \xi \rangle \Psi^{-2} = e^{\lambda - (2,2)},$$

keeping the notation $e^{\lambda} = |\xi|$ of (1.10). From the Fefferman–Phong inequality (Proposition 2.1), we deduce

PROPOSITION 2.3. *Let the weight functions Ψ, ψ satisfy* (2.12). *Let*

$$(2.14)\qquad a(x, \xi) \in S^{\lambda}_{\Psi,\psi}$$

be a positive scalar function and suppose

$$(2.15)\qquad \operatorname{Re} a(x, \xi) \geqslant L(\xi) \beta(\xi).$$

Suppose $L(\xi) \geqslant L_0 > 0$ and

$$(2.16)\qquad L(\xi) \in S^{(2,2)}_{\Psi,\psi}.$$

Then for $v \in C_0^\infty(R^n)$,

$$(2.17)\quad \operatorname{Re}(a(X, D)v, v) \geqslant \left\| L(D)^{1/2} \beta(D)^{1/2} v \right\|^2_{L^2} - C \left\| \beta(D)^{1/2} v \right\|^2_{L^2}.$$

If, in addition,

$$(2.18)\qquad L(\xi) \geqslant C \langle \xi \rangle^{\varepsilon} \quad \textit{for some } \varepsilon > 0,$$

then

$$\text{(2.19)} \qquad \operatorname{Re}(a(X,D)v,v) \geqslant \tfrac{1}{2}\left\|L(D)^{1/2}\beta(D)^{1/2}v\right\|_{L^2}^2 - C_N\|v\|_{H^{-N}}^2.$$

PROOF. The hypotheses yield

$$\text{(2.20)} \qquad e(x,\xi) = a(x,\xi)\beta(\xi)^{-1} \in S^{(2,2)}_{\Psi,\psi}$$

in light of (2.13). Since $e(x,\xi) \geqslant L(\xi)$, we have $e(x,\xi) - L(\xi)$ positive and in $S^{(2,2)}_{\Psi,\psi}$, so Proposition 2.1 implies

$$\text{(2.21)} \qquad \operatorname{Re}(e(X,D)u,u) \geqslant \operatorname{Re}(L(D)u,u) - C\|u\|_{L^2}^2.$$

Since

$$\text{(2.22)} \qquad e(X,D) - \beta(D)^{-1/2}a(X,D)\beta(D)^{-1/2} \in OPS^0_{\Psi,\psi},$$

if we set $u = \beta(D)^{1/2}v$, we deduce (2.17) from (2.21). From here (2.19) is an obvious consequence of (2.18).

In case we are studying Airy operators, with $\Psi = \langle\zeta\rangle|\xi|^{1/3}$, we have

$$\text{(2.23)} \qquad \beta(\xi) = \langle\zeta\rangle^{-2}|\xi|^{1/3}.$$

Suppose

$$\text{(2.24)} \qquad L_{ab}(\xi) = |\xi|^a\langle\zeta\rangle^b,$$

with

$$\text{(2.25)} \qquad 0 < a \leqslant 2/3, \qquad -3a/2 < b,$$

and

$$\text{(2.26)} \qquad \langle\zeta\rangle^{b-2} \leqslant C\langle\xi\rangle^{2/3-a},$$

which amounts to assuming

$$\text{(2.27)} \qquad \text{either } b \leqslant 2 \quad \text{or} \quad b > 2 \quad \text{and} \quad b \leqslant 3(1 - \tfrac{1}{2}a).$$

One example is

$$\text{(2.28)} \qquad L(\xi) = |\xi|^{1/3}\langle\zeta\rangle^{5/2};$$

another is

$$\text{(2.29)} \qquad L(\xi) = \Psi(\xi) = |\xi|^{1/3}\langle\zeta\rangle.$$

In case (2.24) we have

$$\text{(2.30)} \qquad L_{ab}(\xi)\beta(\xi) = |\xi|^{1/3+a}\langle\zeta\rangle^{b-2},$$

so from (2.19) we have

PROPOSITION 2.4. *Suppose* $a(x,\xi) \in \mathscr{A}^{+,1}$ *or, more generally,* $a(x,\xi) \in S^{\lambda}_{\Psi,\psi}$, *with* Ψ,ψ *given by* (1.9). *Suppose*

$$\text{(2.31)} \qquad \operatorname{Re} a(x,\xi) \geqslant C_0|\xi|^{1/3+a}\langle\zeta\rangle^{b-2} = C_0K_{ab}(\xi),$$

where a and b satisfy (2.25)–(2.27). *Then for* $v \in C_0^\infty(R^n)$,

$$\text{(2.32)} \qquad \operatorname{Re}(a(X,D)v,v) \geqslant \tfrac{1}{2}C_0\left\|K_{ab}(D)^{1/2}v\right\|_{L^2}^2 - C_N\|v\|_{H^{-N}}^2.$$

In particular, if

$$\operatorname{Re} a(x,\xi) \geqslant C_0|\xi|^{2/3}\langle\zeta\rangle^{1/2} = C_0 K(\xi), \tag{2.33}$$

then

$$\operatorname{Re}(a(X,D)v,v) \geqslant \tfrac{1}{2}C_0\left\|K(D)^{1/2}v\right\|_{L^2}^2 - C_N\|v\|_{H^{-N}}^2; \tag{2.34}$$

and if

$$\operatorname{Re} a(x,\xi) \geqslant C_0|\xi|^{2/3}\langle\zeta\rangle^{-1} = C_0\kappa(\xi), \tag{2.35}$$

then

$$\operatorname{Re}(a(X,D)v,v) \geqslant \tfrac{1}{2}C_0\left\|\kappa(D)^{1/2}v\right\|_{L^2}^2 - C_N\|v\|_{H^{-N}}^2. \tag{2.36}$$

In fact, (2.33)–(2.36) are imbedded in the following family of results. We can consider

$$L(\xi) = L_b(\xi) = |\xi|^{1/3}\langle\zeta\rangle^b, \qquad -1/2 < b \leqslant 5/2, \tag{2.37}$$

in which case

$$L_b(\xi)\beta(\xi) = |\xi|^{2/3}\langle\zeta\rangle^d, \quad d = b-2, \qquad \text{so } -5/2 < d \leqslant 1/2. \tag{2.38}$$

We see that if $a(x,\xi) \in \mathscr{A}^{+,1}$ and

$$\operatorname{Re} a(x,\xi) \geqslant C_0|\xi|^{2/3}\langle\zeta\rangle^d = C_0\kappa_d(\xi), \qquad -5/2 < d \leqslant 1/2, \tag{2.39}$$

then

$$\operatorname{Re}(a(X,D)v,v) \geqslant \tfrac{1}{2}C_0\left\|\kappa_d(D)^{1/2}v\right\|_{L^2}^2 - C_N\|v\|_{H^{-N}}^2. \tag{2.40}$$

Note that in Proposition 2.3 hypothesis (2.18), together with the fact that the selfadjoint parts of $a(X,D)$ and $a(x,D)$ differ by an element of $OPS_{\Psi,\psi}^{\lambda,-(2,2)}$, implies that one can replace $a(X,D)$ by $a(x,D)$ in (2.19) and, hence, also in (2.32), (2.34), (2.36), and (2.40).

It is convenient to have the following consequence of Proposition 2.3 when $L(\xi)$ does not satisfy (2.18) but does dominate a large constant for $|\xi|$ large.

PROPOSITION 2.5. *Keep hypotheses* (2.14)–(2.16), *but replace* (2.18) *by*

$$L(\xi) \geqslant L_0 \quad \textit{for } |\xi| \textit{ sufficiently large}, \tag{2.41}$$

where $L_0 > 0$ *is sufficiently large compared to appropriate seminorms on* $a(x,\xi)$. *Then*

$$\operatorname{Re}(a(X,D)v,v) \geqslant \tfrac{1}{2}\left\|L(D)^{1/2}\beta(D)^{1/2}v\right\|_{L^2}^2 - C_N\|v\|_{H^{-N}}^2, \tag{2.42}$$

with a similar estimate holding for $a(x,D)$

PROOF. Hypothesis (2.41) makes this an immediate consequence of (2.17).

With this result we can get the following analogue of Proposition 2.4 in the limiting case $a = 0$. Thus we consider

$$L(\xi) = L_b^{\#}(\xi) = \langle\zeta\rangle^b L_0, \qquad 0 \leqslant b \leqslant 3. \tag{2.43}$$

In this case

$$L_b^{\#}(\xi)\beta(\xi) = |\xi|^{1/3}\langle\zeta\rangle^d L_0, \qquad -2 \leqslant d \leqslant 1, \tag{2.44}$$

so we have

PROPOSITION 2.6. *Let $A \in \mathscr{A}^{+,1}$ or, more generally, let $A \in OPS^{\lambda}_{\Psi,\psi}$, with Ψ, ψ given by* (1.8)–(1.9). *Suppose that for a sufficiently large* constant L_0,

$$\operatorname{Re} A(x,\xi) \geqslant L_0|\xi|^{1/3}\langle\zeta\rangle^d = L_0\omega_d(\xi), \tag{2.45}$$

at least for $|\xi|$ sufficiently large. Assume

$$-2 \leqslant d \leqslant 1. \tag{2.46}$$

Then

$$\operatorname{Re}(a(X,D)v,v) \geqslant \tfrac{1}{2}L_0\left\|\omega_d(D)^{1/2}v\right\|^2_{L^2} - C_N\|v\|^2_{H^{-N}}, \tag{2.47}$$

with a similar estimate for $a(x,D)$.

3. Microlocal hypoellipticity; general results. We will begin this section with some hypoellipticity results that can be obtained by elementary means, particularly by examining the behavior of commutators of operators, before getting to results that make use of the Fefferman–Phong inequalities. The following is an illustrative example of an elementary hypoellipticity result. We consider operators acting on $\mathscr{D}'(M)$, M a compact manifold.

PROPOSITION 3.1. *Let $P \in OPS^1_{1,0}(M)$. Suppose P is invertible on $\mathscr{D}'(M)$ and its inverse $T = P^{-1}$ has the property that for some $a > 0$,*

$$T\colon H^s(M) \to H^{s+a}(M) \quad \textit{for all } s \in R. \tag{3.1}$$

Then T is microlocal; i.e., for all $u \in \mathscr{D}'(M)$,

$$\mathrm{WF}(Tu) \subset \mathrm{WF}(u). \tag{3.2}$$

Hence, P is hypoelliptic.

Since the proof will involve only looking at commutators of T with pseudodifferential operators, it is useful to abstract the setting as follows. We say

$$T \in \mathscr{O}(m) \Leftrightarrow T\colon H^s(M) \to H^{s-m}(M) \quad \text{for all } s \in R. \tag{3.3}$$

Then with $\operatorname{Ad} P(T) = [P,T] = PT - TP$, we say

$$T \in \mathscr{O}(m,\rho) \Leftrightarrow T \in \mathscr{O}(m) \text{ and } \operatorname{Ad} P_1 \cdots \operatorname{Ad} P_k(T) \in \mathscr{O}(m-k\rho) \quad \text{for all } P_j \in OPS^0(M). \tag{3.4}$$

The relevance of these classes to microlocal behavior is provided by

LEMMA 3.2. *Suppose $\rho > 0$. Then any $T \in \mathscr{O}(m,\rho)$ is microlocal.*

PROOF. It is easy to see that property (3.2) is equivalent to the property

$$\psi(x,D)T\varphi(x,D) \in OPS^{-\infty} = \mathscr{O}(-\infty) \tag{3.5}$$

whenever $\psi(x,\xi)$, $\varphi(x,\xi)$ belong to $S^0(M)$ and have disjoint conic supports. Let $\varphi_1(x,\xi) \in S^0$ have conic support disjoint from $\psi(x,\xi)$ but be equal to 1 on the conic support of $\varphi(x,\xi)$. Then $\varphi_1(x,D)\varphi(x,D) = \varphi(x,D) \mod OPS^{-\infty}$, so

$$\begin{aligned}\psi(x,D)T\varphi(x,D) &= \psi(x,D)T\varphi_1(x,D)\varphi(x,D) \mod OPS^{-\infty} \\ &= \psi(x,D)[T,\varphi_1(x,D)]\varphi(x,D) \mod OPS^{-\infty},\end{aligned} \tag{3.6}$$

since $\psi(x,D)\varphi_1(x,D) \in OPS^{-\infty}$. However, (3.4) implies

$$[T,\varphi_1(x,D)] \in \mathcal{O}(m-\rho,\rho). \tag{3.7}$$

An iterative argument produces the desired conclusion (3.5).

Consequently, Proposition 3.1 is a special case of the following result.

PROPOSITION 3.3. *Let* $P \in \mathcal{O}(1,\rho)$, $\rho > 0$. *Suppose* $P^{-1} = T \in \mathcal{O}(-\sigma)$ *with* $1-\rho < \sigma \leqslant 1$. *Then* $T \in \mathcal{O}(-\sigma, \rho+\sigma-1)$ *and, hence, T is microlocal.*

Note that Airy operators of order m satisfy

$$\mathcal{A}^{+,m} \subset S^m_{1/3,0} \subset \mathcal{O}(m,1/3). \tag{3.8}$$

Typical hypoelliptic operators like $\Lambda^{2/3}\Phi$ have inverses in $OPS^{-2/3}_{1/3,0} \subset \mathcal{O}(-2/3)$, and the inequality $1-\rho<\sigma$ breaks down. However, Proposition 3.3 can be strengthened to a result in which special consideration is paid to the behavior of first-order commutators. It follows from (1.15)—(1.16) that

$$A \in \mathcal{A}^{+,m},\ P \in OPS^0 \Rightarrow [P,A] \in OPS^{m-2/3}_{1/3,0} \subset \mathcal{O}(m-2/3,1/3). \tag{3.9}$$

This motivates the following definition. Let $0<\rho\leqslant\tilde\rho$. We say

$$T \in \mathcal{O}(m,\rho,\tilde\rho) \Leftrightarrow T \in \mathcal{O}(m,\rho) \text{ and } [P,T] \in \mathcal{O}(m-\tilde\rho,\rho), \quad \text{for any } P \in OPS^0, \tag{3.10}$$

Thus the content of (3.9) is

$$\mathcal{A}^{+,m} \subset \mathcal{O}(m,1/3,2/3). \tag{3.11}$$

We give a proof of the following result, which contains Proposition 3.3.

PROPOSITION 3.4. *For* $0<\rho\leqslant\tilde\rho\leqslant 1$ *let* $P \in \mathcal{O}(1,\rho,\tilde\rho)$. *Suppose* $P^{-1} \in \mathcal{O}(-\sigma)$. *If* $1-\tilde\rho<\sigma$, *then* P^{-1} *is microlocal. In fact, if*

$$1-\tilde\rho < \sigma \leqslant 1-\tilde\rho+\rho, \tag{3.12}$$

then

$$P^{-1} \in \mathcal{O}(-\sigma, \tilde\rho+\sigma-1). \tag{3.13}$$

PROOF. Let $T = P^{-1}$ and $P_1,\ldots,P_k \in OPS^0(M)$. An inductive argument shows that $\operatorname{Ad} P_1 \cdots \operatorname{Ad} P_k(T)$ is a sum of terms, each of which is a product of j terms like

$$\operatorname{Ad} A_{1l} \cdots \operatorname{Ad} A_{\mu_l l}(P) \in \mathcal{O}(-\tilde\rho+\rho-\mu_l\rho), \qquad 1 \leqslant l \leqslant j, \tag{3.14}$$

where A_{pl} are rearrangements of $P_1,\ldots,P_k$, and $j+1$ terms T. Here

$$\mu_1 + \cdots + \mu_j = k. \tag{3.15}$$

It follows that Ad $P_1 \cdots$ Ad $P_k(T) \in \mathcal{O}(\nu_k)$, with

$$\nu_k = (j+1)(-\sigma) + (1+\rho-\tilde{\rho})j - \rho \sum_l \mu_l$$

$$= (1+\rho-\tilde{\rho}-\sigma)j - \sigma - \rho k \leqslant (1-\tilde{\rho}-\sigma)k - \sigma,$$

granted (3.12). This is equivalent to assertion (3.13); therefore the proof is complete.

The classes $\mathcal{O}(m, \rho)$ and $\mathcal{O}(m, \rho, \tilde{\rho})$ were defined in [**29**, §8], where the special case of Proposition 3.5,

$$P \in \mathcal{O}(1, 1/3, 2/3),\ P^{-1} \in \mathcal{O}(-2/3) \Rightarrow P^{-1} \in \mathcal{O}(-2/3, 1/3), \tag{3.16}$$

was discussed. In [**29**] this was applied to the study of transmission problems for the wave equation in situations where the interface between two media is diffractive on each side. A similar argument, figured in the result of [**22**], which states that $\Phi + p(x, D)$ is hypoelliptic if $p(x, D) \in OPS^{1/3}$ has principal symbol taking values in a half-space in $\mathbf{C}$ intersecting a conic neighborhood of the third quadrant only at the origin. This is a special case of a result discussed in §4 of this paper.

The appropriate hypothesis on P^{-1} might be verified in different ways, depending on the circumstances. One general method to establish such a result on P^{-1} is to use the sharp Gårding inequality of Hörmander–Lax–Nirenberg. In this fashion one can obtain the following result.

PROPOSITION 3.5. *Let $p(x, \xi) \in S^1_{1,0}(M)$. Suppose, for some $\rho \in (0, 1]$,*

$$\operatorname{Re} p(x, \xi) \geqslant C_0 |\xi|^\rho \tag{3.17}$$

for $|\xi|$ large. Then $p(x, D)$ is hypoelliptic with parametrix in $\mathcal{O}(-\rho, \rho)$.

We remark that if (3.17) is assumed to hold for some $\rho > \frac{1}{2}$, we get $p(x, \xi)^{-1} \in S^{-\rho}_{\rho, 1-\rho}$, and the pseudodifferential operator calculus yields a parametrix in $OPS^{-\rho}_{\rho, 1-\rho}$. Also, if $p(x, \xi)$ is real and (3.17) is assumed, then $p(x, \xi)^{-1} \in S^{-\rho}_{1/2+\rho/2, 1/2-\rho/2}$, and we get a parametrix with symbol in this class for any $\rho \in (0, 1]$.

PROOF. We apply the sharp Gårding inequality to $\operatorname{Re} p(x, \xi) - C_0 \langle \xi \rangle^\rho$, positive and in $S^1_{1,0}(M)$, to obtain

$$\operatorname{Re}(p(x, D)u, u)_s \geqslant C\|u\|^2_{s+\rho/2} - C'\|u\|^2_s, \qquad u \in C^\infty(M), \tag{3.18}$$

where $\| \ \|_s$ denotes the Sobolev norm on $H^s(M)$. In view of Poincaré's inequality we deduce

$$\operatorname{Re}(p(x, D)u, u)_s \geqslant \tfrac{1}{2} C\|u\|^2_{s+\rho/2} - C_N\|u\|^2_{s-N}. \tag{3.19}$$

From (3.19) we deduce the global a priori estimate

$$\tfrac{1}{2} C\|u\|^2_{s+\rho/2} \leqslant \|p(x, D)u\|^2_{s+\rho/2} + C'_N\|u\|^2_{s-N}, \qquad u \in C^\infty(M). \tag{3.20}$$

We can refine this to a global regularity result:

$$u \in \mathcal{D}'(M),\ p(x, D)u \in H^{s-\rho/2}(M) \Rightarrow u \in H^{s+\rho/2}(M). \tag{3.21}$$

To obtain this let $\varphi_\varepsilon(x, D)$ be a Friedrichs mollifier, i.e., a family of operators in $OPS^{-\infty}$ with the following properties:

$$\varphi_\varepsilon(x, \xi) \text{ is bounded in } S^0_{1,0}(M) \text{ for } 0 < \varepsilon \leqslant 1, \tag{3.22}$$

and

$$\varphi_\varepsilon(x, \xi) \nearrow 1 \text{ as } \varepsilon \downarrow 0, \qquad \varphi_\varepsilon(x, \xi) \geqslant 0. \tag{3.23}$$

Then apply the sharp Gårding inequality to $\varphi(x, \xi)^2(\operatorname{Re} p(x, \xi) - C_0\langle\xi\rangle^\rho)$ to obtain

$$\operatorname{Re}(\varphi_\varepsilon(x, D)p(x, D)u, \varphi_\varepsilon(x, D)u)_s \geqslant C\|\varphi_\varepsilon(x, D)u\|^2_{s+\rho/2} - C'\|u\|^2_s, \quad u \in C^\infty(M), \tag{3.24}$$

with constants independent of ε, which implies

$$\begin{aligned} C\|\varphi_\varepsilon(x, D)u\|^2_{s+\rho/2} &\leqslant \delta\|\varphi_\varepsilon(x, D)u\|^2_{s+\rho/2} \\ &\quad + (1/4\delta)\|\varphi_\varepsilon(x, D)p(x, D)u\|^2_{s+\rho/2} \\ &\quad + C'\|u\|^2_s, \qquad u \in C^\infty(M), \end{aligned} \tag{3.25}$$

with constants independent of ε. Now given $u \in \mathscr{D}'(M)$, $p(x, D)u \in H^{s-\rho/2}(M)$, we must have $u \in H^\sigma(M)$ for some σ. Suppose $\sigma \leqslant s$. Now we use (3.25) with s replaced by σ. By continuity the estimate extends to all $u \in H^\sigma(M)$ with constant independent of ε. Thus

$$\|\varphi_\varepsilon(x, D)u\|^2_{\sigma+\rho/2} \leqslant K < \infty \tag{3.26}$$

for $\varepsilon \to 0$ in this case. Taking $\varepsilon \to 0$ yields $u \in H^{\sigma+\rho/2}$. Repeating this argument, we get $u \in H^{s+\rho/2}$, and the global regularity result (3.21) is proved. With only a little extra work we could microlocalize the arguments above and obtain microlocal hypoellipticity in that fashion, but instead we apply Proposition 3.4.

To set things up to apply Proposition 3.4, it is convenient to add a positive elliptic element $R_N \in OPS^{-N}_{1,0}$ to get

$$P_1 = p(x, D) + R_N, \tag{3.27}$$

so that for P_1, we have

$$\operatorname{Re}(P_1 u, u)_{L^2} \geqslant \tfrac{1}{2}C\|u\|^2_{H^{\rho/2}}$$

as well as the analogues of (3.19) and (3.21) for all s. Hence, the map

$$P_1\colon C^\infty(M) \to C^\infty(M) \tag{3.28}$$

is injective and has closed range, and so its adjoint $P_1^*\colon C^\infty(M) \to C^\infty(M)$ is also. Furthermore, P_1^* enjoys the regularity property (3.21), so P_1^* is injective on $\mathscr{D}'(M)$ and has closed range on this space. Hence, P_1 is invertible on $C^\infty(M)$ and $\mathscr{D}'(M)$, and the regularity property (3.21) yields $P_1^{-1} \in \mathscr{O}(-\rho)$. Thus Proposition 3.4 gives $P_1^{-1} \in \mathscr{O}(-\rho, \rho)$. Now

$$P_1^{-1}p(x, D) = I - P_1^{-1}R_N = I - S_N, \tag{3.29}$$

where $S_N \in \mathcal{O}(-\rho - N, \rho)$, so the Neumann series produces a left parametrix

$$(I + S_N + S_N^2 + \cdots)P_1^{-1} \in \mathcal{O}(-\rho, \rho) \tag{3.30}$$

for P. This proves Proposition 3.5.

A natural generalization of Proposition 3.5 to Beals–Fefferman classes is

PROPOSITION 3.6. *Let Ψ, ψ be a Beals–Fefferman pair of weight functions, and let*

$$p(x, \xi) \in S_{\Psi,\psi}^{(1,1)}. \tag{3.31}$$

Suppose that, for some $\rho > 0$,

$$\operatorname{Re} p(x, \xi) \geqslant C_0|\xi|^\rho \tag{3.32}$$

for $|\xi|$ large. Then $p(x, D)$ is hypoelliptic.

Note that (3.31) and (3.32) together imply

$$\Psi(x, \xi)\psi(x, \xi) \geqslant C'\langle\xi\rangle^\rho; \tag{3.33}$$

i.e., in the terminology of [**1**], Ψ, ψ must be a "strongly coercive" pair of weight functions. We remark that if $p(x, \xi)$ in Proposition 3.6 is real valued, then $p(x, \xi)^{-1} \in S_{\Psi',\psi'}^{(-1,-1)}$, where $\Psi' = \Psi(p/\Psi\psi)^{1/2}$, $\psi' = \psi(p/\Psi\psi)^{1/2}$, and $p(x, D)$ has a parametrix with symbol in this class.

Proposition 3.6 is best proved without recourse to Proposition 3.4 by localizing the estimates used to prove global regularity results analogous to (3.21). Since such an argument will be used in the proof of Proposition 3.8, we omit the details here. Propositions 3.5 and 3.6 hold for $K \times K$ systems of operators, where $\operatorname{Re} p(x, \xi)$ is interpreted as the selfadjoint part $\frac{1}{2}(p(x, \xi) + p(x, \xi)^*)$. The following results are proved only for scalar operators, since the Fefferman–Phong inequalities have been proved only for scalar operators. The following result improves Proposition 3.5.

PROPOSITION 3.7. *Let $p(x, \xi) \in S_{1,0}^2$ be scalar. Suppose that for some $\rho \in (0, 2]$, $\operatorname{Re} p(x, \xi) \geqslant C_0\langle\xi\rangle^\rho$ for $|\xi|$ large, and assume* $\operatorname{Im} p(x, \xi) \in S_{1,0}^1$. *Then $p(X, D)$ and $p(x, D)$ are hypoelliptic.*

Note that Proposition 3.5 applies here only for $\rho \in (1, 2]$. Proposition 3.7 is a special case of the following improvement of Proposition 3.6.

PROPOSITION 3.8. *Let Ψ, ψ be Beals–Fefferman weight functions, and let*

$$p(x, \xi) \in S_{\Psi,\psi}^{(2,2)} \tag{3.34}$$

be scalar. Suppose that for some $\rho > 0$,

$$\operatorname{Re} p(x, \xi) \geqslant C_0\langle\xi\rangle^\rho \tag{3.35}$$

for $|\xi|$ large, and assume

$$\operatorname{Im} p(x, \xi) \in S_{\Psi,\psi}^{(1,1)}. \tag{3.36}$$

Then $p(X, D)$ and $p(x, D)$ are hypoelliptic.

We prove the following result, which generalizes Proposition 3.8 by relaxing the condition (3.36) on $\operatorname{Im} p(x, \xi)$. The relaxation is probably not important in the study of classical operators, as in Proposition 3.7, but it is important in the study of Airy operators.

PROPOSITION 3.9. *Let Ψ, ψ be strongly coercive Beals–Fefferman weight functions, $p(x, \xi) = a(x, \xi) + ib(x, \xi) \in S^{(2,2)}_{\Psi,\psi}$, scalar. Assume that for large $|\xi|$,*

$$a(x, \xi) \geqslant K(\xi) \geqslant C_0\langle \xi\rangle^{\rho} \tag{3.37}$$

for some $\rho > 0$, where we suppose $K = e^{\kappa} \in S^{\kappa}_{\Psi,\psi}$. Suppose, furthermore, that for any $\varphi(X, D) \in OPS^0$,

$$W(X, D) = \varphi(X, D)[b(X, D), \varphi(X, D)],$$

which belongs a priori to $OPS^{(1,1)}_{\Psi,\psi}$, satisfies the condition

$$|W(x, \xi)| \leqslant C_1 K(\xi)\langle \xi\rangle^{-\gamma} \tag{3.38}$$

for some $\gamma > 0$. Then $p(X, D)$ is hypoelliptic.

Note that if $b(x, \xi) \in S^{(1,1)}_{\Psi,\psi}$, then $W(x, \xi) \in S^{0}_{\Psi,\psi}$. Thus the hypothesis here on $\operatorname{Im} p(x, \xi)$ is more general than that of Proposition 3.8. We assume we are working on $M = \mathbf{T}^n$.

PROOF. If we have open conic sets $\Gamma_1 \subset \Gamma_0$, let $\varphi_j(x, \xi) \in S^0$ be supported in Γ_j, with $\varphi_0(x, \xi) = 1$ on a conic neighborhood of $\overline{\Gamma}_1$. The Fefferman–Phong inequality gives

$$\|K(D)^{1/2}u\|^2 \leqslant \operatorname{Re}(p(X, D)u, u) + C\|u\|^2. \tag{3.39}$$

Apply this to $\varphi_1(X, D)u$ to get

$$\begin{aligned}\|K(D)^{1/2}\varphi_1(X, D)u\|^2 \leqslant{}& \operatorname{Re}(p(X, D)\varphi_1(X, D)u, \varphi_1(X, D)u)\\ &+C\|\varphi_1(X, D)u\|^2.\end{aligned} \tag{3.40}$$

Hence, with $p(X, D) = P = A + iB$,

$$\begin{aligned}\|K(D)^{1/2}\varphi_1(X, D)u\|^2 \leqslant{}& \operatorname{Re}(\varphi_1(X, D)Pu, \varphi_1(X, D)u)\\ &+\operatorname{Re}([A, \varphi_1(X, D)]u, \varphi_1(X, D)u)\\ &+\operatorname{Re}(i[B, \varphi_1(X, D)]u, \varphi_1(X, D)u)\\ &+C\|\varphi_1(X, D)u\|^2.\end{aligned} \tag{3.41}$$

Since $\operatorname{Re}[A, \varphi_1(X, D)] \in OPS^{0}_{\Psi,\psi}$, we have

(3.42)

$$\begin{aligned}\|K(D)^{1/2}\varphi_1(X, D)u\|^2 \leqslant{}& \operatorname{Re}(\varphi_1(X, D)Pu, \varphi_1(X, D)u)\\ &+C\|\varphi_0(X, D)u\|^2\\ &+|(W(X, D)\varphi_0(X, D)u, \varphi_0(X, D)u)| + C_N\|u\|^2_{-N},\end{aligned}$$

where $W(x,\xi)\in S^{(1,1)}_{\Psi,\psi}$ is assumed to satisfy (3.38). Using

$$\begin{aligned}|(\varphi_1(X,D)Pu,\varphi_1(X,D)u)| &\leqslant \|K(D)^{-1/2}\varphi_1(X,D)Pu\|\,\|K(D)^{1/2}\varphi_1(X,D)u\|\\ &\leqslant \tfrac{1}{2}\|K(D)^{-1/2}\varphi_1(X,D)Pu\|^2+\tfrac{1}{2}\|K(D)^{1/2}\varphi_1(X,D)u\|^2,\end{aligned}$$

we obtain

$$\begin{aligned}\tfrac{1}{2}\|K(D)^{1/2}\varphi_1(X,D)u\|^2 \leqslant{}& \tfrac{1}{2}\|K(D)^{-1/2}\varphi_1(X,D)Pu\|^2\\ &+|(W(X,D)\varphi_0(X,D)u,\varphi_0(X,D)u)|\\ &+C\|\varphi_0(X,D)u\|^2+C_N\|u\|^2_{-N}.\end{aligned}\tag{3.43}$$

If we apply the Fefferman–Phong inequality to

$$\operatorname{Re}\big(C_1K(\xi)\langle\xi\rangle^{-\gamma}-e^{i\theta}W(x,\xi)\big)\geqslant 0 \quad \text{in } S^{(2,2)}_{\Psi,\psi},\tag{3.44}$$

we obtain

$$|(W(X,D)u,u)|\leqslant C_1\|K(D)^{1/2}u\|^2_{-\gamma/2}+C\|u\|^2.\tag{3.45}$$

Consequently, (3.43) yields

$$\begin{aligned}\tfrac{1}{2}\|K(D)^{1/2}\varphi_1(X,D)u\|^2 \leqslant{}& \tfrac{1}{2}\|K(D)^{-1/2}\varphi_1(X,D)Pu\|^2\\ &+C_1\|K(D)^{1/2}\varphi_0(X,D)u\|^2_{-\gamma/2}\\ &+C\|\varphi_0(X,D)u\|^2+C_N\|u\|^2_{-N}.\end{aligned}\tag{3.46}$$

We can generalize (3.46) from L^2-norm estimates to H^s-norm estimates if we conjugate all operators by Λ^s with symbol $\langle\xi\rangle^s$. Thus let $\tilde\varphi_j(X,D)=\Lambda^s\varphi_j(X,D)\Lambda^{-s}$, still essentially supported in Γ_j with $\tilde\varphi_0=1 \bmod S^{-\infty}$ on a conic neighborhood of $\overline{\Gamma}_1$, and let $\tilde P=\Lambda^sP\Lambda^{-s}=\tilde p(X,D)$. Note that

$$\tilde p(x,\xi)=\tilde a(x,\xi)+i\tilde b(x,\xi),\tag{3.47}$$

with

$$\tilde a(x,\xi)=a(x,\xi)-\{b,\langle\xi\rangle^{-s}\}\langle\xi\rangle^s \mod S^0_{\Psi,\psi}\tag{3.48}$$

and

$$\tilde b(x,\xi)=b(x,\xi)+\{a,\langle\xi\rangle^{-s}\}\langle\xi\rangle^s \mod S^0_{\Psi,\psi},\tag{3.49}$$

where $\{\ ,\ \}$ is the Poisson bracket. We have $\{c,\langle\xi\rangle^{-s}\}\langle\xi\rangle^s\in S^{(1,1)}_{\Psi,\psi}$ for any $c\in S^{(2,2)}_{\Psi,\psi}$. Thus the hypotheses of the proposition are valid for $\tilde P$ if and only if they are valid for P. Hence, we have the Sobolev estimates

$$\begin{aligned}\tfrac{1}{2}\|K(D)^{1/2}\tilde\varphi_1(X,D)u\|^2_s \leqslant{}& \tfrac{1}{2}\|K(D)^{-1/2}\tilde\varphi_1(X,D)Pu\|^2_s\\ &+C_1\|K(D)^{1/2}\tilde\varphi_0(X,D)u\|^2_{s-\gamma/2}\\ &+C\|\tilde\varphi_0(X,D)u\|^2_s+C_N\|u\|^2_{s-N}\end{aligned}\tag{3.50}$$

for any $u \in C^\infty(M)$. Now replace $\varphi_j(x,\xi)$ by families of smoothing operators defined as follows. Let $\chi(t) \in C_0^\infty(R)$ be equal to 1 on $[-1,1]$, 0 outside $[-\frac{3}{2},\frac{3}{2}]$. Let

$$(3.51)\qquad \varphi_1^\tau(x,\xi) = \chi(2\tau|\xi|)\varphi_1(x,\xi), \qquad \varphi_0^\tau(x,\xi) = \chi(\tau|\xi|)\varphi_0(x,\xi).$$

Then with $\tilde{\varphi}_j^\tau = \Lambda^s\varphi_j^\tau\Lambda^{-s}$, we have

$$(3.52)\qquad \begin{aligned} \tfrac{1}{2}\left\|K(D)^{1/2}\tilde{\varphi}_1^\tau(X,D)u\right\|_s^2 \leqslant \tfrac{1}{2}&\left\|K(D)^{-1/2}\tilde{\varphi}_1^\tau(X,D)Pu\right\|_s^2 \\ &+C_1\left\|K(D)^{1/2}\tilde{\varphi}_0^\tau(X,D)u\right\|_{s-\gamma/2}^2 \\ &+C\left\|\tilde{\varphi}_0^\tau(X,D)u\right\|_s^2 + C_N\|u\|_{s-N}^2. \end{aligned}$$

This is a priori valid for $u \in C^\infty(M)$, but it extends by continuity to any $u \in H^{s-N}$. The constants are independent of $\tau \in (0,1]$. Letting $\tau \downarrow 0$, we get the regularity result

$$(3.53)\qquad \begin{aligned} &u \in \mathscr{D}'(M),\, u \in H^s \text{ on } \Gamma_0,\, K(D)^{1/2}u \in H^{s-\gamma/2} \text{ on } \Gamma_0,\, K(D)^{-1/2}Pu \in H^s \text{ on } \Gamma_0 \\ &\qquad \Rightarrow K(D)^{1/2}u \in H^s \text{ on } \Gamma_1. \end{aligned}$$

Replacing $\Gamma_1 \subset \Gamma_0$ by an appropriate nested sequence of cones, we improve this to

$$(3.54)\qquad u \in \mathscr{D}'(M),\, K(D)^{-1/2}Pu \in H^s \text{ on } \Gamma_0 \Rightarrow K(D)^{1/2}u \in H^s \text{ on } \Gamma_1.$$

This microlocal regularity result proves the proposition.

Let us note that condition (3.38) is implied by the hypothesis

$$(3.55)\qquad \left|\nabla_\xi b(x,\xi)\right| + \langle\xi\rangle^{-1}\left|\nabla_x b(x,\xi)\right| \leqslant C_2K(\xi)\langle\xi\rangle^{-\gamma}.$$

We now generalize Proposition 3.9 by allowing more general orders. For convenience we restrict our attention to weight functions of the form

$$(3.56)\qquad \Psi = \Psi(\xi), \qquad \psi = \psi(\xi).$$

Let $\mu \in \mathscr{O}(\Psi,\psi)$; suppose $e^\mu \in S^\mu_{\Psi,\psi}$ and $\mu = \mu(\xi)$. Now suppose

$$(3.57)\qquad p(x,\xi) = a(x,\xi) + ib(x,\xi) \in S^\mu_{\Psi,\psi},$$

with a and b real, and

$$(3.58)\qquad a(x,\xi) \geqslant C_0L(\xi)e^{\mu-(2,2)} = C_0K(\xi),$$

where

$$(3.59)\qquad L(\xi) \geqslant \langle\xi\rangle^\rho$$

for some $\rho > 0$. Suppose $L(\xi) = e^{l(\xi)} \in S^l_{\Psi,\psi} \subset S^{(2,2)}_{\Psi,\psi}$. Suppose, furthermore, that

$$(3.60)\qquad \left|\nabla_\xi b(x,\xi)\right| + \langle\xi\rangle^{-1}\left|\nabla_x b(x,\xi)\right| \leqslant C_2K(\xi)\langle\xi\rangle^{-\gamma}$$

for some $\gamma > 0$. Now consider

$$q(x, \xi) = p(x, \xi)\beta(\xi)^{-1} \in S^{(2,2)}_{\Psi,\psi}, \tag{3.61}$$

where

$$\beta(\xi) = e^{\mu-(2,2)}. \tag{3.62}$$

Note that

$$q(X, D) = \beta(D)^{-1/2} p(X, D)\beta(D)^{-1/2} \quad \operatorname{mod} OPS^0_{\Psi,\psi}. \tag{3.63}$$

If we separate $q(x, \xi)$ into real and imaginary parts

$$q(x, \xi) = a^{\#}(x, \xi) + ib^{\#}(x, \xi), \tag{3.64}$$

then (3.58) implies

$$a^{\#}(x, \xi) \geqslant C_0 L(\xi) \geqslant C_0\langle\xi\rangle^{\rho}. \tag{3.65}$$

Note that

$$(\partial/\partial x_j)b^{\#}(x, \xi) = \beta(\xi)^{-1}(\partial/\partial x_j)b(x, \xi),$$

while

$$(\partial/\partial \xi_j)b^{\#}(x, \xi) = \beta(\xi)^{-1}(\partial/\partial \xi_j)b(x, \xi) - b(x, \xi)\beta(\xi)^{-2}(\partial\beta/\partial\xi_j).$$

We see that one cannot deduce from (3.60) that $b^{\#}(x, \xi)$ satisfies the hypotheses of Proposition 3.9 without imposing the stronger hypothesis that

$$b(x, \xi) \in S^{\mu-(1,1)}_{\Psi,\psi}. \tag{3.66}$$

In this case we can apply Proposition 3.9 to deduce that $q(X, D)$ and $p(X, D)$ (by (3.63)) are hypoelliptic. However, since we do not want to assume (3.66), we must argue further. In light of (3.65) one can apply the Fefferman–Phong inequality to get

$$C_0\|L(D)^{1/2}u\|^2 \leqslant \operatorname{Re}(q(X, D)u, u) + C\|u\|^2, \tag{3.67}$$

and, in light of (3.63), if we set $v = \beta(D)^{-1/2}u$ we have

$$C_0\|K(D)^{1/2}v\|^2 \leqslant \operatorname{Re}(p(X, D)v, v) + C\|\beta(D)^{1/2}v\|^2. \tag{3.68}$$

Taking φ_0, φ_1 as in the proof of Proposition 3.10, and replacing v by $\varphi_1(X, D)v$, we have

$$\begin{aligned} C_0\|K(D)^{1/2}\varphi_1(X, D)v\|^2 \leqslant{} & \operatorname{Re}(p(X, D)\varphi_1(X, D)v, \varphi_1(X, D)v) \\ & + C\|\beta(D)^{1/2}\varphi_1(X, D)v\|^2; \end{aligned} \tag{3.69}$$

hence,

$$\begin{aligned} C_0\|K(D)^{1/2}\varphi_1(X, D)u\|^2 \leqslant{} & \operatorname{Re}(\varphi_1(X, D)Pu, \varphi_1(X, D)u) \\ & + C\|\beta(D)^{1/2}\varphi_1(X, D)u\|^2 \\ & + \operatorname{Re}([A, \varphi_1(X, D)]u, \varphi_1(X, D)u) \\ & + \operatorname{Re}(i[B, \varphi_1(X, D)]u, \varphi_1(X, D)u), \end{aligned} \tag{3.70}$$

in parallel with (3.41). This time

$$\operatorname{Re}[A, \varphi_1(X, D)] \in OPS^{\mu-(2,2)}_{\Psi,\psi}, \tag{3.71}$$

so

$$|\operatorname{Re}([A, \varphi_1(X, D)]u, \varphi_1(X, D)u)| \leqslant C\|\beta(D)^{1/2}\varphi_0(X, D)u\|^2 + C_N\|u\|^2_{-N}. \tag{3.72}$$

On the other hand, by hypothesis (3.60), if

$$W(X, D) = \varphi_1(X, D)[B, \varphi_1(X, D)], \tag{3.73}$$

we have not only $W(x, \xi) \in S^{\mu-(1,1)}_{\Psi,\psi}$, but also

$$|W(x, \xi)| \leqslant C_0 K(\xi)\langle\xi\rangle^{-\gamma}. \tag{3.74}$$

Write

$$W(x, \xi) = V(x, \xi)\beta(\xi), \qquad V(x, \xi) \in S^{(1,1)}_{\Psi,\psi}, \tag{3.75}$$

and recall $K(\xi) = L(\xi)\beta(\xi)$, so

$$|V(x, \xi)| \leqslant C_0 L(\xi)\langle\xi\rangle^{-\gamma}; \tag{3.76}$$

hence

$$\operatorname{Re}\big(C_0 L(\xi)\langle\xi\rangle^{-\gamma} - e^{i\theta}V(x, \xi)\big) \geqslant 0, \quad \text{in } S^{(2,2)}_{\Psi,\psi}. \tag{3.77}$$

Thus

$$|(V(X, D)u, u)| \leqslant C_0\|L(D)^{1/2}u\|^2_{-\gamma/2}. \tag{3.78}$$

Since

$$V(X, D) = \beta(D)^{-1/2}W(X, D)\beta(D)^{-1/2} \quad \operatorname{mod} OPS^{(-1,-1)}_{\Psi,\psi}, \tag{3.79}$$

we get

$$|(W(X, D)v, v)| \leqslant C_0\|K(D)^{1/2}v\|^2_{-\gamma/2} + C\|\beta(D)^{1/2}v\|^2. \tag{3.80}$$

Now (3.72) and (3.80) applied to (3.70) give

$$\begin{aligned}\|K(D)^{1/2}\varphi_1(X, D)u\|^2 \leqslant{} & C\|K(D)^{-1/2}\varphi_1(X, D)Pu\|^2 \\ & + C\|\beta(D)^{1/2}\varphi_0(X, D)u\|^2 \\ & + C\|K(D)^{1/2}\varphi_0(X, D)u\|^2_{-\gamma/2} + C_N\|u\|^2_{-N}.\end{aligned} \tag{3.81}$$

Next we can get H^s estimates by conjugating with Λ^s with symbol $\langle\xi\rangle^s$. The analogue of (3.47)–(3.49) shows that P satisfies hypotheses (3.57)–(3.60) if and only if $\Lambda^s P \Lambda^{-s}$ does, so we obtain

$$\begin{aligned}\|K(D)^{1/2}\tilde{\varphi}_1(X, D)u\|^2_s \leqslant{} & C\|K(D)^{-1/2}\tilde{\varphi}_1(X, D)Pu\|^2_s + C\|\beta(D)^{1/2}\tilde{\varphi}_0(X, D)u\|^2_s \\ & + C\|K(D)^{1/2}\tilde{\varphi}_0(X, D)u\|^2_{s-\gamma/2} + C_N\|u\|^2_{s-N}.\end{aligned} \tag{3.82}$$

Replacing φ_j by φ_j^τ, given by (3.51), and arguing as before, we deduce the regularity result

$$(3.83)\quad u \in \mathscr{D}'(M),\ K(D)^{-1/2} Pu \in H^s \text{ on } \Gamma_0 \Rightarrow K(D)^{1/2} u \in H^s \text{ on } \Gamma_1.$$

Let us state formally the result we have proved.

PROPOSITION 3.10. *Let* $p(x, \xi) = a(x, \xi) + ib(x, \xi) \in S^{\mu}_{\Psi,\psi}$. *Suppose*

$$(3.84)\qquad a(x, \xi) \geqslant C_0 K(\xi),$$

with $K(\xi) = L(\xi)e^{\mu-(2,2)}$ *and* $L(\xi) \geqslant \langle \xi \rangle^{\rho}$ *for some* $\rho > 0$. *Suppose* $L(\xi) = e^l \in S^l_{\Psi,\psi} \subset S^{(2,2)}_{\Psi,\psi}$. *Suppose, furthermore, that for some* $\delta > 0$,

$$(3.85)\qquad |\nabla_\xi b(x, \xi)| + \langle \xi \rangle^{-1} |\nabla_x b(x, \xi)| \leqslant C_2 K(\xi) \langle \xi \rangle^{-\delta}.$$

Then $p(X, D)$ *is hypoelliptic, and the regularity result* (3.83) *holds.*

Now we specialize to the weight functions Ψ, ψ that arise in considering the Airy operators

$$(3.86)\qquad \Psi = |\xi|^{1/3}\langle \zeta \rangle, \qquad \psi = 1.$$

Recall from §1 that $\mathscr{A}^{+,1} \in S^{\lambda}_{\Psi,\psi}$, where

$$(3.87)\qquad e^{\lambda} = \langle \xi \rangle.$$

Also recall that for $p(x, \xi) \in \mathscr{A}^{+,1}$,

$$(3.88)\qquad \nabla_\xi p \in S^{\lambda-\gamma-(1,1)}_{\Psi,\psi},$$

where

$$(3.89)\qquad e^{\gamma} = \langle \zeta \rangle^{-1/2} |\xi|^{1/3}.$$

Our result is

PROPOSITION 3.11. *Let* Ψ, ψ *be the weight functions* (3.86). *Let*

$$(3.90)\qquad c(x, \xi) = a(x, \xi) + ib(x, \xi) \in S^{\lambda}_{\Psi,\psi}$$

with a and b real valued. Suppose

$$(3.91)\qquad a(x, \xi) \geqslant C|\xi|^{1/3+\rho}\langle \zeta \rangle^{-1/2}$$

for some $\rho > 0$ *and*

$$(3.92)\qquad \nabla_\xi b(x, \xi) \in S^{\lambda-\gamma-(1,1)}_{\Psi,\psi},$$

with e^{γ} *given by* (3.89). *Then* $c(X, D)$ *is hypoelliptic. If* (3.91) *holds on some conic open set* Γ, *then* $c(X, D)$ *is hypoelliptic microlocally in* Γ.

PROOF. In order to apply Proposition 3.10, we must have $a(x, \xi) \geqslant K(\xi)$, where $K(\xi)$ satisfies two constraints that become (since $e^{\lambda}\Psi^{-2} = |\xi|^{1/3}\langle \zeta \rangle^{-2}$), respectively,

$$(3.93)\qquad K(\xi) \geqslant C_0 |\xi|^{1/3+\delta}\langle \zeta \rangle^{-2} \quad \text{on } \Gamma$$

for some $\delta > 0$, and $e^{\lambda-\gamma-(1,1)} \leqslant |\xi|^{-\delta} K(\xi)$ for some $\delta > 0$ on Γ. Since

$$(3.94)\qquad e^{\lambda-\gamma-(1,1)} = |\xi|^{1/3}\langle \zeta \rangle^{-1/2}$$

in this case, the second constraint is

$$K(\xi) \geqslant C_1|\xi|^{1/3+\delta}\langle\zeta\rangle^{-1/2} \quad \text{on } \Gamma. \tag{3.95}$$

We see that this constraint is stronger than (3.93), and applying Proposition 3.10 completes the proof.

Next we consider a refinement of the notion of wave front set beyond that achieved with elements of OPS^0. For the moment let Ψ, ψ be any strongly coercive pair of Beals–Fefferman weight functions. Pick $\varphi(x, \xi) \in S^0_{\Psi,\psi}$ and let

$$\Gamma_0 = \{(x, \xi) \in T^*M \setminus 0 \colon \varphi(x, \xi) > 0\}.$$

In fact, for τ near 0 let

$$\Gamma_\tau = \{(x, \xi) \colon \varphi(x, \xi) > -\tau\}. \tag{3.96}$$

We assume the Γ_τ "fan out" as τ varies in the following sense. Suppose there is a $\tau_0 > 0$ such that, for any $\tau \in (-\tau_0, \tau_0)$ and any $\sigma \in (0, \tau_0]$, the following condition holds. Namely, there are $\varphi_1(x, \xi)$ and $\varphi_2(x, \xi)$ in $S^0_{\Psi,\psi}$, supported respectively in $\Gamma_{\tau+\sigma}$ and Γ^c_τ, such that

$$\varphi_1(x, \xi)^2 + \varphi_2(x, \xi)^2 \geqslant 1 \quad \text{for } |\xi| \geqslant 1. \tag{3.97}$$

In such a case we say Γ_0 is *admissible* (relative to $S^0_{\Psi,\psi}$). For admissible $\Gamma_0 \subset T^*M \setminus 0$, we say $u \in \mathscr{D}'(M)$ is in H^s in Γ_0 if $\varphi(X, D)u \in H^s(M)$ for all $\varphi(x, \xi)$ in $S^0_{\Psi,\psi}$ supported in Γ_0. A cover of $T^*M \setminus 0$ by admissible sets $\Gamma_1, \dots, \Gamma_K$ will be called *regular* if any $u \in \mathscr{D}'(M)$ that is in H^s on each Γ_j belongs to $H^s(M)$. For this property to hold, it is sufficient that there are $\varphi_j(x, \xi) \in S^0_{\Psi,\psi}$, supported in Γ_j, such that $\varphi_1(x, \xi)^2 + \cdots + \varphi_K(x, \xi)^2 \geqslant 1$ for large $|\xi|$. In particular, whenever Γ_0 is admissible and Γ_τ is given by (3.92), it follows that the cover of $T^*M \setminus 0$ by Γ_0 and $\Gamma^c_{-\sigma}$ is regular for small positive σ. More generally, we have

LEMMA 3.12. *Let $\Gamma_1, \dots, \Gamma_K$ be admissible sets embedded in families $\Gamma_{j\tau}$, as above. Suppose that for some $\sigma > 0$, $\Gamma_{1,-\sigma}, \dots, \Gamma_{K,-\sigma}$ cover $T^*M \setminus 0$. Then $\Gamma_1, \dots, \Gamma_K$ is a regular cover.*

PROOF. It follows from the admissibility criteria that for $1 \leqslant j \leqslant K$ there exist $\varphi_j(x, \xi) \in S^0_{\Psi,\psi}$ supported in Γ_j and $\geqslant 1$ on $\Gamma_{j,-\sigma}$. If the $\Gamma_{j,-\sigma}$ cover $T^*M \setminus 0$, it follows that $\varphi_1(X, D)^2 + \cdots + \varphi_K(X, D)^2 \in OPS^0_{\Psi,\psi}$ is elliptic; therefore if $u \in \mathscr{D}'(M)$ and each $\varphi_j(X, D)u$ belongs to $H^s(M)$, then certainly $u \in H^s(M)$.

If $a(X, D) \in OPS^\mu_{\Psi,\psi}$ for some $\mu \in \mathscr{O}(\Psi,\psi)$, and if Γ is admissible relative to $S^0_{\Psi,\psi}$ (or perhaps relative to some other symbol class $S^0_{\Psi',\psi'}$), we say $a(X, D)$ is hypoelliptic microlocally in Γ provided that whenever $u \in \mathscr{D}'(M)$ and $a(X, D)u = f$ has the property that f is in H^s in Γ^b for all s, then so does u for any admissible $\Gamma^b \subset \Gamma$ (relative to $S^0_{\Psi',\psi'}$). The following result refines Proposition 3.10:

PROPOSITION 3.13. *Let Ψ, ψ, Ψ', ψ', be strongly coercive weight functions of the form* (3.56), *and suppose $S^0_{1,0} \subset S^0_{\Psi',\psi'} \subset S^0_{\Psi,\psi}$. Let $p(x, \xi) = a(x, \xi) + ib(x, \xi) \in S^\mu_{\Psi,\psi}$ with a and b real valued. Let Γ_0 be admissible (relative to $S^0_{\Psi',\psi'}$), $\sigma > 0$, and suppose that for large $|\xi|$,*

$$a(x, \xi) \geqslant K(\xi) \quad \textit{on } \Gamma_\sigma, \tag{3.98}$$

where we assume $K(\xi) = L(\xi)e^{\mu-(2,2)}$, $L(\xi) = e^{l} \in S^{l}_{\Psi,\psi} \subset S^{(2,2)}_{\Psi,\psi}$, *and*

$$L(\xi) \geqslant C_0\langle\xi\rangle^{\delta} \quad \text{on } \Gamma_\sigma. \tag{3.99}$$

Also suppose

$$(\psi')^{-1}|\nabla_\xi b(x,\xi)| + (\Psi')^{-1}|\nabla_x b(x,\xi)| \leqslant C_2 K(\xi)\langle\xi\rangle^{-\delta} \quad \text{in } \Gamma_\sigma. \tag{3.100}$$

Then $p(X, D)$ *is hypoelliptic microlocally in* Γ_0. *We have the regularity result*

(3.101)

$$v \in \mathscr{D}'(M),\ K(D)^{-1/2}\varphi(X,D)p(X,D)v \in H^s \Rightarrow K(D)^{1/2}\varphi_0(X,D)v \in H^s$$

if $\varphi, \varphi_0 \in S^0_{\Psi',\psi'}$, *with* φ_0 *supported in* $\Gamma_{-\sigma}$ *for some* $\sigma > 0$, $\varphi = 1$ *on* $\Gamma_{-\sigma/2}$.

The proof is the same as that of Proposition 3.10.

4. Hypoellipticity of Airy operators. The principal goal of this section is to give a proof of the following conjecture of Melrose and Sjöstrand [**22**]. Suppose that

$$A(x,\xi) = |\xi|^{2/3}\Phi(\zeta) + p(x,\xi), \qquad p(x,\xi) \in S^1, \tag{4.1}$$

and the principal part $p_1(x,\xi)$ of $p(x,\xi)$ takes values in the complement of a conic neighborhood of the third quadrant in $\mathbf{C}$. We recall that $\Phi(\zeta)$ takes values in the open first quadrant, so that if

$$B(x,\xi) = A(x,\xi)^{-1}, \tag{4.2}$$

we have the estimate

$$|B(x,\xi)| \leqslant C|\xi|^{-2/3}\langle\zeta\rangle^{-1/2} \tag{4.3}$$

for $|\xi|$ large. In this case it was conjectured that $A(X, D)$ is hypoelliptic, and we prove this. Another proof has been proposed by Eskin [**4**]. We follow [**4**] in considering microlocal hypoellipticity in a conic neighborhood of (x^0, ξ^0) in two separate cases:

$$(\partial/\partial x_n)p_1(x^0,\xi^0) = 0, \tag{4.4}$$

and

$$(\partial/\partial x_n)p_1(x^0,\xi^0) \neq 0. \tag{4.5}$$

We can assume $\xi_n^0 = 0$, since otherwise this is a standard ellipticity result; for the same reason we can also suppose $p_1(x^0, \xi^0) = 0$. Our treatment of case (4.5) follows [**4**] in outline, but it differs in the technical details of applying sharp Gårding inequalities; our treatment of case (4.4) will be completely different from [**4**].

To start our analysis let us note to what symbol class $B(x, \xi)$ must belong.

LEMMA 4.1. *Under hypotheses* (4.1)–(4.3) *we have*

$$B(x,\xi) \in S^{-2/3}_{1/3,1/3}. \tag{4.6}$$

More precisely,

$$B(x,\xi) \in S^{\gamma-\lambda}_{\Psi,\psi^\#} \subset S^{-2\lambda/3}_{\Psi,\psi^\#} \tag{4.7}$$

with

$$\Psi(x,\xi)=|\xi|^{1/3}\langle\zeta\rangle,\qquad \psi^{\#}(x,\xi)=|\xi|^{-1/3}\langle\zeta\rangle^{1/2}. \tag{4.8}$$

Since it is equally easy to analyze all real powers of such a symbol, we prove the following, more general, result.

LEMMA 4.1′. *Suppose that $A(x,\xi)\in\mathscr{A}^{+,1}$ or, more generally, $A(x,\xi)\in S^{\lambda}_{\Psi,\psi}$, where $e^{\lambda}=\langle\xi\rangle$ and Ψ,ψ are the weight functions* (1.8)–(1.9); *also suppose $\nabla_\xi A(x,\xi)\in S^{\lambda-\gamma-(1,0)}_{\Psi,\psi}$, as in* (1.16). *Suppose the estimate*

$$\left|A(x,\xi)^{-1}\right|\leqslant C|\xi|^{-2/3}\langle\zeta\rangle^{-1/2}=Ce^{\lambda-\gamma} \tag{4.9}$$

holds, and if $r\in R$ is not an integer, suppose $A(x,\xi)$ avoids R^-. Then for $r\in R$,

$$A(x,\xi)^r\in S^{r\lambda}_{\Psi,\psi^{\#}}\quad \text{if } r\geqslant 0,\qquad S^{r(\lambda-\gamma)}_{\Psi,\psi^{\#}}\quad\text{if } r<0, \tag{4.10}$$

with $\Psi,\psi^{\#}$ given by (4.8).

PROOF. We have

$$D_x^\beta D_\xi^\alpha A^r=\sum_{\substack{\alpha_1+\cdots+\alpha_k=\alpha\\ \beta_1+\cdots+\beta_k=\beta}} C\left(D_x^{\beta_1}D_\xi^{\alpha_1}A\right)\cdots\left(D_x^{\beta_k}D_\xi^{\alpha_k}A\right)A^{r-k}. \tag{4.11}$$

To estimate the terms in this sum note that

$$\alpha_j=0\Rightarrow\left|A^{-1}D_x^{\beta_j}A\right|\leqslant Ce^{\gamma} \tag{4.12}$$

and

$$|\alpha_j|\geqslant 1\Rightarrow|A^{-1}D_\xi^{\alpha_j}D_x^{\beta_j}A|\leqslant Ce^{\gamma-\lambda}e^{\lambda-\gamma-|\alpha_j|(1,0)}=Ce^{-|\alpha_j|(1,0)}.$$

Thus a typical term in (4.11) is bounded by

$$Ce^{\gamma|\beta|}e^{-|\alpha|(1,0)}|A|^r, \tag{4.13}$$

since there are at most $|\beta|$ factors of the form (4.12), and in turn (4.13) is bounded by

$$C|A|^r\Psi^{-|\alpha|}\left(\psi^{\#}\right)^{-|\beta|},$$

since $e^{-\gamma}=\Psi^{\#}$. This proves Lemma 4.1′, and Lemma 4.1 is a special case where $r=-1$. We note that estimate (4.13) implies the more refined result that $|A|=e^{\tilde a}$ defines an order function $\tilde a$ such that $A^r\in S^{r\tilde a}_{\Psi,\psi^{\#}}$ for $r\in R$. Note that

$$\Psi(x,\xi)\psi^{\#}(x,\xi)=\langle\zeta\rangle^{3/2}. \tag{4.14}$$

Now the symbol calculus gives

$$B(x,D)A(x,D)=I+E(x,D), \tag{4.15}$$

a priori in $OPS^{\gamma}_{\Psi,\psi^{\#}}$, with

$$E(x,\xi)\sim\sum_{\alpha>0}\left(\frac{1}{\alpha!}\right)D_\xi^\alpha B(x,\xi)\cdot D_x^\alpha p(x,\xi) \tag{4.16}$$

if A is of the form (4.1). Note that

$$D_\xi^\alpha B(x,\xi)\cdot D_x^\alpha p(x,\xi)\in S^{\gamma-(|\alpha|,0)}_{\Psi,\psi^{\#}}. \tag{4.17}$$

Published accounts of the Beals–Fefferman calculus, such as in [1], contain propositions that imply the difference between $E(x,\xi)$ and the sum over $|\alpha| < N$ in (4.16) belongs to $S_{\Psi,\psi^\#}^{\gamma-(N,N)^\#}$, where $e^{(1,1)^\#} = \Psi\psi^\#$. Since, according to (4.14), the pair Ψ, $\psi^\#$ is not strongly coercive, this is not an incisive result. In fact, (4.17) suggests this difference belongs to $S_{\Psi,\psi^\#}^{\gamma-(N,0)}$, and this is correct. Rather than consider details of the Beals–Fefferman calculus, we can justify this assertion as follows. Since $B(x,\xi) \in S_{1/3,1/3}^{-2/3}$, Theorem 2.10 of Hörmander [9] implies the difference considered above belongs to $S_{1/3,1/3}^{1/3-N/3}$, so (4.16) is asymptotic, and hence (4.17) implies the difference does belong to $S_{\Psi,\psi^\#}^{\gamma-(N,0)}$. Since

$$e^{\gamma-(1,0)} = \langle\zeta\rangle^{-3/2}, \tag{4.18}$$

we have (4.17) in $S_{\Psi,\psi^\#}^{0}$ for $|\alpha| = 1$ and in $S_{\Psi,\psi^\#}^{(-j,0)}$ for $|\alpha| = j+1 \geqslant 2$. Thus $E(x,D)$ actually belongs to $OPS_{\Psi,\psi^\#}^{0}$, and modulo $S_{\Psi,\psi^\#}^{(-1,0)} \subset S_{1/3,1/3}^{-1/3}$ its symbol is given by the sum over $|\alpha| = 1$ in (4.16). In fact, we can do better, using the full strength of (1.5); it follows that $D_{\xi_j}B(x,\xi)$ belongs to $S_{\Psi,\psi^\#}^{\gamma-2\lambda}$ if $j \neq n$. This gives

LEMMA 4.2. *Under hypotheses* (4.1)–(4.3) *we have*

$$B(x,D)A(x,D) = I + E(x,D) \in OPS_{\Psi,\psi^\#}^{0}, \tag{4.19}$$

with

$$\begin{aligned} E(x,\xi) &= i(\partial/\partial\xi_n)(\Phi + p^\#)^{-1}\cdot|\xi|^{-2/3}(\partial p/\partial x_n) \mod S_{\Psi,\psi^\#}^{-(1,0)} \\ &= -i\Phi'(\zeta)(\Phi(\zeta) + p^\#(x,\xi))^{-2}|\xi|^{-1}(\partial p/\partial x_n) \mod S_{\Psi,\psi^\#}^{-(1,0)}. \end{aligned} \tag{4.20}$$

We have set

$$p^\#(x,\xi) = |\xi|^{-2/3}p(x,\xi) \in S^{1/3}.$$

Let us denote the last line of (4.20) *by* $E_0(x,\xi)$. *In light of Lemma* 4.1 *we have*

$$\Phi'(\zeta)(\Phi(\zeta) + p^\#(x,\xi))^{-2} \in S_{\Psi,\psi^\#}^{0}, \tag{4.21}$$

while

$$|\xi|^{-1}\partial p/\partial x_n \in S_{1,0}^{0}. \tag{4.22}$$

We are now in a position to prove microlocal hypoellipticity of an Airy operator $A(x,D)$ satisfying (4.1)–(4.3) in a conic neighborhood of a point (x^0,ξ^0) in case (4.4). First, replace $p(x,\xi)$ in (4.1) by $\tilde{p}(x,\xi) \in S_{1,0}^{1}$, which is equal to $p(x,\xi)$ on a conic neighborhood of (x^0,ξ^0), so that $\partial\tilde{p}/\partial x_n$ is sufficiently small; hence, $A(x,D)$ is replaced by $\tilde{A}(x,D)$. Denote by $\tilde{B} \in S_{\Psi,\psi^\#}^{-2\lambda/3}$ and $\tilde{E} \in S_{\Psi,\psi^\#}^{0}$ the symbols obtained by replacing p by $\tilde{p}$ in the construction above. Thus,

$$\tilde{B}(x,D)\tilde{A}(x,D) = I + \tilde{E}(x,D), \tag{4.23}$$

with

$$\tilde{E}(x,D) = \tilde{E}_0(x,D) \mod OPS_{\Psi,\psi^\#}^{-(1,0)}, \tag{4.24}$$

where

$$\tilde{E}_0(x,\xi) = -i\Phi'(\zeta)(\Phi(\zeta) + \tilde{p}^\#(x,\xi))^{-2}|\xi|^{-1}(\partial\tilde{p}/\partial x_n)(x,\xi). \tag{4.25}$$

By the Calderón–Vaillancourt theorem we deduce that, at least after altering by an element of $OPS^{-\infty}$, $\tilde{E}_0(x, D)$ has small operator norm on L^2. Since $\tilde{E}(x, D) = \tilde{E}_0(x, D) + R$ with $R \in OPS^{-1/3}_{1/3,1/3}$, we can write $R = R_1 + R_2$, with $\|R_1\|$ small and $R_2 \in OPS^{-\infty}$; therefore, with $\tilde{E}_1(x, D) = \tilde{E}_0(x, D) + R_1$,

$$I + \tilde{E}(x, D) = I + \tilde{E}_1(x, D) + R_2, \tag{4.26}$$

where $\tilde{E}_1(x, D)$ has small operator norm on L^2 and $R_2 \in OPS^{-\infty}$. Now $I + \tilde{E}_1(x, D)$ is clearly invertible on L^2. We claim it is invertible on H^s for each s, which is not quite trivial, since this operator need not commute with an elliptic operator Λ^s in $OPS^s_{1,0}$. Indeed, as an operator on H^s, $I + \tilde{E}_1(x, D)$ is certainly Fredholm of index zero, since it is equal to $I + \Lambda^s\tilde{E}_1(x, D)\Lambda^{-s} + T_s$, $T_s \in OPS^{-1/3}_{1/3,1/3}$, a compact perturbation of an operator which is invertible on H^s. Thus $I + \tilde{E}_1(x, D)$ is invertible on H^s if and only if its kernel on H^s is 0. Since its kernel on L^2 is zero, we have this for positive s. Hence, $I + \tilde{E}_1(x, D)$ is invertible on H^s for $s \geqslant 0$, and a simple duality argument shows it is invertible on H^s for $s \leqslant 0$ also.

Thus, on each H^s,

$$\left(I + \tilde{E}_1(x, D)\right)^{-1}\tilde{B}(x, D)\tilde{A}(x, D) = I + R^{\#}, \qquad R^{\#} \in OPS^{-\infty}. \tag{4.27}$$

Now a simple modification of Proposition 3.4 shows that

$$W = \left(I + \tilde{E}_1(x, D)\right)^{-1}\tilde{B}(x, D) \in \mathcal{O}(-2/3, 1/3), \tag{4.28}$$

and hence, by Lemma 3.2, W is microlocal. It is a microlocal left parametrix for $\tilde{A}(x, D)$, and hence, on some conic neighborhood of (x^0, ξ^0), it is a microlocal left parametrix for $A(x, D)$. This finishes the analysis of case (4.4).

Next we consider case (4.5). Following **[4]** we note that due to the hypothesis on $p_1(x, \xi)$, if $\xi^0_n = 0$, then

$$(\partial/\partial x_n)\mathrm{Re}\, p_1(x^0, \xi^0) = 0 \Leftrightarrow (\partial/\partial x_n)\mathrm{Im}\, p_1(x^0, \xi^0) = 0 \tag{4.29}$$

(provided $p_1(x^0, \xi^0) = 0$). Thus, in case (4.5) we actually have

$$(\partial/\partial x_n)\mathrm{Re}\, p_1(x^0, \xi^0) \neq 0 \quad \text{and} \quad (\partial/\partial x_n)\mathrm{Im}\, p_1(x^0, \xi^0) \neq 0. \tag{4.30}$$

More precisely, the product of these two quantities must be negative. Let $p_1(x, \xi) = q_1(x, \xi) + iq_2(x, \xi)$, with q_1 and q_2 real valued. The implicit function theorem, together with (4.30), implies that we can find smooth $\alpha_1(x', \xi)$, $\alpha_2(x, \xi)$, homogeneous of degree 0 in ξ, and smooth $\mu_1(x, \xi)$, $\mu_2(x', \xi)$, homogeneous of degree 1, such that

$$\begin{aligned} q_1(x, \xi) &= \mu_1(x, \xi)(x_n - \alpha_1(x', \xi)) \\ q_2(x, \xi) &= \mu_2(x', \xi) + \alpha_2(x, \xi)q_1(x, \xi) \end{aligned} \qquad (x' = (x_1, \ldots, x_{n-1})). \tag{4.31}$$

Also, $\mu_1(x^0, \xi^0) \neq 0$ and $\alpha_2(x^0, \xi^0) \neq 0$. Following **[4]**, we have

LEMMA 4.3. *The negativity of the product of the terms in* (4.30) *implies*

$$\mu_2(x', \xi) \geqslant 0 \quad \textit{and} \quad \alpha_2(x, \xi) < 0 \tag{4.32}$$

near (x^0, ξ^0).

PROOF. In fact, at $x_n = \alpha_1(x', \xi)$ and, in particular, at (x^0, ξ^0),

$$(\partial/\partial x_n)q_1(x, \xi) = \mu_1(x, \xi) \quad\text{and}\quad (\partial/\partial x_n)q_2(x, \xi) = \alpha_2(x, \xi)\mu_1(x, \xi):$$

if this product is negative, it forces $\alpha_2 < 0$ at such a point. To see that $\mu_2(x', \xi) = q_2(x, \xi) - \alpha_2(x, \xi)q_1(x, \xi)$ is $\geqslant 0$, since it is independent of x_n, it suffices to show it is $\geqslant 0$ at $x_n = \alpha_1(x', \xi)$. Suppose that $\pm\mu_1(x, \xi) > 0$. Let $\pm(x_n - \alpha_1) > 0$ tend to zero. Now our hypothesis yields

$$q_1(x, \xi) < 0 \Rightarrow q_2(x, \xi) > 0 \quad\text{and}\quad q_2(x, \xi) < 0 \Rightarrow q_1(x, \xi) > 0.$$

But by the first half of (4.31) we have $q_1(x, \xi) < 0$ here, so $\mu_2(x', \xi) = q_2 - \alpha_2 q_1$ is a sum of a positive term and a term tending to zero. This proves the lemma.

Now, following [**4**], if we write

$$A(x, \xi) = |\xi|^{2/3}\Phi(\zeta) + p(x, \xi) = N_1 + q_1 + i(N_2 + q_2) + r(x, \xi), \tag{4.33}$$

with $r \in S^0_{1,0}$ and

$$N_1(\xi) = \operatorname{Re}|\xi|^{2/3}\Phi(\zeta), \qquad N_2(\xi) = \operatorname{Im}|\xi|^{2/3}\Phi(\zeta) \in S^\lambda_{\Psi,\psi}, \tag{4.34}$$

we have, mod $S^0_{\Psi,\psi}$,

$$\begin{aligned}\operatorname{Re}(-\alpha_2(x, \xi) - i)A(x, \xi) &= -\alpha_2 N_1 + N_2 - \alpha_2 q_1 + q_2 \\ &= -\alpha_2 N_1 + N_2 + \mu_2 \geqslant -\alpha_2 N_1 + N_2,\end{aligned} \tag{4.35}$$

and so, by (4.32) and the behavior of $\Phi(\zeta)$—namely,

$$\begin{aligned}\operatorname{Re}\Phi(\zeta) &\geqslant C\langle\zeta\rangle^{1/2}, \qquad \zeta \geqslant 0, \\ \operatorname{Im}\Phi(\zeta) &\geqslant C\langle\zeta\rangle^{1/2}, \qquad \zeta \leqslant 0,\end{aligned} \tag{4.36}$$

—we have

$$\operatorname{Re}(-\alpha_2(x, \xi) - i)A(x, \xi) \geqslant C|\xi|^{2/3}\langle\zeta\rangle^{1/2} \tag{4.37}$$

for large $|\xi|$ on a conic neighborhood of (x^0, ξ^0). Thus

$$F(x, \xi) = (-\alpha_2(x, \xi) - i)A(x, \xi) \in S^\lambda_{\Psi,\psi} \tag{4.38}$$

satisfies the conditions of Proposition 3.11, and we deduce that $F(x, D)$ is hypoelliptic microlocally near (x^0, ξ^0). In fact, we have the microlocal regularity result

(4.39)

$$u \in \mathscr{D}'(M),\ K(D)^{-1/2}\varphi(X, D)F(x, D)u \in H^s \Rightarrow K(D)^{1/2}\varphi_0(X, D)u \in H^s$$

for $\varphi(x, \xi)$, $\varphi_0(x, \xi) \in S^0$, supported in a conic neighborhood of (x^0, ξ^0), with $\varphi = 1$ on supp φ_0, where

$$K(\xi) = |\xi|^{2/3}\langle\zeta\rangle^{1/2}. \tag{4.40}$$

Now since

$$F(x, D) - (-\alpha_2(x, D) - i)A(x, D) \in OPS^{\lambda-\gamma-(1,1)}_{\Psi,\psi}, \tag{4.41}$$

it follows that the regularity result (4.39) also holds if $F(x, D)$ is replaced by $(-\alpha_2(x, D) - i)A(x, D)$. This in turn immediately gives the same microlocal regularity result for $A(x, D)$ and completes the proof of the following result.

PROPOSITION 4.4. *Consider the symbol*

$$A(x, \xi) = |\xi|^{2/3}\Phi(\zeta) + p(x, \xi), \qquad p(x, \xi) \in S^1. \tag{4.42}$$

Suppose $p_1(x, \xi)$ takes values in the complement of a conic neighborhood of the third quadrant in **C**. *Then $A(x, D)$ is hypoelliptic microlocally in a conic neighborhood of $\xi_n = 0$.*

If one considers, for example, an oblique derivative boundary condition for the wave equation with a real vector field, then one considers a symbol $A(x, \xi)$ of the form (4.42), where $p_1(x, \xi)$ is purely imaginary. Thus the hypothesis of Proposition 4.4 does not hold. Some propagation of singularities results will be contained in the results of §5, but we will record one microlocal regularity result here.

PROPOSITION 4.5. *Consider a symbol of the form* (4.42) *with $p_1(x, \xi)$ purely imaginary. Then $A(x, D)$ is hypoelliptic, microlocally, in any set*

$$\xi_n > C|\xi|^{1/3+\varepsilon}. \tag{4.43}$$

PROOF. By (4.36) we see that in this region $|A(x, \xi)^{-1}| \leqslant C|\xi|^{-2/3}\langle\zeta\rangle^{-1/2}$, and, hence, the proof of Lemma 1.1 shows that $A(x, \xi)^{-1} = B(x, \xi)$ belongs to $S^{\gamma-\lambda}_{\Psi,\psi^\#}$ microlocally in this region. But in this region, by (4.14), the weight functions Ψ, $\psi^\#$ are strongly coercive, so arguments parallel to the analysis of (4.15)–(4.17) show $A(x, D)$ has a microlocal parametrix in the region (4.43). The proposition is proved.

We conclude this section with some remarks about the possibility of utilizing Lemma 4.1′ to analyze square roots of certain operators with positive symbols dominating $C\langle\xi\rangle^{2/3}\langle\zeta\rangle^{1/2}$. Suppose $A(x, \xi) \in \mathscr{A}^{+,1}$ or, more generally, with Ψ, ψ, is of the form (1.8)–(1.9),

$$A(x, \xi) \in S^{\lambda}_{\Psi,\psi}, \qquad \nabla_\xi A(x, \xi) \in S^{\lambda-\gamma-(1,0)}_{\Psi,\psi}, \tag{4.44}$$

with

$$A(x, \xi) \geqslant C|\xi|^{2/3}\langle\zeta\rangle^{1/2} = Ce^{\lambda-\gamma} = CK(\xi). \tag{4.45}$$

Recall that by Lemma 4.1′ for $r > 0$,

$$A(x, \xi)^r \in S^{r\lambda}_{\Psi,\psi^\#}. \tag{4.46}$$

Better estimates on the derivatives of $A(x, \xi)^r$ are possible for $r > 0$. Suppose $0 < r < 1$. We have

$$|\nabla_\xi A^r| = r|A|^{r-1}|\nabla_\xi A| \leqslant Ce^{(\gamma-\lambda)(1-r)}e^{\lambda-\gamma-(1,0)} = Ce^{r(\lambda-\gamma)}(\Psi)^{-1} \tag{4.47}$$

and

$$|\nabla_x A^r| = r|A|^{r-1}|\nabla_x A| \leqslant Ce^{(\gamma-\lambda)(1-r)}e^{\lambda} = Ce^{r(\lambda-\gamma)}(\psi^\#)^{-1}. \tag{4.48}$$

From here we may readily sharpen Lemma 4.1′ to obtain

LEMMA 4.6. *Under hypotheses* (4.44)–(4.45), *if* $0 < r < 1$ *we have*

$$\nabla_\xi A^r \in S^{r\lambda - r\gamma - (1,0)}_{\Psi,\psi^\#}, \qquad \nabla_x A^r \in S^{r\lambda - r\gamma - (0,1)^\#}_{\Psi,\psi^\#}, \tag{4.49}$$

and hence

$$\nabla_x \nabla_\xi A^r \in S^{r\lambda - r\gamma - (1,1)^\#}_{\Psi,\psi^\#}. \tag{4.50}$$

Here the natural notation is

$$e^{(k,l)^\#} = (\Psi)^k (\psi^\#)^l. \tag{4.51}$$

Now, under hypotheses (4.44)–(4.45) let us set

$$B(x,\xi) = A(x,\xi)^{1/2}. \tag{4.52}$$

Use the Weyl calculus to form $A(X,D) \in OPS^\lambda_{\Psi,\psi}$ and $B(X,D) \in OPS^{\lambda/2}_{\Psi,\psi^\#}$. In view of Lemma 4.6 we obtain

$$B(X,D)^2 = \tilde{A}(X,D), \tag{4.53}$$

where

$$\tilde{A}(x,\xi) = A(x,\xi) + R(x,\xi), \tag{4.54}$$

and all the terms in the expansion of $R(x,\xi)$ belong to $S^{\lambda-\gamma-(2,2)^\#}_{\Psi,\psi^\#}$ or better. Propositions of **[12]** imply only that $R(x,\xi) \in S^{\lambda-(2,2)^\#}_{\Psi,\psi^\#}$, a situation we discussed in the proof of Lemma 4.2. In this case we cannot deduce from **[9]** that

$$R(x,\xi) \in S^{\lambda-\gamma-(2,2)^\#}_{\Psi,\psi^\#}, \tag{4.55}$$

and we omit the details of a proof of (4.55), since it is not central to this paper. Note that

$$e^{\lambda-\gamma-(2,2)^\#} = |\xi|^{2/3}\langle\zeta\rangle^{-5/2}. \tag{4.56}$$

If we applied this construction instead to $A(x,\xi) - \frac{1}{2}CK(\xi)$, we would have $B(X,D)^2 = \tilde{A}(X,D)$, with $\tilde{A}(x,\xi) = A(x,\xi) - \frac{1}{2}CK(\xi) + R(x,\xi)$, and, hence,

$$A(X,D) \geqslant \tfrac{1}{2}CK(D) - R(X,D). \tag{4.57}$$

As (4.56) shows, there exists a constant $M < \infty$ such that

$$\|R(X,D)u\| \leqslant M\|K(D)u\|. \tag{4.58}$$

The operator $R(X,D)$ is not strongly dominated by $K(D)$ in the sense that one has

$$\|R(X,D)u\| \leqslant C'\|K(D)u\|_{-\varepsilon}, \tag{4.59}$$

so these estimates by themselves do not imply semiboundedness of $A(X,D)$, such as we proved in Proposition 2.4 (especially (2.33)–(2.34)). In some cases it is possible to arrange M in (4.58) to be *small* (with perhaps a term of the form (4.59) thrown in); some examples of this phenomenon have arisen in work of Eskin **[3]**.

5. Propagation of singularities. We obtain some general results on the microlocal behavior of the singularities of solutions u to

$$Pu = f, \tag{5.1}$$

where P belongs to a Beals–Fefferman class, $P = p(X, D) \in OPS^{\lambda}_{\Psi,\psi}$. Since in the classical case it is typical to normalize P to have order 1, we here take

$$e^{\lambda} = \langle \xi \rangle. \tag{5.2}$$

As in the latter part of §3, we microlocalize in sets that are admissible relative to some class $S^0_{\Psi',\psi'}$, where

$$S^0_{1,0} \subset S^0_{\Psi',\psi'} \subset S^0_{\Psi,\psi}. \tag{5.3}$$

We begin by deriving the basic commutator identity used in the proof of propagation of singularities for classical operators. If C is a microlocalizing operator, and if

$$P = A + iB \tag{5.4}$$

with A and B symmetric, we have

$$(CPu, Cu) = (ACu, Cu) + i(BCu, Cu) + ([C, A]u, Cu) + i([C, B]u, Cu);$$

hence,

$$\operatorname{Im}(CPu, Cu) = \operatorname{Re}(\{(1/i)C^*[C, A] + C^*BC + C^*[B, C]\}u, u). \tag{5.5}$$

In the case $P \in OPS^1$ one assumes B is bounded from below. In certain situations it might be desirable to utilize some of the positivity of B; suppose

$$(Bv, v) \geqslant (B_0 v, v) - C_0\|v\|^2, \qquad B_0 \geqslant 0. \tag{5.6}$$

Then using Cauchy's inequality $|(CPu, Cu)| \leqslant \delta^2\|Cu\|^2 + (1/4\delta^2)\|CPu\|^2$, we obtain from the basic commutator identity (5.5) the following basic commutator inequality:

$$\operatorname{Re}\big(\{(1/i)C^*[C, A] - (\delta^2 + C_0)C^*C\}u, u\big) + (C^*B_0Cu, u) \leqslant (1/4\delta^2)\|CPu\|^2 + |(Wu, u)| \tag{5.7}$$

where

$$W = \operatorname{Re} C^*[B, C]. \tag{5.8}$$

Note that if $C \in OPS^0$, $B \in OPS^1$, and both are selfadjoint, then $W \in OPS^{-1}$.

Now in the classical case if we want to show that u is in H^s on a conic neighborhood of some curve γ in $T^*M \setminus 0$, we construct $C \in OPS^0$ and $\varphi(X, D) \in OPS^0$ such that $\varphi(x, \xi) \geqslant 1$ on γ and

$$\operatorname{Re}\{(1/i)C^*[C, A] - (\delta^2 + C_0)C^*C\}(x, \xi) - \varphi(x, \xi)^2 \geqslant -C|\xi|^{-1} \tag{5.9}$$

for large $|\xi|$. This is arranged by having the symbol of C satisfy a differential inequality along the integral curves of $H_{a(x,\xi)}$. One then has to cut off the symbol of C near one endpoint of γ, since otherwise it would explode. Thus (5.9) would no longer hold everywhere, but we would have

$$\operatorname{Re}\{(1/i)C^*[C, A] - (\delta^2 + C_0)C^*C\}(x, \xi) - \varphi(x, \xi)^2 + E(x, \xi)^2 \geqslant -C|\xi|^{-1}$$

for a certain $E(x, \xi) \in S^0$ supported on a conic neighborhood of the endpoint of γ toward which H_a points. Thus we could apply the sharp Gårding inequality, since the symbol of the left side belongs to $S^0_{1,0}$, to obtain

$$\|\varphi(X, D)u\|^2 \leqslant \operatorname{Re}\big(\{(1/i)C^*[C, A] - (\delta^2 + C_0)C^*C\}u, u\big) + \|Eu\|^2 + C\|u\|^2_{-1/2}.$$

From this and (5.7) one easily obtains that $u \in H^s$ on γ, given that $Pu \in H^s$ on γ, an integral curve of H_a, and $u \in H^s$ on $\operatorname{supp} E$. This is the classical propagation of singularities argument of **[11]**, which we want to generalize.

In light of the discussion of the classical case, let us rewrite the basic commutator inequality (5.7) as

(5.10)
$$\operatorname{Re}\big(\{(1/i)C^*[C, A] - (\delta^2 + C_0)C^*C\}u, u\big) + (C^*B_0Cu, u) + (E^2u, u) \leqslant (1/4\delta^2)\|CPu\|^2 + |(Wu, u)| + \|Eu\|^2$$

and arrange to apply a sharp Gårding inequality to the left side of (5.10). In our first generalization we make the following five hypotheses on the size of various operators involved.

First, we recall our basic setup:

(5.11)
$$P = A + iB \in OPS^{\lambda}_{\Psi,\psi}, \quad C \in OPS^{0}_{\Psi',\psi'}, \quad A, B, C \text{ symmetric}, \; C(x, \xi) \geqslant 0.$$

Next, we suppose that there is a $K(\xi) = e^{\kappa} \in S^{\kappa}_{\Psi,\psi}$ such that

(5.12)
$$[C, A] \in OPS^{\kappa}_{\Psi,\psi}.$$

Furthermore, we suppose (5.6) holds with

(5.13)
$$B_0 = B_0(\xi) = e^{\beta} \in S^{\beta}_{\Psi,\psi}$$

and, for some $\varepsilon > 0$,

(5.14)
$$W = \operatorname{Re} C[B, C] \in OPS^{\beta-\varepsilon\lambda}_{\Psi,\psi};$$

finally, we assume

(5.15)
$$E(x, \xi) \in S^{\kappa/2}_{\Psi,\psi}.$$

Regarding the function $K(\xi) = e^{\kappa}$, we need

(5.16)
$$e^{\beta-\varepsilon\lambda} \leqslant Ce^{\kappa}$$

and

(5.17)
$$e^{\kappa} \leqslant C\langle\xi\rangle^{-\varepsilon}B_0(\xi)e^{(2,2)} = Ce^{\beta-\varepsilon\lambda+(2,2)}.$$

We also restrict our attention to weight functions Ψ, ψ of the form

(5.18)
$$\Psi = \Psi(\xi), \qquad \psi = \psi(\xi).$$

Now we have the following general result, which leads to analysis of propagation of singularities.

PROPOSITION 5.1. *Let* Ψ, ψ *be strongly coercive weight functions of the form* (5.18), *and let* $P = A + iB$, C, E *satisfy* (5.11)–(5.15), *where* κ *satisfies* (5.16)–(5.17). *Suppose* $\varphi(x, \xi) \in S^0_{\Psi', \psi'}$ *and*

$$\operatorname{Re}\{(1/i)C[C, A] - (\delta^2 + C_0)C^*C\}(x, \xi) + E^2(x, \xi) \geqslant \varphi^2(x, \xi) - C_2\langle\xi\rangle^{-\varepsilon}C'(x, \xi)B_0(\xi), \tag{5.19}$$

where $C'(x, \xi) = 1$ *on* $\operatorname{supp} C \cup \operatorname{supp} \varphi$, $C' \in S^0_{\Psi', \psi'}$. *Then we have the a priori inequality*

$$\|\varphi(X, D)u\|^2 + \|B_0^{1/2}Cu\|^2 \leqslant K_1\|CPu\|^2 + K_2\|B_0^{1/2}C'u\|^2_{-\sigma} + K_3\|C'u\|^2_{-\sigma} + \|Eu\|^2 + K_N\|u\|^2_{-N} \tag{5.20}$$

for $u \in C_0^\infty$ *with some* $\sigma > 0$.

PROOF. First we note that (5.14) implies

$$|(Wv, v)| \leqslant C_1\|B_0^{1/2}v\|^2_{-\sigma} + C_2\|v\|^2_{-\sigma}, \tag{5.21}$$

with $\sigma = \varepsilon/2$. If we set

$$T = \operatorname{Re}\{(1/i)C[C, A] - (\delta^2 + C_0)C^2\} + E^2,$$

we have $T \in OPS^\kappa_{\Psi, \psi}$; hence, using (5.16) we obtain

$$T^\# = T - (\varphi^2 - C_2\langle\xi\rangle^{-\varepsilon}B_0) \in S^\kappa_{\Psi, \psi}$$

with symbol $\geqslant 0$. Let $L(\xi) = e^{\kappa - (2,2)}$, so $T^\# L(\xi)^{-1} \in S^{(2,2)}_{\Psi, \psi}$ is $\geqslant 0$. Thus we can apply the Fefferman–Phong inequality to get

$$\begin{aligned}\operatorname{Re}(T^\# u, u) &= \operatorname{Re}\left(L(D)^{-1/2}T^\# L(D)^{-1/2}L(D)^{1/2}u, L(D)^{1/2}u\right)\\ &\geqslant -K\|L(D)^{1/2}u\|^2.\end{aligned}$$

In view of hypothesis (5.17) we deduce

$$\|\varphi(X, D)u\|^2 \leqslant \operatorname{Re}(Tu, u) + K_1\|C'u\|^2_{-\sigma} + K_2\|C'B_0^{1/2}u\|^2_{-\sigma} + K_N\|u\|^2_{-N}. \tag{5.22}$$

This combined with (5.10) and (5.21) proves the proposition.

We note that, basically, Proposition 5.1 was obtained by taking each ingredient in the well-known analysis of (5.10) in the classical case and generalizing in a natural manner so as to make full use of the Fefferman–Phong inequalities, where the analysis in the classical case makes only weak use of a sharp Gårding inequality. Also, we have added one extra ingredient, a potential to make heavier use of positivity of B. We postpone for a short while replacing hypothesis (5.19), which involves operator products such as $(1/i)C[C, A]$, by symbols such as $C(x, \xi)\{C, A\}$. First we crystallize somewhat Proposition 5.1 as follows. We suppose the symbol classes to which the commutators $[C, A]$ and $[C, B]$ belong are just those dictated by the symbol calculus for $S^*_{\Psi, \psi}$; therefore no assumptions are made about $\nabla_\xi A$ and $\nabla_\xi C$ being of lower order than implied by (5.11), and we may as well suppose $\Psi = \Psi'$ and $\psi = \psi'$. In other words, instead of hypothesizing (5.12), (5.14), and (5.15), we merely work with the general containments

(5.23) $$[C, A] \in OPS_{\Psi,\psi}^{\lambda-(1,1)}, \qquad W \in OPS_{\Psi,\psi}^{\lambda-(2,2)},$$

and we pick

(5.24) $$E(x, \xi) \in S_{\Psi,\psi}^{(\lambda-(1,1))/2}.$$

Thus, in order for the hypotheses of Proposition 5.1 to be in effect, we need

(5.25) $$e^{\lambda-(1,1)} \leqslant Ce^{\kappa}, \qquad e^{\lambda-(2,2)} \leqslant Ce^{\beta-\varepsilon\lambda},$$

as well as estimates (5.16)–(5.17), which are

(5.26) $$C_1 e^{\beta-\varepsilon\lambda} \leqslant e^{\kappa} \leqslant C_2 e^{(2,2)+\beta-\varepsilon\lambda}.$$

Note that

(5.27) $$\operatorname{Re}(1/i)C[C, A](x, \xi) = C(x, \xi)\{C, A\}(x, \xi) \quad \operatorname{mod} S_{\Psi,\psi}^{\lambda-(3,3)}.$$

In view of the second inequality in (5.25), the remainder can be absorbed in the last term in (5.19), so we have

COROLLARY 5.2. *Let* $P = A + iB \in OPS_{\Psi,\psi}^{\lambda}$ *as in* (5.11). *Suppose there exist* $e^{\kappa} \in S_{\Psi,\psi}^{\kappa}$, $e^{\beta} \in S_{\Psi,\psi}^{\beta}$, *such that for some* $\varepsilon > 0$ *estimates* (5.25)–(5.26) *hold, and also* (5.13) *holds. Let* φ, $C' \in S_{\Psi,\psi}^{0}$ *be as in Proposition* 5.1, *let* $E(x, \xi)$ *satisfy* (5.24), *and suppose*

(5.28) $$C(x, \xi)\{C, A\}(x, \xi) - (\delta^2 + C_0)C(x, \xi)^2 + E(x, \xi)^2 \geqslant \varphi(x, \xi)^2 - C_2\langle\xi\rangle^{-\varepsilon}C'(x, \xi)B_0(\xi).$$

Then we have the a priori inequality (5.20). *In particular, if we set*

(5.29) $$e^{\kappa} = e^{(2,2)+\beta-s\lambda}, \qquad s \geqslant \varepsilon,$$

then, as long as $B_0(\xi) = e^{\beta} \in S_{\Psi,\psi}^{\beta}$, *we can deduce these conclusions, provided that*

(5.30) $$e^{\lambda-(2,2)} \leqslant Ce^{\beta-\varepsilon\lambda};$$

i.e., provided that

(5.31) $$B_0(\xi) \geqslant C|\xi|^{1+\varepsilon}(\Psi\psi)^{-2}$$

and s is chosen small enough that

(5.32) $$\langle\xi\rangle^{s} \leqslant C\langle\xi\rangle^{\varepsilon}(\Psi\psi).$$

As we can see, unless $\Psi\psi \geqslant C|\xi|^{(1+\varepsilon)/2}$, we must make some nontrivial use of positivity assumptions on B_0 in order to use this method to obtain estimates leading to propagation of singularities. In particular, for our Airy operators, where

(5.33) $$\Psi = |\xi|^{1/3}\langle\zeta\rangle, \qquad \psi = 1,$$

$\Psi\psi = |\xi|^{1/3}\langle\zeta\rangle$ fails to satisfy such an estimate, so positivity of B_0 must play an important role.

COROLLARY 5.3. *Let* $P = A + iB \in S^{\lambda}_{\Psi,\psi}$, *with* Ψ, ψ *given by* (5.33). *Suppose* (5.6) *holds with* $B_0(\xi) = e^{\beta} \in S^{\beta}_{\Psi,\psi}$ *and*

$$B_0(\xi) \geqslant C|\xi|^{1/3+\varepsilon}\langle \zeta \rangle^{-2}. \tag{5.34}$$

If C, C', φ, E *are as in Corollary* 5.2, *then the symbol inequality* (5.28) *implies the a priori estimate* (5.20).

Let us look at a special class of operators to which Corollary 5.3 applies. Let

$$P = i(N + iQ), \tag{5.35}$$

where

$$N(x, \xi) = |\xi|^{2/3}\Phi(\zeta) \tag{5.36}$$

is essentially the Neumann operator for a diffractive boundary problem as described in the introduction. Suppose

(5.37) $Q(x, \xi) \in S^1$ and $Q_1(x, \xi)$ is the real-valued principal symbol.

Such an operator arises in the study of oblique derivative problems for the wave equation, the well-posedness of which was studied in [**13**]. We show how Corollary 5.3 yields some of the propagation of singularities results of [**3**]. In order to write $P = A + iB$, we break up $\Phi(\zeta) = A'(\zeta)/A(\zeta)$ into its real and imaginary parts:

$$\Phi(\zeta) = \Phi_1(\zeta) + i\Phi_2(\zeta). \tag{5.38}$$

Now, at least for $A = A_+$, we have, for real ζ, $\Phi(\zeta)$ taking values in the first quadrant in **C**, more precisely that

$$\Phi_1(\zeta) \geqslant C\langle \zeta \rangle^{-1} \quad \text{if } \zeta \leqslant 0, \tag{5.39}$$

$$\Phi_1(\zeta) \geqslant C\langle \zeta \rangle^{1/2} \quad \text{if } \zeta \geqslant 0, \tag{5.40}$$

while

$$\Phi_2(\zeta) \geqslant C\langle \zeta \rangle^{1/2} \quad \text{for } \zeta \leqslant 0, \tag{5.41}$$

but

$$0 < \Phi_2(\zeta) \leqslant C_N\langle \zeta \rangle^{-N} \quad \text{for } \zeta \geqslant 0. \tag{5.42}$$

A proof of these elementary estimates is given in [**23**, Appendix A]; see also [**3**]. With P given by (5.36) we have

$$P = -\Lambda^{2/3}\Phi_2 - Q + i\Lambda^{2/3}\Phi_1 = A + iB, \tag{5.43}$$

where $\Lambda^m \in OPS^m$ has symbol $|\xi|^m$. In particular, modulo an unimportant zero-order term,

$$B(x, \xi) = B_0(\xi) = |\xi|^{2/3}\Phi_1(\zeta), \tag{5.44}$$

and it follows from (5.39) that

$$B_0(\xi) \geqslant C|\xi|^{2/3}\langle \zeta \rangle^{-1}, \tag{5.45}$$

an estimate which is stronger than (5.34). Of course, for $\zeta \geqslant 0$ we have an even stronger estimate—namely,

$$B_0(\xi) \geqslant C|\xi|^{2/3}\langle \zeta \rangle^{1/2} \quad \text{for } \zeta \geqslant 0. \tag{5.46}$$

We remark that by virtue of the results of §1, $B_0(\xi)$, given by (5.44), satisfies the condition $B_0(\xi) = e^\beta \in S^\beta_{\Psi,\psi}$.

In order to deduce propagation of singularities results from Corollary 5.3, we need to produce a rich family of nonnegative symbols $C(x, \xi)$ (together with φ, C', and E) such that the symbol inequality (5.28) holds. We follow Eskin [3] in constructing $C(x, \xi)$ in the form

$$C(x, \xi) = \kappa(x, \xi'; N_2), \tag{5.47}$$

where

$$N_2(\xi) = |\xi|^{2/3}\Phi_2(\zeta). \tag{5.48}$$

As in (5.38), $\Phi_2 = \operatorname{Im}\Phi$. As $\zeta \to -\infty$, $\Phi_1(\zeta) \sim 0$ and $\Phi(\zeta) \sim i\Phi_2(\zeta)$. The function $\kappa(x, \xi'; \nu)$ will be C^∞ in all its arguments, for $|\xi'|^2 + \nu^2 \neq 0$, and homogeneous of degree zero in (ξ', ν). We have

$$\begin{aligned}\{C, A\} = \{N_2, \kappa(x, \xi'; N_2)\} + \{Q, \kappa(x, \xi'; N_2)\} \\ + (\partial\kappa/\partial\nu)(x, \xi'; N_2)\{Q, N_2\}.\end{aligned} \tag{5.49}$$

We make the notational convention that the Poisson bracket $\{F, \kappa(x, \xi'; \nu)\}$ regards ν as a parameter and applies only x and ξ' derivatives to κ. Note that

$$\begin{aligned}H_{N_2} &= |\xi|^{2/3}\Phi_2'(\zeta)H_{|\xi|^{-1/3}\xi_n} + \Phi_2(\zeta)H_{|\xi|^{2/3}} \\ &= |\xi|^{1/3}\Phi_2'(\zeta)H_{\xi_n} + |\xi|^{-1/3}\left[\tfrac{2}{3}\Phi_2(\zeta) + \tfrac{1}{3}\zeta\Phi_2'(\zeta)\right]H_{|\xi|}.\end{aligned} \tag{5.50}$$

The construction of $\kappa(x, \xi'; \nu)$ involves looking at the analogue of (5.49) with C, A replaced by $C^\#$, $A^\#$, obtained by replacing N_2 by

$$N_2^b(\xi) = \sqrt{|\xi|(-\xi_n)} = |\xi|\sqrt{\alpha}, \tag{5.51}$$

where we set

$$\alpha = -\xi_n/|\xi|. \tag{5.52}$$

We have

$$\begin{aligned}\{C^\#, A^\#\} = \{N_2^b, \kappa(x, \xi'; N_2^b)\} + \{Q, \kappa(x, \xi'; N_2^b)\} \\ + (\partial\kappa/\partial\nu)(x, \xi'; N_2^b)\{Q, N_2^b\}.\end{aligned} \tag{5.53}$$

Note that $H_{N_2^b} = -(1/2\sqrt{\alpha})H_{\xi_n} + (\sqrt{\alpha}/2)H_{|\xi|}$, so

$$\begin{aligned}\{C^\#, A^\#\} = \frac{1}{\sqrt{\alpha}}\Bigg[\frac{1}{2}\frac{\partial\kappa}{\partial x_n} - \frac{1}{2}\frac{\nu^2}{|\xi|^2}H_{|\xi|}\kappa - \frac{1}{2}\frac{\partial Q}{\partial x_n}\frac{\partial\kappa}{\partial\nu} \\ - \frac{1}{2}\frac{\nu^2}{|\xi|^2}(H_{|\xi|}Q)\frac{\partial\kappa}{\partial\nu} - \frac{\nu}{|\xi|}H_Q\kappa\Bigg]\Bigg|_{\nu = N_2^b}.\end{aligned} \tag{5.54}$$

Note that the quantity in brackets is just a *smooth* vector field applied to κ! Call it X. It has the property that the coefficients of $\partial/\partial x_j$ are homogeneous of degree 0 in (ξ', ν), and the coefficients of $\partial/\partial\xi_j$ and of $\partial/\partial\nu$ are homogeneous of degree 1 in (ξ', ν). All of these coefficients are C^∞ for $|\xi'|^2 + \nu^2 \neq 0$. Note that the coefficient of $\partial/\partial x_n$ in X is nonzero, in fact close to $\frac{1}{2}$, at least if $|\nu|/|\xi|$ is small. If γ is an orbit of X on which X is nonvanishing, then, as usual, one can construct $\kappa_1 \geqslant 0$, supported on a small conic neighborhood of a segment $\tilde{\gamma}$ of γ, with $X\kappa_1 \geqslant 0$, strictly positive on $\tilde{\gamma}$, and $\kappa_2 \geqslant 0$ such that $X\kappa_2 \geqslant 1$ on supp κ_1, except near one endpoint of $\tilde{\gamma}$; then taking $\kappa = \kappa_1 e^{\lambda\kappa_2}$ yields

$$X\kappa \geqslant \lambda\kappa. \tag{5.55}$$

Take $\lambda \geqslant \delta^2 + C_0 + 1$, with $\delta^2 + C_0$ as in (5.28). Consequently,

$$\{C^{\#}, A^{\#}\} \geqslant (\lambda/\sqrt{\alpha})C^{\#}. \tag{5.56}$$

We now show that, with κ so prescribed and C given by (5.47), the symbol inequality (5.28) holds. To do this we take (5.49), plug in (5.50), and produce a formula whose similarity to (5.54) we exploit. We obtain

$$\begin{aligned} \{C, A\} = -2|\xi|^{1/3}\Phi_2'(\zeta)\Bigg[&\frac{1}{2}\frac{\partial\kappa}{\partial x_n} - \frac{1}{2}\frac{\nu^2}{|\xi|^2}H_{|\xi|}\kappa - \frac{1}{2}\frac{\partial Q}{\partial x_n}\frac{\partial\kappa}{\partial\nu} \\ &-\frac{1}{2}\frac{\nu^2}{|\xi|^2}(H_{|\xi|}Q)\frac{\partial\kappa}{\partial\nu} - \frac{\nu}{|\xi|}H_Q\kappa\Bigg]\Bigg|_{\nu=N_2} \\ &+R_1 + R_2 + R_3, \end{aligned} \tag{5.57}$$

where

$$R_1 = |\xi|^{-1/3}\left[\tfrac{2}{3}\Phi_2(\zeta) + \tfrac{1}{3}\zeta\Phi_2'(\zeta) + \Phi_2(\zeta)^2\Phi_2'(\zeta)\right]H_{|\xi|}\kappa, \tag{5.58}$$

$$R_2 = |\xi|^{-1/3}\left[\tfrac{2}{3}\Phi_2(\zeta) + \tfrac{1}{3}\zeta\Phi_2'(\zeta) + \Phi_2(\zeta)^2\Phi_2'(\zeta)\right](H_{|\xi|}Q)\partial\kappa/\partial\nu, \tag{5.59}$$

and

$$R_3 = \left[1 - 2\Phi_2(\zeta)\Phi_2'(\zeta)\right]H_Q\kappa. \tag{5.60}$$

To analyze R_j first note that

$$H_{|\xi|}\kappa \in S^0_{\Psi,\psi}, \quad (H_{|\xi|}Q)\partial\kappa/\partial\nu \in S^0_{\Psi,\psi}, \quad H_Q\kappa \in S^0_{\Psi,\psi}. \tag{5.61}$$

As for the other factors in (5.58)–(5.60), we note that cancellations inherent in our setup involving $\Phi_2(\zeta) \sim (-\zeta)^{1/2} + \cdots$ as $\zeta \to -\infty$ give

$$\rho_1(\zeta) \sim \begin{cases} a_{11}(-\zeta)^{-1} + a_{12}(-\zeta)^{-5/2} + \cdots & \text{as } \zeta \to -\infty, \\ b_{11}(-\zeta)^{1/2} + b_{12}(-\zeta)^{-1} + \cdots & \text{as } \zeta \to +\infty, \end{cases} \tag{5.62}$$

for

$$\rho_1(\zeta) = |\xi|^{1/3}\left[\tfrac{2}{3}\Phi_2(\zeta) + \tfrac{1}{3}\zeta\Phi_2'(\zeta) + \Phi_2(\zeta)^2\Phi_2'(\zeta)\right] \tag{5.63}$$

and

$$\rho_2(\zeta) \sim \begin{cases} a_{21}(-\zeta)^{-3/2} + a_{22}(-\zeta)^{-2} + \cdots & \text{as } \zeta \to -\infty, \\ b_{21} + b_{22}(-\zeta)^{-3/2} + \cdots & \text{as } \zeta \to +\infty, \end{cases} \tag{5.64}$$

for

$$\rho_2(\zeta) = [1 - 2\Phi_2(\zeta)\Phi_2'(\zeta)]. \tag{5.65}$$

Consequently, it is clear that $R_1 + R_2 + R_3$ is dominated by $\langle\xi\rangle^{-\varepsilon}B_0(\xi)$ for some $\varepsilon > 0$ in light of (5.45)–(5.46). We now look at the main term on the right side of (5.57).

This term is the product of the factor in brackets, which has been arranged to be $\geqslant \lambda C(x, \xi)$, except near one endpoint p_0 of a segment of an integral curve of X in (x, ξ', ν) coordinates, and the factor $2|\xi|^{1/3}(-\Phi_2'(\zeta))$. Now it is known that

$$-\Phi_2'(\zeta) > 0 \quad \text{for all } \zeta \in \mathbf{R}. \tag{5.66}$$

In fact, the decomposition $\Phi(\zeta) = \Phi_1(\zeta) + i\Phi_2(\zeta)$ can be written

$$\Phi(\zeta) = F'(\zeta)/F(\zeta) + c_0 i/2F(\zeta)^2, \tag{5.67}$$

where $c_0 > 0$ and $F(\zeta) = |A(\zeta)|$ for ζ real. Thus $\Phi_2(\zeta) = c_0/2F(\zeta)$, so (5.66) is equivalent to $F'(\zeta) > 0$, which is equivalent to positivity of $\Phi_1(\zeta)$, a phenomenon we have already mentioned. Again we refer to Appendix A of **[23]** for details on this. In view of the asymptotic behavior of $\Phi_2(\zeta)$, (5.66) implies

$$-\Phi_2'(\zeta) \geqslant C_1\langle\zeta\rangle^{-1/2} \quad \text{for } \zeta \leqslant 0, \tag{5.68}$$

though $\Phi_2'(\zeta)$ is rapidly decreasing as $\zeta \to +\infty$. Thus, the main term on the right side of (5.57) is greater than or equal to

$$C_1\lambda|\xi|^{1/3}\langle\zeta\rangle^{-1/2}C(x, \xi) \quad \text{for } \zeta \leqslant 0, \tag{5.69}$$

where λ can be taken as large as desired.

We have assembled all the ingredients needed to establish the symbol inequality (5.28). It is natural to check (5.28) in three separate regions:

$$-\xi_n \geqslant C_0|\xi|^{1/3+\varepsilon}, \quad |\xi_n| \leqslant C_0|\xi|^{1/3+\varepsilon}, \quad \xi_n \geqslant C_0|\xi|^{1/3+\varepsilon}. \tag{5.70}$$

These regions may also be defined by $-\zeta \geqslant C_0|\xi|^\varepsilon$, $|\zeta| \leqslant C_0|\xi|^\varepsilon$, and $\zeta \geqslant C_0|\xi|^\varepsilon$, respectively. The region $-\xi_n \geqslant C_0|\xi|^{1/3+\varepsilon}$ includes the "hyperbolic region" for the diffractive boundary problem giving rise to the Airy operator $P = i(N + iQ)$. In this region the quantity (5.69) is greater than or equal to $C_2\lambda C(x, \xi)$ except near p_0, and if we pick λ sufficiently large and set $\varphi(x, \xi) = C(x, \xi)$, then one can pick $E \in S_{\Psi,\psi}^{(\lambda-(1,1))/2}$ supported near p_0 such that (5.28) is satisfied in this region. The region $|\xi_n| \leqslant C_0|\xi|^{1/3+\varepsilon}$ is a "subconic" neighborhood of the grazing set $\xi_n = 0$. In this region $\{C, A\}(x, \xi)$ is controlled by (5.66) and estimates of R_j, but by virtue of (5.45)–(5.46), $C(x, \xi)^2 = \varphi(x, \xi)^2$ is dominated by $\langle\xi\rangle^{-\varepsilon}B_0(\xi)$ in this region, so again (5.28) is satisfied. Finally, in the region $\xi_n \geqslant C_0|\xi|^{1/3+\varepsilon}$, since $[C, A]$ belongs to $OPS_{\Psi,\psi}^{\lambda-(1,1)}$ for any $C \in OPS_{\Psi,\psi}^0$, and $e^{\lambda-(1,1)} = |\xi|^{2/3}\langle\zeta\rangle^{-1}$, estimate (5.46) shows that $C[C, A]$ and C^2 are dominated by $\langle\xi\rangle^{-\varepsilon}B_0(\xi)$ for any choice of $C(x, \xi)$ (not necessarily satisfying (5.47) and (5.56)). Thus (5.28) holds in the third region. The fact that $C(x, \xi)$ can be chosen arbitrarily in this region means that the estimates given by Corollary 5.3 imply hypoellipticity of P in this region, but of course we have already seen in Proposition 4.5 that this microlocal hypoellipticity is an elementary consequence of the Beals–Fefferman symbol calculus.

By the usual process of replacing $C(x, \xi)$ by a family of smoothing operators, one elevates the a priori estimate (5.20) to a microlocal regularity theorem. As this procedure and its geometrical interpretation are standard, we conclude here our study of propagation of singularities.

REFERENCES

1. R. Beals, *A general calculus of pseudodifferential operators*, Duke Math. J. **42** (1975), 1–42.

2. R. Beals and C. Fefferman, *Spatially inhomogeneous pseudodifferential operators*, Comm. Pure Appl. Math. **27** (1974), 1–24.

3. G. Eskin, *Initial-boundary value problems for second order hyperbolic equations with general boundary conditions*, I, J. Analyze Math. **40** (1981), 43–89.

4. ______, *Initial-boundary value problems for second order hyperbolic equations with general boundary conditions*, II, Preprint.

5. C. Fefferman, *The uncertainty principle*, Bull. Amer. Math. Soc. **9** (1983), 129–266.

6. C. Fefferman and D. Phong, *On positivity of pseudodifferential operators*, Proc. Nat. Acad. Sci. U.S.A. **75** (1978), 4673–4674.

7. ______, *The uncertainty principle and sharp Gårding inequalities*, Comm. Pure Appl. Math. **34** (1981), 285–331.

8. ______, *Symplectic geometry and positivity of pseudodifferential operators*, Proc. Nat. Acad. Sci. U.S.A. **79** (1982), 710–713.

9. L. Hörmander, *Pseudodifferential operators and hypoelliptic equations*, Singular Integrals, A. P. Calderón (Editor), Proc. Sympos. Pure Math., Vol. 10, Amer. Math. Soc., Providence, R. I., 1967, pp. 138–183.

10. ______, *Fourier integral operators* I, Acta Math. **127** (1971), 79–183.

11. ______, *On the existence and the regularity of solutions of linear pseudodifferential equations*, L'Enseign. Math. **17** (1971), 99–163.

12. ______, *The Weyl calculus for pseudo-differential operators*, Comm. Pure Appl. Math. **32** (1979), 359–443.

13. M. Ikawa, *Probleme mixte pour l'equation des ondes*. II, Publ. Res. Inst. Math Sci. **13** (1977), 61–106.

14. M. Imai and T. Shirota, *On a parametrix for the hyperbolic mixed problem with diffractive lateral boundary*, Hokkaido Math. J. **7** (1978), 339–352.

15. K. Kubota, *A microlocal parametrix for an exterior mixed problem for symmetric hyperbolic systems*, Hokkaido Math. J. **10** (1981), 264–298.

16. H. Kumano-go, *Algebras of pseudo-differential operators*, J. Fac. Sci. Univ. Tokyo Sect. IA Math. **17** (1970), 31–50.

17. P. Lax and L. Nirenberg, *On stability for difference schemes; a sharp form of Garding's inequality*, Comm. Pure Appl. Math. **19** (1966), 473–492.

18. A. Majda and M. Taylor, *The asymptotic behavior of the diffraction peak in classical scattering* Comm. Pure Appl. Math. **30** (1977), 639–669.

19. R. Melrose, *Microlocal parametrices for diffractive boundary value problems*, Duke Math. J. **42** (1975), 605–635.

20. ______, *Airy operators*, Comm. Partial Differential Equations **3** (1978), 1–76.

21. ______, *Equivalence of glancing hypersurfaces*, Invent. Math. **37** (1976), 265–291.

22. R. Melrose and J. Sjöstrand, *Singularities of boundary value problems*, II, Comm. Pure Appl. Math. **35** (1982), 129–168.

23. R. Melrose and M. Taylor, *Boundary problems for wave equations with grazing and gliding rays*, Monograph (in preparation).

24. ______, *Near peak scattering and the corrected Kirchhoff approximation for convex obstacles*, Adv. in Math. (to appear).

25. J. Miller, *The Airy integral*, Cambridge Univ. Press, 1946.

26. M. Taylor, *Pseudodifferential operators*, Princeton Univ. Press, Princeton, N. J., 1981.

27. ______, *Grazing rays and reflection of singularities for solutions to wave equations*, Comm. Pure Appl. Math. **29** (1976), 1–38.

28. ______, *Propagation, reflection, and diffraction of singularities for solutions to wave equations*, Bull. Amer. Math. Soc. **84** (1978), 589–611.

29. ______, *Diffraction effects in the scattering of waves*, Singularities in Boundary Value Problems, Reidel, Boston, 1981, pp. 271–316.

30. ______, *Airy operator calculus*, Microlocal analysis, Contemporary Math., Vol. 27, Amer. Math. Soc., Providence, R. I., 1984, pp. 169–192.

STATE UNIVERSITY OF NEW YORK AT STONY BROOK

ABCDEFGHIJ–AMS/AP–898765